BOSTON STUDIES IN THE PHILOSOPHY OF SCIENCE

VOLUME XXVIII

SCIENCE IN FLUX

SYNTHESE LIBRARY

VOLUME 80

BOSTON STUDIES IN THE PHILOSOPHY OF SCIENCE

EDITED BY ROBERT S. COHEN AND MARX W. WARTOFSKY

VOLUME XXVIII

JOSEPH AGASSI

SCIENCE IN FLUX

D. REIDEL PUBLISHING COMPANY

DORDRECHT-HOLLAND / BOSTON-U.S.A.

Library of Congress Cataloging in Publication Data

Agassi, Joseph.
　　Science in flux.
　　(Boston studies in the philosophy of science ; v. 28)
(Synthese library ; v. 80)
　　　Includes indexes.
　　　1.　Science—Philosophy.　I.　Title.　II.　Series.
Q174.B67　vol. 28　[Q175]　　　　501s　[501]
ISBN 90–277–0584–4　　　　75–12926
ISBN 90–277–0612–3 (pbk.)

Published by D. Reidel Publishing Company,
P.O. Box 17, Dordrecht, Holland

Sold and distributed in the U.S.A., Canada and Mexico
by D. Reidel Publishing Company, Inc.
306 Dartmouth Street, Boston,
Mass. 02116, U.S.A.

Printed in The Netherlands by D. Reidel, Dordrecht

To Sir Karl Popper

in gratitude
with admiration and dissent

None of us is blessed with the knowledge of the absolute truth. Even our science is not blessed with the absolute truth. It is based on common sense, which is to say, that it is ever changing and that it is content gradually to replace big mistakes by smaller ones.

(ALFRED ADLER, *The Science of Living*, 1929, p. 36)

EDITORIAL PREFACE

Joseph Agassi is a critic, a gadfly, a debunker and deflater; he is also a constructor, a speculator and an imaginative scholar. In the history and philosophy of science, he has been Peck's bad boy, delighting in sharp and pungent criticism, relishing directness and simplicity, and enjoying it all enormously. As one of that small group of Popper's students (including Bartley, Feyerabend and Lakatos) who took Popper seriously enough to criticize him, Agassi remained his own man, holding Popper's work itself to the criteria of critical refutation.

Agassi's range is wide and his publications prolific. He has published serious studies in the historiography of science, applied sociology (on Hong Kong with I.C. Jarvie), foundations of anthropology, interpretive scientific biography (Faraday), Judaic studies, philosophy of technology (which Agassi pioneered, particularly in distinguishing it from the philosophy of science), as well as the many works on the logic, methodology, and history of science. Even as we go to press, Agassi's works are appearing; we append an imperfect and selected bibliography.

For Agassi, the test of relevance is whether something is interesting. Eschewing the outward accoutrements of 'serious scholarship' when they mask dullness or lack of imagination, Agassi nevertheless pursues a constructive path in his many essays – but not a straight line. Digression often becomes the most direct means to bring a point home. When Agassi is puzzling, perverse, provocative, it is part of the argument. His form of argument is not 'analytic' in the simple sense of a sequence of statements in logical form, but rather a dialectic where questions and answers interweave in a complex fabric. Yet, clarity is one of his goals. And his complaint is that he means exactly what he says, in the simplest way, but that readers and commentators often refuse to take him at his word. A kind of tribute to his provocativeness is the (private) punning comment once made by an eminent philosopher of science: "Cet Agasse m'agace".

We have happily suffered the benefits (and enjoyed the irritations) of Professor Agassi's collaboration in the Boston Colloquium for the

Philosophy of Science for more than a decade. He has been a dedicated and generous critic, affectionate friend, and receiver of criticisms. He has, above all else, never let philosophy become mere technique. These essays convey, we think, the seriousness, the disturbance and the intellectual vivacity of a skeptical and rationalist spirit.

Boston University Center for the   R. S. COHEN
Philosophy and History of Science,   M. W. WARTOFSKY
May 1975

TABLE OF CONTENTS

SELECTED BIBLIOGRAPHY OF JOSEPH AGASSI

'Duhem versus Galileo', *Brit. J. Phil. Sci.* **8** (1957).

'A Hegelian View of Complementarity', *Ibid.* **9** (1958).

'Koyre on the History of Cosmology', *Ibid.* **9** (1958).

'Corroboration versus Induction', *Ibid.* **9** (1959).

'Epistemology as an Aid to Science', *Ibid.* **10** (1959).

'Methodological Prescriptions in Economics', with K. Klappholz, *Economica,* 1959.

'How are Facts Discovered', *Impulse* **10** (1959), 1-3.

'Jacob Katz on Jewish Social History', *Jew. J. Soc.* **1** (1959).

'Methodological Individualism', *Brit. J. Soc.* **11** (1960).

'The Role of Corroboration', *The Australasian Journal of Philosophy* **39** (1961), 8–91.

'An Unpublished Paper of the Young Faraday, *Isis* **52** (1961).

'Between Micro and Macro', *Brit. J. Phil. Sci.* **14** (1963).

'Empiricism without Inductivism', *Philosophical Studies* **14** (1963), 85–86.

Towards an Historiography of Science, Beiheft 2, to *History and Theory,* March 1963 (117 pp.).

'The Nature of Scientific Problems and Their Roots in Metaphysics', in *The Critical Approach, Essays in Honor of K. R. Popper,* ed. M. Bunge (New York and London: Free Press and MacMillan, 1964), pp. 189–211.

'The Confusion Between Physics and Metaphysics in Standard Histories of Science', in *Ithaca: Proceedings of the Xth International Congress for the History of Science,* ed. H. Guerlac (Paris: Hermann, 1964), pp. 231–238, 249–250.

'Sensationalism', *Mind* **75**, (1966), 1–24.

'The Confusion between Science and Technology in Standard Philosophies of Science', *Tech. and Culture* **7** (1966), 348–66.

'Planning for Success: a Reply to Professor Wisdom', *Tech. and Culture* **8** (1967), 78–81

The Continuing Revolution (New York: McGraw-Hill, 1968).

'Science in Flux: Footnotes to Popper', in *Boston Studies in the Philosophy of Science,* Vol. III, eds. R. S. Cohen and M. W. Wartofsky (Dordrecht and Boston: Reidel, 1968), 293–323.

'On Novelty', *International Philosophical Quarterly* **8** (1968), 442–63.

'Precision in Theory and in Measurement', *Phil. Sci.* **3** (1968), 287–90.

'The Logic of Technological Development', in *Akten des XIV. Internationalen Kongresses für Philosophie,* Vol. 2 (Vienna: Herder, 1968), 134–37.

'Comments: Theoretical Entities Versus Theories', in R. S. Cohen and M. W. Wartofsky, (eds.), *Boston Studies in the Philosophy of Science,* Vol. 5 (Dordrecht, Holland and Boston: Reidel, 1969), 457–59.

'Replies to Diane: Popper on Learning from Experience', *Am. Phil. Quart.,* Special Monograph Series, No. 3 (1969), 162–70.

'On Privileged Access', *Inquiry* **12** (1969), 420–26.

'Unity and Diversity in Science', in R. S. Cohen and M. W. Wartofsky, (eds.), *Boston Studies in the Philosophy of Science,* Vol. 4 (Dordrecht and Boston: Reidel, 1969), 463–522.

'Can Religion go Beyond Reason?', *Zygon* **4** (1969), 128–68.

'Positive Evidence in Science and Technology', *Phil. Sci.* **37** (1970), 261–70.

'Duhem's Instrumentalism and Antinomism', *Ratio* **12** (1970), 148–50.

'Positive Evidence as Social Institution', *Philosophia,* Israel **1** (1970), 261–70.

Faraday as a Natural Philosopher (Chicago and London: The University of Chicago Press, 1971).

'What is a Natural Law', *Stud. Gen.* **24** (1971), 1051–56.

'The Standard Misinterpretation of Skepticism', *Phil. Stud.* **22** (1971), 49–50.

'Imperfect Knowledge', *Phil. Phenomenol. Res.* **32** (1972), 465–77.

'When Should we Ignore Evidence in Favor of a Hypothesis?', *Ratio* **15** (1973), 183–205.

'Random versus Unsystematic Observations', *Ratio* **15** (1973), 111–13.

'Testing as a Bootstrap Operation', *Z. Allg. Wiss.* **4** (1973), 1–25.

'Criteria for Plausible Argument', *Mind* **83** (1974), 406–16.

'Modified Conventionalism', in P. A. Schilpp, (ed.), *The Philosophy of Sir Karl Popper* (La Salle, Ill.: Open Court, 1974), pp. 693–96.

'On Pursuing the Unattainable', in R. J. Seeger and R. S. Cohen, (eds.),

Philosophical Foundations of Science (*Boston Studies in the Philosophy of Science* vol. XI), Reidel, Dordrecht and Boston, 1974, pp. 431–44.

Towards a Philosophical Anthropology. Van Leer Jerusalem Foundation Series (Atlantic Highlands, N. J.: Humanities Press, 1975).

'Assurance and Agnosticism', to appear in A. Michelos *et al.*, (eds.), *PSA 1974* (*Boston Studies in the Philosophy of Science,* Vol. XXXII), Reidel, Dordrecht and Boston, 1975.

'The Lakatosian Revolution', to appear in P. Feyerabend *et al.*, (eds.), *Essays in Memory of Imre Lakatos* (*Boston Studies in the Philosophy of Science,* Vol. XXXIX), Reidel, Dordrecht and Boston, 1975.

'Can Adults become Genuinely Bilingual?', to appear in A. Kasher, (ed.), *Language in Focus: Foundations, Methods and Systems; Essays Dedicated to Yehoshua Bar-Hillel* (*Boston Studies in the Philosophy of Science,* Vol. XLIII) Reidel, Dordrecht and Boston, 1975.

Science in Flux (*Boston Studies in the Philosophy of Science,* Vol. XXVIII) Reidel, Dordrecht and Boston, 1975.

PREFACE

The present volume is a collection of essays selected and arranged in the hope of presenting a comprehensive view of science, from three different levels of abstractness. On the factual level I suggest that no factual information is final and that the authority of factual evidence can be overruled in diverse ways. On the scientific level, I suggest, scientific theories are testable explanations which conform, or which are hoped to conform, to a general theory of the world. On the metaphysical level, I suggest, competing general theories of the world are metaphysical yet they are used to generate competing scientific research programs and competing scientific theories. On each level the social aspects of science are emphasized. And it is on the social level that the requirement for positive evidence signifies: an innovation that may be a public hazard – like a new machine or a new pill – has to be tested before its launching is permitted, according to legally specifiable standards.

The point of departure of this volume is the philosophy of Sir Karl Popper. His criticisms of his predecessors, even his characterization of the various schools of thought, is largely endorsed – somewhat streamlined, and with some inconsistencies removed. His major inconsistency, I think, is between his praise of the pre-Socratic philosophers, whose metaphysical views he erroneously takes to be scientific, and his requirement for maximum empirical testability or refutability. This requirement he uses as the major weapon for ousting metaphysics. And though later in life his opinion of metaphysics has been much more favorable he never withdrew the requirement for maximal refutability.

The demand seems to me to be exaggerated in two ways. We do not always prefer the most refutable view; sometimes we prefer views which better conform to our metaphysics. Nor do we always know whether our views are testable or not.

This is a matter of principle concerning the onus of proof of rationality. Popper requires that we be able to prove our rationality. I think it better to indicate why we think this or that is irrational and allow all else to enjoy the benefit of doubt. The only exception, to repeat, concerns public safety, where the public demands minimum standards of examination of whatever anyone wishes to launch in the public domain.

What this amounts to is the proposal to liberalize our scholarly and scientific standards, and at least recognize that repeatedly scientists employ a laxer standard than the ones preached by rationalist philosophers. They reject the authority of facts; at times they offer *ad hoc* excuses with which they hope to shelve difficulties for a time; often they treat hopes as if they were fulfilled, which, in my opinion, is the hallmark of pseudo-science. Newcomers to science are bewildered by the discrepancy between stringent standards and lax conduct; when they realize what happens, rather than lower the standards and make them more reasonable, they blame their elders and peers for their violation of a sacred code. All this is painful and quite unnecessary.

What I say may be true or not. The fact is, we do not know what is rationality, and so it is advisable, I think, to take rationality always on a common-sense level, as attempts to reach given goals, including the immediate goal of scientific communication and cooperation. And any stringent requirement of rationality may be erroneous and detrimental to cooperation.

It was my intention to finish this volume with an essay on the limits of reason, and show that reason is too limited to investigate its own limits, that we can rationally study the rational method only within limits. But I got stuck. A leading advocate of this view in our century is S. Hugo Bergman (whose student I was privileged to be in Jerusalem before I went to London where I soon became a student of Popper), and Bergman was the student of the works of Solomon Maimon, the great skeptic and critic of Kant. I have not yet had the occasion to study the works of Bergman and of Maimon with sufficient care. This, however, I would say now about skepticism in general and the limits of reason in particular.

In a sense skepticism concedes too much to justificationism: the justificationist has dismissed existing standards of rationality as question-begging and sought the final and unquestionable ones. The skeptic accepted the dismissal but declared the justificationist's quest hopeless. Rather, it is more reasonable to observe the existence of questionable and question-begging standards and watch them improve and try to help improve them. The final essays in this volume are but an elaboration on this proposal. Whether my proposal is ultra-skeptical or non-skeptical is hard for me to say. It certainly observes that there are limits to reason; it also suggests hopes for improvability of rationality. (Popper allows both the limits of reason and the progress of science – but not, to my knowledge, progress in the theory and practice of rationality as such. It was Maimon, and he alone, who ventured to take Kant's theory of rationality as a hypothesis, and all hypotheses, even Newton's or Kant's, as capable of being surpassed.)

It is always an embarrassment for a philosopher to place his view within a context, since naturally he takes as his context the pillars of wisdom familiar to all, thus inviting either an unfair comparison or a conceited suggestion that the comparison becomes his standing. Yet he has to do it, one way or another, to assist his reader's orientation. For my part, I should like to consider the following comparison – fair or scandalous – because it poses a problem which I feel very strongly. Consider the following parallel:

Newton: Kant: Maimon: Einstein: Popper: Agassi

and assume, for a while, that my comparing Popper with Kant is apt. (Popper already has explained it, and the following pages will, I hope, further justify it.) The reason I compare myself with Maimon is not only that my critique of Popper is quite the same as Maimon's just critique of Kant – skeptical and metaphysically oriented – but also that it renders my venture so very Quixotic: he was a man of genius of the kind I cannot possible claim, yet he failed so utterly he was almost entirely forgotten, more remembered as an eccentric Jew (he would not have minded that) than as a forceful critic and clear thinker (which would have struck him as disappointing, I suppose). My problem, then, is this. What chance do I have in advocating both skepticism and metaphysics if even Maimon has failed?

I suppose I should say, first, we all do what best we can, in the duty of hope; and second, that my chances are objectively much brighter than his, since the dogmatism of Newton has largely been replaced by Einstein's toleration and skepticism, and since the romantic gush of the nineteenth century has been checked with the result that irrationalism today is not that rampant. This lucky streak, the objective advantages I have, may make up, I hope, for the obvious defects of the present volume.

Boston, Massachusetts, Thanksgiving, 1974
Tel Aviv, Israel, Passover, 1975

ACKNOWLEDGEMENTS

Most of the material in this volume has been published elsewhere, as follows:

Chapter 1: Previously unpublished.
Chapter 2: Robert S. Cohen and Marx W. Wartofsky (ed.), *Boston Studies in the Philosophy of Science*, vol. 3, Reidel and Humanities, Dordrecht and New York, 1968, pp. 293–323.
Appendix: *Australasian Journal of Philosophy* **39** (1961), 8–91.
Chapter 3: *International Philosophical Quarterly* **8** (1968), 442–63.
Appendix: *Impulse* **10** (1959), 1–3.
Chapter 4: *American Philosophical Quarterly*. Special Monograph Series, No. 3 (1969), 162–170.
Appendix: *Philosophical Studies* **14** (1963), 85–86.
Chapter 5: *Mind* **75** (1966), 1–24.
Appendix: *Inquiry* **12** (1969), 420–426.
Chapter 6: *Ratio* **15** (1973), 183–205.
Appendix: *Ratio* **15** (1973), 111–113.
Chapter 7: *Zeitschrift für allgemeine Wissenschaftstheorie* **4** (1973), 1–25.
Appendix: *Philosophy of Science* **3** (1968), 287–290.
Chapter 8: Previously unpublished.
Appendix: Previously unpublished.
Chapter 9: in Mario Bunge (ed.), *The Critical Approach, Essays in Honor of Karl Popper*, Free Press and Macmillan, New York and London, 1964, pp. 189–211.
Appendix: *Studium Generale* **24** (1971), 1051–1056.
Chapter 10: Paper read at the Canadian Philosophical Association annual meeting in 1969; to be published in *Philosophical Forum*.
Appendix: Previously unpublished.
Chapter 11 and the Appendix: H. Guerlac (ed.), *Ithaca, 1962: Proceedings of the Xth International Congress for History of Science*, Hermann, Paris, 1964, pp. 231–238, and 249–250.

Chapter 12: *Technology and Culture* **7** (1966), pp. 348–366.
Appendix: *Ibid.* **8** (1967), 78–81.
Chapter 13: *Philosophy of Science* **37** (1970), 261–270.
Appendix: *Ratio* **12** (1970), 148–150.
Chapter 14: *Philosophia*, Israel **1** (1970), 261–270.
Appendix: *Akten des XIV. Internationalen Kongresses für Philosophie*, vol. 2, Herder, Vienna, 1968, pp. 134–137.
Chapter 15: *Philosophy and Phenomenological Research* **32** (1972), 465–477.
Chapter 16: *Mind* **83** (1974), 406–416.
Appendix: *Philosophical Studies* **22** (1971), 49–50.
Chapter 17: An extract of three pages was published in P. A. Schilpp (ed.), *The Philosophy of Sir Karl Popper*, Open Court, La Salle, Illinois, 1974, pp. 693–96.
Appendix: Previously unpublished.
Chapter 18: Robert S. Cohen and Marx W. Wartofsky (eds.), *Boston Studies in the Philosophy of Science*, vol. 4, Reidel and Humanities, Dordrecht and New York, 1969, pp. 463–522.
Appendix: Previously unpublished.
Chapter 19: *Zygon* **4** (1969), 128–168.
Appendix: Previously unpublished.
Chapter 20: Paper read at the Philosophy of Science Association biennial meeting in 1974; to be published in *PSA 1974* (*Boston Studies in the Philosophy of Science*, vol. 32), Reidel, Dordrecht and Boston, 1975.

I am grateful to all the editors and publishers concerned for their permission to republish.

A number of colleagues, friends, and students, have helped me writing the following essays – in discussion about the topics and about the presentations, and in reading and correcting the manuscripts extensively. I can only mention here those who did it at great length, namely, my wife Judith, W. W. Bartley III, R. S. Cohen, I. C. Jarvie, Tom Settle, Kenneth Topley and J. W. N. Watkins. I am grateful to them all. Though my long and close association with Sir Karl Popper was over by the time I wrote these essays, I think I owe him special gratitude for all his help.

CHAPTER 1

A PROLOGUE: ON STABILITY AND FLUX

When addressing the question of stability and of instability a philosopher may be talking on two very different planes or levels, as it were. The deep and the superficial, or common-sense. (Strictly, the superficial may be different from the common-sense, but let us ignore this.) He may speak superficially or common-sensically of relative stability and instability, and he may speak metaphysically of stability inherent in things he considers, of ultimate stability, as it were, or of ultimate instability. It is my wish to stress that when I speak of the stable and the unstable in science I speak on the superficial or common-sense level, not on the metaphysical level. The metaphysical situation concerning science is rather obvious – or at least has been so since Einstein: science, like all human endeavour, is essentially or inherently unstable; and this makes the discussion of its possible stability on the metaphysical level rather a bore. It is much more interesting to ignore these inherent qualities on the metaphysical level and speak superficially on the common-sense level of what is more stable and what is more ephemeral in science, and why.

We have here a classical question, concerning the stability inherent in the universe, in things in general, inanimate or animate. And we have here a new question, concerning the stability inherent in science, in a specific product of the human mind, perhaps also a specific social institution. Of course the question of stability in general is deeply linked with the question of stability in science – both because science is a part of the universe and because science is intended to mirror the universe in some sense or another. Yet the two questions, of stability in things and of stability in science, are quite different. The question, is the world of science stable? – is metaphysically uninteresting and common-sensically very interesting. The question, is the world of things stable? – is metaphysically very interesting indeed, and there is a vast interesting literature dealing with it. Common-sensically it is not interesting in its generality: some things are more stable than others and that is all there is to it. Only by making the common-sense question more specific, e.g., why is chalk more stable than cheese? can we render it more interesting and proceed with scientific investigations into it.

Let us, then, glance first at the metaphysical question, is the world of things stable? and then move to the common-sense question, is the world of science stable?

On the metaphysical level it may be claimed that everything in the universe radiates stability, is imbued with stability; that reality and stability are one. The opposite metaphysics sees in reality one and only one characteristic, instability, change, flux. And there is the middle view, or myriads of middle views, of inherent stability interplay with inherent flux which together keep the world going.[1]

Let us consider then, these three views: the one of utter permanence, shared by Parmenides, Spinoza, Einstein, and Schrödinger, which I shall call the Parmenidean; and the one of utter flux, shared by Heraclitus, Hegel, Marx, and (the young) Wittgenstein, which I shall call the Heraclitean; and the moderate view. The moderate view may have a great many variants, such as Plato's view of the world of permanence (the Ideas) and the (observable) world in flux, and such as Laplace's, which distinguishes between permanent laws and initial conditions which depend on no law (though not contingent since they depend on laws and previous initial conditions, of course). The moderate view is not to be confused, however, with the common-sense everyday unthoughtful attitude, which, like the moderate view, assumes certain things to be more ephemeral and others more permanent. For, it can be easily shown that the healthy common-sense view does not and need not distinguish between the-permanent-as-far-as-we-are-concerned and the permanent-proper, the common-sensically permanent such as the sun and the metaphysically permanent such as Democritus's atom. Moreover, it is not to be supposed for one moment that even the most ardent advocate of one or the other of the two extreme views, who opposes the moderate view proper, ever doubted the common-sense moderate view. Yet, one may not conclude from the previous statement that the disagreement between the two extremes and the moderates proper have no relevance to the moderate common-sense view. Let us take this point slowly.

Perhaps for psychological reasons, perhaps for social reasons, perhaps for intellectual reasons, we all seek the more stable in our environment; I think this is essential to the struggle for survival. It is also a part of the struggle for survival that a certain stability is assumed or taken for granted or treated unquestioningly. The child trusts his parent, the bread-winner trusts the forces of nature, the voter his social milieu. Well, not quite: the child has neurotic fears, the farmer and hunter alike perform rituals to secure the return of spring, and the voter may consider going to the barricades. These expressions of fear and anxiety may cause a breakdown or a breakdown may give rise to them. And the struggle for survival may be lost. At such junctures the Heracliteans begin to be heard by the public; all of a sudden their fantastic un-common-sense view starts sounding frighteningly common-sensical; the conviction concerning their insanity is insanely shaken. At such times Heracliteans may, as they did, influence common-sense profoundly and irreversibly. At precisely such moments, Parmenideanism may be heard too, and sound deliciously com-forting, and similarly have a profound and irreversible influence on common-sense. Its effect may be Utopian and Platonist, or, to the con-trary, resigned and Hindu; but it then becomes a part of the common man's heritage, and thus a part or an aspect of his common-sense.

All this is said within a rather common-sense framework. Within that framework, we tend to agree, there is the recommendation not to judge things in moments of stress. It is much better to deliberate questions before they become pressing, before we are psychologically tuned to certain kinds of solutions to them to the extent that our judgement gets limited and clouded, not to say blinkered and prejudiced. Now, if we want to view matters in common lights and in cool temper, what we would want to know first is, how do the three schools, and especially the two extreme ones, account for the common phenomenon of relative stability in some spots and of relative flux in other spots? The fact is that they cannot. But it is equally a fact for all three schools: the world presents us with too many unexplained phenomena, so that even the very best view we have of it is no more than a very rough and sketchy program of explanation, with only a very minor part of it actually executed, not any real explanation of the world. Nevertheless, we may insist, at least the moderates have a rough sketch of explanation. Is it not the fact that the extremes are barred even from the hope of ever being able to offer even a sketch of an explanation? This, we remember, is the link between the question of stability in the world and that of stability in science.

It so happens that the extremists did offer programs, and ones which did develop into important ideas in spots. It was Emile Meyerson who claimed, in his *Identity and Reality*, that all physical science stems from a Parmenidean impulse. Even Stuart Hampshire, who once followed Wittgenstein in considering metaphysics nonsense, admitted at that period that Spinoza's Parmenideanism was very fruitful in the development of the sciences. Similarly, Russell has confessed that he was attracted to mathematics out of a Parmenidean impulse, and that he is a Spinozist at heart. (Aren't we all?)

Now it may be claimed, and with much justice, that there is a world of difference between Parmenideanism proper and the Parmenidean impulse: it is one thing, fairly within common-sense, to search for stability, and quite another, crazy idea, to claim that everything in the world reflects stability and no flux whatsoever. Also, it may be claimed, and again with much justice, that the motive of the search, even the search itself, may be quite irrelevant to our valuation of the finding. Heraclitus called Pythagoras a charlatan and an ass; unable to dismiss this charge, we try to ignore it: we merely divorce his philosophy from his mathematics and ignore the former. We likewise divorce Russell's motives from our admiration of the *Principia Mathematica*. Indeed, motives come in only as either a historical curiosity (Russell) or as an explanation for people's dogmatism (Pythagoras). But once we call someone a dogmatist it matters little how we explain his dogmatism – at least in comparison with our (right or not, but) tragic decision to have no intellectual commerce with him.

Here, however, there is one very practical common-sense point. Parmenides may have been utterly crazy, but it was well known in antiquity that he and his school were master-dialecticians, the last people to be called dogmatic. Indeed, unlike the Pythagoreans, the Parmenideans had no tribalist school and when their master's doctrines were effectively criticized they did not try, as the Pythagoreans did, to retain the school by a semblance of a doctrine. To generalize: however extremist one's views or impulses, rationality need not be damaged as a result. Indeed, even a moderate view may become dogmatic. Even the advocacy of dialectics may become a school's dogma.[2]

The contemporary view of science as anchored in stable empirical evidence, is a paradigm of a common-sense moderate view. Its Parmenidean impulse is well checked; indeed, most of its advocates are avowed opponents of all metaphysics, extremist metaphysics in particular. Yet I fear I see no way of avoiding the conclusion that the post-Einsteinian followers of this doctrine – inductivism – are plainly dogmatists. Speaking within the common-sense moderate view or within the metaphysical moderate view, as you wish, we cannot deny that the Parmenidean impulse plays a great role, that the search for the stable is valuable. But whereas science searches for the stable in the world, the inductivist philosophers search for the stable in science; they cannot find it, but no rational argument will dissuade them from continuing the search.

The inductivist view of Man resembles those faintly ludicrous statues of the Stalin era, of a man standing, legs slightly apart, knees locked, shoulders stretched backward, chin slightly up, and gaze fixed on the horizon. To me mankind seems, especially on the common-sense level, to be much more of a clown walking blindfolded on a tightrope – with the difference that the clown is exceptionally aware of his predicament. The world may be stable – our environment (social as well as physical) is plainly unstable, even hostile.

If the Parmenidean impulse is so very essential for survival, and especially for a balanced childhood, the Heraclitean impulse is equally valuable, especially for the adult who has to create for his youngsters the necessary security. Particularly when a philosopher comes searching for stability in science, he may be disillusioned in the face of a major scientific revolution and become panicky and anti-scientific, and even anti-rational. To me, the fact that science is in flux is so obvious that it is hard to take seriously the inductivist – whose main aim is to explain the relative stability and steady growth of science – 'facts' which he considers equally obvious.

This is not to say that there are no elements in science that reveal a greater degree of stability than others. And the interesting task of identifying those elements and explaining their relative stability seems to me to be extremely interesting. This is the task undertaken, and hopefully executed in some small measure, in the various essays in the present volume. I do not here take sides in the metaphysical controversy concerning stability; rather, at the common-sense level I view science as in flux, though with some elements in it less stable and some more stable. So, from the start, I take it for granted that induction is not serious, and try to explore the picture of science in somewhat more detail from a non-inductivist, common-sense, moderate viewpoint. The best non-inductivist view seems to me to be that of my teacher, Sir Karl Popper, which I take as my point of departure.

The stabilizing, the relatively stable, factors in science seem to me to be both the metaphysical foundations of science and the social institutions of science, especially of scientific opinion. And, of course, the metaphysical foundation of science is itself an institution. I shall discuss here an instance which I avoid elsewhere in this volume – the institution of the mathematical tools of science.

The traditional literature considers mathematical statements as true beyond any shadow of doubt. There was an older skeptical tradition which raised arguments from computational errors, lapses of memory, etc. Traditional epistemology as well as its critics systematically ignore these arguments, viewing them as outside the universe of discourse of traditional epistemology. I do not think this was an error or a lapse into dogmatism. Nevertheless, let us consider what universe of discourse such funny arguments might belong to.

What book, will you declare, contains only truths? If you are a trained philosopher you may, after some hesitations, to be sure, accept a volume of elementary mathematical tables as such. The volume contains thousands of statements on every page, and it is to be doubted that any user can check more than a fraction of these. Even if all users of the same texts came together to compare notes, the checking would not amount to much. One might think of a fiendish enemy having replaced all the mathematical tables in a given metropolis by slightly erroneous ones. The reader can easily imagine a science-fiction book hinged on such a story, and the reader familiar with the rules and conventions of science-fiction can write such a book within a few days. This science-fiction work will remain science-fiction as long as the social conventions and the situation it describes differ from ours; we can also write a science-fiction work about situations similar to ours, but cast in a different set of conventions; we doubt, however, that we can alter our situation, stick to our conventions, yet obtain a science-fiction work: we expect that the story will show how our conventions will readjust the situation, and the story will then qualify as fiction, not as science-fiction. Hence, when I trust my mathematical tables I am implicitly trusting certain social institutions: I assume that they are stable enough to eliminate wrong tables. If you do not like this argument as it is not prosaic enough, think of all possible misprints which may occur in one slim volume and of the enormity of the task of proof-reading and checking a book. Consider the fact that you trust your volume to contain no misprints because it has been published by a reputable firm, and my argument concerning stable institutions is translated from science-fiction to a much more prosaic context.

Why are such discussions outside the universe of discourse of traditional epistemology? Because they introduce sociological factors. The universe of discourse of traditional epistemology is not sociological but psychological. The skeptical criticism of the traditional view of mathematical certainty is likewise psychological, but it makes psychological assumptions very different from those of the traditional epistemologist. Personal errors, forgetfulness, etc., the traditional epistemologist assumes, can in principle be rectified and hence cause no problems of principle. For the skeptical criticism to cut through this dismissal, the skeptic needs a better psychology, not a new sociology; or else he needs attack the psychologistic framework within which epistemology is couched and replace it with a sociologistic framework.

In this volume I follow the traditional attack on traditional epistemology on these lines: it has been shown that the psychological assumptions incorporated in traditional epistemology are erroneous. It has been shown that a scientist depends on his society, that the new epistemology has to be sociological rather than psychological. This much is in the universe of discourse of the traditional criticism of traditional epistemology. And in this respect it is to this tradition that the present essays belong. But this volume is not another contribution in the same vein. Rather, I wish to stop the trend: we have wasted too much time and effort criticizing traditional epistemology, and merely on the sociological assumption that as long as a thesis is popular it deserves further criticism. This thesis can be refuted by evidence from the social history of science. We need a social set to help us study our problems, but a small set who know the errors of the old view and the new problems, who are ready to push ahead amongst themselves, will do a better job of ousting the error than hundreds of critics-proselytizers. The thinkers of the Enlightenment went too far in suggesting that we ignore error altogether and merely let new light dispel old darkness; but the opposite view seems to be an extremist error too. One may ask where exactly is the happy middle? I do not know, but clearly, it is where enough stability exists to insure continuity of investigation. There are many new problems to be discussed in a new sociologically oriented method, which may lead to new ideas. Incidentally, the new ideas may attract the multitude much better than a wide front of criticism. For instance, why do we trust our standard mathematical tables? Do these contain errors to any serious degree? Presumably not, and this is no mean achievement; how was it made? Can we hope to introduce it to other, and much poorer branches, such as chemistry and biology (not to say philosophy)? More important, what determines a universe of discourse? Who decides what is a good question? How?

Such questions will be in the centre of the epistemology of the future from whose viewpoint this volume will be an obsolete curio, a mere adumbration.

The present volume, then, has an in-built obsolescence. This can be shown in a few ways. Let me use one more. The position of this volume is allegedly a purist Popperian. But Purist Popperism is a contradiction in terms: purism is an instrument of stagnation, and quick and rapid change leave no time for much purification. I think Popper has trained his disciples to preach purified Popperism, and I think this is unhealthy. I try to make my purism as critical of the doctrine originally presented as possible. But enough of that. More interesting problems lie ahead.

NOTES

[1] See Karl R. Popper, *Conjectures and Refutations*, London and New York, 1963 and 1964, Chapters 2 and 5. Popper's views are repeated in Stephen Toulmin, 'The End of the Parmenidean Era', in Y. Elkana (ed.), *The Interaction Between Science and Philosophy*, Humanities Press, Atlantic Highlinds, N.J., 1974, pp. 171–184. See also my comments there, 191–3.
[2] See Daniel E. Gershenson and Daniel A. Greenberg, 'The "Physics" of the Eleatic School: A Reevaluation' in *The Natural Philosopher*, Vol. 3, 1964, pp. 99–111. A most remarkable paper.

CHAPTER 2

SCIENCE IN FLUX

Footnotes to Popper *

To what extent and in what respect is science intellectually valuable? This is a controversial matter. What is hardly disputed is that what is alterable in science is of mere ephemeral value; and what is valuable in it is that which is more universal and permanent, that which is more solid and lasting. One of the very few philosophers who oppose this accepted view is Sir Karl Popper. In his view, science is so valuable because of its open-mindedness, because any of its achievements may at any time be given up and newer achievements may be hoped for to replace the relinquished ones. Science, says Popper, is at constant war with itself, and it progresses by revolutions and internal conflicts.

Popper's philosophy is far reaching in its consequences and thus very challenging and most interesting. It should be particularly interesting to examine its claims and compare what it describes as science with the phenomenon of science, in its history and in its contemporary manifestations. In particular, it may be seen that there exists a certain degree of stability in science, and one may query how does Popper's theory of science in flux account for the relative stability which science manifests. Possibly the observed stability refutes Popper's theory. Possibly, the observed stability is neither a refutation of Popper's theory nor is it explained by it. Possibly the observed stability is explained, but only with the aid of certain minor *ad hoc* additions to the theory. Possibly the theory of science in flux explains the relative stability in some more satisfactory manner. (These, of course, are all the possible relations between a theory and a fact.) The view which will be presented here is two-fold. First, Popper himself accounts for the relative stability of science only with the aid of a few very unsatisfactory *ad hoc* assumptions; and this renders his theory much less exciting and attractive than it looks at first. Second, these assumptions are not necessary at all for Popper's theory of science. In as much as one may report an observed stability in science (rather than express wishful thinking), one may consider it from the Popperian view as some rather regrettable sluggishness. Alternatively, one may view it as the stability of the social institutions which administer and apply science, rather than the stability of science itself.

The attitude from which the present discussion is launched is somewhat logically austere, then. Popper's fundamental assumptions are presented as a whole theory of science, and they are examined as such. If Popper's fundamental assumptions will not work as a complete theory, then, doubtlessly, one may try to add epicycles to Popper's cycles and see if these help. Alternatively, one may prefer to destroy the whole machine and search for a totally new one. Although thus far this has not been found necessary, it may yet be; the addition of epicycles, however, is worse than scrapping the whole machine.

Let me, then, introduce the chief ideas of Popper's philosophy, and discuss the additional assumptions of it later on. Had I started in a traditional manner, I would have the choice between the inductive and the deductive modes of presentation: either starting from the detailed facts on which Popper's philosophy is based, or starting from the fundamental axioms from which it follows. Being a Popperian, I shall proceed dialectically: let me first introduce the intellectual background to the problems which Popper has attempted to solve, then his problems and his solutions, and then related problems which may be of some interest in our context.

I. EINSTEIN HAS UPSET THE VIEW THAT SCIENCE IS STABLE

A philosopher has strong reasons to fumble and hesitate when he starts. In particular, he knows how much where he ends depends on where he begins: and where he begins may be all too arbitrary. Existing canons of science insure the exclusion of too much arbitrariness, or at least the minimizing of its effects if it has intruded. Existing canons of the arts permit, and even encourage, certain kinds of arbitrariness, but on certain conditions which may be very constraining at times. How much arbitrariness is allowed in philosophy? What place does philosophy allot to idiosyncrasy? These philosophical questions are difficult and the division over them is fundamental. Often the first step a philosopher takes is the answer he gives to these very questions. His first step, then, is at once much too difficult and much too final. Moreover, to tell you of my own idiosyncrasy from the start, the two existing alternatives, the two currently popular answers to the problem of arbitrariness in philosophy, seem to me to be most unsatisfactory. Briefly, the tradition of modern philosophy, stemming from Bacon and Descartes, emphasized and stressed the wish to build a scientific philosophy, an objective philosophy, a philosophy with no arbitrariness and no idiosyncracy. Its chief pride was science, and it took science as the model for rationality in general, indeed, for humanity at large; what traditional philosophy shared with religions which are not esoteric, is the idea that man is superior to animals in his ability to choose intelligently. This ability of rational choice they equated with objectivity and with science.

The rebellion against traditional philosophy started already in the 19th century; but it became prominent only in our century. It was not serious, nor was it taken seriously by the intellectual leadership of the time. It is no accident, perhaps, that most of the rebels advocated rather esoteric alternatives to science. Nietzsche and Kierkegaard, who are nowadays considered pioneers in the new movement, considered themselves journalists rather than philosophers; other pioneers of that movement were writers and artists like Dostoevskij and Wagner, and nationalist politicians of all sorts. The very first phase of the rebellion against scientific philosophy is known as romanticism; it embraced certain backward philosophical schools, as well as certain experimental literary and artistic schools, and some daring political movements. As a philosophy of art, romanticism gained grounds in the last century even within traditional philosophy; even John Stuart Mill, a pillar of scientific philosophy who saw man as a primitive learning machine, was romantically inclined when he spoke of self-expression of the individual in the arts. In epistemology, metaphysics, and even political philosophy, the popularity of the esoteric amongst intellectuals began when science showed itself to be as much in flux as other human phenomena – if not more. When the scientific revolution shook the intellectual world, esotericism emerged as an intellectual power *par excellence*. Nationalisms of all sorts and advocacies of esoteric, or semi-esoteric, religions became serious contenders; not due to their own merits, but because scientific philosophy started to decline. In his *Religion and Science* (chapter on mysticism) Russell records with chagrin the fact that in the 'thirties many men of science endorsed, and publicly advocated, some sort of non-scientific philosophy or another. This is one of the unintended consequences of the scientific revolution of the 20th century in general, and of Einstein's work in particular.

To be more precise, it is not so much the occurrence of revolutions in science, the fact that science is in flux, that created the major change in the philosophical scene; rather, what has happened is that suddenly the fact that science is in flux ceased to be a secret. When some Christian theologians were ready to admit that Christian doctrine is changeable, since the Bible is not infallible, they effected more radical a break with their tradition than the break created by the Renaissance protestant theologians who claimed that they were merely purifying the tradition and thus returning to the original unpolluted teaching. In science, things are not as bad, but prior to Einstein much effort did go into the rewriting of history in concealment of the fact that science is in flux.[1] The great agitation in the whole philosophical world which accompanied the revolution in chemistry of the late 18th century can be explained as the outcome of the first open revolution within the ranks of science: the disagreements on fundamentals between such established scientists as the phlogistonists and the anti-phlogistonists must have been as shocking to scientists as the split between Stalinists and Trotskyites was to communists. But, like the Trotskyites, the phlogistonists were declared enemies, traitors, etc. Then came the revolution in optics of the eighteen-twenties, the overthrow of Newton's theory of light. Some writers of that period expressed feelings close to hysteria, others close to witch-hunting; but things were patched up rapidly, and as the Newtonians had not put up a serious fight they could switch positions quietly. There is evidence, however, that after the revolution in optics, even Laplace was a broken man. When the conservation of force was announced, it was declared romantic *Naturphilosophie*: both Oersted and Helmholtz were suspected as romantics. In his obituary notice on his best friend, Michael Faraday, August de la Rive rejected Faraday's views mainly, if not only, because they looked to him idealistic. But this was forgotten too. In all the changes, switches, surreptitious changes, and the rewriting of the history of philosophy and of science, the ideal of science and its rationality as unalterable was upheld, and the standard instance for the equation of science with rationality and of both with stability was Newtonian mechanics. Sir John Herschel, William Whewell, John Stuart Mill – these were the leading 19th-century methodologists, and they repeatedly insisted that Newtonian mechanics, at least, is entirely perfect, is in no need of improvement or modification to the slightest, and cannot be superseded *even in being explained by a still deeper or more general theory*: Newtonian mechanics was the last word on all the questions it answers – and in every sense in which it could be construed as the last word. When Maxwell tried a field interpretation of Newtonian action-at-a-distance and found this impossible, he declared he could not understand his own result and that he was rather baffled. He must have known Faraday's suggestion that Newton's theory needs modification since the action at a distance is not possible. But daring and revolutionary as he was, Maxwell could not entertain the possibility of modifying Newtonianism – not even as a mere exercise. (Later in life, as Poynting observes, he became much bolder; but he died young and so he never returned to that exercise.)

And then came Einstein, and he openly and frankly declared science to be fallible. And people argued about Einstein in street corners and boulevards, in beerhalls and saloon bars, in salons and coffee-houses; not so much out of interest in atomic energy or in the curvature of space as out of interest in the intellectual role science was going to play from then on: was it going to retain its hegemony? If not, are we going to return to religion? If yes, how and why?

And the romantics now came for the first time with a serious argument against the rationalists. The argument has been very forcefully presented by Michael Polanyi, I think, in his *Logic of Liberty* and elsewhere. Some people believe in science, some people believe in the Bible, he said. It all depends on your starting-point, which is arbitrary and idiosyncratic; it depends on the tradition you align yourself with, and on the people you associate with. In science there is no objective standard of rationality; taste, style, idiom, and the like, matter as much in science as in the arts or in religion. Commitment is arbitrary; and where you stand is a matter of initial commitments.

The upholder of the objectivist tradition thus under attack, the old-fashioned rationalist epistemologists, themselves felt that some concession to the opponent was to be made. They admitted that those whose foot-steps they were following had exaggerated their claims for the stability of science; and hence they had also exaggerated their claim for the objectivity and rationality of science; and when the exaggeration was found out, so the story continues, those unable to discriminate threw out the baby with the bath-water, the exaggerations concerning rationality, with rationality pure and simple. Science, some of the old-fashioned philoso-

phers now say, is not certain but only highly probable. So, we may live to
see a revolution in science, a modification, or even a rejection, of some
well-established theory in science. But this is neither too likely nor too
unsettling: after all, Einstein's work is but a slight modification of
Newton's, and as modified Newton's work still stands, as safe as ever (if
not safer). It is out of all proportion to pin on such a rare and small need
for modification and improvement a licence for all arbitrariness and
caprice.

This is, at least, how Lord Rutherford saw the situation when he
delivered his somewhat cryptic presidential address to the British Association for the Advancement of Science in 1913, when he declared the view
that Einstein had overthrown Newtonianism to be a popular prejudice,
since, on the contrary, Einstein had "broadened its basis". How popular
Rutherford's view is nowadays I do not know, but I think it is very popular
indeed amongst those philosophers who try to retain the old standards in
minimum modifications, as well as amongst biologists. Amongst physicists
the more popular view is that of science as a branch of formal mathematics, or perhaps of applied mathematics, and hence as utterly certain.
View Newtonian mechanics as a system of total differential equations, and
you have no philosophical problems left. What is obviously common to
those who view scientific theories as empirically probable and those who
view them as mathematically certain is the view that the value of science
rests in its stability.

II. THE EMPIRICAL SUPPORT OF SOME SCIENTIFIC THEORIES
REQUIRES EXPLANATION

This is the way I see the two traditions in contemporary philosophy,
roughly the conventionalists, instrumentalists, pragmatists, positivists,
and analysts on the one side, and the irrationalists, including the existentialists, on the other; this is how I saw them ever since I became somewhat familiar with the contemporary philosophical scene (I do not claim
much familiarity with it). I have rejected the irrationalists offhand,
partly because of my faith in rationality and in science, partly because
they recommend arbitrary esoteric religions, from one of which I had
just run away. The irrationalist school I rejected after understanding it
or at least after feeling that I had understood it; the rationalist school

simply did not make sense, to me. Amongst my teachers in science some believed that science is a branch of applied formal mathematics, that the formulas of science are stable and hence important, whereas their interpretations are ephemeral and hence of little value. One of my teachers quoted in class with great approval Philipp Frank's comparison of applied mathematics to sewing machines and the interpretations of its formulas to clothings which are subject to the whims of fashion. I could not believe that men of science would debunk science so freely and voluntarily. To be sure, the mathematical relations between Einstein's formulas and Newton's formulas are mathematically not very interesting, and yet for the technicians they are very important and quite satisfactory. But the intellectual value of the Einsteinian revolution, the new picture of the world that he has created, the exciting questions about the nature of space and time – all these I could not believe my teachers were giving up or viewing as mere fashion. They insisted that the interpretation of a formula is never of great importance; in particular, they stressed, the interpretation of the formulas of quantum or wave mechanics are not important. These formulas, it is well known, are open to both the wave interpretation and the corpuscular interpretation, but either interpretation is rather problematic. Is the lack of interest in them but an attempt to ignore a problem? Max Born, in his attack on Schrödinger of 1953, has claimed that the problem belongs to the philosopher, not to the scientist. The problematic interpretations constitute answers to the question what is, really, an electron? This question, said Born, is entirely metaphysical, and hence outside the domain of science. My teachers in science went further. They said metaphysics always lags behind science, and so it is better ignored. This argument seemed to me to be besides the point, and dangerously erroneous. It is very difficult to develop a scientific theory, and the process of development may start with a new metaphysics which slowly becomes more scientific. It is easy to be tough toward theories which are as yet unsuccessful; it is harder to try and be appreciative of half-baked ideas.[2]

So much for the formalist instrumentalist philosophy of science; as to the inductivist philosophy of science, it looked to me even more puzzling than the formalist philosophy. Its main theses, you remember, are two. First, that scientific theories are highly probable so that they are not likely to be overthrown. I have heard Born estimate in a public lecture

the likelihood of a revolution against the quantum-theory at odds of one
against one thousand or thereabout. Second, that if, *per improbabile*, a
scientific theory is rejected, its main features are salvaged and reincorporated into the new theory. The first point, concerning the probability of
scientific theory, fills the literature. The second, concerning modification,
is passed over rather glibly, and for a good reason. What was considered
the main features of a theory prior to the revolution, turns out after the
revolution to be the less significant features of it. The indivisibility of
atoms, surely the central element of Dalton's theory, has not survived
the revolution, and the same can be said about the constancy of mass and
the ability of forces to act at a distance in relation to Newtonian mechanics.
If we observe closely any salvage operation, we shall clearly see that what
is salvaged is first, all the facts which the old theory had explained, and
second, some of the formulas of the old theory which are yielded from
the new theory as first approximations. So, at heart, the salvage operation
is explicable by a formalist instrumentalist philosophy, not by an inductivist philosophy; but the inductivist philosophy yields only high probability,
whereas the formalist instrumentalist philosophy yields utter certainty!
What use, then, do we have for the theory of probability?

The answer to this question concerns positive evidence. If the formal
theory is certain, its domain of applicability is not. Indeed, one may view
a revolution in science as a revision of the views of the domain of applicability of a given theory. This view was advanced by Duhem, and recently
repeated by J. B. Conant and Thomas Kuhn. What they all tend to pass
over with no comment, is the fact that when ascribing a domain of applicability to a formula we are making a hypothesis, and that the hypothesis
may be supported by striking positive evidence. For example, the domain
of applicability of Maxwell's equations depended on whether he has
described correctly the relations between the speed of light in vacuum and
the ratio between the units of electricity and of magnetism, and whether
he has described correctly the relations between dielectricity and re-
frangibility, etc. On all these points Maxwell got answers from his theory
that were later spectacularly supported by experiment. Why? How? Was
he a magician? Was he a prophet? Was he lucky? These questions are most
intriguing and the answer to them is less easy to come by than one imagines
when one simply rejoices in the success of science instead of asking how
it comes to be.

The fact that scientific theories are so spectacularly supported by positive evidence was explained by Bayes, Laplace, Whewell, Helmholtz, and many other thinkers, on the following lines. There are two possible explanations of the existence of positive evidence: first, that the theory it comes to support is true, and second, that though false, the theory agrees with facts by accident. The more improbable the evidence is, they said, the less likely it is that it fits the theory by accident and hence the more likely it is that the theory is true.

This argument, whether correct or not, belongs to the theory of probability or of games of chance or of relative frequency. Therefore, if the argument is correct, then he who endorses only theories which fit the facts well will be less often in need of changing or modifying his view than he who endorses testable theories which were not yet empirically tested; and the more improbable the evidence in favour of given theories, the less likely it is that they be overthrown or modified. This is the corollary from the probability theory of positive evidence which, historically, gave it its great importance.

Now, this corollary explains why in the 19th century there was little discussion about the difference between certainty and high probability. If we wish to be right as often as possible we aim at high probability, preferring certainty; but we do not complain about having attained only high probability any more than the owner of a vending machine whose machine accepts true coins and rejects false ones correctly with the exception of one in one hundred or so. Subsequent to this, we may see that the modern idea that the Einsteinian revolution has proved that we can aim only at high probability but not at certainty, is not as much of a change as was claimed. Further, the spirit of the 19th-century philosophy of science viewed any need for modification as a defect, quickly to be eliminated. It fully agreed with the probability theory of positive evidence which speaks only of the likelihood of a theory to be true or false, not to be in need of big modifications or small ones. Once modifiability is introduced, there is no knowing how the probability theory requires from us to treat positive evidence and its import. It is thus not surprising that when speaking of modifications philosophers seldom speak of the probability of the need for modification. The classical theory of probability, at least, applies only when every theory which is in need of modification is considered to be false. Classically, then, no matter what probability a theory has, it also is either true or false: amongst the probable theories, the majority are true; and amongst the improbable ones, the majority are false. It is this classical rational readiness to judge the truth of hypotheses by the strictest standards that many contemporary philosophers dislike. These contemporary philosophers wish to use probability not in the sense of likelihood to be absolutely true but in some weaker sense of truth; probability is often viewed today as truth-surrogate.[3]

 This change is not very surprising: the corollary to the classical theory which tells us that he who believes positive evidence is seldom in error, is false. Take any 19th-century thinker who believed only well-supported theories – William Hyde Wollaston should qualify very well – and ask how many of his beliefs are literally true. The answer is, of course, that none of them will pass today without some modification. This is, I suppose, what became obvious to some philosophers thanks to Einstein, and this is why they have tried to provide a new sense to the notion of the probability of a hypothesis, as I said. For my part, I still wonder, how were false theories such as Maxwell's so fantastically supported by empirical evidence? Was it really a most unlikely coincidence? It is very tempting to say that a miracle happened – perhaps to reward men of science for their efforts, perhaps to encourage them to make further efforts. How else can we explain the facts?

 To my surprise I find this a most unpopular question. Even Popper seldom faces it. Most scientists and most philosophers just rejoice in having positive evidence; perhaps they think that asking questions spoils the fun. A young colleague of mine has quite recently chosen, for some purpose or another, to discuss cases of happiness which are obvious and undiluted; he chose meeting one's mother in the airport, and the confirmation of one's hypothesis. When I read that I was grieved; I can well see how obvious and undiluted a case of happiness the meeting of one's mother after a long absence may be; but is it so incredible that one may hold a false hypothesis and tragically have it empirically supported again and again just when it would be so wonderful if he had given it up and tried new avenues? Is confirming one's hypothesis really as straightforwardly desirable as meeting one's mother after a long absence? Undoubtedly, people do get impressed by some positive evidence, and rightly so; also, it is a fact that they would all too often put their faith in a theory impressively supported by evidence; but is it so obvious that they are right? Once people believed in miracles as a matter of course; once people believed in the written letter as a matter of course; nowadays the desirability of believing in theories with empirical evidence supporting them is a matter of course. What of it? At all times and places some people believed in some omens and signs or others; does this prove anything? Is one faith so obviously much better than the others? Before we can leave this question, we may find it not so outrageously eccentric that someone wishes to know why is empirical support more imposing than miracles. But I guess I am crying out from the depth of my personal prejudices, which I am quite ready to expose.

III. THE DESIRE FOR STABILITY MAKES US SEE MORE OF IT THAN THERE IS

In my adolescence I found myself having Jewish religious education. I soon had doubts which led me to philosophy, and philosophy led me to science; by the time I entered the university I decided to study physics in order to have a better understanding in philosophy. Unfortunately for me, it is no longer customary to teach physics to budding philosophers, but rather to budding nuclear engineers and their like. I must have been quite a pest to my teachers. When they told me that I should postpone arguing for a few years and do my homework first, I retorted saying that I had heard the same proposal from my teachers in theology. When they said that the theories they were teaching me fit the facts admirably well, I said this had been the case with theories of earlier generations which are by now superseded. When they said all that physical theory has to do is fit the facts well, I asked what is so great about such a fitting, which they found all too infuriating to answer. When they said science was useful, I said I had not come to a university to master a trade; when they said it was beautiful, I said they meant it was arbitrary and not necessarily true; and I wanted nothing short of the truth. The fact remained. I wanted some metaphysical enlightenment, they wanted tools of a very respectable trade. I was cantankerous, and they wanted to be left in peace; and as long as they were doing their job reasonably well they felt entitled to it. My main teacher was a quantum physicist of some moderate reputation in the world, and his lack of interest in Einstein's work, his readiness to pay all the lip-service to complementarity as long as he could ignore it – all this shocked me. One remark I heard in a lecture on applied mathematics impressed me greatly: "One revolution per one generation or two is quite enough; we do not want too many of them." I had to leave my professors in peace, then, as indeed they were entitled to, but I remained cantankerous. I found soon, what I knew all the time, that others were even less willing to argue than my physics teachers. I soon found that I had to debate with various people, often students like myself and young academics, the pros and cons of debate. I should mention here Shmuel Ettinger, at least, from whom I have learned to contrast dialectics with apologetics. This is only one point, yet a surprisingly important and useful one.

But before going further into views on dialectics, I wish to stress one point. The main reason why I have later changed my attitude towards my teachers in science is not so much the realization that I had made much too great a demand on them in a rather adolescent fashion. The main reason why I have learned to appreciate them is that they did not try to satisfy my demand in an adolescent fashion. In the last century people like myself, who asked questions similar to mine, received an answer which entirely or almost entirely satisfied all or most of them. My teachers lived in the post-Einsteinian era. They did not know how to answer me, but they knew what answer will not do, and openly explained why. This seems to me rather of some significance. All too often we ignore Einstein's impact. Arthur Koestler, for instance, who is a remarkably sensitive intellectual journalist, has said recently that Einstein did not yet make an impact on man's changing view of the universe, because relativity still is not public knowledge. Koestler's oversight is the oversight of the best intellectual public which he has observed with his keen journalistic eye. Like so many philosophers and scientists, he has overlooked the fact that so much of the new science was made possible when the old attitude towards science had been demolished by the Einsteinian revolution, and this demolition surely is public knowledge. Consider, for instance, E. A. Burtt's *Metaphysical Foundations of Modern Physical Science*. This work contains a critical assessment of Newton's theories, the metaphysical ideas on which they depend, the tyrannical authority Newton had on his followers, etc. It is hard to imagine that Burtt could have written all this prior to the Einsteinian revolution. Yet Burtt was not aware of such an indebtedness, though he is quite willing to admit is as a possibility.[4] Whatever problems we have, even if they are not new, Einstein has made both the answers to them and the debates and arguments concerning these answers, quite new. This point may not be very striking; indeed, I think it will be taken for granted and agreed upon without much ado. Yet the fact remains that though Einstein's impact in this way will be acknowledged without much effort, Einstein's influence on the general outlook of the 20th-century man (other than his contributions to atomic warfare) is still viewed as minute. This may be due to mere oversight or to a simple inconsistency. But it need not be; one may explain this as the outcome of viewing debates and arguments, criticism and clearing the grounds, as of secondary importance in intellectual life, as mere preliminaries to the real work. My own opinion, for what it is worth, is not only that Einstein has had a wide and great influence. In addition, by having had such a wide influence, he has shown how important argument is; but on this score his impact is not as large as it might be. And so, in my view, the Einsteinian revolution may be pushed further and have a much greater general impact than it has had hitherto. But, if this needs saying, my view here is very much of a minority opinion. The majority, even amongst men of science, are still ambivalent, at best; even those who enjoy debate and argument, often tend to view such activity with some measure of distaste, as something not very useful and below a gentleman's station.

The best way to win an argument, goes a popular nasty quip, is to start with the right position. What position is right, of course, we do not know or else we would not start a debate; when we know, we all agree about what we know. Debate, I heard since my childhood, is the symptom of ignorance, and this is why politicians have debates but scientists agree amongst themselves. I hated this idea before I could say why; later I came to think it is the same idea as: clean people do not have to take a bath, only dirty people bathe regularly. And so my counter-move was to show that politicians seldom debate, even though they often give the impression that they do, by conjuring imaginary opponents and knocking them down with great ease. It was harder to show that men of science constantly dissent, and debate, and examine, and cross-examine, and re-examine – always ready to consider. There is, to be sure, as great a semblance of agreement amongst scientists as there is a semblance of debate amongst politicians. And this semblance of agreement deserves some examination, both in order to show that it is not correct and in order to explain its persistence. That it is not correct is not so easy to show; the task looks rather baffling. It was with me somewhat approaching a religious conviction. I drew great encouragement from the little I knew about Einstein, and from remarks of his which expressed the critical attitude which I was after. But much remained baffling and obscure. Much of this obscurity and bafflement is presented in an impressive story which I have recently heard from I. B. Cohen. It is about a classics professor and a philosophy professor strolling on the lawns of a small American college – I forget the names – one lovely afternoon a generation ago. "You may find me old-fashioned", said the classics professor, "but I confess, I do not believe the germ theory of disease." "Old-fashioned, my dear fellow, not at all", answered the philosophy professor, "I wouldn't call it old-fashioned, but plainly ignorant."

The jibe at the classics professor, which for all I know was well deserved, and the fact that a philosopher delivered it, raise this lovely anecdote to the level of a fable: like a fable in Christian mythology it comes to drive home an obvious truth: in science there are no fashions, only ignorance can be eccentric in any way. But take the germ theory of disease; ask yourself what the definite article in the descriptive phrase "the germ theory of disease" stands for. If you conduct an empirical research, as I have done, you will find that most people misuse the definite article here. (Philosophers have tried hard to find misuses of words, but they centred attention on such recondite words as 'exist' and 'God'; the misuse of the definite article, and of rather common names, so well-known to school teachers, has thus far escaped their notice.) You will find that with very few exceptions, when people use definite articles and proper names of scientific ideas, they are not able to identify what they so describe or name. I find this fact most amazing and disturbing, and I wish to report it as an empirical finding, and as one to which I was led by the influence of Popper's philosophy. For a time I was an avid student of the interpretations of quantum theory. I finally gave up hope to understand the literature, being convinced that there is no body of knowledge as the one labeled by so many physicists by the title 'classical physics', that no one knows what *the* Copenhagen interpretation of quantum theory is, and that there is no proof answering the description 'von Neumann's proof'. What even such a common name as 'Newtonian mechanics' stands for is vague. It is fairly clearly – though not too clearly – identifiable, if the class of people who use it has a fairly homogeneous background. To some of the better contemporary authors, this expression denotes the theory discussed in the early part of Einstein's classical *The Meaning of Relativity*; to others, including my own teachers, something much more similar to what was presented by the late 19th-century Cartesians – Thomson and Tait – and positivists – Mach and Poincaré; to most historians of science, the phrase denotes the ideas expounded by Laplace, and to erudite historians of science, the ideas expounded by Newton. But let us return to the germ theory of disease. If you take a sample of people who use it, laymen, professional scientists, or medicine men, you will get answers vague or divergent. The germ theory of disease was originated by Pasteur and contested by Koch. In case you are interested, Koch won in the debate, but his ideas have meanwhile been superseded too. No one in his senses believes the germ theory of disease today: we all are all too familiar with the fact that some diseases are caused by very different factors – genetic factors, nutritional factors, malfunctioning of internal organs, whether caused by virus infection or not – and some diseases defy all etiology (cancer being one of these puzzles). So what do people mean when they profess belief in that theory? At best they mean to express their opinion that syphilis and malaria, for instance, and sometimes pneumonia, are caused by germs. If they are ignorant enough, they also think that diphtheria too is caused by germs (rather than by the toxin that certain germs secrete) or even that rabies is caused by germs (rather than by viruses). Anyhow, I think it is a fact that all well-informed people agree that malaria is caused by germs and scurvy by malnutrition of sorts (vitamin C deficiency). Does this not prove that in science some agreement exists?

It does. And, indeed, there is some agreement almost anywhere – between scientists and even between politicians, between friends and even between foes; but the stress on agreement between scientists as opposed to the disagreement between politicians, the claim that agreement in science is good and argument is a symptom of ignorance – this claim is very far from having been established. We may ask: why do people agree so much? Why in particular do scientists agree? Why do they agree so widely on what are the facts of the matter and on what are the theories the facts support and on our reasonableness in accepting supported theories?

Let us pause here for a moment and reflect on what I have presented

thus far. What I have tried to offer is an admittedly subjective picture of
the intellectual background in which I found myself as a student, and
which has made my encounter with Popper's philosophy, and with the
man himself, the most crucial influence on my outlook. But let there be
no mistake: my own experiences of before having met Popper are here
cast so very much in a Popperian frame as to be highly suspect historical
records. It is a well-known fact that we see facts from given perspectives,
so that any different perspective makes us see the same facts differently;
but as I have not tried to set a historical record here, I need not bother
about historiographic problems here. What I have tried to show, is how
different the situation looks from different starting-points. Most philosophers take it for granted that science is agreement and agreement is good;
their problem is to justify a principle which they think is responsible for
the agreement. My own starting-point was so very different. Science had
disappointed me, but I refused to give up interest in it. Consequently I
was prepared to doubt most of what scientists say about science. Consequently, my problem was how to explain unanimity in science. How do
you explain the fact that one Albert Einstein, a lad in his mid-twenties,
unsuccessful in his search for an academic career, could upset all well-
established scientific doctrine and become the new leader of science? The
traditional answer is, of course, that his doctrines were empirically
proven. So were the doctrines of Newton. Can we never learn from experience? Is science a Latin-American dictatorship?

Popper's theory of science as disagreement rather than as agreement
may offer new answers to such questions. I shall now very briefly sketch
his theory, say how much of the situation it explains, and how we may
push the explanation further.

IV. POPPER'S THEORY PRESENTS SCIENCE AS AN ENDLESS SERIES OF DEBATES

Popper's *Logic of Scientific Discovery* comes to solve the problem of
demarcation of science, and the problem of induction. Customarily,
scientific theories are viewed as stable or near-stable – as true, as essentially
true, or as highly probable; consequently, by the customary view scientific
theories are less prone to upset and overthrow than pseudo-science and
superstition. Popper, by contradistinction, views scientific theories as

more prone to overthrow, as more ephemeral, than non-scientific doctrines – at least in the sense that when we try to render a scientific doctrine less prone to overthrow we thereby make it less scientific and more pseudo-scientific. So much for the problem of demarcation. The problem of induction has many formulations and variants, and Popper chose the following: how do we learn from experience? The usual answer to this question is that experience leads to stable beliefs, to the adherence of established theories. Popper's answer, by contradistinction, is that learning from a piece of experience is the very act of overthrowing a theory with the help of that experience.

Popper's view, then, is that a theory is scientific if and only if it can be overthrown with the help of experience, and that we gain theoretical knowledge from experience when and only when such revolutions occur. This raises a few problems, chiefly the problems of the stability of science. What causes research to coordinate? What makes us accept empirical evidence? What causes unanimity in science and technology? What is the role of positive evidence in science and technology?

Popper's theory of positive evidence is the better-known part of his teaching. People often identify it with his theory of learning from experience: they take it for granted that he joins the multitude in viewing learning from experience as finding empirical support to our theories. Those who think so are advised to glance at his book[3] and find there with very little effort that he declares learning from experience to be not by positive evidence but by negative evidence. What Popper initially said about positive evidence in his early days concerns not the role of positive evidence, but the nature of positive evidence. He said: If we search for positive evidence, the evidence will not be scientific, the only way to find positive evidence recognizable by science is by looking for negative evidence. If you look for negative evidence, says Popper, you may find it, or else you will find positive evidence; if, however, you look for positive evidence, you will find only positive evidence, but which is of no scientific value. The question was immediately raised: If you want positive evidence, why not look for it directly, why all this round-and-about way? The answer is, of course, that we benefit from negative evidence, since only by it do we learn from experience; so we should look for negative evidence if we wish to improve.

Some people think Popper cannot seriously mean what he says here.

Some think he is being perverse or masochistic. But what do these people say when they find Socrates make these same remarks, as for instance in. *Gorgias* and *Greater Hippias*? Well, many scholars think that when speaking so Socrates is being ironical, by which they mean tongue-in-cheek and even sarcastic; others think that on these occasions he is merely being educational. But in *Phaedo*, which is likely to be a rather near-stenographic record, Socrates says: If all your life you have backed the wrong horse, you are better off switching horses in your last hour, than not. Dividing people into black and white is always silly, to be sure, but I confess to the folly of dividing people into those who see Socrates' point and to those who shrug a shoulder when told of it.

Science, then, is to Popper a special case of Socratic dialogue, with experiment and observation offering new arguments, or new empirical criticisms, to use Faraday's idiom. And when an attempt at empirical criticism misfires the result is positive evidence. What, however, is the role of positive evidence? Let us take seriously Popper's theory that science comprises of series of conjectures plus the refutations of some of them and nothing else. We are then bound to say that whatever role positive evidence may play, it cannot play any role *qua* positive evidence. It may stimulate the invention of a conjecture and it may be not only a positive support of one conjecture, but also a refutation of another. This, and no more. To show that such must be the case, let us construct a thought-experiment: consider a universe in which science is almost like ours, with series of conjectures, some of which are tested, but in which, by luck or otherwise, every test is successful; that is to say, in that universe every test refutes a theory. Query: does that universe have science proper or not? Popper's answer must be in the affirmative, which proves that to him, positive evidence *qua* positive evidence plays no essential role in science.

I have put this point to Popper a few years ago. His present view is this.[5] He says a good theory should be not only capable of being refuted, but also it should not be refuted too soon: we want positive evidence before we get negative evidence. We want positive evidence so as to be assured, he adds, that knowledge grows. So now, it seems, Popper has changed his view of science from a theory of conjectures and refutations to a theory of conjectures, corroborations, and refutations. In a footnote, however, he denies that any change has occurred when he says: "I feel greatly indebted to Dr. Agassi for drawing my attention to the fact that

I have previously never explained clearly the distinction between" the desirability of having refutable theories and the desirability of their not being refuted in the very first test. So Popper says he has not changed his mind, merely explained clearly what he had explained earlier. It is not that Popper had not seen that the requirement for testability is different from the requirement for corroboration, of course; it is merely that he had overlooked, he says, the need to explain clearly the other requirement. As I say, either Popper assigns no value to positive evidence *qua* positive evidence, or he is in the same boat as the inductive philosophers who cannot bring positive evidence to support their theories of positive evidence. Popper himself, in the same place, admits that much, when he concedes to me that his view contains "a whiff of verificationism here", but he thinks I am worse off without it, since Popperism without it contains "a whiff of instrumentalism".

Here I disagree on general methodological grounds. Pure Popperism may be instrumentalism, or even worse; modified Popperism, whether Popperism plus the addendum that it is nice to have positive evidence from time to time, or Popperism with any other addendum, may be quite true for all we know. But the rules of the game forbid putting addenda without extensive study. This may be easily seen when we take other case-histories.

P. W. Bridgman was shocked by the Einsteinian revolution, and felt that the role of the philosopher is to insure the prevention of such mishaps in the future, by offering maximum stability to science. And he tried to reform science by introducing the maxim that all concepts of science should be operationally understood or else expurgated; e.g. 'length' should mean the measuring of length. Bridgman's philosophy is admirable, not only in its immense tenacity, but also in its being an intended reform of science rather than the usual philosophical approval of the scientists. In his tenacity and reformism Bridgman did not shrink from rejecting general relativity because according to it the measuring of a length is not identical with the finding of that length, but, at best, with approaching that length asymptotically. Einstein himself was sympathetic to Bridgman's views, which in part he is the coauthor of, and he conceded that had operationalism worked in all other instances of science except general relativity, Bridgman would have had a somewhat significant point against it. But Bridgman's original program does not work at all. He himself has conceded that much when he admitted that the list of operations on which

science rests is larger than he had originally asserted: he had to add to
his initial list of operations, which contained measuring, observing, and
the like, the operations he called paper-and-pencil operations, and which
we normally call thinking and calculating.

There is little doubt that though Bridgman's philosophy had initially
raised much opposition, practically all its original opponents dropped all
their objections after Bridgman has inserted his correction. Indeed, they
lost all interest in it. The reason for this is obvious: though the corrected
version of Bridgman's philosophy is true, it is no answer to Bridgman's
original quest. He had developed an austere program for science in order
to keep it absolutely stable. He then relaxed his stringent requirements,
and the desired stability is all gone. We are now where we were before
Bridgman started, though made wiser by his failure. His correction of
the failure, however, is hardly more than an admission of failure, though
in a face-saving manner he has pretended to be merely modifying it.

Lest anyone think that failure is loss of face, that saving face is in
any way useful, let me mention one great failure in the history of 20th-
century philosophy, which is also one of the most monumental achievements of the century. I am speaking of logicism, namely the view that
mathematics is a part of logic. The idea is ancient, indeed a cornerstone
of pre-Socratic philosophy. It became so important when modern logicians, Frege and Russell in particular, showed both how much of mathematics is a part of logic, and also, that not all of it is. I shall leave this
point at that, referring the interested to the most critical and most appreciative recent paper by Alonzo Church on the subject.[6] What I wish to say
briefly here, is that Popper has a program akin to logicism and that he
has deviated from it as Russell did, but not in the same manner, since
Russell was aware of his failures. Popper's program belongs not to the
philosophy of mathematics, but to the philosophy of science. And his
program was to reduce methodology to dialectics. What is dialectics is a
problem too (akin to the logicists' problem, what is logic). For the time
being let us assume that dialectics is the act of process of Socratic debate,
where the immediate aim is to show one's interlocutor to be in error, and
where the very final end is the attainment of the truth. In particular,
dialectics does not lead to knowledge; its final end, to use Socrates' own
metaphor, is the twilight-zone between knowledge and ignorance, where
even if we do possess the truth we cannot prove it.

Assume, then, that science is that kind of dialectic process. How much of science makes sense, then? This is a tricky question, since, if some of science makes no good sense according to this view, we may follow one of two opposite alternatives. The one alternative is to give up this view. The other is to reform science. Indeed, if Popper were not a reformer like Bridgman, but a justifier like Reichenbach, I doubt that he would be as interesting as he is. Yet there is a risk here of becoming a dogmatic Popperian and of always blaming science when it fails to stand up to the Popperian standards!

Indeed, this is true. Let us, then, explore this danger of dogmatism and see what it amounts to. Let us, in other words, first see how much of science is made sense of, and how much is left out, if we take Popper's program in its rigorous version.

V. POPPER MAKES ADDITIONAL ASSUMPTIONS

One additional assumption I have mentioned is the assumption that we need positive evidence from time to time. There is another additional assumption here, which I think Popper is ambivalent about: he sometimes implies it, sometimes its rejection. The assumption is that positive evidence renders a hypothesis credible. I do not know what this has to do with dialectics at all, and I do not know what is Popper's theory of belief. A third assumption is that factual evidence must be believed, or else we may all become dogmatic. I dislike this assumption, even though it is tempered by the dialectical assertion that factual evidence too may be criticized and shown to be in error. In *Gorgias* we find Socrates handicapped by his interlocutors' rather dogmatic resistance to the need to admit the truth of some factual information. I think he managed very well without complaining that in this manner they were breaking a rule of the dialectic procedure. In the history of science we also find men of science refusing to accept factual evidence merely because it conflicts with their views. Dirac's case is one well-known instance.

It seems to me that originally Popper tried to make his theory of science as formal and rigorous as he could. He required that of all criticizable hypotheses that hypothesis should be taken up first, which is most highly criticizable. There is clear reason for this additional requirement in the theory of science as critical debate. C. S. Peirce had already made the same proposal, but he, at least, made it on the ground of economy, which may indeed be justified occasionally, as we shall see. It is not clear by itself we must consider first the most criticizabie theory available. Nor is it clear by itself why we should accept every evidence against the theory if we cannot refute that evidence, and accept the theory if it is well supported by positive evidence. Why do we need so rigid a view of science? Is it a corollary of the view of science as a special case of dialectics?

Not at all. The main problems concerning science, the problem of demarcation and of induction, are solved by the very idea of Popper's program, and are indeed solved in his classical *Logic of Scientific Discovery* prior to any discussion of degrees of testability, of the empirical basis of science, or of corroboration. What do these additonal ideas come to explain? What problems do they come to solve?

The roles which the additional assumptions come to play are two: first to insure that the game of dialectics is played fairly, and second to explain the apparent relative stability of science and the coherence of research in various fields and within any one field. Let us take the first poiɴt first. Popper seems to fear that dogmatism may creep into science through the back-door, by the choice of barely testable auxiliary hypotheses, by doubting unpalatable factual evidence, and by hosts of other shifty techniques and clever twists.

In an attempt to block all attempts at dogmatism Popper has entirely isolated the open-minded from the possibly dogmatic and the dogmatic. I do not like this even as a limiting and ideal case. Why should we always be able to discern the possible apologist and dogmatist from the critically-minded and open-minded? Why not allow all practices other than the clearly apologetic ones, and leave it at that? Popper's claim that unless we have a criterion for acceptance and rejection of factual information we may permit dogmatism is correct, but his conclusion that we should therefore have such a criterion is a *non sequitor*, and his criticism of those who have failed to do so is invalid. Rather, we may try to explain how was it possible to develop science in relative freedom from apologetic disputes and other dogmatic practices, even though there was no good criterion of acceptance and rejection of factual information?

This is one of the many problems concerning the relative stability and coherence within science which Popper may have tried to explain. For instance, people prefer theories which are based on many facts and which are simple. Popper claims the first preference to be a misstatement for the preference for theories of high explanatory power. And he also claims that high explanatory power, as well as simplicity, are monotonous functions of high testability. In other words, Popper claims that given the preference for a high degree of testability, all accepted proper preferences can be explained. My point here is not merely that Popper is in error here, but also that his idea is an auxiliary hypothesis which amounts to giving up the whole program.

Whether the program has to be given up I do not know, but I think it is worth further examination. I think it is clear that even if Popper's theory worked well, one might well question his additional requirements and auxiliary hypotheses – even refute them by historical evidence. The question is, has the program thus far failed?

Thus far, I think it has not. It seems clear that much as he fought against his narrow positivist environment Popper was not entirely immune to its influence – which should not be so surprising, especially in view of the fact that his positivist colleagues regularly accused him, as they still do [7], of exaggerating differences with them so as to stand out more distinctly. If one wishes to rectify Popper's program, one must start from the very start, from Popper's view of dialectics and of rationality, indeed, and expurgate it of the traits of positivism which are manifest almost everywhere in his writings, at least to a person as allergic to positivism as some of us are.

VI. RATIONALITY IS A MEANS TO AN END

Theories of knowledge and theories of rationality since antiquity centered on one major problem. It is the problem of the foundation of knowledge and of rationality. In Indian geography, you remember, the earth was made to rest on an elephant, the elephant on a tortoise, and so on; yet the suspicion remained that below it all was a fathomless abyss. An infinite regress to a bottomless pit where, as Democritus tells us, truth resides, is equally discomforting. How can we find a foundation of knowledge and show that truth does not forever reside in a bottomless pit? So was the problem set in the 18th century – by Diderot, by Kant, and by others – and so it remained at least until the end of the 19th century, as long as all important thinkers were deeply convinced that truth is not forever hidden in a bottomless pit but was made manifest through science which is stable and which rests on stable rational foundations. Many thinkers rejected existing views on what the foundation was, or rather on what the foundation of the foundation was. Although science was manifestly well-founded, they found the foundation of the foundation annoyingly elusive, and they sometimes viewed this fact as a scandal in philosophy, to use Whitehead's apt phrase. Yet as long as Newtonian mechanics was there – unmovable and as stable as ever – there was no doubt that the foundation was there too. So stable was the view that Newtonian mechanics was stable, that it spilled over well into the Einsteinian era. In Wittgenstein's *Tractatus* of 1921, for instance, it is taken for granted that, as everybody knows, Newtonian mechanics is unshakeable. But Einstein's impact was soon felt, and the generation of philosophers who have heard of him in their youth took account of him – and, as I said, in two diametrically opposite ways. The irrationalists concluded from the Einsteinian revolution that science is in flux and hence not intellectually satisfying, and hence philosophically uninteresting. The rationalists who followed Einstein faithfully in physics, tried to minimize the Einsteinian revolution by pooh-poohing some of the ideas which Einstein overthrew and by declaring that the other ideas which Einstein overthrew were soon reincorporated in physics after minor surgery and face-lifting. As to the foundation of science, they saw little need for modifying the old-fashioned theory of the foundation of science. They merely changed the idea that science can attain certitude, with the idea that it can attain near-certitude, that certainty can be approached only asymptotically, by ever increasing the probability of our theories. Of course, this left the problem of the foundation of science untouched, as the infinite regress is hardly different when the foundation of science is certainty or the near-certainty of high probability. Indian geography would be no different either, if the earth stood on a three-legged tortoise – or was it an elephant – instead of a four-legged one. But to come back to the impact of Einstein.

What is common to both new schools of thought in the post-Einstein era, the pro-science and the anti-science schools, is the idea that if science is totally unstable, then it is intellectually of little value. This is why the one school sees it as intellectually valuable and near-certain or highly probable, and the other as highly alterable and thus as intellectually valueless. The premiss common to both – that if science is in flux, then it is intellectually valueless – was contradicted by Popper, who considers the proposal of theories and the successful criticism of them both as the method of science and as the method of rational thinking – of intellectual development and intellectual achievement. Though he insists that the aim of science is truth, he has given up all hope or aspiration for certitude or high probability or any other stabilizing factor.

To avoid misunderstanding let us stress that the word 'criticism' is here used in its commonest and broadest sense in which opinions can be criticized – whether in every-day use or scientific use. In addition, it should be noted, practically all rational philosophers have stressed the value and usefulness of criticism. Yet they almost all viewed criticism – the elimination of error – as a preliminary to the positive advancement of the understanding. Popper almost alone, and alone in our century, has claimed that criticism belongs not to the *hors d'oeuvre*, but to the main dish.

Popper has initially thought that all criticism is either empirical or logical. This led him to view moral proposals and metaphysical and theological statements beyond the realm of rational debate. This was much better than the positivistic view of such statements as meaningless, but still not good enough, as Popper himself now stresses. His development came through his study of the ancient problem of rationality, and the contemporary irrationalist critique of classical rationalism. He claimed in his *Open Society* that indeed rationality has an irrational basis, but that it is most rational to minimize that basis. The minimum, he claims, is the Socratic assumption that we can learn from our mistakes. Once this idea was assumed, its very austerity ousted the idea that criticism in morality and metaphysics is impossible, and in the English edition of his *Logic of Scientific Discovery* of 1959 Popper makes this correction.

This is a nice case of how austerity of assumption may make one more broad-minded, contrary to what may perhaps be intuitively expected. Moreover, quite apart from whether we should be open-minded or not, we may find it very exciting to view science as a special case of Socratic dialogue, to wit the case of Socratic dialogue in which the criticism is empirical. But the theory of Socratic dialogue as presented in Popper's *Open Society* (Chapter 24) is in need of modification, since in order to be effective, criticism must be selective and discriminating; and since in order to be properly selective and discriminating, one needs some criteria for selectivity and for discrimination. Popper himself may have noticed the need for modification, and perhaps this is why he spoke of the need for orientation of rational thinking towards specific problems. He may have noticed that one has to speak of the aim of science as something more specific than the aim of Socratic dialogue in general. For my part I do not think so. I do not think that Popper himself thinks his own view of rationality as presented in his *Open Society* is in any need of modification. But it is. The view advocated by Popper is: rationality is critical debate; the modification suggested here is: rationality is not any critical debate, but only that which is oriented towards a specific goal as well as might be reasonably expected. To show that the modification is not vacuous, you may consider scholasticism or Talmudism, which is definitely dialectical but, one might claim, not very rational. The modification proposed here takes care of such cases. The modification, I dare say, is very much in the spirit of Popper's philosophy, but I do not think Popper's own. If it turns out to be rather insignificant or trivial, I gladly claim authorship of it; if, however, it turns out to be even half as interesting and rich in implications as I think it is, I must set the historical record straight and report it as the result of a group-effort, to which Popper, Bartley, and all other known disciples of Popper belong.

Consider, again, scholasticism. Doubtlessly, it was critical towards its own minor assumptions and defended its major ones by any amount of auxiliary hypotheses. Here, then, Popper's proposal to step up the degree of testability of the assumptions under consideration is justified by the desire to avoid stagnation. It is far from obvious, however, that in all cases Popper's proposal is equally welcome. One may imagine cases, and even look for historical cases, where it was more profitable to consider a less testable but more promising ideas in terms of offering stimulus to thought or a unifying theme to diverse scientific problems, a kind of world-hypothesis or a metaphysics. If need be, one may wish that after some time effort be made to render the interesting new idea testable. As long as it is interesting and engaging, I do not like a Popperian policeman to tell me that it is still metaphysical and untestable, and that therefore I should leave it alone.[2]

The modification here proposed solves other problems as well. Assuming that rationality is solving certain problems as best possible and subjecting these solutions to critical debate, one may characterize science as a proper special case of rationality, while using the following two criteria. First, the problems of science are related to explanations of facts, or more generally, to comprehension of the world. Second, the criticism of science is empirical. Popper himself has characterized science by the second criterion alone. Namely of empirical criticizability. He thought, I understand, that when we take care of the second criterion, that will also take care of the first: he thought, in other words, that good explanations are well open to empirical criticism, and *vice versa*. On this he is in error, but we can easily see that on the correction of Popper's view of rationality and on holding to his view of science as a special case of rationality, this error gets eliminated rather naturally.

To show, again, that the modification is not vacuous, let us consider examples. Ambroise Paré refuted many medical superstitions of his day (such as the alleged healing effect of consuming bits of Egyptian mummies). It was indeed because many superstitions were refuted in the Renaissance that refutability came to be considered a vice rather than a virtue. Compare this with the – admittedly regrettable – irrefutability (or seeming irrefutability) of Einstein's unified field theory or of Faraday's adumbration of it!

Popper likes to liken scientific refutability with the ability of democracies to overthrow their governments with relative ease – which he even elevates to a definition of democracy. But, again, his doctrine seems wanting, as Judith Agassi has pointed to me, and in its very theory of rationality: Popper's theory of democracy has no reference to ends; it is not only the ability to overthrow governments with ease, but also the ability to thereby achieve better conditions for the furthering of certain aims, which should count: and these aims in themselves should be democratic in another sense, such as having a society of relatively independent-minded and relatively responsible and relatively educated and free citizens. There were periods in history where governments in a given country were overthrown in succession but for no avail and leading but to frustration (e.g. post-war France). Again, we see, the modification of Popper's general view of rationality itself may suggest an improved version of a specific theory of his – in this case of his theory of democracy.

The interesting task, now, is to follow through Popper's idea of science again as a special case of rationality. It may turn out that this idea presents science as being much more in a state of flux than it really is, that the existing consolidating factors in science must be either erroneously overlooked by that theory or taken account of by adding to it extra assumptions, or by making some additional rules, perhaps, in order to render Socratic dialectics into the dialectics of science. If it turns out that one may explain the existing stability of science only by additional hypotheses, one may well prefer to condemn the stability than to add the auxiliary hypotheses. The question then will be: How shall we decide between these two, and can there be a rational method of arbitration? For my part, I think some of the stability may be explained by reference to the aim of science and our inability to execute them fast enough, some of the stability may be explained as the unwelcome intrusion of dogmatism and pomposity into science. The latter case – of intrusion – is very useful to study and experiment with, but it is not terribly interesting, except in that Popper's theory does have practical reformist consequences. The idea that much of the stability is due to certain human sluggishness turns out to be more interesting, I think.

Assuming, for instance, that the faith in positive evidence is some residue from religion, we may try to explain it rather than worship it. Different cases of positive evidence require different kinds of explanations, perhaps, or at least allow for different kinds of explanation. Einstein explained the positive evidence of Newton's theory of gravity by his theory of gravity, plus claims such as that the field of the sun's gravity is not too strong, etc. Doubtlessly, the positive evidence is thus explained partly as luck, partly as coming near enough to the truth. If Einstein's theory turns out not to be the truth we may, indeed, feel the need for a different explanation of the positive evidence in question. One way or another, we can see that the more imaginative we are the more we may design new ways of viewing positive evidence. No doubt, the positive evidence concerning the success of different theories is even today radically different; the evidence for Rutherford's theory of the atom, for instance, is often viewed as a kind of a fluke. How we explain positive evidence, then, much depends on our theory of the world in general.

The aim of science, to be sure, is the true explanation of phenomena – but only partly so. It does not attempt to cover all the phenomena at once, but it does attempt to provide the true world-view, the true metaphysical blueprint of the universe. Science, accordingly, not only proposes explanations to be examined, but also starts with metaphysical theories within which to incorporate scientific theories in manners which can be critically examined as well. The metaphysical frameworks may be criticized and rejected; they also provide some stabilizing elements as they are less easily criticizable than scientific theories and thus less ephemeral. They are less ephemeral because of our lack of ability to render them testable, at least not so quickly. Hence the stability is due to some defects! Similarly, we do not have to accept empirical evidence as authoritative. The aim of science is to have empirical information explained, but we may explain information as false, reinterprete information as merely approximation to the facts in order to render it consistent with our theories; and so on – as long as we are not too apologetic in doing so, that is to say as long as we present our views because they may be interesting, and as long as we remain open to criticism, there is nothing wrong in our stubborn refusal to toe the line. But this is easier said than done. In particular, we may always reinterpret corroborating evidence in a manner which will render it more as criticism than as corroboration, but the effort is not small! How much we are able to reinterpret and review depends on our imagination and ingenuity.

To take an example from a current situation, many dislike general relativity and they try to offer alternatives to that theory, which should explain at least all the facts which that theory explains; as that theory is corroborated, their task is not so easy, and general relativity still lacks serious competition. To show how differently corroboration may look relative to other ends, take corroborations in technology, where positive evidence has to be acquired by a public organization prior to the implementation of a technical innovation, such as civil jet-plane service. The positive evidence does not prove the organization's decision to be correct; it does, however, prove it to be responsible. That in spite of all positive evidence in favour of a new technique, that technique may be disastrous is well known from quite a few dramatic cases. There is no doubt, also, that severe testing does eliminate some of these techniques before they even enter the market. Assuming, then, that the end of testing in technology is to eliminate some undesirable technical innovations, the value of corroboration is of evidence of having done one's fair share in avoiding catastrophe. How different is the corroboration of some new aerodynamic formula in an airplane plant from the positive evidence in favour of general relativity is intuitively obvious; the explanation of this difference, I suggest, is in terms of the different cases of rationality involved, namely, the different purposes at hand.

Here I have touched upon the topic of the social stability of science as applied science or technology. There are other stabilizing factors in the social setting of science, such as the fact that school-teachers cannot be up-to-date in science but must follow certain requirements of school-boards. The requirements of school-boards, the requirements for industrial standards of severity of tests, etc., are set by social conventions. And though social conventions are indeed alterable, they must be stable to some extent. But this is no reason to discourage fluidity in scientific research: our lack of imagination is a sufficiently strong factor in preventing high speed. Indeed, when we look at the geniuses of science, at the enormity of their ideas, at the fruitfulness and ease with which they can turn out new theories and new criticisms, we may well see that science would be very different if we all were such geniuses. Philosophy, said William Gilbert, is but for the few. This is no longer true, and Gilbert would have enjoyed seeing the thousands of scientists who live today. They are not all Gilberts, to be sure, and their work is less revolutionary than his. But even small thinkers can try to appreciate and to criticize and to imitate big thinkers, and there is no need to fear that if all scientists were Einsteins, science may become chaotic. Perhaps there would be chaos in the arts, too, if we had many Beethovens or Cézannes; but we do not preach stability in the arts. If and when we shall have too many Einsteins, or Beethovens, we may tackle some problems that this may give rise to. Meanwhile, we can well try to appreciate the revolutionary effect of some scientific geniuses. I have chosen to conclude with one very brief example, from the work of Faraday.

Michael Faraday, who interpreted Newtonian mechanics as an approximation to a future gravitational field theory, admitted that he was rejecting the theory which had been as well corroborated, and as impressively as could be. He rejected it on metaphysical grounds and he argued very rationally about it. He invented a few ideas about gravitational fields and refuted them all. He accepted the refutations but not as refutations of his metaphysics. At least not at the time. He was a worshipper of facts, yet he was able to dismiss some information as false by his very fact-worshipping, because he declared that these reports were interpreted and he accepted as phenomenological only such reports as were neutral between his opponents' views and his. One might object that such behaviour is dogmatic; Faraday must have worried about this objection, since he answered it: he said his opponents could not reinterpret his evidence against them half as well as he could do this for them, and even he could not do this for them half as well as he could for himself. Finally, then, the advantage of his field-metaphysics was that it was the most imaginative and provocative available.

Perhaps nobody wishes to curb the imagination of men of science; I do not know. It is my impression that imagination, critical debate, interest in science, all these beautiful things are still viewed with suspicion, and for many people science and scholarship must be tedious and dull and pedestrian before they can appreciate it. Be that as it may, the theories of science which have gained traditional approval are certainly constraining, and more often than not they obviously fail to do justice to the imagination and to criticism. Popper's theory of science as imagination checked and stimulated by criticism is at least in intention so very wonderful. I have tried to examine how much of science it makes sense of without additional constraints and *ad hoc* hypotheses. If, contrary to my view, Popper's program cannot be executed in its purity, we may try something entirely different, with the hope that we shall not do injustice to the value of creativity in free imagination and rigourous criticism.

Boston University

REFERENCES

* I am indebted to my wife Judith, to William W. Bartley, III, and to Robert S. Cohen, for their patient reading of many drafts, and making many corrections and suggestions.
1 See my *Towards an Historiography of Science*, The Hague 1963.
2 See my 'The Nature of Scientific Problems and their Roots in Metaphysics', in *The Critical Approach: Essays in Honor of Karl Popper* (ed. M. Bunge), New York 1964.
3 K. R. Popper, *The Logic of Scientific Discovery*, New York 1959.
4 I owe this to a private conversation with Burtt.
5 K. R. Popper, *Conjectures and Refutations*, Chapter 10, 'Truth, Rationality, and the Growth of Scientific Knowledge', § 5.
6 Alonzo Church, 'Mathematics and Logic', in *Logic, Methodology, and Philosophy of Science* (Proceedings of the 1960 International Congress) (ed. E. Nagel, P. Suppes, and A. Tarski), Stanford, Calif., 1962.
7 See Rudolf Carnap's 'Reply to Critics', in *The Philosophy of Rudolf Carnap* (ed. P. A. Schilpp), Evanston 1964.

APPENDIX: THE ROLE OF CORROBORATION
IN POPPER'S PHILOSOPHY

The following is a comment on D. Stove's review of Popper's *Logic of Scientific Discovery*, 'Critical Notice', *The Australasian Journal of Philosophy* **38** (1959), 173–187. I find that review a forceful expression of a very typical reading of Popper's view, namely, reading it as yet another theory of justifying scientific theories by favorable evidence. Indeed, in his review Stove discusses almost only one point, Popper's theory of corroboration. He finds in this theory two elements. The first is Popper's emphasis on the significance of the sincerity of the attempts to refute existing scientific hypotheses. This element Stove considers to be psychological and irrelevant to the problems at hand. The second element is Popper's view according to which a hypothesis is corroborated only when the (sincere) attempts to refute it have resulted in failure. This Stove considers a rather traditional solution to the problems at hand, and one which may be criticized by traditional arguments, as well as by new arguments which Popper himself states in his book.

I have used the phrase "the problems at hand" twice already; I could not be more specific as I am not quite certain what are the problems which Stove has in mind. I find it somewhat bewildering that Stove does not say more explicitly what problems Popper claims to be discussing, what problems Popper's theory of corroboration is intended to solve, or what problems Stove thinks this theory ought to solve. I wish, therefore, to supplement Stove's interesting but seriously limited review on these points as best I can. I shall explain why I think that his criticism is entirely valid, and that it is in no way a criticism of the book he was reviewing (except, perhaps, for a criticism of some ambiguity on Popper's part).

1. *Popper's Problems and Central Tenets.* Popper claims to have solved two traditional philosophical problems: the problems of induction (how do we learn from experience?) and of demarcation between science and non-science (by what criterion do we decide which hypothesis is scientific?). His solutions are these: (1) We learn from experience by repeatedly positing explanatory hypotheses and refuting them experimentally, thus approximating the truth by stages. (2) Those hypotheses are scientific which

are capable of being tested experimentally, where tests of a hypothesis are attempts to refute it.

(This is the core of Popper's book, as he himself claims, and as, I should think, is rather obvious anyhow. In this core there is nothing explicit about corroboration, namely about the failure to refute hypotheses; but only of success in refuting some hypotheses, which success is alleged to constitute learning from experience. And Popper's criterion of demarcation does not distinguish between refuted and corroborated hypotheses. By Popper's criterion even hypotheses which were amply refuted by experience are fully entitled to the honorary status of scientific hypotheses.)

Stove's review relates almost entirely to Popper's theory of corroboration; he devotes two or three rather short paragraphs to the core of the book (pp. 173–174), without relating this core to the theory of corroboration in a clear and explicit manner. Moreover, in these two or three paragraphs he commits some significant errors of presentation.

For example, he attributes to Popper the view that "there is no problem at all about induction" although Popper stresses the opposite, namely that Hume's criticism, which establishes the invalidity of inductive inferences, leads to a few genuine philosophical problems, one of which is the problem of induction, namely, 'how do we learn from experience?'. Stove, apparently, translates 'how do we learn from experience?' into 'how do we know which of our hypotheses is true or at least probable?'. For Popper this traditional translation is a mistake. Stove, however, does not notice this, and having attributed the wrong question to Popper, he thereby attributes to him the wrong answer, to wit, the view that a well-corroborated theory is likely to be true.

2. *The Traditional View of Scientific 'Success'.* Undoubtedly, the attitude towards Einstein's general theory of relativity changed dramatically with the result of Eddington's observation of the total eclipse, which agreed with the theory rather well. Such a result is viewed by physicists as a great "success", a "verification", a "proof", or "confirmation". Most physicists, I contend, display great pleasure when confronted with "verifications" and give them much publicity, while (with the important modern exception of the discovery of Lee and Yang) the refutation of hypotheses is displeasing to them and is toned down or even pooh-poohed.

If we accept Popper's view, then it seems that we have to view the scientists' traditional delight in "success" and dismissal of "failures" as their misunderstanding of their own activities, or rather as their method of advertisement which need not be taken seriously. And yet, Popper claims, there is an important element in "success".

3. *A Descriptive Theory of 'Success'.* The empirical support which a theory gains from experience, or its "success", or "confirmation", or "verification", is a measure of its having stood up to severe tests, or, in Popper's terminology, its high degree of corroboration. This is the whole of Popper's doctrine of what "success" is. I wish to stress that the theory according to which "success" is failure to refute a hypothesis does not entail that one "success" leads to another. Nor does it entail that "success" is a Good Thing. Much less does it include the view, sometimes stated explicitly, and more often implicitly, that theoretical science is a Good Thing because it is the body of "successful" theories.

4. *Corroboration and Eliminative Induction.* Most men of science, and almost all philosophers, think that "success" is a Good Thing, while Popper thinks that "failure" is a Good Thing. Now, many people have stated before that the refutation of errors is a contribution to learning. But they usually agreed that this is only so because some people have preconceived ideas which are mistaken and which are obstacles to learning and therefore have to be refuted; they usually agreed that, had people been cautious and slow to advance hypotheses and to commit themselves to them, the drudgery of refutation could be greatly reduced. Some people, in particular the great though neglected philosopher, William Whewell, claimed that, as it is most unlikely that one should hit upon the true hypothesis before hitting upon false ones, refutation is a necessary preliminary to any discovery of a true law of nature. Yet even Whewell did not think that refutation is good in itself, but rather good as a preliminary which is a useful means for the discovery of the truth. This theory, that we discover the truth through the elimination of errors, is known as the theory of eliminative induction. In some versions of this theory, refuting errors does not necessarily lead to the truth, but, in any case, to hypotheses with high probability or high credibility. In Popper's terminology Whewell's theory can be put (in a somewhat improved fashion) thus: a well-corroborated theory is true, and its

corroboration is its verification. The somewhat watered-down version, then, would be this: a corroborated theory is probable or credible.

This last doctrine has often been attributed to Popper.[1] And time and again he has claimed that this is not his view. In Popper's view no degree of corroboration of a hypothesis can secure that it will not be refuted in the next test; no degree of corroboration of a hypothesis makes it even slightly more probable that it will not be refuted in the next test. Corroboration implies neither verification nor any increase of probability.

To put this somewhat differently, we can neither hope to escape error, nor to make error less likely – not even in a limited field of research. All we can hope is that we shall eliminate some errors, and replace them by other, smaller errors. This is the main point where Popper differs from the eliminationist inductivist. He views as ends what they view as means.

Strangely, Stove knows this point, but he cannot quite take it seriously. Dismissing this point as inessential to Popper's doctrine, he (rightly) finds Popper's doctrine to be essentially in the eliminationist inductivist tradition; but his dismissal of this point can be explained as his inability to grasp Popper's novel and revolutionary idea.

5. *On the So-Called Rational Degree of Belief.* The problem of induction – how do we learn from experience? – Popper has tried to solve *not* by his theory of corroboration but by his theory of gradual approximation to the truth by repeatedly making explanatory hypotheses and refuting them experimentally. Another problem is, how can we avoid teaching false views or at least diminish our liability to assert mistaken views? This question, says Popper, has only one answer: the less you say the less likely you are to err. But science is an attempt to say more and more about the world, so that those engaged in science should have no fear of asserting an erroneous view, but they should do the utmost to encourage criticism. The last traditionally important question is, what theory should we believe, and why? This question, which is central in the inductivist tradition, is hardly discussed by Popper. A glance at the index to his book will reveal how little he says about beliefs, and reading these passages on belief will reveal that most of them are critical of traditional views rather than constructive.

The faith in science is a faith in a certain open-mindedness and detachment of belief – even to the extent of toning down beliefs and viewing them as entirely private affairs. This is the opposite of the view that science tells us what are the true objects of belief, be those the true laws of nature or the most probable among the known hypotheses in the light of the known factual information.

In the beginning of his 'Critical Notice' (p. 174) Stove correctly attributes to Popper the view that empirical science can lead to disbelief, but not to belief. From the middle of his 'Critical Notice' onwards, however, Stove identifies Popper's view with eliminative inductivism, whose major aim is to prescribe beliefs in probable hypotheses or in verified laws of nature. Take away from eliminative inductivism this prescription of beliefs, and you do get Popper's view. But then the problem arises, how do we gain theoretical knowledge from experience? The eliminative inductivist's answer to this question is, of course, that learning from experience is the same as finding which is the most probable hypothesis – the very answer which Popper rejects. His answer is that we learn from experience by refuting our hypotheses and inventing new explanatory hypotheses, and refuting these again, thus achieving better and better approximations to the truth.

6. *Knowledge and Learning.* The problem of induction concerns the question of how we learn from experience, a problem which belongs to the theory of learning, or methodology. Traditionally, however, philosophers have mainly concentrated on the problem of whether or not we have knowledge based on experience, a problem which belongs to the theory of knowledge, or epistemology. There is a good reason for this: if we could show that we have knowledge based on experience, then we could thereby show that indeed we do learn from experience. But all attempts to show that we have knowledge based on experience have so far failed. One of Popper's major revolutionary approaches was to go back from epistemology to methodology. Popper takes for granted (as an empirical fact, if you will) that we do learn from experience, and he asks by which manner, in what way, we learn. It turns out that it is easier to discuss methodological problems than epistemological problems. This is why Popper has so little to say on epistemology though he says quite a lot on methodology.

I am afraid that Stove has completely missed the point. He claims (pp. 178, 180) that the difference between having invented a hypothesis to explain a set of facts and having discovered these facts as corroborations to that hypothesis is "just nil". In other words, he claims, no matter how we have arrived at our knowledge, it is the state of knowledge that matters. And he says further that discussing this question of how we arrived at our knowledge is a psychological matter. Thus, he claims in effect, there is room only for epistemology and for psychology, but not for methodology, which is outside these two fields.

Now the situation is this. It was Popper who strongly emphasized, in opposition to many inductivist philosophers (including Bacon, Newton, and Mill), that from the point of view of assessing our present-day knowledge, the question of how we have arrived at that knowledge is entirely irrelevant. But though epistemologically irrelevant, this question is methodologically highly relevant. Here Popper argues not that it is irrelevant whether or not hypotheses are derived from facts, but rather that it is most often not the case that hypotheses are found by looking at facts; rather they are found by trying to solve concrete problems. This is not psychology. Similarly, though epistemologically it is irrelevant to ask whether we found facts by opening our eyes wide enough, methodologically it is most important to notice that opening one's eyes does not lead to the discovery of new facts, and that it is easier to arrive at new facts by trying to refute our present hypotheses. Psychologists may ask whether we like to refute our pet hypotheses; methodologically what matters is that we learn by doing so. Similarly, from the epistemological viewpoint there is no difference between having discovered facts before or after explaining them. This last assertion contradicts the central doctrine of Whewell. According to Whewell a hypothesis is verified only when it explains a new fact which it was not intended to explain. Now intentions and chronology are irrelevant to the question of whether the hypothesis at hand is true. And the fact that we are often more impressed with corroborations than with explanations is an irrelevant psychological fact. Now that we have agreed (Stove and myself) that corroboration is irrelevant to the appraisal of our state of knowledge, and that the fact that corroborations are impressive is a psychological irrelevancy, the question to be asked is whether corroboration constitutes progress of knowledge and, if so, why.

7. *Learning by Corroboration.* The significance of corroborating evidence is twofold; first, it is in certain respects new evidence, and secondly, it illustrates the high explanatory power of the corroborated hypothesis. I shall not discuss the fact that high explanatory power is considered valuable, because Popper discusses this point at great length. I should only argue that the fact that we pay much attention to a corroborated hypothesis can better be explained by the desire for explanatory power than by the desire for credibility. A corroboration of a refuted hypothesis is pointless from the credibility viewpoint. And yet some corroborations to already refuted hypotheses were very important, because, I suggest, they increased their explanatory power. In this way they rendered it necessary to build future hypotheses in a manner which would yield the refuted but corroborated hypotheses as a first approximation. (This is Bohr's correspondence principle translated into Popper's system.) It is easy to say, after the event, that the refuted hypothesis was important because it explained much, and not because it was corroborated; the fact remains that we learned that the hypothesis explained much by testing it further and by obtaining corroborations as the results of these tests.

Thus, we want to know not only whether a hypothesis is true or false, but also to find out, as sharply as possible, the limits of its explanatory powers, both from within (corroboration) and from without (refutation).

There is a third argument in praise of corroboration which I put diffidently because, although it is quite simple and may be of some importance, I find it difficult to construct an example for it. Imagine first a development in which a hypothesis A is refuted by the fact a, a hypothesis B which explains both A (as an approximation) and a, but is refuted in the first test by the fact b, and a hypothesis C which explains both B (as an approximation) and b. Imagine a second process, starting with the same hypothesis A and its refutation a, being followed by the hypothesis C which is later corroborated by the fact b. Here, not the corroboration, but the skipping of the stage B, is what makes the second process more rapid than the first. The corroboration is the result of the rapidity, and not *vice versa.*

To conclude, the value of corroboration lies in the discovery of the corroborating facts, in the discovery of as many facts as possible which are explicable by an existing theory, and in its being a result of rapid progress.

8. *Is Corroboration Really Necessary?* So far I have argued that the corroboration of a theory is enlightening – though less enlightening than its refutation – even if it is not really necessary. Stove, rooted in his inductivist tradition, takes it for granted that factual knowledge which is acquired after inventing a hypothesis could have been acquired beforehand. This is clearly not so with refuting evidence, which is observed as a result of a long chain of deductions of a prediction from a hypothesis. But since, as Stove observes, the corroboration of a hypothesis is a refutation of a previous alternative to it, perhaps corroborating evidence could be observed before the invention of the hypothesis it corroborates. I do not wish to decide this problem, but merely to point out that while Stove knows the answer to it as a matter of course, I find it worthy of a critical discussion.

It is a famous idea, already used by Galileo against inductivism, that we see what we think we ought to see, that we interpret facts in the light of theories. This idea, that experience usually tells us only what we have already thought out for ourselves, makes it particularly difficult to see how we can learn from experience. Both Galileo and Popper have claimed that we can escape this limitation (to some extent) by being extremely critically minded. But Popper, at least, admits that this is easier said than done, and even when we are on the alert we often tend to see things as we expect them to be.

Because of this, Popper suggests, it is preferable to have a number of alternative hypotheses, and to design crucial experiments between them. In this case one hypothesis may be a useful instrument with which to refute another, and in the process of using it we may corroborate it.

There exist striking historical examples to this effect. The deviation of Mercury from the path prescribed to it by Newton's mechanics was not viewed as a refutation of that hypothesis. After all, a few apparent deviations had occurred before, and these were later satisfactorily accounted for without involving overthrow of that hypothesis. Moreover, only after Einstein had explained Mercury's deviation along new lines was a similar, though smaller, deviation discovered in other planets, including Earth, in accord with Einstein's hypothesis. A debate is going on at present as to whether the other two corroborations of Einstein's hypothesis are not in fact refutations of it; I suggest that an alternative to Einstein's hypothesis which would be in better agreement with these facts may alter the widely accepted attitude towards them (especially if that alternative be corroborated). So at least in some instances one may doubt Stove's assumption that corroborating evidence could be discovered prior to the hypothesis it corroborates, at least in the sense that the evidence is viewed entirely differently when it is a corroboration, in the sense that what might be considered as small deviations or even statistical errors are now viewed as important facts.

But, we should notice, there exist important refutations which were never considered as corroborations. The Michelson-Morley experiment and the Lee and Yang experiments are famous examples of this.

It is obvious, since essentially we cannot predict the result of a test, that we cannot have a satisfactory explanation of corroborations in general. Nevertheless, I wish to express my profound sense of puzzlement at the incredible corroborations which some theories have received, in the past and in our own times. I cannot escape the feeling that it is as if a deity paid us a premium for any good explanation and let it be well corroborated before it be refuted. To use the inductivist language, I am willing to bet at any odds that if a reasonably good solution should be found to any of the major problems in contemporary physics, it would be well corroborated within a short time, and with much greater ease than it would be later refuted. I cannot support this feeling, which, of course, may be totally pointless. And I do not think that Popper's philosophy so far explains the amazing fact, if it is a fact, that so much corroboration is to be found in the history of science, although, I think, Popper's view is the only reasonable explanation of the value of corroboration.

But, I should add here, in his later work[2] Popper has claimed that corroboration is important, apart from being enlightening as discovery and increasing the explanatory power of a hypothesis, and is even essential to science – as encouragement to our research. I confess that I do not quite see this point, and I would here join Stove in asking what is the essential difference between corroboration and explanation. If psychological factors are ignored, and if the corroborating facts would have been found before the hypothesis which they corroborate, why, then, is corroboration essential?

9. *The Indispensability of the Corroboration of Factual Reports.* As Popper has pointed out, one kind of corroboration is essential for science. It is the

corroboration of factual evidence, or rather of its spatio-temporal universalization, which is often called "a general fact" or "an observable" or "a generalization". We know that only repeatable facts are considered in science, and repeatable in the sense of corroborable, not in the sense in which the reports about flying saucers are constantly flowing in. The reason for this is that a fact is important when it contradicts a theory. And it is important to have them corroborated in order to make it simpler to accept the reports as those of general facts rather than to explain them away in some *ad hoc* fashion.

Popper has no wish to explain the fact that the generalizations we propose are often corroborated: it is the task of specific scientific hypotheses to explain specific generalizations which are corroborated (and thereby to explain the fact that they are corroborated). This would not satisfy an inductivist, since such specific explanations are dubious; he wants certitude or at least high probability. But he is asking for the impossible. All one can say in a general manner is that if we had no corroborable general facts we would not have science as we know it. But a world with no general facts is, perhaps, one in which even life is impossible.

10. *Conclusion.* Stove takes it for granted that Popper's theory of testing is a preliminary to his theory of corroboration, and that this latter theory is a solution to the problems: how do we know and what should we believe? But Popper tries to solve the problem: how do we learn? His answer is: by criticizing our errors. The idea that anything we say can be a subject for a critical examination is the core of Popper's philosophical attitude. Stove views Popper's recommendation of the critical attitude as a part of his theory of corroboration, and he tries to see whether it is a necessary or an eliminable part of it. He is thus putting the cart before the horse. Popper takes the critical attitude as fundamental. Corroboration, according to him, is one sort of happening in the history of science which results from this attitude and to which, in turn, this attitude should be applied. Stove takes it for granted that, to Popper, a corroborated theory is corroborated because it is true or likely to be true or credible. As I understand it, Popper's philosophy contains the idea that we should take notice of a well-corroborated theory and try to explain the fact that it was corroborated – and a variety of explanations may be available, each of which should be critically examined. Undoubtedly, Popper's

philosophy is connected with a long-standing tradition; but it is the critical tradition of Galileo and Boyle, of Kant and Whewell, rather than the inductivist tradition of Bacon, Newton, and Mill.

NOTES

[1] See von Wright, *Logical Problem of Induction*, second edition, Oxford 1957; H. G. Alexander, *BJPS* **10** (1959), 234; and S. F. Barker, *Induction and Hypothesis*, Cornell 1957; I. Levi, *Gambling With Truth*, N.Y., 1967; this is by no means an exhaustive list.
[2] *Conjectures and Refutations*, Chapter 10. See below.

CHAPTER 3

ON NOVELTY

... as in that rule of arithmetic, ... *regula falsi*, ... so in physiology it is sometimes conducive to the discovery of truth, to permit the understanding to make an hypothesis.... [By] examining how far... that hypothesis [goes], the understanding may, even by its own errors, be instructed. For it has truly been observed by a great philosopher [Bacon], that truth does more easily emerge from error than confusion.

BOYLE, 1661

... the worth of a hypothesis lies mainly in that by a manner of *regula falsi* it leads us always nearer to the truth.

MACH, 1863

... The process of scientific discovery is cautious and rigorous, not by abstaining from hypotheses but by rigorously comparing hypotheses with facts, and by resolutely rejecting all which the comparison does not confirm.

WILLIAM WHEWELL, 1858

... the way to get at the merits of a case is not to listen to the fool who imagines himself impartial, but to get it argued with reckless bias for and against. To understand a saint you must hear the devil's advocate....

SHAW, 1907

There is a saying that was attributed to Louis Agassiz (no relation of mine) to the effect that every new idea is first declared contrary to reason, and then contrary to religion, and when it overcomes these two obstacles it is declared to be old hat. Popper's philosophy is just now passing from the second to the third stage – the stage of being old hat. Be that as it may, allow me to record, as an observation, that a number of distinguished philosophers, particularly of science, now seem to be preparing the grounds for this transition: they are playing amongst themselves a game

Note. The mottos are from Robert Boyle, *Certain Physiological Essays* (1661) Proëmial Essay; E. Mach, in *Oesterreichische Zeitschrift für praktischen Heilkunde* 9 (1863), 366; W. Whewell, *Novum Organum Renovatum*, 1858, Of Certain Characteristics of Scientific Induction, Aphorism IV; G. B. Shaw, Preface to his *The Sanity of Art* (1907).

that I both witnessed and heard reported in private conversations any number of times – the game of showing that every new idea that you can quote from Popper has been previously published by others. Various writings of various thinkers, whether scientists or philosophers, of the last three centuries, avail themselves of this game. It is not hard to find in them passages with striking resemblance to important remarks of Popper. His refutability criterion for the scientific character of theories, which is perhaps the best known idea of his, can be found in previous writings in one version or another. Even what I consider his most important idea, his idea that scientific theories are series or stages of approximations to the truth, is not new. I have found expressions of it in the works of Priestley in the eighteenth century, Laplace in the early nineteenth century, Ludwig Boltzmann at the turn of the century, in an appendix to Duhem's *Aim and Structure of Physical Theory*, and even in rather obscure and almost entirely unknown passages, in a few books, and in an article by Bertrand Russell where he contrasts Newton's theory with Einstein's.[1] The American philosopher Max Otto, in a discussion of scientific method, quotes a remark made by a working geologist, Lewis G. Westgate, incidentally to a geological discussion; in this quotation Popper's whole philosophy of science may be declared to be stated in capsule form.[2] So I think the game of finding quotations from writers previous to Popper is not difficult to play, and the results are largely successful. I would even go further and say that many of the remarkably striking technical ideas of Popper's – his ideas of simplicity, of degrees of testability, and other technical ideas – may be found in embryo form in works of earlier writers, particularly William Whewell and Charles Saunders Peirce. Peirce was an unsystematic philosopher, perhaps, but also brilliant and versatile. Perhaps Popper is the exact opposite of Peirce, because the most characteristic thing about him is the systematic and logical way he presents his ideas and fully works them out. But I am anticipating myself.

I. ON THE NOVELTY OF IDEAS IN GENERAL

To begin with, the criterion of novelty which stands behind the game I have just described is not so very good. It is rather uncritical to accept as a matter of course that if you can quote every idea of a thinker from previous thinkers then he has nothing new to add. Let me criticize this criterion first, not only in order to show that by a better criterion of novelty, Popper's are novel; I shall later argue that Popper has also invented new criteria of novelty – for scientific facts and theories – of great importance.

It is not easy to criticize criteria. Imagine an aesthetic criterion, for instance, criticized by the production of counter-examples. These counter-examples ought to be beautiful, but not by the criterion under scrutiny. The holder of the criterion will, of course, declare them not beautiful; if you say he is doing so *ad hoc*, and in order to evade criticism, he will answer that he is doing so since he is told to do so by a criterion that he is systematically and consistently applying regardless of whether he is under critical fire. And so, as long as one employs one's criterion systematically and consistently one is immune to criticism. But we expect beauty to do more than comply to a criterion, namely to impress us one way or another. And so the safest manner to criticize the criterion is to appeal to its open-minded adherents by a barrage of diverse counter-examples in the hope that some of these examples will move some of the adherents, and that they be open-minded enough to see this as criticism.

Here, then, is my barrage of diverse historical cases, some of which might impress the reader as cases of striking novelty or surprise of one kind or another, and all of which are in open conflict with the accepted standards of novelty.

At the very least there is that thing which we all know as putting 2 and 2 together and miraculously getting 5. In this case, the two twos are old, but the putting them together is possibly new. I shall mention one simple example; bifocal spectacles. Before Benjamin Franklin invented the bifocals many old people used to have two pairs of glasses. This is indeed why Benjamin Franklin made two pairs in one frame, and this was evidently an innovation – one of which he was rather proud.

Let me mention another simple example to show how strange novelty can be, and how narrow it is to approach novelty with certain set criteria in order to deprecate a thinker. James Watt improved the steam engine and made it much more marketable than before. It was alleged that James Watt had applied Joseph Black's theory of latent heat to the improvement of the steam engine. James Watt was a friend of Joseph Black, but in the face of this allegation he found it necessary to speak for the truth. He admitted that he owed much to Joseph Black, particularly in his calculations; but, he said, the major innovation, the invention of the cooling chamber, was based on the 'old established fact' of condensation of steam in cold. The major improvement of Watt's engine over Newcomen's was that whereas Newcomen's had one chamber in which to produce alternately heat and cold, Watt designed a two-chamber system, one chamber permanently hot, one permanently cold. In brief, he discovered the exhaust system of the combustion engine.

Let me mention one more example of an acknowledged great novelty, which in a sense is very old: it is the idea known as logicism, or the theory that mathematics is the branch of logic. This theory not only is ancient, but it was extensively discussed by Peirce. Yet we all ascribe logicism to Frege and Russell and Whitehead, because they took the idea seriously enough to try it out in all detail. Here we come again to the idea of being systematic. Somebody may be original in being systematic; even if he is not successful in his effort to be systematic. One of the important results of Russell's study is his failure in his systematic application of the idea of logicism to current mathematics. To a smaller extent, the same can be said of Jacob Burkhardt's studies of the Renaissance: the view of that period as that of the ideal of the universal man, is, of course, part and parcel of that period; how much so, remained for him to work out in all its detail. Just being systematic and the way in which one systematizes, may be novel even though the program of systematizing may be very much in the air. And the significance of that innovation is not simply a matter of success or failure either: Russell's venture, at least, is known to be great in its very failure. To this too, we shall return when we examine Popper's philosophy of science.

The examples cited thus far suffice to show that our naive and intuitive criteria of novelty are quite objectionable – even on intuitive grounds. Indeed, one facet of real novelty is that it may break and shatter old criteria of novelty and suggest that we think again about what novelty is. More generally, a novel idea causes some rethinking about novelty or about logic or about research. Consider, for instance, the work of August Kekule – the invention of the benzene ring. It is hard to realize how important this innovation was unless one knows some history of the chemistry current in the mid-nineteenth century, and in particular, the background of Kekule's discovery. For, the novelty of Kekule's idea was in Kekule's approach or even in his choice of a problem.

In the middle of the nineteenth century, organic chemistry started to develop with the theory of radicals. Chemists then looked at certain groups of atoms, such as one carbon and three oxygens, as if they played the role of one single atom; and the idea was that it is easier to shift such a group from one compound to another than to break it into its constituent atoms. Radicals were things between atoms and molecules. This was the great breakthrough in organic chemistry just before Kekule entered the scene. At that time people were working on every kind of problem which could be attacked that way. One problem that could not be attacked that way was how to explain the constitution of benzene, since this chemical has hydrogen and carbon atoms in equal numbers and thus shows no radical. Most chemists just laid it aside because they had enough work applying the theory of radicals. Often in the history of thought a trail blazer invents a new scheme of ideas, such as the radical theory, and many of his followers try to apply the idea in various ways. Working on radicals, chemists were too busy to bother about benzene. Kekule was first and foremost the man who saw the significance of a neglected problem. He soon added the benzene ring to the list of radicals but it changed the nature of radicals altogether: from then on a radical could be changed by replacing parts of it by different atoms!

My next example relates novelty to problems even more strikingly. It is from the work of Theodore Herzl, the founder of the Zionist movement. His major work is *The Jewish State*, subtitled 'a new solution to the Jewish problem.' The preface starts with the claim that the author's main idea is very old and is to be found in the ancient Jewish prayerbook. This is very interesting: a new solution, but an old idea. The old idea could not have arisen as a solution to any problem that Herzl had in mind, because the chief problem that he had in mind was new. He was chiefly concerned, of course, with the Jewish problem. The Jewish problem as Herzl understood it was something that did not exist before the French Revolution. There were other Jewish problems, we may well suppose, which existed in different times, but this is a different matter. Here again, I am applying a technique which is not new, but which springs from Popper's philosophy most forcefully. It is not enough to refer to a new idea: we have to show that it is a solution to some problem if we wish to stress its novel character. In all the literature of Zionism, for example, this point is eloquently absent. Let me present it, then.

Herzl saw the Jewish problem in a way which is partly very nice and partly not so nice, but all round in a new way. The nice part of it is that he saw the Jewish problem not as the problem of the Jews, but the problem of the whole of Western Europe, Jewish and Gentile alike. And that is as nice as the current claim that the negro problem in the U.S.A. is not a problem of the so-called American negro, but the problem of all the Americans, discriminators and discriminated alike. The not so nice part of Herzl's problem is that the Jewish problem, according to him, was how to stop the Jews from flocking to the socialist movements and other trouble-making revolutionary organizations. This is as unpleasant as the claim that the negro problem ought to be solved in order to stop the race riots. In other words, Herzl recognized that Jewish rebels were often people who had become rebels out of a just grievance; but it neither occurred to him that other rebels could have just grievances, nor did he think (in view of the Dreyfuss affair) that Jews could be compensated in Europe. In particular, and this *is* the unpleasant part of his teaching, he did not stress that grievances, rather than the violence they lead to, have to be rectified. In as much as he would have blamed the cause of the grievance, namely, antisemitism proper, for the inability of the Jews to settle down comfortably in Europe, he would be addressing the perennial Jewish problem, not the new one as he saw it. And so, in his very narrow approach he discovered a distinctly new problem, one which could not have existed before the revolutionary turmoils struck Europe. So here is a new problem to which the old idea is offered as a new solution.

An old idea as a new solution is another kind of putting of the 2 and 2 together, the putting of a new problem and an admittedly old idea together. The novelty is not of the idea, then, but of declaring the idea to be a solution. Supposing (with Popper) that all important ideas are solutions to some problems, one can see novelty in regrouping problems and solutions. In science this is common – as the transfer of a thermo-dynamic solution to electrostatics (Kelvin). In our instance, the old idea comes from the Jewish prayer book, the new problem comes from the Western European political situation, and now Herzl comes and puts them together. This I think is a novelty, and it is a novelty that should make us reconsider our criteria of novelty.

All this may sound somewhat scholarly and abstract. Herzl's solution was old: I am suggesting, with him, that it is new because the problem it answers is new; why should this suggestion be accepted? Why should we not join those who (applied old criteria and) ridiculed his claim for novelty by quoting the Jewish prayerbook? The answer can only be given in detail, and again, by employing a Popperian criterion of novelty: Herzl's solution led to consequences which were rejected by all his immediate predecessors. The prayerbook led no one to organize a new institution which claimed to represent those Jews who wished to return and for the purpose of political negotiations with governments as an equal amongst equals. And the ability of the Zionist movement to negotiate with the big governments is rooted in the recognition that they were a party in the situation – that the Jewish problem was their problem too, not only the problem of the Jews. Thus, the novelty of the solution can be shown in its new corollaries in the new problem-situation.

The retort to this answer may be: let us credit Herzl with the novel idea of organizing Jews to go to Israel, and never mind how novel was his idea of return which made him organize the return. One may support this argument by historical evidence; it mattered little to Herzl whether Jews should leave Europe to go to Isreal or to Uganda, and indeed, in one session of the Zionist Congress he succeeded getting passed a resolution to go to Uganda rather than to Israel.

This retort is false. The idea of organizing a Jewish movement to leave Europe and go to Israel was not novel with Herzl; indeed, in addition to its adumbration a few times, by various Utopian and semi-Utopian thinkers, politicians, and writers, there was already such an organization forming in East Europe. The East European Jews soon joined Herzl and viewed him as their intellectual and political leader. Even when they left the Zionist Congress meeting, when the Uganda resolution had passed, with tears in their eyes, they clung to him as their leader. The Jewish problem in the East, where anti-Semitism was state policy, was much more traditional than in the West; the Eastern Jews did not think Herzl understood the Jewish problem in the East; but they admired the West and viewed him as their spokesman there.

Even when you have a problem in one hand and a solution in another, then, you may get new insights and new practical ideas by putting the two together. It is no accident that Herzl achieved things others failed to achieve; it is no accident that he achieved new results in a field in which he was most unsuitable to act – he being a naive dreamer and a sentimental journalist, and the field being harsh power politics and the inflexible institutions of political intrigue. Nor is it accidental that his successor, Weitzmann, was a leading politician and diplomat with no proper political backing even in the post World War I arena which was almost destroyed by political cynicism.

Leaving political history, then, we may easily generalize the point at issue in more than one way. Discovery of problems, even discovery of the relative significance of problems, is sometimes a significant novelty. It follows immediately that declaring a known but unappreciated solution to a given problem to be important may change our view of the field in which the problem occurs and thus may lead to a series of major discoveries; it may then be itself considered a discovery which renders one of many given problems the central problem in that field.

Let us, then, take examples from diverse fields to show that a change of emphasis may be a discovery.

If we take Lord Kelvin's view of what the central problems of physics of the turn of the century were, we acknowledge his contribution, even though he neither created these problems nor advanced (or even outlined) any cogent or interesting solution to them (indeed he was highly conservative with regard to the desirable or even possible solutions). Similarly, Galileo's very perception that the Copernican Revolution is a revolution not only in astronomy but at least also in mechanics and in theology – this perception is itself novel. (He was anticipated here, in part, by Copernicus, Brahe, and Kepler; still his ideas are novel – like Frege's and Russell's – in their very extension.)

Here we arrive at perhaps the most important kind of novelty, and the one most easily amenable to dismissal by quoting predecessors. It is doubtless true that Locke was not the first to advocate toleration, yet he was the first to base a political philosophy on the principle of toleration not merely as a moral principle and political desideratum, but also as a prime political principle. Hence, all he said on toleration has predecessors, yet his political philosophy, his accent on toleration, does not. Similarly, though Locke spoke of substance as something which always eludes us, he cannot be called more than a minor predecessor of Kant; he did not approach Kant's theory of the thing-in-itself because he did not bother seriously to examine the significance of the elusiveness of the substance he passingly admitted was elusive.

This brings us to the novelty – the striking novelty, I think – of Popper's philosophy of science – even on the assumption that every specific idea of Popper's in the field is not novel. In line with Kant's critical idealism and its novelty, I shall now expound Popper's critical realism and show its novelty. Critical realism is the view – definitely not original with Popper – that any view of the world, being of the world, is realistic, is a set of assertions about the thing-in-itself,[3] yet the elusiveness of the thing-in-itself insures that our assertions about it may be false and perhaps even validly criticized. Apply this to science and you get Popper's philosophy of science. In order to show its novelty let me contrast it now with the philosophies of science preceding it.

To conclude this section, I have illustrated the narrowness of existing criteria of novelty by showing cases where novelty is manifest not by the introduction of new pieces of information or unheard of hypotheses, but by new uses of old materials. These were matching old pieces in new ways, particularly new matchings of problems and solutions; similarly, they were cases of shifts of emphasis, concerning centrality of problems and of modifiability of tenets accepted as more stable and less likely to be modified than others in the field.

In all this I stressed the centrality of problems: to understand a thinker we have to know his problems.

II. SCIENCE AND TRUTH

I come now to the problems that Karl Popper attacked when he started his philosophical researches. The problems were very well known. By now there is a tendency to declare them rather insignificant, but in the first quarter of the century they were very much alive and they had much to do with what is known as the crisis in physics. The crisis in physics was once something everybody talked about. Now this isn't so anymore; in fact mentioning the crisis in physics is quite uncommon in the current literature. The crisis in physics was the inability of physical theory to live up to its promise. The promise was that all predictions that were deduced from that theory would come true. It is indeed amazing how well physical theory, especially in Newton's mechanics, had stood up to the promise. It is amazing how many predictions based on Newton's theory looked false but upon further examination either were found to be true or turned out to be due to inexact calculating rather than to mistakes in the theory. There was, for instance, a calculation which showed deviations of Saturn and of Jupiter from the orbits calculated from Newton's law, and people thought maybe this is the end of Newton's law, but Laplace showed that the calculations weren't good enough, and that if you used better ones you got the right results. The other deviations were found, and Adams and Leverrier suggested it was a mistake to think that there are only seven planets in the solar system, that Newton's theory can be saved by the assumption of an additional planet. The additional planet was soon found. God obviously planted that planet up there in order to make Newton's theory right. Finally, one of the deviations of the facts from Newtonian mechanics, the motion of Mercury's perihelion, was just unwilling to play ball; people tried to postulate new planets, or a halo around the sun, but to no avail – until Einstein came and changed the theory itself. By now we may be used to the idea of approximating truth by modifications – by a series of approximations to it. At the time it was a great shock; the very idea that Newton's theory may be superseded was shocking. A wonderful philosophical literature – the conventionalist-instrumentalist literature – arose; just before the Einsteinian revolution and as a result of the crisis, a literature of which perhaps I should speak a little bit. The last famous conventionalist was A. S. Eddington, whose wonderful style and clarity are more famous than his own contribution to conventionalist philosophy. So let me present his views first.

Consider the case of abstract painting. One is often disturbed when one looks at an abstract painting and sees nothing recognizable there. That this disturbance is rather misstated or misplaced can easily be shown: a Mondrian abstract painting does not irritate, but is usually found uninteresting, whereas a Picasso is found quite irritating: "I cannot see three musicians there"; "this is not how a guitar looks," etc. Indeed when one looks at Picasso one cannot fail to recognize that it is, all too often, a gross distortion of a picture of three musicians, of a guitar, etc.

So much for the most central and well-known facts about abstract painting. The word 'abstract,' one should note, has two very different meanings. Etymologically there is little difference between 'extract' and 'abstract'; yet from antiquity it was in accord with common use that though anything may be extracted, only an essence may be abstracted; thus, though any idea or paragraph may be extracted from a book, only its central idea or theme may be properly *ab*stracted. Induction was the – proper or improper – method of *ab*straction, i.e. of extraction of essences from facts. In accord with this use the abstract is real, indeed even more real than the concrete from which it was abstracted (since the concrete is a mixture of essence and accident); yet in some context clearly the contrast abstract-concrete means unreal-real. Even if one does not believe in induction from facts, as long as one accepts the idea of theories as delivering essences – or forms (for non-empiricists prefer 'form' to its near-synonym 'essence') – that is, as long as theories are regarded as realistic, as having genuine referents in the world, we may still speak of the contrast abstract-concrete but not to equate it with unreal-real. Yet the latter is the artistic contrast between abstract-representational. Traditionally, the abstract represents the form and the concrete the appearances. But once we disclaim all ontological status to our theories, axioms, or definitions, they become parallel to the contrast abstract-representational. When we contrast abstract painting with more traditional paintings, we stress its being non-representational.

This, indeed, is why Mondrian does not annoy viewers as much as Picasso: Mondrian does not even seem to be representational, whereas Picasso often offers us distorted representations. Many art critics have tried to fight this bias by saying that truthful representation is best done by cameras, and that ever since the camera was invented the painter was not required to make likenesses. Therefore, the argument goes, when we look at a Picasso we shouldn't try to find there a representation, a reproduction of an object; we should see other things there, such as the interplay of color and areas and the like; hence, when we look at Picasso's 'Three Musicians,' for instance, we should forget that we see three musicians there.

Obviously, this is an exaggeration. It is, I think, impossible to forget that 'Three Musicians' is in effect a distorted representation of three musicians – even were the title removed; since we can see one blow a distorted wind-instrument and one play a distorted string-instrument, and a piece of music writing. When we see this picture we automatically interpret it – we read a representation into it. So the art critic's suggestion not to read any interpretation into an abstract painting is not applicable. To this the art critic replies by way of improving his position: never mind whether it is a representation or not, he says now, it doesn't matter what

interpretation you read into it. What matters, he still insists, is the interplay of color, area, etc. If there is an interpretation, well and good, if
there isn't an interpretation, well and good – the most important thing is
how the picture affects you. If, however, the picture affects you adversely
because you hold an interpretation of it which makes it affect you adversely, then blame your interpretation and not the artist's work. In
particular, when the representation you read into the picture is a falsehood
or a distortion, either you should not mind it or reject it in an effort
to enjoy the painting. Thus, instead of objecting to interpretation in order
to stress abstractness, it is better to merely play down the value of interpretation. It is important to note that we may justly say that Picasso's
'Three Musicians' is, indeed, a representation (whether we like it or not
and in spite of the fact that the interpretation is largely in the eye of the
beholder), and as a representation it is a very bad one; we hang it on the
wall not because it is a representation, however, but because it is an
excellent work of abstract art.

Now here 'abstract' is again used as opposed to 'representation,' yet
we apply both predicates to the same work! – the same work, that is, in
two different interpretations: logic is not thereby violated.

Since the present discussion does not concern aesthetics, I am not
going to examine this last theory; rather, I have introduced it here as a
perfect analogy to Eddington's theory of science. The conventionalist
philosophers of science declared that those who brought about the crisis
in physics were mistaken in considering scientific theory as representation
of reality, a xerox-copy of laws from the pages of the book of Nature.
Scientific theories, they said, are not laws of Nature, but abstract equations, implicit definitions, and other highly sophisticated pieces of intellectual machinery. Science is a very rarified system which is not
representative of the world; that is, it is not given to interpretation. Hence,
all questions rooted in the quest for correct interpretation may be dismissed as naive. They are questions based on the expectation that science
be representative, and the expectation is out of place – as the crisis in
physics shows. The crisis in physics, in other words, is not a scientific
crisis, but a philosophical crisis. It is the crisis caused by the attempt to
see in science representation. It is true that reports of observations in
science are representations of concrete empirical facts. It is quite true that
reports of experiments in science are likewise representation. Let us say

that reports of experiment and observation in science ought to be true, but anything more abstract in the sense of not-concrete and not directly observational is to remain abstract in the sense of non-representational, in the sense of not read as true-or-false in any sense. Now we all read interpretations into abstract science. We all have some representation of abstract scientific entities like atoms, electrons, curved space. When we speak of space curvature, we actually see a curved space like a curved railway track or a piece of metal curved by the strong man in the circus. Once we represent or concretize our scientific theories in such a manner, we cannot but view them as true-or-false in the sense that reports of observations are true-or-false. Being scientific, and presumably true-or-false, they are naturally considered as true rather than as false; too much is expected of them, and then the expectation is disappointed, and this is the crisis in physics. To avoid crises, say the conventionalists, you should avoid all interpretations – thus avoiding all representation and thus, further, avoiding truth-value. The purpose of science, conventionalists say, is to lump together experiences in a neat way – handy, serviceable, even aesthetically pleasing. It is a neat package and there is no more to it – especially no additional meaning, no independent interpretation above and beyond its strictly observational content. Most conventionalists are therefore simply opposed to all interpretations. Eddington alone has a more liberal conventionalist view; he has no objection to interpretations on the understanding, or on the condition, that they are strictly personal. Indeed, he adds, you cannot avoid interpreting a scientific theory. If you enjoy your interpretation, very good; if not, you may ignore it; if you are troubled about it, this is your private trouble. The main point to stress is that interpretations are not important. So much for Eddington's view.

Now I have no strong objection to all this, even though I do not agree. My heart is not there, but so far not so bad. I now come to something which is really very bad, and it is this. You have a painter who paints the model. If the painting comes recognizably close to the image of the person who sat as a model, the painter considers it a portrait and may try to sell it either to that person or to the friends of that person. But close or not, he cannot declare his portrait to be non-representational: this is the game, heads, I win, tails, you lose – which Galileo already denounced. I think one has to make up one's mind whether one does care about representation or not – whether one willfully distorts, or intends to represent, or is indifferent to representation.[4] What matters, then, is that we declare our aim, rather than fit the aim to the achievement. We should not shoot the target by first shooting and then describing the circles of the target around the bullet wherever it happens to have hit. This would be considered dishonest by most people. I contend that most amateur philosophers of science – whether scientist or other – are doing just this.

Scientists do wish to see scientific theories as representational. To view scientific theory as a mere convenience, as presenting known experience neatly but offering no new and deeper understanding of experience, is a bit demeaning to the stature of science. The deprivation of scientific theory of any depth has already been done when conventionalism starts. Perhaps the most important conventionalist philosopher is Henri Poincaré. He was attacked for demeaning science, and as a reply he wrote a very charming book called *The Value of Science* – the points of which in my opinion and in the opinion of many others are rather dubious. There is no escape from the fact that if scientific theories are mere convenient definitions – mere instruments to aid the memory and prediction and for aesthetic gratification – then the aspirations of scientists like Copernicus, Kepler, Galileo, Newton, and others, cannot relate to science as we know it. And therefore, most innovating scientists do like to consider their theories as representational and hopefully as true representations. It is when they get into a crisis that they declare their intentions to have been more modest, and usually only in retrospect. In other words, our theories may be true or false, on condition that they are true. If they are not true, we better say that they are not meant to be considered as either true or false, but abstract. Now to say that they are true or false on condition that they are true is, of course, unfair; and this is why I put it that way. I don't think anybody before Popper noticed it, but it is there in the literature when scrutinized carefully. Take for instance, the article in the *Encyclopedia Britannica* on mechanics, which was written by a mechanical engineer and a professor in the University of London. He starts the article by saying that there are two ways of looking at Newton's theory: either as the truth, as a theory verified by experience, or as merely a convenient instrument of calculation and prediction. I think this is unfair. There is another possibility that at least should be stated, and it is that Newton's theory is a false theory. Most scientists and many philosophers of science get indignant when you suggest that Newton's theory is false because the word 'false' arouses in them hostile feelings and they conclude that you suggest that Newton was not a respectable scientist. But you can explain – to a scientist even, though not always – that you have no such intentions. Once a person is willing to view Newton's theory as one which may either be true or false most likely that person will agree that it is false. Nevertheless, the fact remains that scientists find it hard to consider the possibility that Newton's theory is false. And this is a point which is noteworthy. What causes this difficulty? Why is it that calling Newton's theory false angers scientists so?

Somehow, says Popper, it is traditional to equate falsehood with sinfulness for the simple reason that traditionally it is taken for granted that falsehood may be avoided. Consider any religion, Catholicism or Protestantism, and you can see why all falsehood, to some degree or other, is viewed by them as sin. If you want to avoid all sin all you have to do is repeat what the party organizer, perhaps the party manual or the Bible, or the priest, tells you. What you are thus told comes *ex-cathedra*, and therefore, cannot be erroneous. Therefore if you just tow the line you too cannot err. The ability to avoid error leads almost inevitably to the equation of sin with error. Now, in the development of modern philosophy and methodology of science there was an attempt to break away from the bonds of the Church. One of the tools by which the Church was attacked was this. Aristotle's theory may be true and it may be false, but it is not the priests' task to declare whether it is true or false; it is a matter for scientific investigation. This is particularly the point made by Galileo who said that the priests who say otherwise were saying so out of political interest and purely for political purposes. But one can go further. The Renaissance thinkers who had a lasting influence did go further: they said that all error, whether of Aristotle or of anyone else, can in principle be avoided. Descartes in particular says that God does not lie to us, because he is benevolent; hence, if we act properly, if we use the right method, we cannot err. If I do everything the best way and yet I err, then God is a deceiver, which is absurd. Hence there exists a way, the best way, which guarantees the avoidance of all error; hence when I err I do not use the right way even though it is available; hence when I err it is my refusal to use the right way which is the cause of it; which refusal is sheer perversity; error and sin are thus equated.

This is the doctrine that truth is manifest. It was advocated by Sir Francis Bacon, and adopted by Descartes. Bacon said truth is manifest through plain facts; Descartes said through plain reasoning. Though Descartes added to our certainty that truth is manifest, it was Bacon's doctrine that became the official doctrine of the scientific community. And Bacon's doctrine of error was at least as influential as his doctrine of truth. Indeed, his views on how truth is achieved were widely ridiculed in the nineteenth century; but his views of error were then considered platitudes. They still are widely accepted.

Bacon's chief problem was, what caused the decline of antiquity and the descent of the Dark Ages? His answer was, Aristotelianism, which was a self-perpetuating error. All errors are self-perpetuating and so the first task of a true scientist is not the search for truth – this is his second task – but the avoidance of error; once one has erred one is disqualified as a seeker of truth.

Bacon's theory explains how error comes about and perpetuates itself even though truth is manifest. It is the same as the Old Testament doctrine of bribery. The Old Testament says, thou shalt not take bribes, for bribes blind the eyes of the just. Whether one intends to pervert justice or not, once one takes bribes one is in no position to prevent perversion any more. When one offers a hypothesis, says Bacon, one does so in an attempt to avoid the labour of searching for the truth because one is impatient to achieve renown.

But one cannot achieve renown by admitting that one's views are false, or even by admitting that one's views may be false. Hence, in order for one's hypothesis to deliver the goods – to earn one fame in the learned community – it has to be expounded authoritatively, *ex cathedra* and with spurious proofs; and all arguments against it have to be dismissed offhand and with hostility. And so, the motive which makes one advance any hypothesis is the motive which makes one defend it through thick and thin. And so hypotheses are bound to become prejudices and blinkers.

It is a fact that Newton advanced his own theory of gravity as an absolute truth. It is also a fact that Newton's own claim cannot be accepted today; we can say that his theory is a good approximation to Einstein's theory of gravity, or we can say it is neither true nor false; we cannot agree that it is the absolute truth. Can we therefore call Newton prejudiced? I do not think anyone would view this proposal as reasonable. And yet, we feel, calling Newton's theory of gravity false is calling him prejudiced. Here, then, we are plainly inconsistent in that we accept Bacon's theory of error, but only in part.

It is doubtless the case that some errors are prejudices, and perpetuate themselves in the manner described by Bacon. But to say that all erorrs are prejudices is an error, and one which may become a prejudice.

To conclude, once we overcome Bacon's doctrine of prejudice we may view science as a portrait of nature – true or false – or perhaps abstract. Once the three alternatives are put in this novel way, the problem of choice has radically changed. Whether Popper's views are true or false, he radically altered the philosophy of science.

III. POPPER'S VIEW OF SCIENCE

Bacon said, you cannot criticize a man and admire him. This view of Bacon was once very popular. In his obituary of David Hume, his friend Adam Smith says, I will not speak here of his philosophy since those who endorse it admire it and those who reject it despise it. Nowadays we think differently; most philosophers agree with Kant's appraisal; they admire yet reject Hume's philosophy. Even within science, the idea of respectful criticism has gained ground in many quarters; but there is a severe restriction here which must be carefully considered. It is easier to criticize a layman than a scientist, a fool than a wise man; even scientists are open to criticism, then, but up to a point.

What is this point? The American philosopher, Max Otto, has referred (see note 2 above) to the fact that a working geologist, Lewis G. Westgate, had held this: there is value in publishing criticizable hypotheses in the hope of eliciting criticism, because that will be useful in the progress of science, as it opens the road to better hypotheses. Westgate was even aware of fighting a very popular prejudice. As late as in 1940 he had to state:

It may seem ungracious to pick on the mistakes of our predecessors and to drag them out for public discussion. But the theory of scientific method is one thing and its practice is often quite different. And it may be that by the examination of its faulty working and the errors which follow, we can best avoid errors ourselves. The progress of scientific discovery is more important than the concern of the individuals. The earlier investigators [Agassiz] were pioneers. We honour them as the founders of the science [of geology].[5]

This passage is very interesting. It clearly identifies the popular view that public criticism is disrespectful and useless. It opposes the popular view by showing what the use is of public criticism and by paying homage to those criticized. In this very homage Westgate puts the limit: these people were pioneers, and could not but err; we should learn to avoid error. We can avoid error, because meanwhile we have achieved verified theories.[6]

Westgate's view is influenced by that of T. C. Chamberlain, who, in turn, was influenced by William Whewell, the little known but highly influential Cambridge methodologist and historian of science of the second quarter of the last century. Whewell considered experimentation *a* method of criticism, and he considered that failed criticism led to the conviction that the object of the failed criticism is true.

Popper agrees with Whewell that experiment is criticism, but he refuses to see any validation in the failure of criticism. Most philosophers who study Popper's view consider it to be a modification of Whewell's. Whereas Whewell viewed validation as conclusive, they think that Popper has claimed such validation to be tentative. To correct this impression we must return to the problem which Whewell's theory of validation has come to solve and see whether Popper solves it in a similar manner or not. What problem, then, does the theory of validation come to solve? The problem of induction, of course. What is the problem of induction?

The problem of induction is peculiar: solutions to it are offered and hotly debated, but it is seldom stated and still less discussed. There is no easy way of learning from the vast literature on the problem what the problem is; one may even claim that it is not possible to arrive at a concise statement of the problem, because in different periods one may have to state it differently. It seems strange that there is a problem so very important that dozens of books and papers in the learned periodicals appear yearly in attempts to offer new solutions to it and to debate existing solutions to it, yet there is practically no discussion of the problem itself, its formulation, its significance, etc. Alfred North Whitehead has said this problem is a scandal in philosophy – a remark often quoted in the literature. Why is it a scandal? Is it a tough question? Is it soluble? If not, what are we to do about it?

Here is another novelty in Popper's philosophy. The novelty may be stated in a popular saying: a problem well-stated is a problem half-solved.

Old as the saying is, it has not led a single philosopher to present fully the problem of induction – a task which could easily fill a volume or two of very interesting material. On this point, I regret to stress, Popper has failed too. In his classical work, *The Logic of Scientific Discovery*, Popper briefly states the problem and he does not discuss it at all. His statement is technical and unusual (though it was usual in the eighteenth century), with the result that quite a number of philosophers have misread his solution as a solution to the problem of induction as they see it, not as he sees it.[7]

The problem of induction is not a problem at all; it is a collection of problems with varying intellectual backgrounds; in different contexts different parts of the collection gain significance. Let me illustrate this with one example.

The way most philosophers of science would – and do – flippantly present the problem of induction is, how does experience validate a hypothesis? This may be reformulated as, how does a scientist choose one hypothesis out of a few competing hypotheses in the light of experience? To claim that the two above formulations are of one and the same problem amounts to two highly questionable assumptions which may, in the manner indicated here, be smuggled into philosophy before the debate even starts. The first is that scientists choose validated hypotheses. This amounts to saying science contains no error; even if the validated hypothesis turns out to be not so good, at the time it was the best validated it was rational to choose it. The second smuggled assumption is that validation is done through competition – an assumption rejected even by some who implicitly accept it!

Suppose a hypothesis can never be validated in any way. What of it? This question is the same as, why is the problem of induction as formulated here of any importance? If it is not important, then, surely, Whitehead has exaggerated in considering its not having yet been solved a scandal in philosophy.

The importance, we may say, is this. Let us assume that scientists do choose to endorse new hypotheses to believe in and to act upon. Now, if the choice is rational it must be justified; hence, no validation of new hypotheses, no rationality but mere alteration of fashion. This view of the importance of induction amounts to equating rationality with learning – an appealing proposal, but surely not one to smuggle through by implicitly identifying the two formulations (in the previous paragraph) of the problem of induction. Yet this, I contend, is precisely how the proposal is made in a number of significant volumes on the theory of induction.

It is an amusing fact that Popper's explicit stress on the theory of rationality as the theory of the growth of knowledge has impressed his fellow-philosophers; it has impressed them, I suppose, precisely because on this very point they are in profound agreement with him. On this point they do not say, we have hinted at this idea before, we can quote William Whewell on it (assuming they know who Whewell is), and such deprecatory remarks. Here they feel free to sympathize with Popper. But what if scientific hypotheses cannot be validated? If they can be validated we have (let us say) learning and rationality – but if we do not have validation is it necessarily the case that we do not have learning?

There is an optical illusion of this kind in many domains. For instance, one may need money for bail, and very badly; but if one simply cannot mobilize it, one may, but often fails to, think of other ways to get bail, e.g. bonds. Whenever a problem turns out to be too difficult, one may dig deeper and find the problem giving rise to it and try to solve it differently.

The problem of induction, on a deeper level, is the problem of learning from experience, and the problem of rational choice of hypotheses; but these cannot be solved by validation – at least not according to Popper who agrees with Hume and others on this. Take the problem of learning: how do we gain theoretical knowledge from experience? That is, suppose we do learn from experience certain facts, and suppose we take this as unproblematic. Suppose also that sometimes in science experience sheds light on theoretical questions – not by validating theories, but in some other way. Which way?

At last we have come to the problem of induction – or half of it (we have dropped rationality for a while) – in a formulation favourable to Popper.

The problem of induction is, how do we gain theoretical knowledge from experience? The usual view of this problem, we have seen, is that it is the same as, how does experience tell us which theory to endorse? How does experience validate a theory? Whewell accepts all this, and tells us to examine carefully any theory by testing severely its corollaries, and to believe that theory which stands up to criticism. Whewell adds that such a theory will remain validated forever and this part of his theory we may simply ignore. After ignoring this part, does Popper agree with this paragraph?

Not at all. Popper denies, first and foremost, that the problem of induction (how do we gain theoretical knowledge from experience?) is the same as the problem of choice of hypothesis (which theory should I choose in the light of experience?) or that these are the same as the problem of validation (how does experience validate the hypothesis we choose?). For the last problem (how does experience validate a hypothesis?) Popper's answer is totally negative; experience never validates a hypothesis. As to the problem of choice of a hypothesis, Popper says, experience never forces us to choose a hypothesis, but at most permits the choice of a hypothesis; we are not allowed to choose a refuted hypothesis, and if we choose an unrefuted hypothesis but refuse to test it for fear that it be refuted we are hardly better off; but we are only allowed, not forced, to choose a hypothesis which we test but fail to refute. Popper denies that the problem of validation (how does experience validate a hypothesis?) or of choice (how does experience direct our choice of a hypothesis?) is the same as the problem of induction (how do we gain theoretical knowledge from experience?). For the problem of induction he offers a solution which is very radical in that it rips the problem of induction off from the problems of choice and of validation.

In brief, Popper's solution to the problem of induction is totally revolutionary *as a solution to that problem of induction*; in other contexts it is all too often stated, even as sheer commonsense: we learn from experience by criticism, by learning from mistakes. Both Whewell and Popper recognize the importance of attempts at criticism and the fact that they are important whether or not they are successful. Yet for Whewell the successful criticisms are preludes to failed criticisms which validate a theory which thus gains supreme importance. For Popper successful criticisms are the junctures at which theoretical knowledge is gained. Popper's theory of the importance of failed attempts to criticize does not come to solve the problem of induction.

We see here the marked novelty. Both Whewell and Popper have the same problem; both speak of unsuccessful and successful criticism. To solve the same problem, Whewell uses the former, Popper uses the latter.

The theory of development of thought through criticism may sound strange, but it is itself not novel at all. It is Socratic, and at least in one field of inquiry it was made by one school of thought the central philosophy of that field. The field is theology and the school is that of negative theology or *theologia negativa*. Popper has now applied it to science and so it is, just like Herzl's solution, a new solution, as a solution to a new problem, but otherwise old. The novelty of it is, indeed, quite obvious: it sounds very strange to the modern ear. It sounds so strange to the modern ear because of the widespread of Baconion ideas. It is the prerogative of God alone to remain hidden forever, said Bacon, promising his reader that Mother Nature must sooner or later reveal her secrets to us if we go about our scientific business the right way, i.e. not offend Her by putting Her in the chains of our preconceived ideas, but flattering Her by taking serious account of Her smallest move. It is well known that in seventeenth century literature Nature often stands for God, even though only Spinoza had the courage to say it out loud. It is surprising that this did not lead at once to a full fledged *scientia negativa*; that such a theory was adumbrated, by Pascal, Boyle, and others, only makes matters even more puzzling.

Take away Bacon's promise that Nature must reveal herself, and you do not as yet get Popper's philosophy: conventionalist philosophers already did that, yet they insisted that science does not err. Take away both Bacon's promise and his contempt for all error, and you can easily render *theologia negativa* into *scientia negativa*. But why should you do that? What problem should this come to solve? The problem of induction.

This can be seen only if the problem of induction is severed from the problems of choice and of validation of science, which the conventionalist failed to do. Consequently, the conventionalists declared that we simply cannot gain theoretical knowledge. Perhaps; but Popper's view may be true, and then perhaps we can.

If you take Popper's view as a serious candidate then you have immediately one serious problem on your hands; what criticism is tribute to the criticized and what not? It is a problem which Popper has studied only by implication. He has repeatedly stated that criticism is a tribute and hence we should choose the best opponent to criticize and criticize him respectfully. But which opponent is the best? Take a significant problem, take a list of its known solutions and try to criticize them

explicitly. Any other solution that is a serious candidate has to be immune to these criticisms and then we may consider it new and undertake to examine it critically.

Apply now this criterion to Popper's solution of the problem of induction. All past solutions are theories of validation, and traditional
criticisms of the theory of induction apply to all theories of validation
as such (any validation is in need of validation). Popper's theory, not
being a theory of validation, is immune to past criticism. Hence it is new.
Is it also true? My answer to that must await another occasion.

This offers us a solution to the problem, which criticism is respectful,
which is not? Any criticism is respectful if it contains recognition of the
value of the criticized theory; a hitherto uncriticized theory can be appreciated because it is hitherto uncriticized; this enhances the value of
criticism as well. We have here also a novel criterion of novelty as was
promised earlier in this discussion. A solution is novel if it is immune to
criticisms of older alternatives to it: a criticism is novel if it hits on as yet
uncriticized solutions. All this is so strikingly simple it needs no elaboration and looks deceptively familiar. In a sense it is; as a solution it is new,
and as a new solution it has strikingly new and interesting applications.

NOTES

[1] B. Russell, 'Had Newton Never Lived', *Radio Times* 15 (1927), 49–50. See also beginning of Ch. 2 of his *Scientific Outlook*.
[2] Max Otto, *Science and the Moral Life*, New York, 1949, p. 156.
[3] Popper tends to view even idealism as a version of realism: in Berkeley's view the
thing-in-itself is God, in Mach's view it is elements of sensation. This is intriguing but
rather problematic both in its paradoxicality and vis-à-vis its historical context.
[4] Representation and portrayal are two very different matters, but we shall not enter
this point here.
[5] *Scientific Monthly* (October 1940), 309.
[6] *Ibid.*, p. 301.
[7] But let it also be said that Popper returns to the problem again and again, and that
parts of his *Conjectures and Refutations* contain more extensive discussions of the
same problem. He has finally tried to do justice to the problem in his *Objective Knowledge*, Oxford 1972, Chapter 1.

APPENDIX: ON THE DISCOVERY OF GENERAL FACTS

The problem to be discussed in this note is: how are discoveries of facts
made, and why do we find it strange that a given discovery was not made

earlier? The three views answering the problem which we consider are as follows:

1. *Francis Bacon:* Discoveries are made accidentally. They were not made earlier because people were misled by their own opinions. Discoverers discard opinions and trust only their senses.

2. *William Whewell:* Discoveries are verifications of new ideas. They could not be made before the hypotheses which they verified were conceived. Discoverers conjecture explanatory hypotheses and try to verify them.

3. *Karl Popper:* Discoveries are refutations of scientific hypotheses. They could not be made before the hypotheses which they refute (and the specific attempts to criticize these hypotheses) were conceived. Discoverers conjecture explanatory hypotheses and try to criticize them.

Long Struggle

For logical reasons these are all the possible cases: if a discovery is neither a verification nor a refutation of a prediction deducible from any existing hypothesis, then by definition it is independent of any existing hypothesis and is thus accidental. Naturally, the views I have mentioned above are deeply connected with much wider issues of logic and scientific method. In particular, they are connected with the issue of theoretical discoveries. I shall not enter these issues, and only discuss discoveries of facts, using the word 'facts' in its ordinary sense, or in the sense of 'generalizations' or 'low-level generalizations' or 'observable facts'. The following example should make this clear. I put forward this example in order to show, in an intuitive way, that even a most simple and obvious fact had to be discovered after a long struggle.

If you have an electric torch and a compass – a toy compass will do – you will hardly find any difficulty in observing that the electric current in the torch deflects the needle of the compass. This is Oersted's effect. Something like it was looked for in the late 18th century and in the early 19th century by many great discoverers. Oersted himself looked for it for over 10 years. When he ultimately found it he was thought to be the greatest discoverer of his age. It is difficult to realize now what an important discovery this was.

The same difficulty arises when we consider even the latest discoveries, though, naturally, in a somewhat less obvious manner. Take the discovery of the non-conservation of parity. Professor Abdus Salam pointed out in his inaugural lecture at the University of London, "that the results could have been discovered 10 years back, for the evidence existed on all photographic plates with π-μ-e decay". Why then were they not observed 10 years earlier?

Bacon's View

This question was already posed in the early 17th century by Francis Bacon. His answer was this. The chief obstacle to discovery is people's self-assuredness in their belief that they know all that there is to know, while in fact they are merely blinded by their prejudices and superstitions. In order to be a discoverer one has to forget all that one was taught and to realize the need for more factual knowledge. A discoverer doubts everything and takes nothing for granted. A discoverer merely observes facts diligently, collecting as many of them as he can. This is all he can do. The rest is up to Mother Nature, or accident, or luck, or call it what you will. As the discoverer does not rely on any opinion but only on what he sees, he cannot foretell what his eyes will tell him. (The theoretician can evolve demonstrable theories out of the discoverer's findings, and he can predict. But this is a totally different matter.)

Bacon has a very strong argument in favour of his view that discoveries are unpredictable. Discoveries, he observed, are surprising. If you could tell people of previous generations about today's discoveries, they would not believe you, or they would consider your tales miraculous. They could not predict the discoveries of today precisely because discoveries are novelties. As a discovery is of something not previously known, it must be independent of all previous knowledge (and thus be accidental).

Discoveries to Timetable

Some discoverers resented the doctrine that no thinking, no theorizing, was involved in the process of discovery. When Bacon's views were most popular – in the early 19th century – some discoverers tried to show that their discoveries went along preconceived plans, by trying to make discoveries in public lectures. Davy wrote in a letter to a friend that he was going to decompose nitrogen – which he (erroneously) thought was a compound – in a public lecture. Oersted made his discovery in a public lecture. Ampère, too, made his discovery shortly afterwards in a public lecture. Yet people remained unconvinced. The fact that Oersted made his discovery in a lecture was taken as evidence that his discovery was accidental, and the fact that Ampère had predicted his discovery was taken as evidence that it was no discovery at all. It was therefore necessary to show the error in Bacon's reasoning rather than to perform a discovery in public. This was done by the great Cambridge philosopher of the 19th century, William Whewell.

Whewell's Qualification

Whewell agreed with Bacon that new discoveries cannot be predicted on the basis of established theories, but he claimed that they must be predicted or else they would not be seen at all. It follows then, that a discovery is predicted on the basis of a new, not yet established theory. An investigator conjectures an explanatory hypothesis and then tests it. He refutes it and starts all over again. This process is repeated many times. Eventually, though very rarely, and even then only with the aid of intuition and luck, he succeeds in verifying a prediction based on a hypothesis, thus rendering that hypothesis into an established scientific theory.

The great merit of Whewell is that while rejecting Bacon's anti-intellectualist view he retained its most significant element – the idea that scientific discoveries are surprising. He even extended the element of surprise to cases which are most obviously excluded by Bacon. An example would be the prediction of Einstein which was surprisingly verified by Eddington (during a total solar eclipse).

Yet Whewell's view does not fit discoveries which do not come out according to plan, and refute seemingly well-established theories in an unexpected manner. I am referring to the famous crisis in physics of the turn of the last century. As atomism was considered to be well established, radioactivity seemed quite perplexing, even to its discoverers. And as both Newton's theory and Maxwell's theory were claimed to have been verified, it was plain that Michelson's experiment should indicate ether drift. But it did not. With this the whole philosophy of verification collapsed. Most philosophers are at present concerned with a rescue operation trying to substitute high probability for verification. The most important exception is Karl Popper, who suggested that we learn from experience by refuting our hypotheses experimentally. Surprising facts are those which, according to existing views, are quite impossible.

Refuted Theories

Thus Popper retains and even fortifies Bacon's idea of discovery as a surprise, while allowing, at the same time, for the case of two investigators discovering independently the same fact (by criticizing the same well-known theory). He also retains Whewell's idea that discovery involves more thinking than observing. Bacon and Whewell had concluded from the idea that discoveries do not follow from previous theories, that they are independent of them. In doing this they just ignored refuted theories. But refuted theories are very important because they led to discoveries which were made as refutations of them. Whewell was right in claiming that a fact must bear some relation to our theories in order to be observed. But the relation need not be that of agreement – if we are critically minded we may see facts which disagree with our anticipations. Eddington's observation was not a verification of Einstein's theory of gravity; Einstein himself (and many people with him) found it very difficult to accept his own theory. But all agree that Eddington's observation refuted Newton's theory which thus should no longer be considered to be true. Bacon was right in claiming that people fail to see a fact (obvious after its discovery) because they are misled by their hypotheses. But he was wrong in believing that we can altogether dispense with hypotheses. We have to eliminate the false hypotheses by refuting each of them and not, as he proposed, by trying to forget all hypotheses.

Oersted's Sudden Thought

Many people (including Oersted's pupil Hansteen) viewed Oersted's discovery as accidental. Others (including Whewell) used this discovery as an example for a planned experiment. The latter party could not explain why it took Oersted over 10 years to make his discovery, and neither parties could explain what is its significance. The answer to these questions can be given only if the discovery is viewed as a refutation. Although the effect can be repeated with any weak battery, Oersted failed again and again to perform it while using stronger and stronger batteries. Why? We know that the only way to fail is to place the conductor perpendicular to the needle of the compass, i.e. in the east-west direction. Oersted, as well as Young, Davy, and quite possibly other investigators, repeated this mistake many times and systematically. Why? From Popper's viewpoint the answer is pretty obvious: it was due to a theory which they all took for granted. For if you assume, with Newton, that all forces are central (i.e. either pull or push), then even if you do not know whether the electro-magnetic force is attractive or repulsive, or where the centres of force in the conductor lie, though you know that it is cylindrically symmetrical, you can see that the best position of the conductor must be in the east-west direction. For as you do not know whether it will make the needle rotate clockwise or anti-clockwise, you must put it in a position where this question matters least – in the east-west direction. When Oersted got no result he knew that he had erred. He suspected that the force was smaller than he had expected and he therefore increased the power of his battery. He ultimately invented a very powerful battery, ordered it to be made and brought to the lecture hall. (He himself could not build a battery; he was clumsy with his hands and very short-sighted.) When even this attempt failed, he gave up. The lecture was over and the audience was about to leave when it dawned on him that possibly the error had been not in the assumption concerning the magnitude of the forces but in the Newtonian assumption of central forces. Once this idea had occurred to him, the discovery was there – before the audience had time to go away he asked his assistant to put the conductor in the north-south direction and the needle moved. The discovery was so shocking that for a few months he did not publish it; he was perplexed and bewildered. But when he published it, he implicitly rejected the theory that the electro-magnetic force was a central force. Meanwhile the news spread, and the attention of the scientific world was diverted from chemistry to electromagnetism. Ampère claimed that all forces must be central, while Faraday followed Oersted. Being Newtonians, most people followed Ampère and did their best to ignore Faraday's ideas. The dispute was settled only in 1905; Newton's doctrine that all forces are central was ultimately given up as all attempts to rescue it (including the ether theory) had failed. This, of course, was the victory of Einstein's theory of relativity.

A Modern Instance

A particularly simple and beautiful modern discovery is Anderson's discovery of the position. Curved lines looking like antennae of some insects are in fact clouds which were created by charged particles moving through cold vapour. These tracks are curved because at the time a magnet was present in the vicinity. Lorentz's theory of force enables us to say that two tracks curve in opposite directions either if the two particles having the same charge move in opposite directions or if the two particles having opposite charges move in the same direction. Before Anderson everyone naturally agreed that two tracks curving in opposite directions are of two electrons which have the same charge and which moved in opposite directions. The reason for it is the hypothesis – call it prejudice if you like – that all electrons are negatrons. The hypothesis that positrons exist, and are created and annihilated together with negatrons, was made by Dirac as a part of a stop-gap hypothesis designed to overcome a severe difficulty in his theory of the electron.

The difficulty exists no more, and the stop-gap hypothesis was never taken seriously anyhow. But the idea of positrons struck Millikan and Anderson and they decided to check the assumption that the two particles which formed oppositely curved tracks moved in opposite directions. Anderson checked the direction by putting a lead plate in the particles' way. Lorentz's theory of force tells us that the more curved path belongs to a slower particle. Assuming that the particle lost rather than gained speed when meeting the obstacle, we can determine the direction of its motion. In this way Anderson refuted the received hypothesis. Refuting also the hypothesis that the positively charged particle was a proton, he was left with the hypothesis that it was a positron. We now know of many other possibilities (mesons) which Anderson did not consider. Thus Anderson proved nothing. But he disproved the previously accepted hypothesis. By now pair creations are repeated much more easily than in Anderson's day, so if Dirac's and Anderson's hypothesis is false its refutation will be quite a revolution of the greatest order of magnitude.

A Rational Attitude

I have discussed these two examples in some detail in order to show that Popper's way of viewing discoveries allows a better interpretation of them. I should stress again that the choice is only between interpreting a discovery to be a verification, or a refutation, or an accident. This is so simply because the relation between two statements (theory and an observation report) is by definition either deducibility (the report is a verification of a prediction based on the theory), contradiction (the report is of a refutation of the theory), or independence (accidental discovery). When we claim that a discovery is accidental we are totally unable, I think, to explain the intellectual significance which a discovery has when it is announced. When we claim that it is a verification of a prediction we can only claim that it is intellectually significant if we can show that it was a verification of a theory. But then we cannot explain why discoveries – like Oersted's or Anderson's – are intellectually significant, although they are verifications of predictions based on theories which were not accepted at all. Nor can we explain the immense impact of experimental refutations like those of Michelson and Morley and of Lee and Yang.

This, however, is only one of the reasons why I accept Popper's view. The main reason why I accept it is that it is inspired by the faith that we learn from criticism, namely, from finding out our own mistakes. Popper has shown the surprising power and fruitfulness of this modest form of rationalism.

CHAPTER 4

REPLIES TO DIANE:
POPPER ON LEARNING FROM EXPERIENCE

Sir Karl Popper has claimed repeatedly that different events are independent of each other in the sense of "independence" used in the theory of probability. That is to say, the probability of a conjunction of two different events is the product of their probabilities. From this it follows that the probability of any event given any set of different events is the same as it was prior to those events having been given. For example, the chance of the next swan being white is not affected by our having seen many swans before, all of which are white. Similarly, Sir Karl has insisted that the initial probability of any universal law is zero. From this it follows that the probability of any universal law is always zero, regardless of how much empirical evidence supports or backs it.[1]

This should lead to a complete breakdown of rational science. If Sir Karl is right, then, it seems, there is no mode of rational choice of scientific theories; and with this the hopes that we do – or at least may – learn from experience must evaporate.

All this worried Diane, and she has asked for a clarification of his view on this point: how does Sir Karl view learning from experience? I have set my clarification in a few preliminary paragraphs which break into an imaginary dialogue between Diane and Sir Karl. Though I do think that the dialogue is faithful in content, and though I have chosen the atmosphere with as much care as I could, I am afraid I have taken a literary license here and there and have made no attempt to make the dialogue faithful in every respect to Sir Karl's characteristic style of conversation. Hence, this work has no claim for great biographical accuracy. Also, the style of the dialogue is less conversational than originally intended, because editors and referees exercise not only philosophical judgments but also artistic and stylistic ones.

Customarily, philosophers of science discuss at length the problem of choice of scientific theories. The words "choose", or "choice," and the frequently used expression "we choose," are ambiguous. By "we" a philosopher often means Einstein and himself. By "choose" he means, *believe to be true.* Thus, "we choose general relativity" usually means Einstein and the philosopher believe that general relativity is true. This, of course, is utterly false: Einstein disbelieved general relativity on account of some metaphysical arguments, and the philosopher all too often does not believe general relativity since he cannot believe a thesis he does not know and all too often he does not know what general relativity says. So there – refuse to understand "we choose" and maybe you will have nothing more to understand or fail to understand.

To set the problem a bit more cautiously. Men of science are often faced with alternative hypotheses, and they try – sometimes successfully – to make an experiment help them choose one, i.e., take one to be the one which they tell their students and lay-audiences to choose, or at least the one which they talk more about. Also, some say, it is the one hypothesis which they train engineers and navigators and their likes to use. This is ridiculous because engineers and navigators use Newton, not Einstein – excepting some nuclear engineers and some space navigators (and even then, they are more likely to use special relativity, hardly ever general relativity).

Never mind. Suppose we have a set of competing hypotheses. (What Makes Newton and Einstein, Lamarck and Darwin, but not Lamarck and Einstein, competitors? If you find the answer keep it in mind.) Suppose under some conditions experience helps us choose. Which conditions? What rules of choice? The process of choice may be called inductive. The rules may be called inductive logic.

Now, the Carnapian, or current, theory may be characterized simply as follows. Inductive logic follows the rules of the calculus of probabilities. The conditions of choice are two: (1) We must set a measure of *a priori* probabilities and dependences. These are not specified by the calculus, and must be added. As we shall see, this is easier said than done. (2) We may now bring relevant evidence to tip probability in favor of one candidate against the rest.

By contrast, and the accent is on contrast, the Popperian theory is this. In science the important process is not at all choice or endorsement but rather criticism or rejection, namely the conclusion that a given theory is unsatisfactory in view of this, that or another specific criticism – usually specific empirical data which (seem to) conflict with the theory.

Q. Do scientists choose?

A. Usually, yes.

Q. Do they have to?

A. Not really.

Q. Is their actual choice rational?

A. Yes.

Q. How so?

A. They do reject unsatisfactory theories, and so what they do not reject they have failed to declare unsatisfactory, i.e., they have corroborated.

Q. Is this a logic?

A. To some extent, though not all the way.

Q. Can we call this logic inductive?

A. Call anything by any name.

Q. Sorry. I mean, is there a problem of choice and is the problem soluble by empirical means and a logic and have you not supplied the logic?

A. No.

Q. How so?

A. It is not the case that scientists have a problem of choice and we philosophers of science offer them tools. Rather, it is the case that scientists do choose. Their choice may be extrascientific or it may be anti-scientific, but it cannot be scientific – it is always metaphysical, though more or less in accord with science (more or less in accord with my theory of corroboration).

Q. The distinction is subtle: you simply refuse to include the metaphysics as a part of the logic of science. Why?

A. Because the problem of induction is not that of how we choose a hypothesis, but that of how we learn from experience. To the second question "How do we learn from experience?" the current answer is by choice. And it is this very answer which leads to the first question "How do we choose?" If you do not ask the second question, or if you do not give the second question the current answer, you need not ever bother with the first question. And, indeed, the current answer is untrue. We learn from experience not by choice, but by rejection. And this answer leads not to the question "how do we choose?" but to the question, "how do we reject?"

Q. If so, why did you, Sir Karl, study choice?

A. Choice does occur, and it is not inductive, as I have tried again and again to show.

Q. Why not inductive?

A. Inductive logic is the logic of learning from experience by choice and it follows the rules of the calculus of probability. Corroboration is somewhat a logic of choice, but not a logic of learning from experience and it does not follow the rules of the calculus of probabilities.

Q. So you agree with Carnap: his logic does and yours doesn't follow the calculus of probabilities, and so they differ?

A. Obviously.

Q. So your views do not compete?

A. Our *logics do differ*; this is why our *views do clash*.

Q. How so?

A. In my view scientists choose in accord with my logic. In Carnap's view scientists choose in accord with his logic. Our logics differ: we both apply them to the same phenomena: therefore, we arrive at a conflict. Q.E.D.

Q. Not so fast. Can you not apply different logics in different ways so as to get the same results?

A. Yes, you can. You are quite right.

Q. In which case there will be no different views?

A. Correct again.

Q. Can this be the case of Popper versus Carnap?

A. No.

Q. Can you prove this?

A. Easily: Carnap's logic as applied by Carnap, when applied to choices in scientific situations leads to some wrong results: Popper's only to right results. Hence they differ. (Even if they are both ultimately mistaken, they still differ because the one is now known to be mistaken but not the other.)

Q. Please spell this out.

A. You are tiresome: all you need to do is read my works, they are crystal clear.

Q. I deny that vehemently. Don't go away! Please explain.

A. As Carnap says in his *Continuum of Inductive Methods*, if the next event depends in no way on the last few observed ones, our last observations are no guide to the choice of a guess on its outcome (by definition of independence). Even the observation of a million white swans does not tell us, in this case, what is the color of the next swan, and *a fortiori*, what is the color of all (unobserved) swans. And therefore, as Carnap notes, probabilistic independence prevents experience from helping us choose. Hence we do not choose by Carnap's rules unless there is dependence. Carnap says we do choose by my rules, hence that there is dependence. Let it be so. What measure of dependence? Should we assume that measure to be high or low? That is to say, at what speed do we learn from past experience? Carnap says that he doesn't know. He wants to consult experience about this. This is funny: he wants to consult experience about the rules for consulting experience, which is either a vicious circle or infinite regress. Also, we may learn fast in one field and slowly in another field. How will Carnap find this out? By experience? But experience may be balanced by unequal distributions of intellectual energy. Carnap correlates intuitively worlds with higher measures of dependence in them, namely worlds with more order in them, with ideal students learning more rapidly from experience, namely quicker to generalize, and less ordered worlds with ideal slow learners. But learning is supposed to tell us chiefly how and to what measure the world is ordered! Moreover, probabilistic dependence leads under the best conditions merely to a choice of *limited forecasts*, but in a *possibly infinite world* (in space or time) *even dependence does not lead to choice or universal statements*: even a big measure of dependence fails to let experience rescue universal statements from their initial zero probability.

Q. But if there is dependence not all universal statements have zero probability.

A. You mean I have made a small logical error?

Q. Did you not?

A. No. There can be dependence in an infinite sequence of event-statements the conjunction of all of which is equivalent to one universal statement. And there can be dependence in an infinite sequence of universal statements; these two possibilities are mutually (logically) independent.

Q. Good Lord!

A. That is right. You try to solve a simple question of choice by a simple calculus of probability, and the calculus rebels and turns up in a really complicated fashion.

Q. Why? Is it as complicated in mathematics?

A. Not in such an annoying way.

Q. Meaning?

A. Questions of dependence and of probability measures are decided in mathematics arbitrarily, just as in the case of arbitrarily different geometries, and their consequences can be studied.

Q. And in statistics, how do we determine which of these arbitrary probability measures to use?

A. Much as we decide which arbitrary geometry to apply.

Q. Namely?

A. We try any of them for size and if it fits –

Q. – we choose them?

A. – we test them –

Q. – and then choose them?

A. – and try to eliminate them –

Q. – and if the tests fail we choose them?

A. – and if the tests fail we *may* choose them.

Q. – Oh!

A. The probability of the inductive philosophers is that of compulsory choice. But there is no compulsory choice. And so philosophers need no probability, and thus no probability measure, and no dependence measure. Rather, for different problems in probability theory we assume alternating solutions involving different measures of probability and of dependence.

Q. So we need not say that all events are *a priori* independent and all hypotheses have zero initial probability!

A. You are right.

Q. So why do you say what we *need* not say?

A. I say what *may* be said.

Q. On what ground?

A. On the ground that experience and the laws of probability make me choose it.

Q. You must be pulling my leg.

A. Indeed, I am.

Q. Why do you tease me? Do you have no heart?

A. I do. It is because I do, and because you try to make me apply induction to my choice of equiprobability of events and to my choice of zero initial probability of hypotheses, that I say what you wish me to say.

Q. Sorry. What criteria, other than inductive, do you employ for the choice of independence and zero probability?

A. I do not fully know, but perhaps simplicity is one.

Q. But how can you call your choice of zero probability and total independence simple if it makes life so difficult; for does it not prevent any further choice, in particular choice of empirical hypotheses or learning from experience?

A. Take care! Now you are committing a small logical error. My choice is legitimate and excludes inductive logic. But it permits, to begin with, learning from experience by rejection and after this, as an option, the choice of hypotheses by corroboration.

Q. And is this inductive or not? I am at a loss.

A. As you yourself say, my dear, zero initial probability of hypotheses plus equiprobability of events prevents inductive logic, but not the logic of rejection and of choice by corroboration.

Q. Can you prove the latter point – on corroboration, I mean?

A. Certainly; with mathematical precision, even.

Q. And is it not less simple to say that learning from experience is by rejection but that choice is corroboration?

A. Less simple than what?

Q. Than saying that *both* learning from experience *and* choice are corroborations?

A. This is an inductive variant of Popperism.

Q. I am delighted. What's wrong with it?

A. It does not work.

Q. Oh! Why?

A. Infinite regress and all that. Good night.

Q. Wait! You didn't say yet. Why do you assume events to be independent?

A. I do not. They interdepend in accord with causal laws.

Q. I mean why do you assume events to be *a priori* independent?

A. I do not. I assume *a priori* that some depend on others some not, in accord with....

Q. Sorry again; why is the probability of events in your system such that they are probabilistically independent of each other?
Is that precise enough for your finicky taste?

A. Almost. You are very acute, if I may compliment you. I say, events are probabilistically independent in the absence of laws.

Q. Thanks. Why?

A. Because laws and only laws are the measures of the mutual dependence of events.

Q. How interesting! And for Carnap both laws of nature and *a priori* probabilities are measures of dependence of events. Is that why you say your system is simpler than Carnap's?

A. This is, roughly, the general idea.

Q. Any thrid alternative?

A. I may not be overjoyed to agree with Carnap, but I am afraid I must. No third alternative is logically possible. Only two inductive systems are possible. Either you assume *a priori* dependence in your system, or you do not. The former is his, the latter is Wittgenstein's and mine. You see, I even have to agree with Wittgenstein.

Q. And Carnap's system is defective, whereas in yours no learning from experience is at all possible!

A. Not so fast or you have again a small logical error on your hands. What you should have said is this. In Carnap's system no kind of learning from experience is possible, and in mine no inductive learning from experience is possible. Good night.

Q. Please wait. But learning from experience by elimination is possible?

A. By elimination of errors, not by confirmation of one view through the elimination of another. Good night.

Q. Please, wait. One last question, please. Just how do we learn from experience by refutation?

A. Good night.

Q. Are you peeved?

A. Yes.

Q. Why?

A. All the time I told you again and again that we learn from experience by refutations and you didn't bother to try to understand.

Q. But I was bothered with another point. Can I bother about all points at once?

A. No. But why bother with inductive logic for so long before even trying to see what I mean?

Q. What *do* you mean by learning from experience?

A. I mean our intellectual horizon widens with a refutation. I mean that a refuting observation report is more theoretically loaded the more abstract and general the theory it refutes. I mean that when a theory is refuted we may see better how far it goes, explain it, and so see perhaps why it goes that far: some breakthroughs of great importance are great refutations.

Q. Is this a theory of breakthrough?

A. No. Breakthroughs are unique; get them anyway you like. If you can't, try refuting an existing theory.

Q. Which one?

A. Good night.

Q. Peeved again?

A. No. Tired. Inductivists think that all repetition is reinforcement. I wonder. Sometimes repetition simply fills one with profound tiredness, if I may make an empirical observation. But I know: philosophers are not supposed to take recourse to experience. Good night.

NOTE

[1] Definition of independence: a and b are independent when and only when $p(a \& b) = p(a) \times p(b)$.

Definition of conditional probability: If $p(b) \neq 0$, then $p(a, b) = p(a \& b)/p(b)$.

If a and b are independent, then $p(a, b) = p(a \& b)/p(b) = [p(a) \times p(b)]/p(b) = p(a)$.

If $p(a) = 0$ then, by the law of monotony, $p(a \& b) = 0$, and $p(a, b) = 0$.

If $p(b) = 0$, then $p(a, b)$ is undefined for most systems (including Carnap's). It is normally assumed in the literature that since b is an observation or observation-report of unique events, $p(b)$ is never zero. This seems to me to be very questionable. Assume $p(b)$ to be zero, and the computing of $p(a, b)$ may be highly problematic.

BIBLIOGRAPHY

Agassi, Joseph, 'The Role of Corroboration in Popper's Methodology', *Australasian Journal of Philosophy* **39** (1961), 82–91. This volume, pp. 40–50.

Agassi, Joseph, 'Discussion: Analogies as Generalizations', *Philosophy of Science* **31** (1964), 351–356.

Bar-Hillel, Y., 'Comments on "Degree of Confirmation" by Professor K. R. Popper', *The British Journal for the Philosophy of Science* **6** (1955–56), 155–157.

Carnap, Rudolph, *The Continuum of Inductive Methods*, University of Chicago Press, Chicago, 1952.

Carnap, Rudolph, *Logical Foundations of Probability*, 2nd ed., University of Chicago Press, Chicago, 1962.

Carnap, Rudolph, *The Philosophy of Rudolph Carnap* (ed. by P. A. Schilpp), Open Court, LaSalle, Illinois, 1963.

Popper, Sir Karl, '"Content" and "Degree of Confirmation": A Reply to Dr. Bar-Hillel', *The British Journal for the Philosophy of Science* **6** (1955–56), 157–163.

Popper, Sir Karl, *The Logic of Scientific Discovery*, Hutchinson, London, 1959.

Popper, Sir Karl, 'On Carnap's Version of Laplace's Rule of Succession', *Mind* **71** (1962), 69–73.

Popper, Sir Karl, *Conjectures and Refutations*, Routledge & Kegan Paul, London, 1963.

APPENDIX: EMPIRICISM WITHOUT INDUCTIVISM

One of the main points of Popper's philosophy is his suggestion that empiricism is distinct from inductivism and therefore that one can be an empiricist without accepting induction. This point seems to me not to have had the notice it deserves.

Consider the following inference:

First premise: We learn from experience.
Second premise: Whatever we learn from experience,
                we learn it by using induction.
Conclusion: We learn from experience by induction.

Evidently the conclusion follows from the two premises. Hume showed that the conclusion is false if by 'learn' we mean 'gain theoretical knowledge'. Let us accept without debate this interpretation and Hume's criticism. As an inference with a false conclusion is either invalid or contains at least one false premise, the following approaches are open to us. (1) We may render the inference invalid. We may do so, with Hume, by reinterpreting 'learn' in the first premise only, or in both premises only, but not in the conclusion, to mean 'acquire habits.' (2) We may deny the truth of the first premise and say that our theoretical knowledge is not derived from experience but is a priori valid. This is the Kantian approach. (3) We may accept the first premise and deny the second premise. This is the Popperian approach. Other approaches are possible, but they were never seriously entertained. The Humean and the Kantian approaches are now in almost universal disfavor.

The Popperian approach raises the question of how we learn from experience if not by induction. Popper's answer to this question may, if it is acceptable, strenthen his rejection of the second premise. In any case, it will show that the second premise can be denied without inconsistency. However we view Popper's answer to this question, we must acknowledge that we have no argument for accepting the second premise. The only argument in its favor (namely that there is no conceivable empirical method other than the inductive method) was superseded by Popper's proposal of his own methodology. This should be admitted even if Popper's methodology is rejected.

In defense of the second premise, one may argue that 'we learn from experience' is synonymous with 'we learn by the use of inductive method.' But then the phrase 'inductive method' can be omitted altogether and instead of using it one can characterize methods which, according to Hume's criticism, are not workable. These are the methods of generalizing from observations, and of rendering some hypotheses more probable than others with the aid of observations. And the question remains, what alternative methods there can be, which may be said to explain the fact – if it is a fact – that we gain theoretical knowledge from experience. Popper's methodology – his view that we refute hypotheses with the aid of observation – may or may not be acceptable. But it has not been shown to be erroneous, while all existing alternatives to it have.

Those who nonetheless reject Popper's view and seek alternatives should provide empiricist but not inductivist views; i.e., ones which are not hit by Hume's arguments.

CHAPTER 5

SENSATIONALISM

Sensationalism is the traditionally important doctrine according to which all our knowledge of the world comes to us through the senses. The chief aim of these pages is to systematize the traditional arguments against sensationalism, to show their incompleteness, and to supplement them with some modern arguments. Round the turn of the century a new version of sensationalism was proposed by Duhem and Meyerson. It is not surprising that only modern criticism meets their version, since they constructed it after they had accepted the traditional arguments against the traditional versions of sensationalism. It will be easily seen that theirs is the last possible version, so that criticizing it may be considered as criticising sensationalism altogether.

There are two traditional divisions of sensationalism, yielding four possible sensationalist schools of thought. The first division is that between the sensationalists who think that informative theoretical knowledge is possible – the inductivists – and the sensationalists who think that informative theoretical knowledge is not possible – the conventionalists. The second division is that between naive and sophisticated sensationalists: the naive sensationalists assert that all (well attested) reports of observation are entirely reliable; the sophisticated sensationalists assert that only some sort of observation-reports are entirely reliable.

	Sensationalism	
	inductivism	conventionalism
Naive	Telesio	Poincaré
Sophisticated	Bacon	Duhem

Both of these divisions of sensationalism, it should be noticed, are exclusive and exhaustive. A complete criticism of sensationalism must be the criticism of all the four versions, or classes of versions, mentioned above. The traditional criticism of sensationalism was the criticism of inductivism and of naive sensationalim: it left room for sophisticated conventionalism, which was later constructed with incredible ingenuity by Duhem and Meyerson. Ingenious as it is, however, sophisticated conventionalism too has to be rejected, and with this sensationalism is completely superseded.

For the sake of simplicity the discussion of the present essay will be confined to our knowledge of the external world. The problem of our knowledge of ourselves will be avoided in order to avoid reference to any psychological theory except perception theory, and in order to avoid exegesies of the classical sensationalist texts. The very famous dictum 'nothing is in the mind which has not been previously in the senses' may be understood – prior to exegesies – to refer solely to our knowledge of the external world. It is therefore understood here in this restricted sense, and criticized together with the four philosophical schools of thought which endorse it.

1. *Sensationalism vs. Theoretical Knowledge*

According to sensationalism all knowledge of the world comes through the senses. This obviously entails that knowledge consists exclusively of observation-reports and statements derivable from them. It is therefore inconsistent with the view that there exists theoretical knowledge about the world, since (by the very meaning of the word "theoretical") theoretical knowledge of the world is that knowledge which is not a derivable from observation-reports alone.

Sensationalists are well aware of this criticism; they view the problem of how to answer it as an integral part, if not the core, of the problem of induction. They vacillate between two alternative answers to this criticism, inductivism and conventionalism.

The inductivist answer to the above criticism is that theoretical knowledge is "indirectly" derived from the senses, being based on observation-reports by induction. This answer dodges the issue: whatever "indirectly" means, induction is either a purely deductive process or not; if it were deductive, then theoretical knowledge would be derivable from observational reports; which is not the case. Therefore whether or not theoretical knowledge is gained by the process of induction, and whatever process induction may be (by the very meaning of the word "theoretical") if theoretical knowledge exists then sensationalism is false.

The inductivist may try another answer: he may claim that although all factual knowledge is derived from the senses, a particular piece of knowledge may at first have to be a conjecture, and only then it has to be verified, or become a result of observation. This answer may be called verificationism, since according to it conjectures of a given kind are capable of verification by observation.

Verificationism is vague and dangerous. It is so vague that one cannot know whether it is a version of sensationalism, and it is dangerous in its raising false hopes. Let us consider its vagueness first. Verificationists often do not tell us whether or not the verified conjecture follows from the reports which have verified it, and whether or not the verifying could be secured without any prior conjecture. Does verificationism include the claim that the conjecture does follow from the reports which verify it, and the claim that in principle these reports could be procured without any previous conjecture? In case it does, verificationism is decidedly sensationalist; in case it denies at least one of these two claims it is decidedly not sensationalist; in case it does not answer the question it is not decidable whether verificationism is or is not a version of sensationalism. Now, in case verificationism is made a version of sensationalism by adding to it a positive answer to the above question, then, according to it, verification leads to not theoretical knowledge. Sensationalist verificationism only states that prior to making an observation (such as that the sun will rise tomorrow, to take Wittgenstein's example) we may, but need not, guess correctly that observation. This is not what is usually meant by "verificationism"; so, verificationism as commonly understood is not at all sensationalist, though, unfortunately, the vagueness of it gives such an impression. Quite apart from this, and more generally, verificationism is dangerous because it raises false hopes. It raises false hopes because it can provide no assurance or guarantee that our conjectures will ever be verified. Sensationalism requires that if there be a guarantee it should be verified; it also requires that there be a guarantee that the guarantee can be verified, and so on; this is the infinite regress argument already discovered by Hume. Moreover, the assumption that a guarantee exists will make verifications of conjectures unnecessary, as will now be shown, and this will prove the incompatibility between verification with a guarantee and sensationalism.

Take first the simplest case of one conjecture and a guarantee that we shall be able to verify that conjecture. As we have a guarantee that this conjecture is true, we need not verify it. Hence, this case is inconsistent with sensationalism. Even when a theory entails that possibly there is a case in which we need not verify a conjecture, that theory is inconsistant with sensationalism which asserts that we cannot ever know the truth of a conjecture prior to its verification. So let us now take different cases of verificationism and show that in each case there is a possibility of its reduction, under some possible circumstances, to the first case above. Take, again, one conjecture, and a guarantee that it be verified, but now replace the full guarantee by a partial one. Now, a partial guarantee may be of a good chance of success, which does not exclude utter failure; such a guarantee is of no avail since in case of utter failure one may claim that the guarantee is valid nonetheless. In order to exclude such a pre-posterous case, one usually claims, a partial guarantee which only assures a chance of success also assures, by the mathematical laws of chance, that on a sufficiently repeated application of a partial guarantee the partial guarantee of success becomes a perfect guarantee for success in some cases (Bernoulli's law). So we can use a partial guarantee only with a set or a class of conjectures, not with a single one. The partial guarantee, then, will be the guarantee that of the given class of conjectures of a given kind, some – at least one – will be verified. This guarantee was presented explicitly by Keynes, who has labelled it 'the principle of limited variety'. This principle is consistent with the conceivable case of refuting all the members of the given class of conjectures except one. Thus, by this principle, it is quite possible that after a certain amount of failure, the perfect guarantee for partial success would become a perfect guarantee for full success. (Thus, Keynes' principle is incompatible with Bernoulli's law!) And this possibility is the possibility of a case in which the process of verification is unnecessary (because it is guaranteed); as a theory which permits this possibility, the principle of limited variety is inconsistent with sensationalism (which does not permit this possibility). Furthermore, any guarantee whatsoever would render at least some questions of fact decidable without observations, namely, such questions of fact as the ones concerning the possibility and nature of human knowledge. Obviously, in this case we cannot claim that the view that human knowledge is possible is based on experience without begging the question and thus running into an infinite regress.

The next retreat of the inductivist would be an attempt to replace the notion of the verification of a conjecture by that of its confirmation, namely, of the verification of some of the consequences of a conjecture. However, the problem reappears. We can have no guarantee for any verification whatever, not even for a verification of a weak consequence of a conjecture. If we had, then, again, we could construct a possible case in which the guarantee will render observation unnecessary – and thus the existence of a guarantee will be inconsistent with sensationalism. Moreover, the margin between the verification of a conjecture and its mere confirmation, namely the unverified consequences of an accepted hypothesis, would be the nonsensational element of the existing body of theory.

Thus, sensationalism is incompatible with the view that informative theoretical knowledge exists, no matter how it was acquired – by verification, by confirmation, or in any other way. As Hume has already shown, inductivism (be it correct or not) fails to reconcile sensationalism with the view that theoretical knowledge exists. (Being both an inductivist and a sensationalist, Hume ended up denying the existence of theoretical knowledge.)

Historically, another limitation of sensationalism was noticed prior to Hume's discovery of its limitation with respect to theoretical knowledge; it is the limitation with respect to theoretical concepts. Since the Middle Ages sensationalists have realized that sensationalism implies that we cannot have theoretical concepts, that all our concepts are either those derived from observations, or their combinations; that even in our wildest imagination we cannot fancy anything but new combinations of old observational material, so that all concepts are, like the concept 'sphinx' (to take Francis Bacon's example), merely combinations of observational concepts. Indeed, this is Hume's starting point. Einstein's and Russell's favourite argument against sensationalism is the essentially Kantian idea that mathematical concepts go very far beyond any past experience, and that some of these concepts are employed very fruitfully in science.

Yet this very criticism gives the clue to the alternative sensationalist view, namely conventionalism. Conventionalism gives great scope to the imagination, and views both mathematics and theoretical science as admirable structures produced by the imagination. But, admitting that theories go beyond experience and remaining true to sensationalism, conventionalism must empty theories of all factual or empirical content. Assuming that theories come from the imagination, and that information comes from experience, conventionalism justly concludes that theories are devoid of any information. If theory is not informative knowledge, what is it? It is, says conventionalism, our way of looking at particular facts, our way of classifying particular observed facts. Like formal mathematics, theoretical science is merely an empty structure to store information in, a way of saying things, a language. Nothing in reality strictly corresponds to abstract or imagined theoretical concepts like 'space curvature' or 'atom'. These words are no more than shorthand symbols with no independent meaning (their meanings are given by implicit definitions), and statements containing them impart no more information than the information procured by sensations alone.

Although conventionalism is much clearer and more coherent a view than inductivism, it was traditionally viewed by men of science as a defeatist position, because the aim of science, it was felt, was not just to replace an unordered or an arbitrarily ordered heap of information by an elegantly ordered yet not richer stock of information. The intuitively accepted view behind the scientific tradition between 1600 and 1900 was that there exists a hidden reality (i.e. hidden from the senses) and that the aim of science is the search for it; to wit, the aim of science is to attempt to discover the laws of nature and not to invent laws of elegant and concise language. Although hardly any of the classical natural philosophers adhered to conventionalism consistently and persistantly, many of them used it as a second best alternative to inductivism; that is to say, when they could not present their theories as inductively based on observation-reports, they chose the alternative sensationalist interpretation of their theories, and viewed them in a conventionalist manner as mere systems of classifications of knonw facts. It seems clear that traditional thinkers confined themselves to inductivism as a rule and to conventionalism as a temporary refuge, because they wished to retain their sensationalism; they wished to retain sensationalism as they considered it the only basis for empiricism; and they wish to retain empiricism as they considered it the only ground for the validity of empirical science. But in this they are mistaken.

2. *Sensationalism vs. Empiricism*

If we assume that informative theoretical knowledge is possible, then we may inquire into the grounds for its validity. The two traditional answers to the problem, what is the ground for the validity of theoretical knowledge, are apriorism and empiricism. Apriorism is the claim that the ground for the validity of informative theoretical knowledge is intellectual, and empiricism that it is sensational. Thus, empiricism is traditionally sensationalist. But sensationalism may be not empiricist at all. Obviously, conventionalism is a version of sensationalism and yet it is not a version of empiricism; indeed, it is neither apriorist nor empiricist: by including the claim that informative theoretical knowledge does not exist, conventionalism avoids giving rise to the problem what is the ground for such knowledge – to which problem both empiricism and apriorism are the traditional alternative solutions. Yet, as conventionalism is a sensationalist view, traditional empiricists prefer conventionalism to apriorism, and were even ready to use it as a temporary refuge when their empiricism was beaten, hoping that with the increase of the amount of factual information they would be able to return to their empiricism in order to find informative theoretical knowledge about the world.

Traditional natural philosophers have always emphasized the significance of empirical theoretical knowledge. The most often quoted passage of Bacon's was his parable of the ant, the spider, and the bee: the empiric or skeptic who has only reports of observed facts is like the ant which only collects; the reasoner or apriorist who has only theories is like the spider which only spins out its own material; the interpreter of nature, the true empirical theoretical philosopher, is like the bee which both collects and adds something of its own to the collected material. (This parable, says Russell, is unfair to the ant who also orders. If so, it is also unfair to the spider who also collects.) In another famous passage Bacon speaks of science as the wedding of the intellect and the senses.

These metaphors conceal a problem. Admittedly the contribution of the senses is empirical. But what is the contribution of the intellect? Is it not the case that the contribution of the intellect is non-empirical? Is not the idea of empirical theoretical science self-defeating?

Bacon must already have been aware of this problem, for he gave an answer to it (in his Preface to *The Great Instauration*). His answer is this. Just as by sensing the rays of light our eyes see things, so, by analogy, by sensing things our intellect sees the laws of nature. This answer is a traditional mystical or intuitionist view which assumes the existence of a mental eye that sees or intuits laws with complete assuredness just as the eye of the flesh sees things with complete assuredness. (The traditional mystic formula is that of the unity of the knower and the known with knowledge; it occurs in a crucial passage of Bacon.)

No wonder that no later empiricist shared this view with Bacon. The problem remains, then: how is empirical theoretical knowledge possible?

The obvious substitute for Bacon's answer is the view that the senses provide the material and the intellect the order. But this answer is the denial of empiricism. It is either conventionalist, if the order which our mind provides is claimed to be merely ours, or apriorist, if that order is claimed to coincide with the order of the world. (Kant seems to have vacillated between these two claims and clearly preferred to leave the choice between them undecided. He has stressed that the order is provided by us, but intentionally, systematically, and persistently left open the question whether this order of ours coincides with the order of the world or not. This point was overlooked by reviewers of his *Critique of Pure Reason*, with the result that he was incensed and wrote his *Prolegomena* – a much easier book to read – to set this point right, as he explains in the preface to that book. Unfortunately, some of his best modern commentators, such as Russell, still do him injustice on this point by ascribing to him answers to questions which he insisted on leaving unanswered, such as, is space real?)

Another empirist answer is this: when the senses make their own contribution they stimulate the intellect to make its own contribution. But this is not to the point: the apriorists themselves, since Bruno and Descartes, have always asserted that the senses may stimulate the intellect; they only denied that the senses are the source of knowledge; they declared that the intellect makes a contribution, and that we can see the independence of the validity, or the self-evidence, of this contribution. Clearly, they would argue, if any part of the contribution of the intellect is logically independent of the senses to any extent apriorism is not excluded, while if the contribution of the intellect logically depends on perceptions it cannot add to the information which can be provided by the senses. The only other alternative is to claim that the contribution of the intellect is not informative, as the conventionalists do. (Incidentally, Wittgenstein claims (*Tractatus*, 6.342) that theoretical knowledge is not informative, but he adds that the existence of theoretical knowledge and its simplicity are informative. This, obviously, suffices to overthrow his sensationalism and to establish him as an apriorist.)

The above discussion may explain the fact that quite a few great thinkers despaired of their empiricism and became apriorists. It is rather cheap, perhaps, to ridicule eminent apriorists (Descartes is the traditional scapegoat) whenever the validity of your own brand of empiricism is challenged. The function of the repeated sneer at apriorism is to drive home the idea that any deviation from narrow sensationalism leads towards apriorism. This idea does not solve our problem, however, but rather sharpens it. For we can state the dilemma in this way: if we do not go beyond sense experience we have no theoretical knowledge of the world, while if we do go beyond it the margin is not contained in sense experience, and is, thus, *a priori*.

This is the logic which led thinkers to abandon empiricism in favour of either apriorism or conventionalism. For according to both these views our present theoretical knowledge necessarily transcends our experience; they differ only as to the question of whether this knowledge is informative (apriorism) or not (conventionalism).

Yet empiricist philosophers who have studied the problem of knowledge have usually stuck to their sensationalism in spite of this refutation. Perhaps they hoped that somewhere a logical error had been committed in the refutation of sensationalist empiricism. They were unable to refute any step of the criticism, but they had a strong argument in favour of the view that a logical error could be found in the criticism. The argument is this. In our ordinary behaviour we show that we consider theoretical science as informative, for we normally rely on theoretical science as informative, for we normally rely on theoretical information. Moreover, we show that this information is indeed connected with experience, for if theoretical information clashes with the information gained by experience we prefer the latter; we accept theoretical information only when it is strongly supported by experience. Furthermore, we gain theoretical knowledge or at least theoretical hints from certain important experiments, like that of Michelson and Morley. In brief, we know that the error is there, since we know that we gain theoretical knowledge from experience.

There is no need to accept the suggestion that there is an undetected error in the criticism, and it is even dangerous to accept such a suggestion since in this act one might very well be giving up one's rationality and hope to learn from criticism. We may even endorse the above criticism as valid while accepting the contention that we gain theoretical knowledge from experience – although it is in no way imperative to accept it. In a way this contention – we gain theoretical knowledge from experience – is indeed the core of empiricism, as it amounts to the rejection of both conventionalism and apriorism. But, strictly speaking, this contention is not empiricist in the traditional sense, at least in that traditional empiricism is a theory of the ground of the validity of existing knowledge, whereas this contention is a theory of learning or of scientific method. And, most important, perhaps, is that traditional empiricism is sensationalist (and thus has to be rejected), whereas this contention – that we learn about reality from experience – is not. Even if we say that all we learn about the world is derived from experience, we need not commit ourselves to sensationalism except when we identify experience with information derived from the senses. Hence, either the identification of all experience with sense experience is an error, or else any kind of empiricism is hopelessly inconsistent: only if we get rid of the identification of experience with sense experience may we retain some modern version of empiricism. Let us now consider the different ways in which experience was identified with sense experience; it will turn out that this identification leads to the surprising conclusion that we cannot describe our experiences, that we never know what experience is.

3. *Sense-Experience vs. Experience*

The identification of experience with sense experience has been done in a naive way and in a more sophisticated way. The naive identification of experience with sense experience is a version of naive realism. It is simply the claim that (when we look carefully) we see things as they are. It was admirably criticized by many modern philosophers from Galileo and Kant to Einstein and Russell. Galileo thought the naive thesis refuted by the fact that when we walk at night we may see the moon jumping like a cat from one roof-top to another. A somewhat less picturesque but more elegant argument against it is, perhaps, Schrödinger's argument (in *Nature and the Greeks*). We see the sun as being not much bigger than a cathedral. Assuming that the sun is as big as we see it, and accepting very simple, and intuitively quite obvious, trigonometrical theorems, we can calculate the distance between the eastern and western positions of the sun and find it to be no more than one day's walking distance.

This argument does not convince the adherents of naive sensationalism. Naive sensationalism carries great force with it. Even if we do not see all things precisely as they are, we all admit that we can see this table tolerably well (arguments from perspective notwithstanding). We admit that endless speculations and disputations will not be useful to determine some question of fact which can easily be determined by plain observation and experient – for instance, the question, what is the color of the table in the next room? Surely that much we all admit, and it is, quite obviously, the core of what the naive sensationalist wishes to assert. This explains the fact that though no serious thinker defends systematically naive sensationalism, many lapse occasionally into this position.

Many historians of physics and a few physicists claim that the medieval scholars were apriorists who would consult Aristotle's and their own reason, rather than observation, in order to determine the number of teeth a mole has. It is not known whether any medieval scholar deserves ridicule as one who never relied on his eyes but always preferred reasoning to plain observation. If such a person existed, however, he must have been most unusual. Undoubtedly, most people agree that it is often preferable to rely on one's eyes than on one's reasoning; and undoubtedly naive sensationalism explains or justifies this preference. (Even apriorists like Descartes and Spinoza endorsed this preference, of course; they offered most ingenious explanations or justifications of this preference, but we need not go into that here, where we take it for granted that apriorism is false.) As naive sensationalism is plainly false, the (correct) frequent preference of observation over reasoning needs a better explanation or justification. We do not rely on the eye of the mind because it may mislead us, and to be fair we should not rely on the eye of the flesh as it may mislead us too. This, it seems, suffices to show that the preference of perception over reasoning is not a matter of reliance. Sensationalists, however, stick to their sensationalism even after they are forced to admit that the senses can, and sometimes do, mislead us. These sensationalists claim that though our senses may mislead us, under certain conditions they do not. This is still unfair to the intellect. If we are willing to rely on the senses, even though they mislead us, after finding the conditions under which they do not, then, in all fairness, we should be also ready to rely on the intellect, even though it misleads us, after finding the conditions under which it does not. Yet according to sensationalism we should sometimes rely on our senses but never on our intellects. When the sensationalist becomes aware of the fact that experience can mislead us, instead of ceasing to rely on experience and reopening the whole problem afresh, he claims that there must exist some kind of experience – pure experience, as he calls it – which cannot mislead us. This is the doctrine of sophisticated sensationalism.

There exist strong versions of sophisticated sensationalism. They specify which, or what kind of, experience cannot mislead us. The weak version is the mere assertion that pure experience exists. This may be the case. Let us examine, first, the strong versions of sophisticated sensationalism, Bacon's and Locke's. Bacon's position is that the theoretical element of experience is what mislead us, not the sensational element of experience. Once we get rid of all our prejudices, of all of our preconceived ideas, we can experience things as they are (in the manner assumed by the naive sensationalist).

Very interestingly, the mere substitution of 'class-prejudice' for 'prejudice' in the statement above yields Marx's view, and of 'neuroses' instead yields Freud's view. Now, Locke's view is that the reliable experiences are the elements of individual sensations, which were later called sense-date; they are pure sensations like the sensations of sounds or the sight of coloured patches. Bacon's doctrine can be shown to be inconsistent, for it is itself a preconceived idea. Locke's view has been experimentally refuted. The identification of patches was shown to be dependent on our knowledge of geometry and perspective, the identification of a colour shown to be dependent on our language, and even the ability to distinguish between two similar sounds depends on theoretical instruction. Thus, although sophisticated sensationalism as such – the view that there exist pure sensations – is irrefutable, its two more substantial versions – Bacon's and Locke's – are logically or empirically refutable, and they were refuted. As to the irrefutable version of sophisticated sensationalism – the bare claim that pure sensations exist – it does not explain the existing preference, in certain cases, of observation over theory: the preference can only be explained as the preference of the reliable if the observation in question can be identified as a pure sensation.

No one doubts that sensations are a necessary part of any experience. The fact of experience is, however, that thus far we cannot specify a type of experience which must be fully reliable; that in particular we are not aware of pure sensations; that there exists no known or reported immediate or direct sense experience, just as there exists no known or reported immediate or direct experience of the electric signals which, according to modern neuro-physiology, sensations consist of. It is a fact of experience that when describing or reporting scientific experiments and observations we very rarely describe or report our sensations. Observation-reports describe facts and their having been observed – not at all the sensations of the observer while he was observing, except in rare instances such as the sensation of electric currents on the observer's tongue.

This fact a sophisticated inductivist will readily admit, and yet he will claim in describing our scientific experiments we do report our sensations, even though indirectly. Here again our inductivist runs against our dilemma, and again he refuses to consider it, being sure that we learn from experience and that experience must be, ultimately, sense experience. Rather than resolving the dilemma he tries to purify a given report of a scientific experiment of its theoretical element and reduce the scientific report to a report about past sensations. Yet when trying to do this he soon uses theories and statements of objective facts rather than reports about sensations. He will justify his use of statements of objective facts by claiming that they were once constructed out of pure-sense-elements – thus assuming what he has set out to prove. He will also justify his speaking of facts by defending naive sensationalism. He will then retreat from naive sensationalism to a sophisticated one, and so on. It is a historical fact that very few thinkers ever tried to show how a given piece of scientific information can be decomposed into, and recomposed from, sense-perceptions; such attempts, notably Laplace's, Mach's and Russell's, were complete failures because of their authors' vacillation between naive and sophisticated views, as well as between inductivist and conventionalist views.

This argument leads inductivists to two characteristic reactions. The one is to try again. The other is to dismiss the whole debate as too sophisticated. In order to show that it is not unnecessarily over-sophisticated I shall take an example of the inductivist's muddled approach to experimental errors.

It is a well-known fact that John Dalton reported having observed the atomic weight of oxygen to be, on the average, near to but slightly above 6.5 and decided that it is actually 7. Obviously, he could not get the result which we have today, namely 16, because he thought that water contains oxygen and hydrogen in equal proportions and not, as we think today, in the ratio of one to two; but better experiments, it is alleged, might have led him to the result 8 rather than to 7. It is therefore unanimously accepted by modern historians of science that Dalton was a bad observer, Dr. Thomas Thomson's personal testimony to the contrary notwithstanding. It is difficult to imagine that a bad observer was the inventor and improver of experimental techniques in weighing gases. If the historians who condemn Dalton were serious about the whole matter they would have tried to repeat Dalton's experiments as his contemporaries did. In this case they would undoubtedly get the same result as Dalton's, just as Dalton's contemporaries did before Davy discovered a better method which yielded the result 7.5.

It is obvious to me that Dalton's result is respectable and yet untrue. He who doubts it will have to apply the same doubt to the results of all nineteenth-century chemical experiments. The best and most precise experiments concerning the atomic weight of chlorine then gave 35.5 as a result, and they were equally mistaken; the atomic weight of chlorine, is much nearer to either 35 or 37 than to 35.5. The naive and sophisticated inductivists alike must fail to explain all this. In order to explain why our predecessors accepted and we reject 35.5 as the atomic weight of chlorine different theories have to be referred to. It transpires, then, that contrary to all we were taught in chemistry classes and in history of science classes and in philosophy degree courses, it was the factual report which has been declared to be false in the light of modern theories. But before trying to defend this startling conclusion I wish to discuss the sophisticated conventionalists' explanation of this situation. For it is the great advantage of sophisticated conventionalist that he handles this situation with great ease.

4. Sensationalism vs. Common Sense

The criticisms of naive sensationalism and of inductivism which I have presented so far do not cause the slightest difficulty to the sophisticated conventionalist. He does not claim that theoretical knowledge is derived from experience but neither does he claim that theoretical knowledge is informative. He therefore can easily reconcile the existence of uninformative theoretical knowledge with sensationalism. He does not claim any observation report is purely sensational, so that he can stick to his reliance on the senses in spite of all the alterations which the observation reports undergo. The major defect of this position seems to lie in the fact that it sounds just too defeatist a position, defeatist both regarding theory and regarding observation. But this is an error: defeatism regarding theory is quite sufficient. Sophisticated conventionalists can argue that even though we cannot separate the sensational element in any observation statement, this element is invariant regarding any translation of a report from one language to another. The nineteenth-century observation report 'the atomic weight of chlorine is 35.5' is not discarded by modern chemists but is translated by them into the twentieth-century language; the translation reads: 'the average atomic weight of terrestrial chlorine is 35.5'. The translation of the report preserves its sensational element. The sensational element has not been rejected, only the theoretical element has been replaced. The twentieth-century report 'the atomic weight of chlorine is 35 or 37' does not contradict the nineteenth-century report 'the atomic weight of chlorine is 35.5': they are cast in different languages, and forgetting this fact we rashly conclude that they contradict each other. Before we can find out whether they are in contradiction or not we must state them both in one and the same language. Now we cannot easily translate the twentieth-century report into the nineteenth-century language, because the later language is better – more elegant – than the older language. So it is more convenient to translate the nineteenth-century report into the twentieth-century language. As we have seen, the translation shows the two reports to be perfectly compatible with each other.

It is essential for this mode of thought that it is both conventionalist – in viewing theoretical science as a mere system of languages – and sophisticated. Had we been able to state one observational report with no theoretical overtones, then the problem which the sophisticated conventionalist has solved would have arisen in a very different manner and his solution to it would be obviously unacceptable. And if we wish to conclude that it is impossible to have purely observational reports, we must assume that even though we can translate a report into many languages without losing or altering its sensational content, we shall never be able to isolate this sensational element entirely. No doubt, had we constructed all the possible languages, and had we then stated one report in all these languages, the 'conjunction' of all these many statements of this one observation report should give us a fair idea of the observation as such. But this is merely a thought experiment: there can be infinitely many languages, or theoretical systems, for any finite set of observation reports to be expressible in. Also, we have seen from the example above, using a better language is easier and more elegant than using an inferior language – even when an observation-report fits both. This may suggest that the simplest, most ideal, language, if it will ever be achieved, will enable us to state pure observation-reports. Duhem arrived at the same idea from another angle: he thought that the ideal theoretical system will also be a true informative picture of the world – from which the above follows.

This discussion seems to me to clarify a number of points. First, it explains why sophisticated conventionalism never was popular: it is somewhat too sophisticated. Second, it explains the modern search after pure observation reports. Any sensationalist alternative to Duhem's and Meyerson's doctrine must contain the claim that we can isolate sense impressions from the theoretical element with which it amalgamates when presented in a scientific report. Yet in order to be convincing one must indicate how this can be done. Now Popper has argued (1935) that, since universal names are dispositional, reports containing them contain predictions, and are thus no pure reports. *E.g.* the report 'here is a glass of water' contains predictions since the glass is breakable or else we would not call it 'glass', and water is decomposable, etc. Hence the immense literature concerning dispositions and dispositional terms which has followed Carnaps' study (1936) of the relations between dispositions and pure observations.

Yet one should notice, perhaps, that the sophisticated view according to which we cannot separate the sense information from the theoretical element in an observation report, though unpopular amongst philosophers, has gained popularity amongst some schools of contemporary psychology. This is so, partly, I suppose, owing to the fact that psychologists cannot evade problems concerning observation reports as easily as other scientists: such troubles are their business. Partly it is due to the influence of Külpe's critical realism. The full discussion of this point is beyond the scope of the present essay; yet this much can, and ought to, be said here. The sophisticated view according to which we cannot separate the sensational and the theoretical element in an observation report is not in itself intolerable; it is intolerable only when we adopt it together with an inductivist or with a conventionalist attitude. For, when we adopt it together with a critical realistic attitude, we merely admit that any observation report must contain some hypothetical element; only when we adopt it together with a conventionalist attitude it turns out that we do not quite know what we are saying. For, according to sophisticated conventionalism, only the sense element is informative, and the sense element is unisolable; hence, according to sophisticated conventionalism, the information contained in a report is unisolable. The analogous realist attitude only entails that the certain and entirely warrantable element of a report is unisolable; namely, that observation reports are never certain.

To make this clearer, consider an observation report stated in court. To say that the judge does not quite know what is the information he receives from a witness is very disquieting. Moreover, it is not at all difficult to imagine, or to draw out of history, a case in which a piece of evidence would condemn the accused when cast within one theoretical system, and acquit him when cast in another. To critical realists this causes no trouble, since they permit error in any observation report. But not so to the sophisticated conventionalist.

Duhem was not unaware of this difficulty, for he tried to solve it. He suggested that naive sensationalism should apply to commonsense situations (such as the one described above) and sophisticated sensationalism to science. His argument in favour of this suggestion is too involved to reproduce here. Nor need it be reproduced. For this division between science and common sense cannot be maintained, especially in the light of modern perception studies, of the Külpe school and its derivatives. Today's common sense, as Maxwell has already claimed, is yesterday's frontier of science.

The sophisticated conventionalist may attempt to answer this criticsim, but the more he will do so the more he will defeat his own purpose, for he is presenting a more and more elaborate theory about science and its role in society – a theory which he must consider as informative, and which entails that it is itself uninformative. As long as we only look at scientific theories we may suggest viewing them as empty – as Duhem does. But we cannot merely suggest to a judge that he view scientific theories as empty; we have to explain to him why he should do so by providing a theory of sorts; and he will rightly apply this theory to itself in order to dismiss it as empty.

This discussion explains, I hope, why Poincaré, by no means an unsophisticated philosopher, preferred naive sensationalism to sophisticated sensationalism; it is untenable to claim that we do not quite know what we say when in ordinary circumstances we state a simple and unproblematic observation report; nor is it tenable to divorce such reports from scientific enquiry. But though Poincaré's rejection of sophisticated sensationalism is well founded in common sense, his acceptance of naive sensationalism was a serious error.

We have now come to the end of the list of traditional alternatives. Those philosophers who pin their hopes on the future success of present-day efforts to discover pure observation reports hope to erect a new inductivist epistemology or a less sophisticated conventionalism than Duhem's. The rest are faced with the choice between apriorism, sophisticated conventionalism (a Kantian vacillation between the two), or the search for a revolutionary approach. The notorious conservatism of the bulk of philosophers (plus the unpopularity of apriorism and sophisticated conventionalism) is my only explanation for the popularity of the search for pure observational reports, for observable hard and fast facts, in the face of the increasing amount of evidence from modern psychology, from modern perception theory, which shows the futility of this search. Psychologists are usually unaware of the philosophical implication of their studies, of the fact that their studies give rise to the need for a new epistemology. But then they are usually not interested in this aspect.

The claim that there are no pure (or 'neat') observation reports is central to Ryle's argument in his *Concept of Mind* (1949). By implication Ryle also rejects sophisticated conventionalism (when he denies the existence of a fundamental difference between common sense and science, pp. 288ff.). He thus faces *the* problem of epistemology, and he is well aware of it: he sketches a programme for a new epistemology (pp. 317–318). This need for a new epistemology is rooted not in Ryle's central doctrine, in his proposed solution to the body-mind problem, but in his revolutionary perception theory. Popper, who dissents from Ryle's solution to the body-mind problem, but shares Ryle's perception theory, had outlined over a decade earlier a similar, if not the same, programme, and also proposed theories which answer the *desiderata* of that programme. It is regrettable that this logic of Ryle's argument has not been clearly seen by the general philosophical public. It is this oversight which is responsible for the popular view of Ryle's doctrine as a version of behaviourism, even though he explicitly rejects behaviourism because it is based on the naive belief in pure observation reports. It is the same oversight which is responsible for the popular identification of Ryle's psychological theory of knowledge as a set of dispositions with Mill's and Schlick's similar epistemological theory of knowledge as the proper procedure of connecting past and future events. (The two sets of theories of knowledge obviously come to solve two quite different sets of problems. The epistemological theory answers questions of status and basis of validity, the psychological theory answers questions of the seat of knowledge and its influence on behaviour. It is regrettable that Ryle's metaphors allow for the confusion of his view with Schlick's.)

Popper's new theory of the status and methods of science is opposed by many philosophers because it entails the non-existence of pure observation reports. This, as we have seen, is a very scanty ground for opposition. Others find it difficult to share his reasons for the acceptability of some observation reports in spite of their inherent uncertainty. It is this last point which I now wish to discuss in some detail.

5. *Explanation vs. Consent*

The whole literature concerning the methods of science seems to be agreed on one point, which I shall now try to criticize. It is agreed amongst philosophers that when it is said that a certain piece of information is scientifically acceptable it is meant that the piece of information in question ought to be accepted as true – to be believed. At least I have never come across any philosopher who has contested this. Popper has stressed that this acceptance must be tentative; but even he agrees that accepting a report is, for the time that it is accepted, considering it to be true. My own alternative is that observation reports ought to be accepted as a task, as something which we should try to explain, and this does not exclude the possibility that we should explain that piece of information as based on an error. This forces us to admit, I shall argue, that the problem of observation, the problem of why an observation report was made, and what is our guarantee that it is true, belongs to science and not to philosophy.

Science deals with factual information, but not with all factual information and particularly not with information concerning miracles. Much has been written about the difference between scientifically acceptable and scientifically unacceptable information, and none of it seems to me satisfactory. Let me first state the difference and then discuss it. The bare facts of the matter seem to be these. In 1627 Galileo ridiculed (*The Assayer*) an argument from an unrepeatable ancient experiment. In 1661 Boyle published an essay 'On the unsuccessful Experiment' (in his *Certain Physiological Essays*) in which he ruled that science has nothing to do with unrepeatable experiments, that if we cannot repeat an experiment which someone claims to have performed we do not have to call him a liar or explain his claim in any other way – we can simply ignore it until it is reported to have been repeated by others. This proposal of Boyle has become a part of the scientific tradition. Although very few philosophers have discussed this situation, every physicist is well aware of it. Yet this situation should have been discussed more often, as it is problematic: the claim that any experiment is repeatable is a mere hypothesis. Boyle himself was extremely worried about this, because he thought that only factual information is certain to some degree ('morally certain'), and that factual information is therefore always to be preferred to a hypothesis with which it clashes. Yet as the rejection of a hypothesis is based on the acceptance of an observation report and the acceptance of the observation report is based on the hypothesis that it is repeatable, it follows that we reject a hypothesis not on the basis of solid facts but on the basis of another hypothesis. Is this not too arbitrary?

That the repeatability of an experiment is hypothetical can be shown by general considerations and by historical example. The general considerations are these: a description of an experiment is a description of the circumstances in which a certain event takes place, and a report of the experiment is the statement that at a certain time and place under the said circumstances the said event was indeed observed. Now many other circumstances were observed at the same time and place, of which there is no record and yet which may be, and sometimes are, essential to the success of the experiment. Thus, the success of the nineteenth-century experiments which made chemists think the atomic weight of chlorine to be 35.5 depended on circumstances which they did not notice but which we can vary today and thus approximate any result between 35 and 37 as we wish.

Boyle was aware of this difficulty. He demanded that we should report as many of the circumstances under which the experiment has been conducted as we can, and that we should vary the circumstances as much as possible. But the more circumstantial the description, the less repeatable is the described experiment; we do not know all the circumstances; we cannot vary all of them; and we cannot even report all of those which we notice. Boyle's last and posthumous publication, *Experimenta et Observationes Physicae,* is burdened with superfluous descriptions of irrelevant circumstances. Yet in his preface to it, which he probably wrote on his death-bed, he expressed the fear that negligently he had omitted some relevant circumstances, thus rendering his own experiments unrepeatable.

The cause of this insoluble problem of Boyle is, I suggest, his rule, according to which whenever a hypothesis and a report of a repeatable experiment contradict each other it is the hypothesis which has to be thrown overboard. To my knowledge nobody has contested this rule. Even Popper, the first philosopher who has stressed the utter and inescapable tentativity of all observation reports, has accepted Boyle's rule. Yet the rule has to be rejected. Here is a historical example of a case in which the rule was at first correctly broken and then mistakenly adhered to.

In 1815–16 Prout published his celebrated hypothesis, according to which the ancient philosophers' primordial matter is identical with hydrogen. According to this hypothesis the chemical atoms are not quite atoms, or indivisible, and their atomic weights must be multiples of the atomic weights of hydrogen atoms, namely whole numbers. Prout's essay is full of experimental evidence, mostly not his own but compiled from the most up-to-date works of the leading chemists of his age. None of these results agreed with Prout's hypothesis very well, and some of them did not agree with it at all. Yet he evidently considered these results as quite encouraging.

A short time later a youngster, Jean-Servais Stas, heard about this hypothesis and, to use his own words, fell in love with it. Like Prout he hoped that with the improvement of the available experimental techniques the results of the measurements of atomic weights would converge towards the results predicted by Prout. Stas soon became the greatest expert in the field. His results did not agree with his expectation, and they broke his heart: he declared that his loyalty to science stood above his loves; consequently he gave up Prout's hypothesis. One may remember that some of the techniques by which isotopes are isolable were available to Stas. Had he insisted that the atomic weight of chlorine cannot be 35.5 he might have suggested that chlorine is a mixture of two physically different though chemically identical substances. But unlike Prout, Stas refused to stick to the hypothesis in the face of known facts in the hope that the facts will adjust themselves to theory rather than the other way round. To be more accurate, he was willing to modify his theory in one way, but one way only. Had atomic weights converged well to even whole-numbers or half-numbers, he would double the conjectured number of hydrogen atoms in any given compound; but chlorine refused to converge even to 35.5.

This example shows that we have to improve upon Boyle's rule. I suggest that Popper's theory allows for a new rule. According to Popper's view scientific theories are explanatory and testable, and the more highly explanatory and testable they are the better. This view seems to me to have gained a sufficiently wide recognition to enable me to use it without any preliminaries. My present discussion, if correct, renders Popper's theory of the empirical basis of science superfluous; this theory (Section 29 of his *Logic of Scientific Discovery*) is perhaps the subtlest and most intriguing part of his study, but it is also unsatisfactory in its very subtlety, and the cause of most of the criticisms and the misunderstandings of his views. I am glad it can be dismissed without any loss.

As Popper has argued, the demand for high testability leads to the demand to exclude the explanation of a series of successful repetitions of an experiment as due to chance. As he has also noticed, the demand for testability justifies the rule according to which unrepeatable experiments should be ignored, since repetitions are a kind of test. This led him to the tacit assumption, which I propose to reject, that results of repeatable experiments must be (tentatively) accepted as true. That he does make this assumption, though tacitly, can be seen in his acceptance of Fries' claim that as the acceptance of observation reports should not be dogmatic it must be justified. This is a sensationalist relic in his theory. It led him to agree with Fries that the attempt at a justification leads either to an infinite regress or to a sensationalism. His own solution to the problem is that although we do not go on for ever testing observation reports by repeating the observation, we can do so when and if challenged. Hence, says Popper, there is an element of dogmatism or conventionalism in the acceptance of the report, since it may be false, as well as a sensational element, as it is causally related to sensations, as well as an element of (potential) infinite regress, since the possibilities of testing it are inexhaustible.

All this can be ignored, I propose. We need speak neither of acceptance, nor of justification of acceptance, of any observation report. We merely have to demand that account be taken of the fact that some observation reports were made repeatedly, and that this fact be explained by some testable hypotheses. The demand to explain given observation reports by highly testable hypotheses entirely suffices. If the most testable hypothesis explains given observation reports while assuming them to be true, which is sometimes the case, we choose that hypothesis. Yet the most testable hypothesis may explain the observation reports as being the results of crude measurements, as Prout's hypothesis did; or as results of sense illusions, as many psychological hypotheses do; or as results of specific initial conditions or specific circumstances, as Einstein's relativity did; or as lies and propaganda – remember the totalitarian scientists! There is no empirical reason to reject such hypotheses on the basis of past experience; rather we go and test them by having recourse to new experiments.

To put it differently, it is not for the general theory of scientific experiment to explain why an experimental report was made, since the possible and even the actual explanations are varied. It is the task of a scientific hypothesis to do this. In particular, we must consider as false Boyle's, Fries', and Popper's view, according to which all (repeatable) scientific observation reports are explained (tentatively or not) as true, and are therefore preferred to hypotheses which conflict with them. Whenever a report is made repeatedly, a scientific hypothesis which explains why it was made is sought for. And of all those specific hypotheses which are found, that one is preferred which is more testable than the others. Thus, when Mercury was reported to deviate from its Newtonian path, a few explanations of it were offered. One explanation of the report was based on the assumption that the observation was inaccurate, *i.e.* that the report was false. Another on the assumption that the initial conditions in the vicinity of the sun are more complicated than previous observers had assumed, *i.e.* that the report was true, but that other reports were false. Both these explanations incorporated Newton's theory of gravity. Yet another explanation of the situation was Einstein's theory of gravity. And the latter was preferred and tested, as it was the most easily testable. The preference for Einstein's theory over the other two alternatives was definitely not based on the fact that the other alternatives incorporated the assumptions that some previous observation reports had been inaccurate: Einstein's theory incorporated the assumptions that practically all observation reports had been inaccurate, of course. If this were not so, practically every observation report would refute the theory which the observation came to test: harly any observation ever fully agrees with the prediction which it comes to test.

The philosophical problem of the acceptability or otherwise of observation reports can thus be entirely ignored by non-sensationalists; no philosophical problem even corresponds to it outside sensationalism. Instead, many scientific problems correspond to it: in each field of enquiry investigators have to explain all repeated observation reports, and they may explain them as true, as approximations, or as sheer fancy. These explanations are not justifications and therefore should be suspected, and therefore should be tested. Indeed, the assumption that a new theory contradicts an older observation report is itself a suggestion of how that theory may be tested, namely by repeating the older observation with a higher degree of accuracy. This has been recently noticed by Popper (in his 'The Aims of Science'). But he did not notice, I think, that this amounts to the admission that observation reports may be accepted as false and that hence the problem of the empirical basis is thereby disposed of, which is my proposed view.

This proposal of mine severs the last connection between the philosophy of experience and sensationalism, by suggesting that philosophy should not include the attempt to discuss the causes of observation reports. Consequently my proposal sounds dangerously idealistic. I wish to argue that, on the contrary, it is the most realistic approach to experiment that has ever been proposed.

6. *The Roots of Scientific Realism*

The chief objection to my view would be that it is idealistic. But it is not idealistic; it leaves it to scientific hypotheses to say whether an observation report is true, near to the truth, or utterly false; it leaves it to scientific hypotheses to say whether a specific observation report was stimulated by sensations emanating from things, by hallucinations, by dreams, or by the desire to achieve fame. My view may sound idealistic, but only because it trusts science to take care of realism; which science is doing very well.

But why is science realistic? The generally accepted answer is that scientists have a metaphysical faith in the existence of things physical. Following Popper I consider this answer as true but unsatisfactory: beliefs may dictate our acceptance of the scientific discipline, but the question is whether this discipline leads to realism, or whether we must add to this discipline a disposition towards realism in order to obtain science as we know it. Clearly this disposition is inessential. It is a simple fact that whether you are a realist or not, you must admit that the method of science alone already pushes you towards handling realistic hypotheses, whether you like them or not, whether you accept them or not.

The reason for this fact is very simple. Idealism is just one way of looking at our experiences, and a way whose importance was immensely exaggerated by contrasting it with all the infinitely many alternatives to it as if it were a contrast between merely two views, idealism and realism. The reason for this exaggeration is, of course, the claim, which I endorse, that sensationalism leads one irresistibly to idealism. But once we ignore sensationalism, idealism becomes one of the very many uninteresting ways in which we may try to account for our experiences. As Lewis Carroll knew, we can say not only that the world is my dream, but also that the world is his dream.

The scientific accounts of experience, then, are realistic plainly because they all differ from one historically famous though unscientific account of our experiences – idealism – an account which leads us nowhere, and which was considered significant because of its close relation to sensationalism. As all versions of idealism are untestable, and scientific theories are highly testable, scientific theories are not idealistic, *i.e.* they are realistic. But is not our predilection for highly testable theories rooted in realism? The answer to the question as I have put it is, No. Science is realistic; more precisely, some versions of realism are scientific; but not all versions of realism are scientific. Realism alone is thus merely the rejection of idealism; it leads no more to science than to animism in any of its most primitive versions. We can be realists without wishing to explain or to test our explanations, but not *vice versa*. Let me show this by the following argument.

One may still feel that my attempt to ignore the general question of why observation reports are made is unrealistic; it may be unrealistic in a somewhat narrower and more naive sense than in the sense of being philosophically idealistic. One may suggest not that the whole world is my dream but merely that the scientific world is a dream. Do I allow for the possibility of a mock-science, of a situation in which some people build laboratories and state observation reports and some people try to explain them, but no one ever bothers to observe?

Let us take this possibility seriously for a moment, although it is puerile. I fear that it may have played an important role in the history of the philosophy of science even though it was never explicitly and carefully discussed (except, perhaps, by Bacon; he warned people against making reports without observing first; which, incidentally, is precisely what he himself did). Let us consider the hypothesis – call it hypothesis B – that there are very few observations and experiments going on anywhere on earth. I contend that at present almost nobody can check more than a negligible fraction of the observation reports which fill the current scientific literature, and that even in one's own field of research one must accept many reports without checking them. Thus, no one can deny hypothesis B on the basis of first-hand knowledge. But anyone can pose many awkward qeustions to those who accept hypothesis B; evading them will render hypothesis B unexplanatory, and attempts to answer them will render it more and more *ad hoc*, *i.e.* less and less testable. Hence, one who accepts Popper's demand for explanation and high testability, will reject hypothesis B. All the other existing approaches will make one feel very disturbed by hypothesis B once one has taken it seriously. Sensationalism forces one to take it seriously.

The whole point of the present discussion can be summed up by stressing the unreasonableness of taking hypothesis B seriulsy on philosophical grounds together with the reasonableness of taking it seriously as a testable explanation of a picture of the situation which we may have, say as a result of a hypothetical victory of Nazism. This is nothing but Popper's revolutionary thesis that the basis of science is social and not psychological.

But why do people observe? Why do they not simply imagine facts? My first answer is that they do, in all earnestness, try to imagine facts, but that their imagination is ludicrously less informative than the imagination of experimental investigators. Not only is the imagination of the author of *Arabian Nights* infinitely inferior to that of Jules Verne; when a cinematic version of a science-fiction novel of Verne is done nowadays, its script-writers have to improve upon his imagination – by using what men of science present, rightly or wrongly, as observed facts! Facts are stranger than fiction. Fiction is a very poor substitute for observation!

But this is only my preliminary answer. I do not wish to imply for one minute that we prefer observation to fiction because it is a better fiction than fiction; nor do I wish to belittle the significance of fiction (including that of Jules Verne) as a stimulus for observation; I only wish to argue that the fear of illusion which has ridden philosophers is rooted in an incredible overestimate of the power of our imagination. The attempt to explain an observation report, whether as a result of observation or as a result of hallucination, shows that we do not think we are so good at self-illusion: otherwise we could explain all observation reports, past and future, as a result of illusions, which is a version of idealism. It is because we do realize the limitation of our imagination that we have to ignore hypothesis *B*. The attempt to explain already implies that we think that we live in a world populated with humans who observe, think, and make statements, often because they think, rightly or wrongly, that they are true. In brief, we observe in order to test, though we do not always succeed. This is why I think that the problem of observation has been overrated: it has been overrated because the significance of the desire to explain or to comprehend has been underrated. The desire to explain, in its turn, has been underrated because the desire for certitude was great, and imaginative explanation is quite a different kettle of fish from certainty of any kind. As the quest for certitude or near certitude has to be abandoned anyhow, and as the demand to present highly explanatory and highly testable theories is realistic enough, we may leave it to science to explain each observation report in the most suitable way without trying to explain, in addition, observation reports as such.

7. Conclusion

We all start from "Naive realism", *i.e.* the doctrine that things are what they seem. We think that grass is green, that stones are hard, and that snow is cold. But physics assures us that the greenness of grass, the hardness of stones, and the coldness of snow, are not the greenness, hardness, and coldness that we know in our own experience, but something very different. The observer, when he seems to himself to be observing a stone, is really, if physics is to be believed, observing the effects of the stone upon himself. Thus science seems to be at war with itself; when it most means to be objective, it finds itself plunged into subjectivity against its will. Naive realism leads to physics, and physics, if true, shows that native realism is false. Therefore naive realism, if true, is false; therefore it is false.

This passage from the beginning of Russell's *An Inquiry Into Meaning and Truth* (pp. 14–15), which has aroused the admiration of Einstein, is the core of Einstein's comments on that book (in *The Philosophy of Bertrand Russell*). Einstein explains there how the desertion of naive realism led to sophisticated sensationalism and thus to idealism as the only alternative to apriorism. 'I am particularly pleased to note', says Einstein in the conclusion of his comments, 'that, in the last chapter of the book, it finally crops out that one can, after all, not get along without "metaphysics" [*i.e.* without unwarranted realism]. The only thing to which I take exception there is the bad intellectual conscience which shines through between the lines.' This 'bad intellectual conscience', to sum-up, is rooted in the following implicit assumptions. First, that there exists only one picture of the world which may be properly viewed as naive realism. Second, that if science explains this naive picture of the world, it ought to accept it as true – which it does not. Third, that science explains not our naive picture, which is false, but reports about our sensations, which are true. In contrast to these tacit assumptions of Russell, and in accord with what I take Einstein's tacit proposed to be, I propose the following view. (1) All pictures of the world which science explains are realistic. (2) All of them are naive to this or that degree. (3) Yesterday's frontier of science is today's rather naive realism. (4) Science is the attempt to explain the existing picture of the world, but this attempt is not based on the adoption of this picture; rather it leads to changes of the picture. (5) As Popper has suggested, science must remain at war with itself if it is to progress.

APPENDIX: ON PRIVILEGED ACCESS

That everyone has some privileged access to some information is trivially true. The doctrine of privileged access is that I am the authority on all of my own experiences. Possibly this thesis was attacked by Wittgenstein (the thesis on the non-existence of private languages). The thesis was refuted by Freud (I know your dreams better than you), Duhem (I know your methods of scientific discovery better than you), Malinowski (I know your customs and habits better than you), and perception theorists (I can make you see things which are not there and describe your perceptions better than you can). The significance of this rejected thesis is that it is the basis of sensationalism and thus of all inductivist and some conventionalist philosophy.

The doctrine of privileged access has two versions; one is commonsense and trivially true; the other is somewhat more philosophical yet still fairly commonsense – and false. The true version is this. Every person has access to some information available to that person alone, and it involves one's self, at least as an eye-witness. There is no doubt that this doctrine is true – at least in the sense that not a single author has ever put forth so much as one paragraph in an attempt to question it. The false version is that every person knows his own self best.

It is hard to say what Ludwig Wittgenstein has meant to say in his denial of the existence of a private language – or even in his very term 'private language'. As is well-known, what he wrote on this is brief and cryptic – and subject to a whole exegetic literature. That literature seems to agree on one and only one point, and it is that Wittgenstein did try to assert the truth of the true version of the privileged access theory and deny the false version of it, and to link both with his theory of language as a social entity, as a way of life. However, in the wake of Judith Jarvis Thompson's critique of some exegetes (*American Philosophical Quarterly*, 1964), I shall leave this point as utterly open as I can.

The false version of the doctrine is the one which has an interesting history which I shall now sketch – and so from now on I shall entirely ignore the true and commonsense version, and identify the doctrine with its false version without fear of confusion.

The doctrine of privileged access is a consequence of a certain theory of knowledge, namely sensationalism or the doctrine that all knowledge – of the world and of one's own self – derives from sensation. The clearest formulation of this doctrine was presented by Rudolf Carnap under the title 'Methodological Solipsism'. This title, unfortunately, has caused confusion, as Carnap narrates in his *Testability and Meaning* (1936); people took him to be advocating solipsim proper. Yet, he insists, all he means is that everything I know about the world – at least about the world of facts – is either what I witnessed directly or what someone else has witnessed and then conveyed to my senses as sound or as the printed page, etc. From this, of course, the doctrine of privilleged access follows at once with the aid of the following premiss. I observe my own inner states directly and at any time I am conscious; I observe my outer states at any time I am conscious; contrastingly, others observe my outer states only when I am within their field of vision, and my inner states even more seldom.

So matters have stood within philosophy for most of the time, especially witin the empiricist school. But things have changed dramatically with the almost simultaneous discovery to the contrary in various fields. The discovery is due to Freud, Duhem, and Malinowski.

Freud was a sensationalist, likewise Duhem; most likely so also was Malinowski. Yet, they all showed that the auxiliary hypotheses necessary to derive the doctrine of privileged access from sensationalism are probably false. This is how the discovery failed to shake sensationalism within philosophy. This is regrettable, I suppose, since sensationalism seems to be false. I must admit that the survival of sensationalism is legitimate from the logical point of view, even though somewhat dogmatic from the methodological one. The original doctrine of sensationalism was important as it gave rise to views about the development of science, but these views included the theory of privileged access. In its modern purified version, sensationalism does not contain the old errors but, likewise, it has no relevance to theories of learning which avoid the old errors. Sensationalism, thus, survives merely as a relic of a glorious past.

Freud's discovery within psychology is very simple. Possibly, your information was valid, you have suppressed it, and I have discovered it with the aid of (psychoanalytic) theory. Hence I know you better than you do. Of course, this assumes the theory to be true. Indeed, but this is not question-begging, because the theory has been empirically verified (claims Freud) by ample previous data. Moreover, I can sometimes refresh your memory or make you release the suppressed memories you have, and thus have my information about you confirmed later by you.

A striking case is that of an analyst who hypnotizes his patient in order to reveal certain repressed data, and who instructs the patient to forget the hypnotic session in order to allow the release of repression to take place in regular analytic sessions (so as to prevent the dangerous side-effects of a sudden self-revelation). Few psychoanalysts take recourse to such a technique and I do not know with what degree of success; but success has been recorded.

More striking, I think, is the case of a mildly hysterical patient whom an analyst can often read like an open book. Why do you resent X? asks the analyst early in the encounter. Why, me? Retorts the hysterical patient with some indignation, claiming privileged access and observing that he has not reported any resentment, least of all toward X.

Most remarkable, be it true or false, is Freud's identification of at least one such unknown object of strong resentment in most male members of the population, namely their fathers. This is part and parcel of Freud's theory of the Oedipus complex.

Since then much more has been recorded, and, in the field of perception-theory proper, refuting the privileged-access theory. Any dentist remembers patient after patient feeling a toothache in the wrong place; children may feel a toothache in the stomach. Sometimes such patients, fervently believing in the theory of privileged access, make such nuisances of themselves that dentists prefer to give them total anaesthesia first so as to prevent argumentation. The dentist may know in advance that the patient has such poor knowledge of himself as to insist on the truth of the theory of privileged access.

In the field of science the theory of privileged access has made people consult scientists about their activities, especially the more successful ones; the assumption is, they could easily answer if they only would. Joseph Priestley reported his discovery of oxygen three times, finding it very instructive for future scientists, especially in view of the fact that some scientists report their discoveries incorrectly. Not that they do not know the truth – Priestley, an eighteenth-century sensationalist, fervently believed in the theory of privileged access. Had scientists reported their discoveries honestly – 'ingenuously' is the word he uses – they would all report the truth, namely that their discoveries had been accidental and that even they themselves were surprised at the time. All too often, however, Priestley charges, scientists invent hypotheses to explain their discoveries and then claim that the discoveries were neither accidents nor surprises, but corollaries to the hypotheses, facts sought for after the hypotheses had been invented. This, says Priestley, they do in order to lend authority to their own hypotheses.

The story of Priestley is but one instance of the application of the theory of privileged access to the history of scientific discovery. Of course, we should not object to scientists narrating their own stories, whether at their own volition, at others' prompting, or as the ritual of the Nobel Prize Acceptance Speech requires.

The question is, How reliable and how generalizable are such reports? Pierre Duhem has argued that for logical reasons the reliance on scientists should not go as far as it all too often does. When narrating a discovery, all too often a scientist applies a theory of scientific discovery to his own case. (For example, Priestley was applying Bacon's theory of the surprise and accident of discovery; cf. *Novum Organum*, I, Aph. 109.) And different scientists hold different theories of science which are not compatible with each other.

Most scientists have believed that theory is based by induction on observation and report their activities as rooted, chronologically, in observation (in the way Priestley did and cajoled others to do). It was the fact that many scientists were inductivists and Duhem's own opposition to inductivism which forced him to deny the privileged access of the scientist even to the scientist's own contribution to science. All this has been put ironically by Einstein in his Herbert Spencer lecture of 1933, using the theory of induction against itself. Do not listen to what scientists say they do, he said, but look at what they do. He adds that having said this he should cancel his own lecture, but that he refuses to do so. All he meant is to deny privileged access, and so his own lecture, too, is merely deprived of that status. Hence, it is not that we should not listen to scientists, but that we need not accept their testimony as the Gospel truth; they have no privileged access.

To return to Priestley for a moment. He was in error in reporting that he discovered oxygen by mere accident. His own narrative enables us to show with no great difficulty (as I have done in my *Towards an Historiography of Science*) that he was refuting Bayen's hypothesis of metallic combustion, or his own improvement on it; and Bayen's hypothesis was an extension of Joseph Black's theory of fixed air and very much debated at the time as described in beautiful and vivid detail by A. N. Meldrum (*The Eighteenth Century Revolution in Science*).

Finally, to Bronislaw Malinowski. All anthropological or ethnological information about China accrued prior to the advent of Malinowski is now considered as a reflection not of the situation described, but of the *naïveté* of the ethnologist who believed the informant, and of the informant's obligation to present facts as in accord with Confucian doctrine and good sense and the like. The idea that China is classless, and that the Chinese examination system for public service is an instrument of equality, which had so much beneficial effect on the West, is one such extreme instance.

Perhaps the discovery of the non-existence of any privileged access to one's own social and cultural heritage should go not to Malinowski alone, but to Durkheim and Malinowski together. Be it as it may, for a time it was the fashion in British social anthropology to discard the declared reason for a custom, or an institution, as a mere veil concealing from its own practitioners its true meaning. Thus E. E. Evans-Pritchard (*The Nuer*) found in feuds between diverse tribes a means of keeping them in contact, quite counter-intuitively. Others found exogamy and incest taboos to be such instruments of contact with others. In his *Lectures on Social Anthropology* (1952), Evans-Pritchard offers a number of intriguing instances, not all from primitive society. Indeed, Durkheim already viewed religion, crime, and even suicide, as society's means of self-preservation regardless of the individuals who happen to be the instruments of society and of their ignorance of the true meaning of their actions as revealed by Durkheim.

One curious application of all this to the history of science is the fairly recent – correct or otherwise – study of the allegedly enormous influence of *Naturphilosophie* on science proper. Since it was regarded as dangerous and detestable to take *Naturphilosophie* seriously, clearly scientists could not easily report having done so – or even know about it. Even when a scientist does, on occasion, make such a confession as Oersted's acknowledgement in his doctoral dissertation of his indebtedness to Schelling, this may be viewed as an exaggeration, or aberration, and thus slowly but nicely sink into oblivion.

So much for the doctrine of privileged access and its overthrow. As I have said, the overthrow does not refute sensationalism. We may maintain that only a trained observer is qualified to observe, and deny in all cases we have thus far discussed that the persons in question were trained observers. On the contrary, the fact that the alleged privileged access has led to false observations can be revealed not by the mistaken observer but by a trained observer – be he a Freud or a Malinowski.

So be it. Query: Is Freud a trained observer of Freud? Malinowski of his own custom? If so, then Freud does have a privileged access to his own mind. Indeed, it is an important point in the study of the status of psychoanalysis even at present; generations after psychoanalysis was developed, criticized, and reformed, some opponents still find this objection to it very forceful and unanswerable. A psychoanalyst can be qualified only after he is successfully analyzed and thus have a privileged access to his own self. He may need another's help in overcoming obstacles, but he has to overcome the obstacles – to acquire full self-knowledge, to possess privileged access to his own self. There is the question as to how Freud did it single-handed, but I shall pass over this question here.

The most important question is not even really whether any psychoanalyst has a privileged access to himself (though, on occasion, it may be important to a patient). The question is broader: Is there a trained observer? Are trained observers really infallible? Are there types of observation which are truly incorrigible and infallible and final? If so, then we may perhaps find out irrefragable facts about minds, cultures, and scientific researches, whether ours or others. If not, then all I say about myself, or about you or him, is to some measure doubtful and hence hypothetical. If so, I cannot claim that I know fully even facts about myself; hence I cannot claim I know myself best; I cannot say that I know best just what I have observed (an expert can, and sometimes does, correct others' testimonies). But then, what is the difference between pure conjecture and a (conjectural) testimony of an eye-witness? Why eye-witnesses at all?

This terrible question pushes us to sensationalism, to the view that there does exist a definable and identifiable class of observation reports which are not in the least hypothetical. We do not know how to search for these, however, except through the avenue of privileged access – of methodological solipsisms – now closed. So we may either search for another avenue in order to give sensationalism substance, or try to offer a new theory of eye-witness testimony to replace sensationalism. This latter avenue has been explored by Popper and by myself – but only very rudimentarily. Many philosophers are still sensationalists and, having no other avenue for locating the class of incorrigible observations, they all too often fall back on the theory of privileged access – hoping to give a new and modified version of it, impregnable to the criticsim outlined thus far and defining, *inter alia*, the concept of trained observation. Thus far there is little or no progress to report. But as long as so many hope for success and work for it, the outcome is not very easily predictable. Some may find one avenue fruitful, others another. Only time and rational argument can tell.

CHAPTER 6

WHEN SHOULD WE IGNORE EVIDENCE
IN FAVOUR OF A HYPOTHESIS?

I. CAN OBSERVATION REPORTS BE REVOKED?

There is a supposition behind the title question: when should we ignore evidence in favour of a hypothesis? The supposition is, sometimes we may do so, sometimes we may not. Some philosophers have explicitly rejected this supposition, others have implicitly rejected it. Let me outline the brief history of this point.

Francis Bacon said the hallmarks of a dogmatist were first, that he is unable to see facts that conflict with his theory, and second, that when others force him to notice them either he denies them or belittles them with the aid of a 'frivolous distinction'.

At roughly the same time Galileo expressed admiration of Copernicus for sticking to his hypothesis in the face of observed facts concerning the brightness of Mars. It is difficult to make much of this fact because, unlike Bacon, Galileo did not explicitly offer a clear-cut methodology within which to fit observation. Whereas Bacon argued that it is better to avoid the formation of hypotheses, since they act as blinkers, Galileo never argued that it is better to avoid observations. Galileo sincerely hoped that the facts would fit his theory perfectly but he felt that a temporary discrepancy did not matter much.

Bacon's point seems common-sense and normal, because in theory tradition goes along with him. But in practice tradition went along with Galileo, not with Bacon. Failure to see this may easily make us misjudge the history of the Copernican movement. Bacon charged Copernicus with trying to replace one dogma for another. Dogmas, we remember, were for Bacon marked by 'frivolous distinction' – the excuses dogmatists make for awkward facts. In the case of astronomy the excuses were epicycles and eccentrics. Bacon was willing to go further. He did not know, he said, whether Copernicus's admission that one planet goes round the earth rather than the sun, was not a case of a dogmatist's inability to accept facts which refute his theory. ('The moon is a planet, the moon goes round the earth, and all planets go around the sun' is clearly a contradiction; making special allowance for the moon may be a frivolous distinction intended to save the theory; Bacon was not sure.) But Bacon's disciples, especially in the Royal Society, were Copernican; they did not like to speak of the deviations of facts from Copernicus's theory – or from Kepler's for that matter. Recently, the tide of opinion has been anti-Copernican: following Derek Price and Arthur Koestler, many have accepted the charge that Copernicus had a measure and measure: he deplored and was scandalized by Ptolemy's use of epicycles yet used them himself; indeed, it is said, he had as many as had the Ptolemaians of his day.

Now Copernicus was definitely not scandalized by Ptolemy's epicycles: he did not oppose the tradition of making temporary allowances for deviation of facts from theory. He deplored the deviations being allowed to remain so long, with not enough effort being made to remove them or to try a new hypothesis. A new hypothesis, he seems to have suggested, could tolerate blemishes during its early days, during its trial period. Both Giordano Bruno and Galileo seem to have said the same, though they were not articulate enough on these matters for me to let things rest here without further exegeses.

The official victory of Baconianism in the scientific societies in the generation following Galileo is perhaps the chief reason for our difficulty in interpreting Galileo. But it would be oversimple to think that Bacon's view was endorsed without qualification.

Robert Boyle offered a crucial qualification. Whereas Bacon counselled avoiding the formation of hypotheses for as long as possible, Boyle was both a Copernican and a Cartesian. He explicitly allowed the formation of hypotheses – at least as candidates for tests and thus as stimuli to experiment and observation. His counsel was rather that wherever a hypothesis and a fact clash, the hypothesis must be given up. Facts and only facts are certain, and they must always be preferred to uncertain hypotheses.

The rule that reports of facts should always make us reject the hypotheses they conflict with, I shall call 'Boyle's rule'. We may use the label 'Locke's rule' to refer to the rule, 'accept as final only observation reports' – though in historical fact it was also put forth by Boyle as a justification of 'Boyle's rule'. And indeed Locke's rule does justify Boyle's rule. Locke's rule is itself justifiable by what we may call 'Bacon's rule' –

which tells us not to accept any statement that is not final, plus the rider that only observation reports and what is properly based on them can ever be final. (Descartes accepted Bacon's rule, but not the rider.)

Bacon's rule is one which many philosophers and many scientists endorse. It is still very popular. I myself often heard it in childhood, in adolescene, and in my university days – usually from science teachers. It is plainly nonsensical. Anyone who thinks he endorses it is forced seriously to question such allegations (which he has not checked with his own eyes) as that he was born to a human female. One does not believe this because one saw cases of human birth, or heard about human births from reliable witnesses. We simply accepted the facts of motherhood before we could understand them, and we never doubt them though we sooner or later learn to doubt other things we heard at that period, concerning Santa Claus and the like. If someone says, or even hints that the story of his birth is on a par with stories of storks bringing babies, he should be rushed to a psychiatrist.

Back to our title question. Unless we reject Bacon's rule, we do not have to take hypotheses seriously, particularly after they have been discovered to clash with observation reports. And unless we reject Locke's rule too, we still cannot take hypotheses seriously because Locke's rule entails Boyle's rule; and Boyle's rule answers our question, 'when should we reject observation reports in favour of a hypothesis?', with the straightforward reply: 'never!'

Yet Locke's rule was a fixture of the scientific tradition until fairly recently. Even Kant agreed with Locke here. Locke considered sense-data the most objective evidence; the basis of science. Kant, to the contrary, viewed them as utterly subjective and unscientific. Yet he fully agreed with Locke about scientific observation reports. These, being cast in a scientific language, are final and undisputable facts of experience.

The exact reversal of Kant's view appears at the turn of the century with Pierre Duhem, the first scientist openly to reject Locke's rule. He agreed with Locke that sense-data are final, but declared their scientific articulation, scientific observation reports, subject to modification together with modification of the very theoretical framework within which they are cast.

Our question, then, only begins to signify at the beginning of the present century. It was not felt strongly, however. First, little obligation was felt that each and every known fact accorded with accepted theories. Freud said it first: it was permissible to overlook some facts for a while. The physicist P. A. M. Dirac stressed that in his work he favoured aesthetic considerations over the close agreement between theory and facts. This, and similar declarations, were certainly not overlooked by philosophers. They lead immediately to the problems put in the title of the present essay. To date, however, the question has seldom been asked – and I know of no other attempt to put it on the agenda so as to raise the question of whether we should reinstate Boyle's rule or not. The issue may be of concern to scientists because Boyle's rule is often applied by editors when rejecting scientific papers proposing hypotheses which conflict with accepted scientific observation reports.

There is no doubt, however, that the question has been raised before. Amongst scientists, Dirac's move is often cited as a repudiation of Boyle's rule. Moreover, Michael Polanyi has repeatedly said this in lectures and publications. Polanyi is philosophically opposed to any rule; he thinks that scientific method is ineffable and its practice is a matter of the intuitions of leading scientists. He therefore avoids answering the question, 'when should we violate Boyle's rule?', by invoking another rule: 'Boyle's rule is and should be violated when leading men of science tell us that it must'.

Finding Polanyi's theory authoritarian, my preference would be to propose, however tentatively, rules to replace Boyle's rule, and then try to examine them. Thus far, there are three or four proposed rules: my own (1966), that of Lakatos (1970), and, by implication, that of Kuhn (1970 or perhaps even 1962). My own proposal, however, was within Popper's philosophy of science. I now prefer Bartley's view of the matter (1968) which puts a much better perspective on it, and makes it much less important than it seemed.

Before discussing the question, 'when should an observation report be rejected in favour of a theory?' we may tackle the more general question, 'when should an observation report be rejected?' Of course, any complete answer to the more general question contains an answer to the more specific question. Yet, in response to a general question one may, and often does, offer an incomplete list of cases and then one may start to discuss each of the specific items on the list in depth while totally ignoring the initial specific item or forgetting to put it on the list. And, when one declares the incomplete list to be complete – either absent-mindedly or for very different reasons – one thereby answers unwittingly the initial specific question.

Thus, the question, 'when do we reject an observation report?' may be answered by the statement, we reject it when we can replace it by a better and conflicting one. For example, we reject empirical assessments of atomic weights when better ones are available. Well and good. Does this mean that this is the *only* occasion on which we reject an observation report? It is hard to say, since the English language is flexible on this point and allows one to read the answer one way or another, depending on context. If strong context-dependence indicates that this is the only way, we can say that the author clearly implies that we may never reject an observation report in favour of a theory except when the theory is backed by a better observation report (and hence Dirac's conduct was reprehensible). Or, context-dependence may go the other way, leaving our initial question unanswered.

The point just made will turn out as crucial at the end of Section V below. Meanwhile we may notice an interesting historical example of context-dependence – the case of Duhem. Duhem clearly said that observation reports are not final and are often in need of replacement. (One cannot say more about this without going into subtle details of Duhem's doctrine. This I am reluctant to do, since the long and detailed study that the subtlety of Duhem requires is too cumbersome to publish.)

Duhem views the set of theoretical statements accepted by scientists at a given time as a system; the system is both theoretical and linguistic: it enables us to store factual information; it is a language in which factual information is stated in neat and compact ways. When too much factual information is stacked, the neatness of the system is overcharged, and the system becomes complex. A change of theory then occurs, which is also, of course, a change of the linguistic system (since the theoretical system is the same as the linguistic system). What Duhem claims is that the factual information is now translated from one language into the other: there is no loss of factual information. Only some old verbal formulation of certain information must indeed be rejected, as the result of the rejection of the theories whose terminology is reflected in them.

This is not to say that all the older formulations of observation reports must be rejected. Indeed, even observation reports which employ modified terms may be retained, even though now their meaning is somewhat altered. An old observation report may have to be rejected *a priori* (if it clashes with a new theoretical statement), or it may be rejected *a posteriori* (as the result of an observation report of a new experiment), or it may be accepted *a priori* (as a part of the new theory) or it may be accepted *a posteriori*, or it may be in doubt pending further experimentation. There are historical examples for each of these cases! One thing can be said generally: once a new theory is accepted, and as long as it is, it forces us to reject some reports when taken literally, yet we can rescue these reports by translation. More cannot be said.

To Duhem, the answer to our question, 'when should we ignore evidence in favour of a hypothesis?' is, in a sense (translation) never, in a sense (transplanting with no alteration of words) always! The chief trouble with this answer is that it is utterly useless, say, for Dirac: it does not tell him whether to be concerned about the discrepancy or not. It may help a historian to look back but not a scientists to look forward.

This concludes the historical survey and brings me to the 1935 views of Karl R. Popper.

II. CAN REFUTATION BE FINAL?

It is hard to present Popper's views, particularly on the topic at hand, without regard to current misunderstandings of them. Besides being a bore, misunderstandings are also fascinating as they are excellent diagnostic tools for elucidating both popular prejudices and the weaknesses of authors who are regularly misunderstood. While I hesitate to forward the hypothesis – I should say, pet prejudice – that it takes two to have a misunderstanding, I do recommend it as a diagnostic tool: persistent misunderstanding especially in the face of much effort by the misunderstood author to improve matters, may indicate that there is something to the claims of the one who misunderstands; and this deserves our attention. Some authors refuse to take a clear-cut stand and express matters simply; possibly because they are not sufficiently clear even in their own minds. In such cases, evidently, persistent misunderstanding may be an incentive to commentators to sharpen and clarify the doctrine on which they are commenting. No matter how vehemently an author denies a thesis imputed to him, then, we may ask why it goes on being imputed to him despite his denials, and what is more, if anything can be done to stop the false imputation. Perhaps nothing. I suggest that often the learned public heeds those who evidently misquote because they are disturbed by what seems to be an inconsistency in the work, especially if it attacks their sacred cows, and their excuse for sticking with their sacred cows is that they are waiting to see the alleged inconsistency cleared up. The author in question has after all taken away from them a useful crutch and not replaced it with another. Often this is the core of the misunderstanding: the author says he has offered an alternative and they do not see this.

The evidently false allegation concerning Popper's philosophy is very strange. It was first made by A. J. Ayer, and it has stuck in spite of many protests and lengthy explanations from Popper.

Briefly, Popper's slogan is, we cannot verify a theory, but we can refute it; whereas logic is useless for verification, it is sufficient for refutation; whereas logic never enables us to justify, it does enable us to overthrow. This is one of the few theses of Popper's philosophy which has been repeatedly quoted and misquoted. Even when repeated without being properly ascribed to Popper, it is easy to detect his hallmark because of a logical mistake he made in this context – one which he has corrected, but which has remained a misnomer for the thesis expounded at the beginning of this paragraph. The error remains widespread.

The doctrine, 'we can use logic to refute but not to verify', is known as the asymmetry thesis, or as the application of the *modus tollens*, or simply as the *modus tollens*. Popper states, in a lengthy footnote to the English version of his *magnum opus*, he was logically confused. Since almost all English-speaking readers are familiar only with the English version of his *Logic of Scientific Discovery* of 1959, and since almost all German readers of his *Logik der Forschung* of 1935 are familiar only with the second and later editions which incorporate translations of the English additions, it is surprising that the learned public sticks so stubbornly to a logical error presented to them along with its correction.

Popper employs deductive logic as a model of science. We begin, according to this model, with universal statements and initial conditions from which we deduce final conditions; the final conditions are either explicanda, old observation reports thus explained, or test statements, new observation statements to be checked and affirmed or their negations affirmed as the case may turn out. Now, let us present universal statements in the modern style, e.g. instead of 'all swans are white' let us say 'for everything, if it is a swan, then it is also white'. The conclusion 'Tom is white' follows from the universal statement just formulated plus the initial condition 'Tom is a swan' by *modus ponens*. Popper repeats in modern language a point clearly made by Bacon: *modus ponens* does not help verify; no matter how many tested swans happen to have come up white, it is still just logically possible to come upon a non-white swan in our next test. Popper – still repeating Bacon – shows how different it all is with the *modus tollens*. The form of the *modus tollens* here should be, 'for everything, if it a swan then it is white', and 'Tom is not white', therefore 'Tom is not a swan'. What Popper had in mind, however, is not at all the conclusion 'Tom is not a swan', but rather the metastatement, 'the observation report "Tom is a white swan", is false' or, more often, 'the hypothesis "all swans are white", is false'. This employs a metalogical rule: the rule of transmisssion of truth from the premises of a valid inference to its conclusion, or, and this is an equivalent rule, the rule of *retransmission of falsity* from the conclusion to the premises of a valid inference, to use Popper's term. The two equivalent metalogical rules, of transmission of truth and of retransmission of falsity, were in the air all throughout the history of logic – or even its prehistory, as they can be seen consciously applied in Plato's early dialogues. Yet as part of the explicit apparatus of logic they date only from Tarski's celebrated studies of the 1930s, and in particular his semantic theory of inference; more precisely, they are fully discussed in Popper's works on logic. Popper's work on scientific method just antedated the introduction of Tarski's work into the German- and English-speaking world, and Popper confused the single valid rule of inference *modus tollens* with the much more general criterion for validity of any rule of inference, namely transmission of truth or retransmission of falsity.

Yet scholars well versed in logic, quite averse to confusing the meta- and the object-language, follow the early Popper here, confuse inferences with conditionals, call the falsification rule by its misnomer *modus tollens*, and make a mess of things. An interesting and unusual symptom of something malignant enough not to be easily eradicated by simple surgery; at the very least the surgeon has to find the whole malignant part and see if he can take it all out before he can extend hope to the patient. I am no surgeon, but such pathology occasionally fascinates me.

Popper's asymmetry thesis (I should say, Bacon's), the thesis of the asymmetry between refutation and verification, is simply this: ordinary logic cannot help us verify a theory but it can help us refute it. The asymmetry is logically so trite, it takes but a glance to see it: from an observation report, which specifies the occurrence of some event in some space-time region (to use Popper's terminology and apparatus), we can deduce the existence of that event somewhere in the universe; we do so by simply replacing the reference to space-time co-ordinates with a more vague expression, such as 'somewhere' or 'there exists', and such. We can verify an existential statement in this manner, says Popper. But we cannot refute such an existential statement – a purely existential statement, as he calls it – as this requires scanning the whole space-time manifold, which may be infinite. We can verify some but not refute any purely existential statements. By a simple logical exercise we can show that this amounts to the following claim – we can refute some but not verify any universal statements. Or rather, instead of 'we' I should write 'observation reports'. Observation reports may refute but cannot verify universal statements; they can verify but not refute existential statements. End of Popper's or Bacon's doctrine of the asymmetry between refutation and verification.

This doctrine of asymmetry was taken to mean that (according to Popper), if we can refute a theory then we can do so once and for all. And, since we can verify its negation, it seems fairly reasonable to say that when we speak of its refutation we have in mind something as final as we mean by 'verification'. Before the reader who wishes to defend Popper says that Popper does not mean anything final by 'verification' either, I want to declare that this move is a sleight of hand which will nourish and sustain the logical confusion as long as Popper's views are remembered. Popper's use of the words 'verify', 'verification', and the like, in many places, quite clearly means a conclusive justification, as contrasted with his use of other, milder words, to denote some inconclusive justification, such as 'probability', 'confirmation' and perhaps even 'corroboration'. It seems clear from Popper's tenth and final chapter on corroboration or how a theory stands up to test, and perhaps also from his fifth chapter on the problem of the empirical basis, that Popper does accept *some* kind of weak support of theories by data, though, of course, he rejects both conclusive support, namely verification, and many inconclusive ones, such as probability and confirmation.

The critical reader may be disturbed by my discussion of conclusive evidence against some reading of a text by Popper, when the text in question is one concerning counter-evidence and the question is, 'is the reading of the text correct which reads it to say that counter-evidence can be conclusive?' I shall not dwell on this point here, though it certainly deserves examination.

The methodical reader may be disturbed by something else: thus far nothing has been said about Popper's view on empirical justification of universal hypotheses, and I have reported that he says that there is none. Yet above it has been affirmed, without so much as an explanation, that Popper *does* think some empirical justification of universal hypotheses is possible – namely corroboration or a hypothesis having withstood a test. Corroboration is, presumably, the empirical basis of a hypothesis or a part of that basis. The methodical reader is not alone here. Practically all commentators on Popper, casual or systematic, sympathetic or hostile, have been puzzled over Popper's position on this point – myself included. My 1959 paper on corroboration presented Popper's views as clear enough. In various papers, before and since, J. W. N. Watkins has also tried to clarify the issue. Watkins's reading of Popper's view of corroboration and my reading of 1959, are quite different, a fact I regret trying to gloss over at the time. This only added to the confusion, because I had inadvertently presented my own view, thinking it was Popper's. Although I checked it with Popper himself, I cannot claim that I had Popper's clear and free endorsement of my reading of his own views.

Personal evidence aside, the fact that many commentators puzzle over Popper's theory of support and the fact that some differ about what it says, the fact that even Popper himself joined the band of commentators, is sufficient evidence to allow us to conclude (here again we have evidence justifying a conclusion though not logically yielding it) that in 1959, in the extended version of his *opus magnum*, Popper's theory of corroboration was not sufficiently clear.

Much emotion and ink were spilled over this issue of asymmetry between verification and refutation. Popper's chief opponents, namely A. J. Ayer, Rudolf Carnap, C. G. Hempel, and Herbert Feigl, all declared that where Popper sees asymmetry, there is in fact full symmetry: there is, in perfect accord with his own doctrine, full symmetry between verification and confirmation on one hand, and falsifications and disconfirmation on the other: as long as an observation report may be withdrawn, as he says it may, we have only strong reasons but not conclusive reasons for rejecting a theory, just as for accepting a theory.

No doubt, this sounds blockheaded in view of the fact that logical verification in science is impossible but logical contradiction is possible. Yet Popper's opponents are right, for the following reasons.

When Popper says that an observation report verifies a purely existential statement he means that we know for sure that *if* the one is true *then* the other is also true, or that *as long as* we accept the one we *eo ipso* accept the other. This is not verification but entailment. Admittedly, it is quite traditional for followers of Bacon's or even of Locke's to say that a theory is verified emprically if an only if it follows from, or is entailed by, a set of observation reports. And Popper here unwisely follows accepted usage and declares all corollaries to observation reports verified. In this sense, the thesis of asymmetry between verification and refutation is trivial and when you tell Popper's detractors that this is what he has in mind – I am reporting now – they look incredulous and say, this is impossible!

After some labour, I discovered that what they mean to say is this. Disreputable philosophers make assertions which have double meanings: one trite and obviously true, one fantastic and (seemingly) obviously false. Under attack they affirm the trite sense but soon after they revert to the other. Surely, Popper's detractors say, I am not imputing this vile technique to him.

Here another difficulty arises: it is taken for granted that corollaries from trite theses are trite. Yet such corollaries may solve serious and troublesome problems! Earlier, in Chapter III above, I ascribed to Popper such significant conclusions from a series of fairly trite assertions. An example may be useful. Obviously, the mere acceptance of an observation report, however tentatively, has some theoretical consequences, such as the rejection, however tentatively, of a scientific theory, however well established. Add to this the rule that one should always accept observation reports until they are refuted – which is Popper's variant of Boyle's rule – and it follows that though observations cannot logically confirm or corroborate theories, they can lead to scientific revolutions; indeed by these rules they must.

I dissent from Popper's version of Boyle's rule. Other rules, it seems to me, including ones accepted by Popper, sometimes permit us to accept a theory – in different senses of accept – and hence to question the asymmetry. Yet there remains the logical fact that an observation report can *entail* a purely existential statement but not contradict one, just as it can contradict some universal statement but not entail one. This hard logical fact (Bacon's and Popper's) remains, and I feel that it is all too easy to ignore it as merely logical. For my part, I think it is both the crux of all critical philosophy, Popper's and my own included, and its rejection or obfuscation creates a barrier between the critical and the uncritical. Attempts to break through the barrier lead only to misunderstandings, as we shall see.

III. A SIMPLE ISSUE OBFUSCATED

The prerequisite of criticism is the ability to recognize plain incompatibility between one's view and a putative counter-example. The critically-minded must recognize a counter-example when he sees one, and on occasion accept one and then change his view.

The prerequisite is not in universal terms. All critically-minded people cannot be required to recognize every incompatibility thrown at them – the exercise may require more mathematics than is at hand, or more leisure than one may wish to consume. Nor can we require that every counter-example which *is* reasonable be *viewed* as reasonable. We do not know what is reasonable. We know that reasonable reports of the midnight-sun in antiquity and of elephants and castles in medieval Europe were quite reasonably dismissed as unreasonable. Similarly, as we know now, Wolfgang Pauli was in error in thinking it unreasonable to try to refute the law of conservation of parity, yet he was then not being unreasonable. Finally, if we insist that reasonable counter-examples be accepted forthwith, then we must declare Galileo, for example, unreasonable.

How hard it is to notice all that may be involved in critical debates is illustrated by Galileo's reversion to Aristotelian ideas after he had overthrown Aristotelianism to his own satisfaction; and Popper's requiring immediate acceptance of observation reports and rejection of the hypotheses they contradict, while approving of Galileo's admiration for Copernicus's refusal to reject Copernicanism in the face of refutation from the observed variability of the brightness of Mars.

So much for the fact that critical people are not always critical – perhaps no one is. More needs to be said about critical-mindedness, as almost anybody accepts criticism sometimes, yet few are critically-minded. I shall not now enter this point, though I consider it important for a philosophy which follows Popper's achievements yet rejects his tenets. He says, critical-mindedness or amenability to criticism is simply the acceptance of counter-examples. I think we need this acceptance only on occasion and there is more to critical-mindedness than that. (He demands too much and too little, as the expression goes.) I wish to discuss those who deny that rationality includes the acceptance of criticism, and those who obfuscate the issue.

Let me begin with Michael Polanyi. His views on the matter are very simple to state but hard to develop. Briefly, he says that some criticism is acceptable, some not, this including observation reports. Which criticism is acceptable is a matter of personal knowledge, of intuition and of immediate feel; it is inarticulate, and subject to no criteria. The leadership of science may reject an observation report off-handedly and stick to a conflicting theory – just like that. They usually know what they are doing, or else they pay the price and lose their position at the top.

I do not like this theory, though on one point it seems fairly near the truth, simply by virtue of the fact that so many professional scientists today are less critically-minded than sheep. By Popper's definition of science most scientists are not busy at scientific activity, since they offer neither (explanatory) hypotheses nor tests. Their minute contributions may be of some empirical and even intellectual value (measuring constants, calculating), but they are devoid of critical qualities.

Polanyi criticizes Popper by giving counter-examples to the equation of science with refutability. He does not offer examples from the sheepish behaviour of most scientists – since this would be question-begging. Rather, he gives cases of *bona fide* important scientific attainments which are either immune to empirical criticism or immunized for-the-time-being. Polanyi makes a good point, but errs in going on to claim that criticism, particularly empirical refutation, *never* takes place in science. Yet this wild-sounding idea is rather impressive: if criticism never has to be accepted as such, but only on certain conditions, then it is not criticism that causes changes of heart as much as the specific conditions which make for the acceptance of the criticism.

Impressive as it is, this idea is an obfuscation. It is true, yet it obscures the simple fact that when a situation is clear enough (there are conditions for clarity and these conditions may be involved) criticisms contradict given theories and force us to give them up. Further, such a development is a liberation. Popper's detractors obfuscate this great idea of his: the overthrow of a theory as a result of observation is liberation achieved by empirical means. Thus, a refuting piece of evidence may be of a great theoretical import.

Hence, Polanyi's criticism is answered: we can say, under certain conditions criticism is better accepted, yet stress the criticism, not the conditions. For the conditions are the clarity and the (hoped for) fruitfulness of the liberating effect of the criticism itself.

To see this, however, we have to see that criticism in general may be fruitful, including empirical refutation, of course. Yet one of the misunderstandings that plague Popper's philosophy is the following idea which is often falsely ascribed to Popper: he allegedly does not see merit in refutations as such but only in their being instruments necessary for the proper corroboration of hypotheses. Popper has repeatedly said both that refutations are desirable and that there can be no corroboration except failed refutations. Now, this easily lends itself to the wrong reading that corroboration is desirable in itself and that refutations, though they should not be resisted, are not desirable in themselves but rather should be admitted as necessary occupation hazard. And this reading makes corroboration desirable though not easy to attain. But this reading is silly: if we hold false theories we may reasonably desire to refute them, not corroborate them. However, Popper's lack of clarity on the role of corroboration makes it easy to read into his works the vulgar theory that corroboration is always good in itself, that every time a scientist corroborates a theory it calls for a celebration. In any case, my concern is less with Popper's writings than with corroboration being an impediment, distracting an investigator from a more fruitful path. There are ample historical illustrations. We owe this insight to Popper's idea that refuting a hitherto unrefuted – perhaps also corroborated – theory represents real progress, true learning from experience.

At one time Imre Lakatos forged a compromise between Popper and Polanyi. According to his collaborator Alan Musgrave, the major idea of Lakatos is that certain ideas comprise the hard core of our research activities, the guiding ideas or regulative ideas of our researches, and for a time at least we protect them against refutations. This is both historically questionable and obfuscating. It assumes that only hard cores are protected, it assumes that all researchers in a field share the same guiding ideas, it assumes that refuted ideas cannot be guiding. There is empirical evidence against each of these assumptions, but if they constitute the hard core of Lakatos's own researches, he may well protect them. This is, then, obfuscation.

The reluctance to admit refutations, which is now shown by Lakatos, a former disciple of Popper, is the first motive behind the earliest responses to Popper's views. Moritz Schlick, the father of the Vienna Circle, called it masochistic. This is not much of an argument, to be sure, as it does not indicate any over-eagerness to take the bitter medicine, let alone to show that it is bitter. The suggestion of overeagerness can be found in two versions, one which plainly violates the thesis of asymmetry and one which is more puzzling and may violate even the law of contradiction.

The first view is that of Hempel, of the symmetry between verification and refutation: both are impossible and must be replaced by the weaker confirmation and disconfirmation. (The word 'disconfirmation' is a neologism.) In order to render a refutation into a mere disconfirmation, however, we must raise doubts as to the truth of some observation reports, whereas in order to render a verification a confirmation we do anything else but raise such doubts. This, then, is a clear case of measure and measure.

The second view is that of A. J. Ayer. Ayer suggested, in a very famous passage of his, that maybe a hypothesis is refuted only once, like having the measles; if so, he added, then it would be a mistake to reject it on the basis of one measly refutation. Taken literally, this seems to be nothing short of quarrelling with logic – and logic seems to me to be as haughty and adamant as ever in declaring a universal statement contradicted by even one true existential statement to be false – irrevocably so. But what Ayer had in mind was perhaps a rule expounded by Newton at the end of his *Opticks*: when an exception to a generalization is discovered, the generalization should be modified to exclude the exception and the modified generalization retained. This, at least, is not against logic. Popper would object only that many qualifications may accrue in this manner and the generalization thus repeatedly qualified may lose its testability, content, and interest. Hence Ayer's objection, when put in a logical manner, is less of an objection than it seems: Popper does not oppose the tactic Ayer seems to suggest except when it is over-used: when the measles becomes more than a mere childhood disease.

But all this is needling. The chief issue at hand seems to amount to the question, do we want finality in science, if not finality in the acceptance of a thesis, at least in its rejection? If not, will it not be faintly ludicrous to reject a thesis and then reject the rejection and re-endorse the thesis? I think this is the root of the trouble. But since examples of resurrected hypotheses are easy to find, the trouble is one which scientists handle, and so philosophers should acknowledge its existence.

There is more to Ayer's objection to Popper than meets the eye, but unfortunately Ayer does not go into much detail. Suppose there is only one non-white swan in the whole universe. In a definite sense, then, the observation of this unique specimen is unrepeatable. The observation will soon enter catalogues of curious events and will be dismissed as one not susceptible to empirical examination. It will thus be ignored by empirical science. Hence, the insistence on the testability of observation reports narrows the class of acceptable reports to those which are repeatable. In such repeatable observation reports the spacio-temporal co-ordinates act as parameters of a sort: we say, we saw a non-white swan at such-and-such time and place, but the specified time and place are not unique, we hope. This, of course, makes all observation reports hypothetical. Even Boyle was aware of – and troubled by – this fact: if all observation reports are hypotheses, why should they always have the upper hand?

The argument I develop from Ayer *prima facie* looks reasonable enough to be problematic if not disturbing. So it may be advisable to go over the ground carefully, familiar or unfamiliar as it may be, and see how much is presupposed in even the simplest experimental reports. It is not necessary to decide whether a scientist's presuppositions are right, but merely what presuppositions are involved.

Let me close this discussion with an example in which we do not doubt that scientists were right yet which is a challenge to the philosopher concerning the presuppositions of empirical evidence. There is a story which may be false – for all I know it is true – yet which is very thought-provoking. In one laboratory, so the story goes, measurements of properties of metals, strength, elasticity, etc., seemed clearly to correlate to days of the week! The evidence was declared in some sense impossible, though it was admitted to be true. It was later found that the pieces of metal suffered from fatigue which disappeared on weekends, when they were allowed to rest. Metal fatigue is now considered an observed fact. So when should we ignore evidence in favour of a hypothesis?

IV. A CRITERION FOR REJECTION OF OBSERVATION REPORTS?

In *The Logic of Scientific Discovery*, Popper endorses the following view of Reininger and Neurath, while cautioning his reader that the error of omission committed by these writers may lead to very grave consequences. Observation reports, or protocol sentences, or basic statements, once affirmed or declared by the world of science to be true, need not always be so affirmed. Popper cautions his reader that unless criteria for the denial of a once affirmed basic statement are given, we are in danger of dogmatism, and in particular of the overthrow of empiricism, since without such criteria anyone is free to deny those basic statements which contradict his pet hypotheses.

If Popper had said no more on this point, one might have raised the following question: 'can we do without criteria for the denial of once-affirmed basic statements and not become dogmatic?' We may have criteria which would disallow the denial of basic statements for the purpose of dogmatizing; we may avoid dogmatizing even without such criteria – say by sheer accident or by grace. Obviously, we may identify the criteria with the demand not to dogmatize: we may permit the denial of basic statements for any purpose and for any reason except the desire to uphold a theory. Our last criteria may not be good; for instance, dogmatism may turn out to be the unintended and unnoticed consequence of some other policy (such as pleasing the grand-old-man). This, however, is the fate of a criterion; moreover, even if criteria help to avoid one kind of dogmatism, adherence to those criteria may amount to some other kind of dogmatism.

The operative word in the preceding paragraph is the word 'MAY': its purpose was to indicate a variety of relevant possibilities. All of these are implictly rejected – or so it seems – by Popper who says that 'Neurath fails to give any such rules and thus unwittingly throws empiricism overboard' (p. 97). Of course there is only a subtle difference between 'allows people to throw empiricism overboard' and 'throws empiricism overboard'. The fact remains, however, that I understand Popper to be advocating the view that we urgently need a rule or criterion for the denial of once-affirmed basic statements so as to avert dogmatism.

Two or three questions press themselves on us here. First, 'is the risk imminent?' Secondly, 'does Popper provide a satisfactory rule?' And thirdly, 'does this not lead to a higher level dogmatism?'

The risk of dogmatism may be understood in a variety of ways. It may mean that unless the legislature provides such a rule, the commonwealth will suffer. This is definitely a misinterpretation. Neurath and Popper are not arguing as legislators, and they know full well that the commonwealth of learning was faring fairly well without any rules such as the ones they were trying to formulate. On the contrary, in a sense Neurath and Popper, when discussing the issue at hand, may be viewed as social anthropologists who observe the commonwealth of learning like Malinowski was observing the Trobriand Islanders. It is a fact, one may hear Popper say, that the commonwealth of learning is non-dogmatic and empiricist, and I wish to find out or make a hypothesis about the rule they are following which keeps them from foundering in the wake of the dogmatist.

This interpretation, though much better than the first, is still not very satisfactory. I think the truth lies somewhere in between the two interpretations but I find it impossible satisfactorily to get matters clear. Therefore I do not know in what sense it is urgent – Popper says it is – to legislate a rule to tell us when we are permitted to deny the truth of a once affirmed basic statement, and why the absence of such a rule not merely permits dogmatism but even leads to it in some sense or another.

V. DOES POPPER OFFER A RULE OF REJECTION?

With this I leave the question, 'do we need a criterion of rejection?' Let us assume now, without debate, that the absence of a rule is deplorable, and ask, 'does Popper provide such a rule in a satisfactory manner?'

Popper's chapter on basic statements (Chapter 5) is quite difficult to follow on this point. The chapter contains admirable and understandable criticism of prevalent doctrines. But when I tried to expound and apply its doctrine of the empirical basis (in Chapter V above) I came in for a lot of criticism from Popper, some of his close associates and former students during the four years between acceptance and publication. Its argument was said by those familiar with Popper's views to be hard to follow. Popper himself found it reprehensible. W. W. Bartley III, one of Popper's most brilliant former students, convinced me that I may be able to clarify my argument against Popper – if I have one, which at the time he vehemently denied – by rereading Popper's text more carefully and systematically. The present chapter follows this suggestion, and also a few of his more detailed ones. My own early reading of Popper on observation reports or basic statements was as follows. There is a rule which tells us when we are permitted to reject basic statements; otherwise we are not permitted to reject any. The rule is this. First, we should state or report as basic statement when we (sufficiently carefully?) observe – or think we observe – what the basic statement affirms. Secondly, we should try to test the basic statement, at the very least by repeating the observation it reports. Thirdly, if the basic statement is refuted (see below) it ought henceforth to be denied and the one refuting it be reported and affirmed in its stead. Fourthly, if the basic statement is not refuted it ought to stay affirmed. Fifthly, anyone who wishes to deny any hitherto affirmed basic statement may not do so except if he has succeeded in his attempts to refute it. One is always at liberty – is even encouraged – to test any basic statement one wishes, and as severely as one knows how.

According to the rule just expounded, participants in the scientific enterprise are always bound to affirm all those basic statements which have stood up to test more recently. They are bound, to be sure, not by natural or psychological necessity, and not by the law of the land, but by rules freely adopted, we may remember, in order to escape dogmatism.

It is hard to say whether the previous two paragraphs paraphrase what is advocated in Chapter 5 of Popper's *Logic of Scientific Discovery*. It once looked to me obvious that in that chapter, in Section 29, this rule was advocated. But Bartley has convinced me, at the very least, that no such rule is explicitly advocated there.

Let us suppose, then, that there is no rejection rule in that chapter.

It is then unsatisfactory, in that it condemns Neurath for failing to provide such a rejection rule while likewise failing to provide one. To be sure, Popper does suggest that after refutation a once-affirmed basic statement should be denied. This, of course, is a very important and far from trivial proposal, since in such cases we have the evidence of the senses for two propositions contradicting each other. But the question remains, 'does Popper suggest that *only* when refuted ought a once-affirmed basic statement to be denied?' If he does say that, then a rule which bars dogmatism (or which is meant to bar dogmatism) has been presented and I shall soon attempt to criticize it by showing that it is a rather dogmatic rule. If he does not say that, then he has not provided a rule which bars dogmatism, after having criticized Neurath for not having provided such a rule. (Here is a case where the reading of a text is highly context-dependent. Yet the text is too brief for a clear-cut decision.)

VI. DO WE NEED A RULE OF ACCEPTANCE OF OBSERVATION REPORTS?

With this, I leave exegesis of Popper for the moment, and move to the question, 'do we have to affirm basic statements affirmed by others?' To be more precise, 'are we obliged, *qua* scientists, to affirm basic statement affirmed by other scientists?' To be still more precise, the obligation referred to in the previous question must read not in a legal or a moral sense but as to whether the rule is recommendable to and/or accepted in the commonwealth of learning.

The question, should we believe hear-say in science, has, to my knowledge, often been raised only to incant the injunction – originated by Bacon – trust only your own senses! As I said this rule is mad; anyone who doubts for a moment that he was born of a human female is mad, regardless of whether he has ever observed birth, human or animal, and regardless of from whom he had heard the facts of life and how reliable his informant happens to be.

Some writers, such as Sir John Herschel, in the opening chapter of his *Preliminary Discourse to the Study of Natural Philosophy* of 1831, stress the opposite, the fact that men of science are *bona fide* credible, that there are few hoaxes in science, i.e. that false observation reports are seldom made fraudulently. Some writers see this as the basis for scientific objectivity. They rightly conclude that good faith is what makes the chief difference between science and pseudo-science, including alchemy, astrology, mesmerism, phrenology, etc. Herschel seems to concur, though only in a passing footnote which is no sufficient indication of his view. Popper, however, certainly does not concur.

It is hard to say what Popper's view is on the place of good faith in the commonwealth of learning. He stresses that objectivity in science is rooted not in the good intentions of the scientist, even when they undoubtedly exist and are operative, but in the institution of intersubjective tests. Yet institutions involve mutual understanding and some measure of trust. It is not that tests are the opposite of trust, of course; indeed, the subtlety of the situation stems from the fact that when we test regardless of trust trust may flourish, whereas, when trust is declared to be exclusive of testing it immediately becomes strained. One may note that this is so regardless of whether the trust or test are of scientific or political information, or personal matters of any sort. Tests are generally not the opposite of trust, and so the question can arise, when we are told of a test, and when we are told that its result supports the theory it came to test or otherwise, do we have to believe the report?

Certainly we can doubt the report in the sense of wishing to repeat the test ourselves; we have seen that. Certainly we may offer internal criticism of the test and say its preparation was not precise enough to procure results that may conflict with the theory in question and so it was not a genuine test. (This point is one of Popper's great contributions to methodology even though he merely sharpened a thesis already offered by William Whewell over a century ago.) But supposing I do not intend to repeat the test and see nothing methodologically wrong with it. Do I have to declare its results true? If so, why?

Suppose I am not allowed to declare the result false and stick to my own hypothesis which contradicts it. Do I have to declare the result true if it corroborates my hypothesis? Do I have to declare it true if it relates to a hypothesis I have no stake in? What is the import of endorsement of an experimental report anyway?

A tacit reply can be found elsewhere in Popper's work: we wish to explain facts, and as we endorse people's reports of facts our stock of reported facts increases and the task we have undertaken – to explain facts – becomes more challenging and interesting.

This reply takes us away from our initial problem. Our initial problem was, when should we prefer a theory to a conflicting report of facts; now we speak of a stock of factual evidence to be explained. Our questions then, can be connected thus. Surely, when we endorse a theory in the face of an observation report we do not explain it. When, in particular, we have rejected an observation report on the basis of later and better ones, we do not want to have it explained by our theories, since only a refuted theory explains a refuted report and we want newer unrefuted theories. Do we, then, have the task of explaining past mistakes? No doubt, we do not want to explain, as physicists or chemists, all sort of superstitions and fables about physics – as this is the task of ethnographers, social historians, and cultural historians. But do we wish, as physicists – not as historians of physics – to explain persistent errors in past physics?

This indicates, I hope, that discussion about science easily spills over into discussion about the nature and role of scientific institutions as institutions, as well as into discussion of how much can we comprehend an activity, science or other, without its history.

Perhaps this is unavoidable. Mario Bunge seems to hold the view that our intellectual activities integrate so deeply into their background that at times all separation looks arbitrary. I have great sympathy with Bunge's philosophy in general and with this view in particular. Yet I think that at times we may avoid big questions even when discussing methodology.

What I tried to show in Chapter V above was that Popper's theory of testability and/or of simplicity may include a solution to the central problem of the present Chapter: If the rejection of a hypothesis is the more testable option then reject the hypothesis; if the rejection of an observation report is the more testable option, then reject the report; in particular, if an observation report is rejected because of a very testable hypothesis, so much the better. Now the way to make the rejection of an observation report based on a highly testable hypothesis is generally very plain: we can try to argue that the observation is inaccurate, that it is only an approximation to the truth. This, then, explains, along Popper's methodology, a few items. First, it explains the fact that at times men of science prefer to reject reports which conflict with their hypotheses, at times they prefer to reject the hypotheses. It explains, second, why a new hypothesis is more often preferred to reports which conflict with it than an old one. It further explains the difference between a new and an old hypothesis by viewing the new hypothesis which conflicts with reports and is then tested (refuted or corroborated) as graduating to the status of an old hypothesis. (The more observation reports we have the more difficult it is to match the testability and simplicity of what goes with them.) It explains, thereby, the role of corroboration in Popper's methodology – and in a manner quite contrary to Popper's since it takes it as enhancing explanatory power, simplicity, and testability, whereas Popper erroneously considers corroboration a value in itself and endorses Boyle's rule. It also explains the fact, or is it an alleged fact, that the hierarchy of hypotheses reported in science text-books consists not of all testable hypotheses which were ever proposed as explanations, but only those which were highly corroborated and thus constitute stages in a simplified history of the science. (This again is contrary to Popper's theory of corroboration since it allows for the corroboration of theories known from the start to be false, such as Bohr's model.) It finally explains the hierarchy as that of explanations of ever increasing levels, of both theories and observation reports as approximations in stages. This, then, makes a simplified history of a science part and parcel of that science.

All this, particularly the idea that an explanation may stand in the relation of logical contradiction to the explained observation report, and the idea of hierarchy of theories as approximations, is fairly Popperian, of course, much in accord with Popper's own 'Three Views Concerning Human Knowledge' and his 'The Aims of Science'. It is also in plain contradiction to the theory of explanation (as deduction) presented in Popper's *Logic of Scientific Discovery*, to the theory of the (compulsory) emperical basis offered there, and even to the theory of degrees of testability (rather than levels of approximation). It seems to me that Popper's refusal to admit past inconsistency in open and clear expression prevents him from seeing the force of his own views – in a streamlined version, as I have presented it here, or in some other way.

Impressive as I find this streamlined version, I nevertheless reject it. Meanwhile Bartley has taught us that testability need not always be interesting. Certainly today's aerodynamics is much more testable yet much less interesting than several branches of today's physics. What Bartley suggests, in his classic paper of 1968, is that it is more important to be able to discuss different possibilities, different frameworks, research programs and procedures, different kinds of hypotheses, compare the preference for an observation report with the preference for a conflicting hypothesis, and so on. As long as things keep moving, and interestingly so, Bartley says, things are going well.

What should we rule about dogmatism, and should we allay Bacon's fear of it, Boyle's fear of it, Popper's fear of it? If Bartley is right then this fear is misplaced. Those who wish to dogmatize will do so; and with little ingenuity circumvent every rule in the book. Better let them dogmatize. The free spirits questing for new ideas and new truths will have little patience with dogmas but may develop much patience with dogmatists.

BIBLIOGRAPHY

Agassi, Joseph, 'Corroboration versus Induction', *BJPS* **9** (1959).
Agassi, Joseph, 'The Role of Corroboration in Popper's Methodology', *Australian J. Phil.* **39** (1961), reprinted here.
Agassi, Joseph, 'Empiricism versus Inductivism', *Phil. Studies* **14** (1963), reprinted here.
Agassi, Joseph, 'Sensationalism', *Mind* **75** (1966), reprinted here.
Agassi, Joseph, 'The Confusion between Science and Technology in Standard Philosophies of Science', *Technology and Culture* **7** (1966), reprinted here.
Agassi, Joseph, 'The Novelty of Popper's Philosophy of Science', *I.P.Q.* **8** (1968), reprinted here.
Agassi, Joseph, 'Popper on Learning from Experience', *A.P.Q.*, Monograph Series No. 3, 1969, reprinted here.
Agassi, Joseph, 'Changing Our Background Knowledge', *Synthese* **19** (1968–9).
Agassi, Joseph, 'Agassi's Alleged Arbitrariness', *Stud. Phil. Hist. Sci.* **2** (1971).
Agassi, Joseph, 'The History of the Royal Society', *Organon* (1971).
Agassi, Joseph, 'Explaining the Trial of Galileo', *Organon* (1972).
Ayer, Sir Alfred, *Language, Truth, and Logic*, Victor Gollancz, London, 1936.
Ayer, Sir Alfred, *The Problem of Knowledge*, Macmillan, London, 1956.
Bartley, III, W. W., 'Theories of Demarcation between Science and Metaphysics', in I. Lakatos and A. Musgrave (eds.), *Problems in the Philosophy of Science*, North-Holland Publ. Co., Amsterdam, 1968.
Bunge, Mario, *Scientific Research*, vols. I and II, Springer, Berlin, Heidelberg and New York, 1967.
Bunge, Mario, (ed.) *The Critical Approach to Science and Philosophy, Essays in Honor of Karl R. Popper*, Free Press of Glencoe, London and New York, 1964.
Carnap, Rudolf, 'Testability and Meaning', *Phil. Sci.* **4** (1937).
Carnap, Rudolf, 'Replies to Critics', in P. A. Schilpp (ed.), *The Philosophy of Rudolf Carnap*, Open Court, La Salle, Ill., 1963.
Duhem, P., *Aim and Structure of Physical Theory*, Princeton U.P., Princeton, 1954.
Feigl, Herbert, 'What Hume might have said to Kant', in M. Bunge (ed.), *The Critical Approach to Science and Philosophy*, London and New York, 1964.
Feyerabend, P., 'Problems of Empiricism', in R. G. Colodny, (ed.), *Beyond the Edge of Certainty*, Prentice Hall, Englewood Cliffs, 1965.

Feyerabend, P., 'Against Method', in *Minnesota Studies in the Philosophy of Science*, University of Minnesota Press, vol. 4.

Hempel, C. G., *Aspects of Scientific Explanation*, Free Press, N.Y., 1965.

Koestler, Arthur, *The Sleepwalkers*, London, 1963.

Lakatos, I., 'Criticism and the Methodology of Scientific Research Programmes', *Proc. Arist. Soc.* **69** (1968).

Lakatos, I., 'Falsification and the Methodology of Scientific Research Programmes', in I. Lakatos and A. Musgrave (eds.), *Criticism and the Growth of Knowledge*, CUP, Cambridge, 1970.

Lakatos, I., 'History of Science and its Rational Reconstruction', in R. Buck and R. S. Cohen (eds.), *Boston Studies in the Philosophy of Science*, vol. 8, Reidel, Dordrecht, 1971.

Musgrave, Allan, Forthcoming contribution to the Bucharest International Conference in the Philosophy of Science.

Polanyi, Michael, *Personal Knowledge, Towards a Post-Critical Philosophy*, University of Chicago Press, Chicago and London, 1958.

Popper, Sir Karl, *Logik der Forschung*, Springer, Vienna, 1935. English Translation, *The Logic of Scientific Discovery*, Hutchinson, London, 1959.

Popper, Sir Karl, *The Poverty of Historicism*, Routledge & Kegan Paul, London, 1957.

Popper, Sir Karl, 'The Aims of Science', *Ratio* **1** (1957).

Popper, Sir Karl, *Conjectures and Refutations*, Routledge & Kegan Paul, London, 1963.

Solla Price, Derek J. de, 'Contra Copernicus: A Critical Re-estimation of the Mathematical Planetary Theory of Ptolemy, Copernicus, and Kepler', in Marshall Clagett (ed.), *Critical Problems in the History of Science*, University of Wisconsin Press, Madison, 1959.

Stove, D., Review of Popper's *Logic of Scientific Discovery*, *Australasian J. Phil.* **38** (1960).

Watkins, J. W. N., 'Confirmation, the Paradoxes, and Positivism', in M. Bunge (ed.), *The Critical Approach*, London and New York, 1964.

Wisdom, J. O., 'Some Overlooked Aspects of Popper's Contributions to Philosophy, Logic, and Scientific Method', in M. Bunge (ed.), *The Critical Approach*, London and New York, 1964.

APPENDIX: RANDOM VERSUS UNSYSTEMATIC OBSERVATIONS

Sir Francis Bacon's first rule of induction was that we must free ourselves of all preconceived notions, conscious or unconscious; and second, that we must observe, and record our observations quite unsystematically.

Bacon stressed that only by following both rules can we achieve a random collection of facts. And, he said, when that collection be large enough to be representative it will be ripe enough to enable us to squeeze theories out of it. His third rule, consequently, is that we must collect a lot before attempting to theorize.

The most obvious criticisms of Bacon – namely that we never have enough data – led thinkers of the more intellectualist methodologies, from William Whewell to Popper, to conclude against unsystematic observations as wasteful and pointless. It mattered to them little that unsystematic observations are not fully random. They did, in criticism of Bacon, declare it impossible to free ourselves of all preconceived notions. They sometimes even went further and concluded that hence even unsystematic observations are not random. Especially in our century they sometimes concluded further that these, therefore, merely reflect the preconceived notions of the observer (Russell, Popper).

Does all this make unsystematic observation valueless? Let us assume without debate, that observation is better systematic of some sort than unsystematic. Suppose, however, a case where systematic observation is precluded on technical grounds. What then? Can we declare that there is no future use in unsystematic observation? Though containing much random (and hence superfluous) information and much biased (and hence distorted) information, it may perhaps be of some future use and so it is certainly better than none, and so when the cost of gathering it is not too high it may be commendable.

Instances abound. Charles Darwin, by no means a Baconian, collected while on the *Beagle* as much information as he could – for an obvious technical reason; he did not hope to arrange a second visit. The same is said by E. E. Evans-Pritchard regarding primitive tribes: they are vanishing and so we must record them pretty fast or not at all. The same was felt by those who recorded conversations with participants in the revolution in quantum theory: these men are dying fast.

No doubt, there are different functions to unsystematic observations. In their very limitation due to the bias of the observers, we may find in them the biases. A psychoanalyst encourages unsystematic observations, or free association, not on their scientific merit, but on the very contrary for their revelation of blind spots and biases. Similarly with anthropologists asking their informants to observe freely.

If this might be so, then bias which cannot be criticized and eliminated, should be given free rein. That is to say, even while making a concession to Bacon and admitting unsystematic observation, we should not go so far as to encourage random observation. Not only is random observation in principle – psychological, methodological, epistemological – impossible; the very attempt to make our observation more random than it naturally is may, indeed, back-fire. The natural unsystematic observation is already so much encumbered with irrelevancy, that the significant in it may – for merely technical reasons – never be retrieved for good use; when the attempt to make the unsystematic observation even more random takes over, the swamp is too much.

To take one interesting example, I quote in full one note, the final one, from Edward Sapir's *Language*, Chapter III, on phonetics.

The conception of the ideal phonetic system, the phonetic pattern, of a language is not as well understood by linguistic students as it should be. In this respect the unschooled recorder of language, provided he has a good ear and a genuine instinct for language, is often at a great advantage as compared with the minute phonetician, who is apt to be swamped by his mass of observations. I have already employed my experience in teaching Indians to write their own language for its testing value in another connection. It yields equally valuable evidence here. I found that it was difficult or impossible to teach an Indian to make phonetic distinctions that did not correspond to 'points in the pattern of his language', however these differences might strike our objective ear, but that subtle, barely audible, phonetic differences, if only they hit the 'points in the pattern', were easily and voluntarily expressed in writing. In watching my Nootka interpreter write his language, I often had the curious feeling that he was transcribing an ideal flow of phonetic elements which he heard, inadequately from a purely objective standpoint, as the intention of the actual rumble of speech.

Sapir's proviso, namely that the recorder have a good ear, clearly indicates the risk that the unsystematic observations of the recorder may be of no interest whatsoever. Yet, clearly, for technical reasons, these are to be preferred to the 'minute' recording of all facts.

There is little doubt that the 'minute' observer is historically a descendent of Bacon even when, as usual, his historical and intellectual awareness is too poor for this fact to be conscious with him. There is little doubt that the very reason Bacon recommended minute observation, namely the attempt to achieve a random collection of facts, is what makes the venture useless. It is obvious that once induction was given up as a bad job the pendulum swang full swing towards the highly directed fully designed experiment – as a test of a theory. There was admittedly some indulgence towards the odd nature-lover who collected specimens but his status was deemed as no higher than that of a stamp-collector. And so a lacuna was created, in which various writers, including Darwin, Evans-Pritchard, and Sapir, can be found. These are not inductivists and they do not preach random observation. But they do see room for unsystematic, though not random, observations, for one reason or another. What I have ventured to claim is that the lacuna was created by the reasonable, yet too violent, response to Bacon. Since Bacon was in error, we can differentiate the unsystematic from the random observation, and reject fully only the latter. There is no risk in becoming Baconian by the sheer advocacy of unsystematic observations – for one reason or another, as the case may be.

CHAPTER 7

TESTING AS A BOOTSTRAP OPERATION
IN PHYSICS

FIRST INTRODUCTION: RELIABILITY IS NOT A MATTER
FOR PURE SCIENCE

The reliability of scientific theories has been taken by most philosophers to be a major contribution of science to humanity. It seems advisable to consider reliability or whatever should replace it, as we should consider all extra intellectual benefits of science, within the realm of scientific technology, and restrict the domain of science as much as possible to enlightenment only.

One may object to the view that reliability belongs to technology by saying that there are two senses of reliability, technological reliability of a theory in its practical application and intellectual reliability which constitutes the observation of the intellectual merit of a theory. I agree that observing the intellectual merit of a theory, as well as its weaknesses, is part and parcel of pure science. If this is what one means by the in- tellectual kind of reliability, then I accept the objection after observing that the two types of reliability have little in common. Those who wish to link closely the two types of reliability will have to say a little more about the nature of the second kind of reliability. This was viewed as the same as credibility, which is different from recognizing merit, as we can easily compare the merit of two views which we reject: the two views have equal credibility, namely zero credibility, but perhaps unequal merit.

Due to the religious ancestry of modern science, it was supposed that when the Copernicans expelled Aristotelianism from the body of estab- lished belief, some other doctrine should have replaced it; it was even alleged by some that the only way to replace Aristotelianism as an estab- lished belief is to replace it by the establishment of another belief; and further, that either the alternative belief is scientifically established, or else we merely replace one dogma by another. There is much sense in this view: such things as established doctrines do exist, now as then; and establishing a doctrine is part and parcel of instituting a number of social practices. It turns out, however, that there is no need for anybody to believe in an established doctrine; nobody asks how many Englishmen believe in the Church of England, for example, before being ready to admit it as a state religion. The field of established belief is a matter of social technology; and so it is at worst a matter of tradition, or caprice, at best of rational or scientific technology; but not of pure science.

One may still press the objection to my view that reliability is a technological matter. Even if scientists do not endorse the established scientific doctrine, they do have their own private beliefs, and these can be enlightened or not. Surely, to that extent the reliability of a scientific doctrine in the eyes of a scientist is a matter of his own enlightenment! Again, I will gladly endorse this view if someone offered a reasonable theory concerning enlightenment and belief. Historically, the philosophers Sir Francis Bacon and René Descartes linked the two most strongly. The link became a part of the scientific establishment with the foundation of the Royal Society in 1660–3, and the recognition of Sir Francis Bacon as its ideologue. Bacon assumed – we know now that he was in error – that a scientist acts on his beliefs at least to the extent that they direct his researches and teachings. There is really little need to indicate that all respectable schools teach doctrines not endorsed by their members. Only doctrinaire Freudians for example ignore all non-Freudians and refrain from teaching extensively all non-Freudian psychology. In physics it is simply unthinkable not to teach the ideas of Newton, Hamilton, Maxwell, and other quite outdated masters. A philosopher not familiar with Greek thought is incompetent even though no one today endorses any Greek ideas as they stand. Nor do people conduct research entirely along their own beliefs; much is directed to examine other people's beliefs. History easily illustrates this. Euler was a Cartesian who contributed to Newtonian mechanics in one way or another. Boscovich was an Aristotelian who saught to compromise Newton and Leibniz, Hamilton was not a Newtonian but a Boscovichian. Maxwell did his main work while trying to reconcile Faraday with Descartes and both with Newton. It is hard to say what each of these people exactly believed, what he believed he was doing, and what he believed was the relation between what he believed in and what he did. Nor are historians of science agreed, either about the facts of the matter, or about how the facts appeared to the researchers involved.

To conclude the criticisms, it is not denied here that reliability matters for pure science. It is only claimed that the two chief roles traditionally ascribed to reliability, are matters for scientific technology. In particular, the reliability of a theory in practical application is clearly not in itself of concern for pure science. Further, the idea that erroneous belief vitiates research is at least doubtful: we do not know what is the right belief, and how erroneous beliefs influence research.

It seems as if the point just made is in conflict with the conclusion of the last paragraph. This is not so. Any adequate theory of belief should take cognisance of the fact that thought influences action and thus research activities. Yet, no doubt, the views here discarded fail to do it. Bacon wanted scientists not to be led astray by false conjecture and so forbade all conjecture until enough facts were collected. Hence Bacon was driven to the view that facts have to be discovered by accident unaided by conjecture. He was in error; any discovery aided by thought shows this quite clearly.

There are three possible relations between a theory and a description of a general fact, deducibility, contradiction, or independence. To insure that true descriptions of facts to be deducible from a theory correctly, the theory should be true; and conjecture has little chance to be true. If a statement of fact contradicts a theory, said Bacon, then accepted theory blinds us to that fact. Hence, false theories are evil and true theories have to be based on facts or come after the facts. The remaining case, then, is of independence: a new fact is independent of any existing theory though it follows from some future theory which will be based on it. The history of science forces us to reject Bacon's views: in historical fact new facts constantly appear contrary to old established beliefs. Let us, then, examine Popper's theory. The body of science was viewed traditionally as the body of established theories. According to Popper the body of science constitutes refutable conjectures. The progress of science Bacon and his followers view as that of the increase of body of established belief, whereas Popper views scientific progress as that of the making of refutable conjectures, and the occasional attempts at refutations of some of them, successful or unsuccessful as the case may be.

At once we can see that thought and action may relate in a manner most obvious yet ingored by those who wish science to provide us with reliability. Namely, science can eliminate unreliability. And, in particular, when the action in question is scientific research, the thought may be a suspicious thought, an ingenious attempt to overthrow current theories, and discovery may be the successful refutation, the finding of the refuting facts.

This view of Popper encounters two difficulties, one classical, often attributed to David Hume, one modern, attributed to Duhem and Quine. Hume's difficulty is, how is a theory ever established, or can a theory ever be established? Duhem's difficulty is, how is a specific theory refuted, or can any specific theory ever be refuted? Let us take Duhem's difficulty first.

SECOND INTRODUCTION: THE DUHEM-QUINE THESIS HAS A NEW SIGNIFICANCE

The Duhem-Quine thesis is, strictly speaking, a point of formal logic, and as such both uncontested and incontestable. It is this: When a conclusion in a valid inference is false then all that logic can tell us in general is that one of its premises is false; logic cannot tell us generally which of the premises is false. On occasion, when we can prove all premises but one to be true, we can, indeed, locate the error in one premiss. This is how we can declare one premiss absurd when doing formal logical exercise. Also, one may notice, when, in a logical exercise, we have a finite set of possible alternative, and we refute each of them except one, then we thereby prove that one. It seems as if we have a perfect symmetry here between proof and disproof in logic. Does this symmetry hold in science?

A scientific prediction is one deduced from an explicit hypothesis, which prediction we can prove or disprove. This is, by definition, a prerequisite of a scientific prediction, and we can thus see a symmetry at on point in science, between proof and disproof. We can see that the symmetry is only limited to this case. For, it seems as if, clearly, when we prove a prediction to be true we do not, thereby, prove the hypothesis from which it follows. This is the difficulty or rather the impossibility, encountered already by Hume if not earlier. There exist no way by which a prediction, if proven, leads to the proof of a hypothesis. In particular, we cannot say, if a prediction follows from two hypotheses, we make another prediction to decide between the two. If a prediction would follow from a finite number of alternative hypotheses than perhaps by a process of elimination we could prove one hypothesis by proving ever so many predictions. But, in logical fact, there are always infinitely many hypotheses from which we could have deduced the same prediction or set of predictions. But at least, it seems, when a prediction was refuted, we could, perhaps, thereby eliminate the hypothesis responsible for the error.

We see here a strange situation. Apparently we have a case of symmetry regarding a prediction: it can be proved if true and disproved if false. Apparently we have a case of asymmetry regarding theory: we can disprove a false theory by disproving a prediction based on it. Yet at once it looks perhaps possible to turn on occasion a disproof into proof and thus restore the symmetry in part.

The erroneous idea that we may have in science a finite set of possibilities which may all be refuted except one, and thereby establish that one – this idea has allured many thinkers. The general idea of a finite set of possibilities is known as the principle of simplicity, or of conformity of nature, or of limited variety. The reason that this idea seems so promising is a simple illusion; thought we have to accept it on occasion, as we must, for example, when making a chemical analysis on the basis of a finite known list of elements. In this case we can employ the procedure proposed, which is known by the name of induction by elimination; when the number of contesting hypotheses is small – two or three – the name given to the process is crucial experiment, or experimentum crucis, which originally seems to have meant the experiment of the crossroad. What Duhem's thesis amounts to, we shall presently see, is that there is no crucial experiment (not even in chemical analysis, he said).

This sounds absurd. It is hardly conceivable that Duhem should prove the impossibility of something which happens regularly in science; when we know that Eddington's eclipse observation was a crucial experiment between the Newtonian and the Einsteinian theories of gravity, when we know that the observations on Brownian movement were crucial between atomism and its opposition, when we can offer lists of crucial experiments in the history of quantum theory, how can Duhem say that these do not exist?

Unfortunately, the label 'crucial experiment' is misleading. We do not know what exactly the alchemists who first used it had in mind; and even the locus classicus, Bacon's use of it (in *Novum Organum, Book II*) is controversial. But Duhem's use is crystal clear, and in his sense, surely there is no crucial experiment. When Eddington decided against Newton he did not prove Einstein right. And so on.

What Duhem said was this. We deduce a prediction from a large set of hypotheses. As we cannot prove which of these hypotheses is the culprit which has led to the false prediction, when we eliminate what we think is the culprit we forfeit our hope for a proof. We may have eliminated the wrong hypothesis. Thus, in chemical analysis we make use of certain hypotheses, often such hypotheses as the one telling us that the chemical atom is stable, or that the inert gases so-called are that, etc.

What Duhem contends here is that when making a deduction from allegedly one hypothesis to one prediction, we actually employ the whole of science. If the hypothesis is concerning gravity we use optical instruments to test it, thereby applying a theory of light and heat at least.

This is an extra-logical component in Duhem's thesis, but it is one which is not worth contesting. We can be finicky and search for, perhaps, even find, perhaps, a prediction which does not involve some part of science or another. The search is not worth persuing. We may well agree with Duhem wholeheartedly that there is no proof in science of any kind, and thus no proof by any crucial experiment; that even disproof is only a very non-specific process since making it specific, i.e. putting the blame on one specific conjecture, is itself not absolutely free of conjecture (i.e. a little conjectural or a little pregnant with conjecture).

Once we are free of the desire to prove we can ask, does Duhem's thesis have any further insight to offer? I think it does. Let us examine Popper's view of science as conjectures and refutations. As it stands, Duhem's thesis has little to offer regarding it: Popper's theory is not concerned with proof or establishment, and is not averse to making any conjecture, including a conjecture which puts the blame for a given refutation on this hypothesis or that. Indeed, any case which proves Duhem's thesis, any historical instance of the overthrow of the wrong hypothesis, will easily accord with Popper's view of the use of conjectures in science.

When the polarization of light through reflection was discovered, in the early nineteenth century, it was deemed the deathblow to the corpuscular theory of light and the proof of the wave theory. It sounds funny but with waves being continuous and particles discrete, it was not interference or even polarization or any continuous aspect of light, but rather the fact that polarization through reflection is discrete, which was so hard for the corpuscularian to explain. In a desperate attempt Jean-Batiste Boit, a celebrated 19th-century man of science, suggested to quantize the spin of the light particle, in order to account for the facts. His suggestion was ignored off-hand, and the wave theory became established. This suggestion is part and parcel of the new quantum theory – of quantum field theory to be precise.

To return to Duhem, if we really employ all of our science in every prediction, how do we come at all to the situation where, normally, we have an idea that we are testing one hypothesis? Granted, that post-hoc we can always say, it is doubtful which hypothesis has collapsed, we can still say, almost always, which one was under attack! How is it that chemical analysts use crucial experiments and doctors use differential diagnoses without hesitation?

When Jean-Servais Stas refuted Prout's hypothesis which says that all atoms except hydrogen are compounds of hydrogen, he did so by showing that the atomic weight of chlorine is 35.55. He was in error. His result does not reflect the falsity of Prout's hypothesis, but the falsity of Dalton's hypothesis which says, all atoms of one element have the same weight. They do not: two atoms of the same elements may have different weights; they are then called isotopes. And when isotopes were discovered, and Aston constructed his masspectrograph and showed that some chlorine weighs 35 and some 37, an attempt was made (by Niels Bohr) to revive a version of Prout's hypothesis (allowing as elementary particles both protons and electrons). It looks as if we again prove the point that we do not know where to put the blame. Yet Stas thought he knew, and every experimentor thinks he knows, which specific hypothesis he is focusing on. How? Why is it that chemical analysts are so often (seemingly) right?

Duhem had no interest in refutation. He exhorted people, as all good philosophers since Bacon did, not to ignore refutations, not to become dogmatists. But the very exhortation shows a mistake: we exhort people to act correctly against their inclination; but we need not exhort those who look for a discovery to pay attention to a refutation any more than we need tell a hunter who looks for meat not to ignore the animal he has shot. If Popper is right and discovery is the same as refutation, the question becomes, how does the experimentor decide which hypothesis to track down? And why do we need the exhortation not to revive a dead game?

For the second question we may offer an answer by indicating a common and standard error. But for the first question we cannot expect a full answer: hunting involves resourcefulness and intuition, it is not as standard as a repeated error and so it has no general explanation. But there are some general and fairly loose rules conducive to, though not insuring, good huntint, and these may go a long way towards an answer. This chapter goes after these rules. Roughly, I shall try to argue, since confirmation is failed refutation, we go on failing until we succeed: we temporarily strengthen one hypothesis and thus think – often erroneousy but well enough for the time being – that we have succeeded at last to refute the previously confirmed theory. When this proves a mistake, the catch proves to be bigger then expected: we have a revolution going.

I. CONVENTIONALISTS AND THE PROBLEM OF INDUCTION

The best known theory of science claims that the task of empirical science is to render some theories reliable by empirical means. This theory, inductivism, is easily refuted by anyone who is willing to examine it, yet most philosophers of science currently engage in defending it, and some waste their time attacking it. Let us, rather, examine what the alternative theories of science tell us about reliability.

There are two extant alternatives to inductivism. The one is conventionalism, endorsed by Poincaré, Duhem, Mach, Dingler, Eddington, and others. The third view is sometimes called criticalism, and includes chiefly Popper, and his followers to this or that degree, and others influenced by his writings or who independently came to similar views.

Conventionalism views scientific theories as implicit definitions, as mathematical frameworks within which to store empirical information. As such, scientific theories are certain by virtue of being definitions. Already Poincaré showed that scientific theories cannot be confirmed by facts. To be able to confirm a theory, he said, you have to be able to refute it; but it is so couched that it can always escape refutation. Take, for example, the law of conservation of energy. It is a part of mathematical theoretical physics which is given empirical content by enumerating the observed kinds of energy. But the list is not complete and so we cannot refute the law: whatever you may conceive of as a refutation of the law I may reconstrue as the discovery of a new kind of energy.

This is a thrilling argument. First, it is, indeed, valid. The only way to refute the law is to create a perpetual motion machine and observe it run, without any sort of refuelling, for ever. No one can stay around long enough to make such an observation, and no one can swear that it worked with no refuelling of any sort. Secondly, the perceptive reader may notice that the attempt to insist that confirmation in science is possible while doing justice to Poincaré's ingenious argument will be hit by Duhem's thesis.

Suppose we agree, as we must, that the energy conservation law as formulated by Poincaré is irrefutable; suppose we also join him in concluding that hence it is not confirmable. Yet suppose we are still looking for a confirmable, and hence refutable, version of the same law. I suppose then, we take the obvious step – the one he indeed mused about – and include in the law the finite list of energy-forms thus far observed, and declare the list complete. Now, we can say, our hypothesis is refutable and hence, hopefully, also confirmable. Alas! says Duhem; you can refute the claim that your list is final, or some other subsidiary claim-concerning energy transport, for example – or even the law of conservation of energy itself; but you will never show for sure that it was indeed this or that hypothesis which was refuted; and there we are right where we started.

Duhem was amazingly systematic, I think, and I find that his inconsistencies can always be eliminated without prejudice to the core of his philosophy. He eliminated all confirmation of scientific theory, he eliminated all belief or reliability from science, and with this all the problems traditionally involved with reliability, yet the price was high; Duhem made science lose its major attraction: for him there is no such thing as pure science; what we call science for him was in part pure mathematics and in part applied mathematics which included empirical generalizations put in mathematical language. As to the reliability of generalizations, this is another matter altogether: they should be severely tested and found reliable before they are accepted. Poincaré, who was more honest about difficulties in his system then Duhem, stressed that therefore he did not succeed to rid science entirely of Hume's problem or the problem of induction; he said it was a vexing problem but we must learn to live with it. He only rid theoretical science, he flet, of Hume's problem, by making it part and parcel of mathematics, pure or applied.

We must conclude, then, that there is a serious flaw in the conventionalist philosophy, which may indeed be filled in future by solving the problem of induction for empirical generalization: how do we make these reliable?

No doubt, quite a few philosophers these days study induction strictly with regard to empirical generalizations. It is not clear, however, whether these philosophers are inductivists or conventionalists, since even conventionalists may acknowledge the existence of the problem of induction in matters experimental. In a recent volume Carl G. Hempel has analyzed Semmelweiss' study of the source and transmission of childbirth fever as a paradigm of proof by elimination of given alternatives, and this looks fairly inductivist. Yet, as long as the proof is not declared conclusive, the conventionalist cannot object. And, indeed, I see little objectionable in Hempel's analysis of Semmelweiss, except perhaps his seeing it as a paradigm. This, I think, is what may, perhaps, pin down Hempel as an inductivist proper even in his latest work. I cannot say.

There is little doubt that the conventionalist faces the problem of induction regarding facts. There is little doubt that the only reasonable solution to this problem is the one offered by Mach: take even an empirical generalization merely as a condensed observation report. The cost of Mach's move is, however, enormous; science loses all its predictive force. That is to say, we can project to the future an empirical generalization, as we can an abstract theory, of course; but in each and every case we have no expectation whatsoever as to the truth of falsity of the prediction. This seems simply empirically false. Wittgenstein said early in the century, that the sun will rise tomorrow is a mere hypothesis; I think we all feel that this is not so. Logic tells us nothing about the truth of an extrapolation, yet we all extrapolate, in fact. Hume said, extrapolaration is a purely psychological matter. We all tend to deny that too. Psychologically speaking, all men extrapolate, of course! Yet not all men extrapolate rationally, and not all men extrapolate irrationally. We cannot use logic alone to demarcate rational from irrational extrapolation; but we will not allow psychology alone to explain extrapolation thereby making all extrapolation equally irrational. Empirically we observe that some extrapolations make more sense than others. We want to know, psychology or no psychology, what it this sense? What is there to use as a means for the demarcation of some extrapolation as more rational or reasonable or sensible than other?

It is hard for me to say what is the message of Nelson Goodman, the famous philosopher and author of the celebrated *Facts, Fiction and Forecast*. The only reasonable question which seems to trouble him is the one presented in the last paragraph, though I cannot say this unhesitatingly. The question I presented there, however, seems to me very reasonable and interesting, and I shall be happy to learn that it is, indeed, his.

The question however, is largely a matter for rational technology. Conventionalists, who are people who think theoretical science is true by definition, are usually also instrumentalists; that is to say they are usually of the opinion that the aim of science is prediction, that all theoretical science is applied mathematics and thus technology. For them, therefore, Goodman's question is crucial. This is, perhaps, why Poincaré said, we remember, that the problem of induction is vexing: he could not shake it off.

II. POPPER IS AMBIVALENT REGARDING GOODMAN'S PROBLEM

For anti-instrumentalists like Popper or like myself it is almost possible to relegate Goodman's question – what distinguishes the rational forecast from the irrational one? – from the realm of theoretical science to the realm of rational technology. Almost. For, of course, the planning of research, even within theoretical science, even within the experimental part of the purest of pure science, is a matter of planning and thus of rational technology, or rational forecast.

This is not to say that Popper does relegate reliability to the domain of rational technology. All in all, I do not think he distinguishes clearly enough between science and technology, except when he criticizes instrumentalism. He does say, in technology we may, for technical reasons, prefer to employ a refuted theory instead of an unrefuted theory, so that technology differs from science, where (refutable but) unrefuted theories are always preferenced to refuted ones. This, however, does not make him raise the question, when is the employment of a refuted theory in technology rational, when not. On the contrary, he either ignores the question of reliability altogether or he tackles it in general within science – not within rational technology as distinct from pure science. Clearly Popper is ambiguous here.

In his chapter on the empirical basis of science in his classical *Logik der Forschung* (i.e. logic of research, which he translated under the title *The Logic of Scientific Discovery*), Popper speaks of empirical information as the basis of science in general. He rules – quite incorrectly, historical examples show clearly – that whenever we have empirical information conflicting with theoretical hypotheses we must relinquish the hypotheses and endorse the information. Otherwise, he says, we may become dogmatic and there must be a rule, he says – quite dogmatically, I think –, to barr dogmatism. (I think rules against dogmatism are all both useless and dangerous.) If we do not like some empirical information we are at liberty and even encouraged to try and refute that empirical information; at times we may even succeed in that and thus bring scientific progress. But until we succeed, he says, we must accept it and reject the hypothesis it contradicts.

Popper, then, accepted facts as a basis, but as a merely temporary basis. He likens the foundations of science not to concrete poured on a rock but to piles drilled in a swamp. I am afraid I am not clear about this metaphor. In particular, since Popper's chapter on the empirical basis discusses mainly refutation, not confirmation, I do not see how this exactly relates to the question of reliability – of theories or of general facts. I suppose that when Popper speaks of the empirical basis of science he does not speak of the reliability of theories. Rather, he speaks there of the growth of science – he says at the end of his chapter "the bold structure of its science's theories rise, as it were, above a swamp. It is like a building erected on piles". As long as he speaks of progress, and as long as he keeps clear the idea that progress is fed on refutations, then, it is fairly clear that he is asking about the reason for the acceptability of refuting information. He says, we cannot prove its truth, but we agree to accept it as true as long as it is not refuted, in order to oust dogmatism.

In so far as this is acceptable, and let us consider it acceptable for now, clearly it is a solution to the problem of induction for empirical generalizations (as scientific information is always generalized: unrepeatable data are not allowed within science) within the field of pure scientific research. The scientist takes a hypothesis to be refuted, he looks for an alternative

to it, on the basis of his blanket agreement to accept as reliable all refuting generalizations. This agreement, says Popper, is made in order to
keep science going: violating it opens the door to policies which may
bring about stagnation through dogmatism.

This is all rather drastic. There is no need to fear dogmatism so: the
dogmatism of Newton, or of Heisenberg, did not kill science. There is
no need for a blanket agreement, since we can report an occasional
reasonable rejection of attested empirical information for the sole purpose of rescuing a theory. Dirac's case is celebrated; but already Prout
faced the same situation when he proposed his hypothesis: many atomic
weights were reported then to be not whole numbers; he got rid of some
of them by better experiments, others by modifying some subsidiary
hypotheses (he replaced the formula for water from HO to H_2O); and
he expressed high hopes about the rest. Stas refuted, or so he thought,
some of the hope; yet the hope was reasonable.

We must, to note the logic of our situation, observe a few trivial points.
Goodman's question is, what differentiates a rational prediction from
an irrational one or, which is the same for our present context, a rational
generalization; and this answer must be confined, as yet, to the domain
of pure science. To be brief, let us introduce Popper's technical term
"corroborated" to mean "well-tested and as yet unrefuted" (the word has
a different sense in legal terminology, and the ordinary sense is either the
legal one of that of confirmation or even strengthening; Poincaré, we
remember, linked confirmation to refutation, the way Popper links corroboration to refutation). Popper's view may be an answer to Goodman's
question, and as follows. It is rational to rely on all and only corroborated
generalizations. If, however, Popper were to allow for the rational rejection of even one corroborated generalization then he would thereby allow
us to raise afresh Goodman's question. It seems to me clear that this is
indeed Popper's view, and that he does answer Goodman's question, all
be it erroneously. But suppose that my reading of Popper is false. The
question, then, is, how else does Popper answer Goodman's question? If
my reading of him is false, I contend, then, he leaves Goodman's question
unanswered.

So much for Popper on the reliability of empirical generalization. As
to scientific theories, Popper says they can be corroborated too: we may
fail to refute them. Literally this is doubtlessly trivially true. Does that

make reliance on them rational? Popper is unclear here. Supposing it is, is that rationality obligatory? Again, no clear answer. Practically all philosophers occupied with some exegesis on Popper's text (except myself) say, yes, Popper does find it rational and hence obligatory to believe a corroborated theory. It seems to me that Popper says, belief in a corroborated theory is rational but not obligatory. This, of course, would be trivial if belief in any unrefuted theory is rationally permissible but not obligatory: this is simply permissiveness. Thus, the view that belief in a corroborated theory is permissible may be an austere view if it implies that it is irrational to believe an uncorroborated view, alternatively, it may be the permissiveness and that would make corroboration rather irrelevant to belief.

To conclude I find Popper's theory of corroboration confusing. It may be read, I suggest, as one or the other theories:

(1) A corroborated theory must be accepted until it be refuted or replaced by a more testable one. That is, assent must be given to a corroborated theory.

(2) A corroborated theory need not be accepted, and any doubtful theory (including a corroborated one) may be accepted. That is, assent is always optional, except in cases of refuted theories.

(3) A corroborated theory need not be accepted, but an uncorroborated (and a refuted) one must be rejected. That is to say, we may always suspend judgment but must suspend judgment on all theory except the corroborated theory.

Each of these theories seems to me objectionable, thought, of course, I find the first and most constraining to be the most objectionable. I can find with ease quotations from Popper supporting each of these three readings, but rather than waste time on hermeneutics I should leave it to him to clarify his position if he is interested. If he's not interested, then, I suggest, hermeneutics is out of place: it is unbecoming to read into a text what an author has left open.

III. BOOTSTRAP OPERATIONS IN TESTING

So let me proceed to my own view of the matter. In general, whether within pure science alone or within science and rational technology taken together, I think corroboration is no matter for belief, yet it has an important role in both. Very briefly, we corroborate working hypotheses, and auxiliary hypotheses, when we calibrate our instruments in order to blunt the edge of Duhem's thesis: calibration has to be successful before tests can proceed, or else we declare the results of the test *a priori* invalid and worthless. And the corroboration with uncalibrated instruments means that the instrument is probably not stable enough and not precise enough to refute the hypothesis tested; and so we use the instruments only if they survive the onslaught, if we corroborate claims for their stability. In other words, we try to make it at least a bit hard to blame the instruments for undesirable results of any test *prior* to that test. This is why we have always an *a priori* idea of which hypothesis to blame if the experiment will refute the prediction it comes to test. But this does not mean that calibration excludes all possibility of blaming our instruments; and so, this does not invalidate Duhem's argument. The reason for this is obvious: calibration is *never* complete; but it explains the fact that Duhem's thesis is so abstract: in concrete situations we take measures to exclude it.

To show that calibration is never complete we may take actual cases where successful calibration not only permits using a Duhemian ploy of saving the tested hypothesis at the expense of the auxiliary hypotheses, but also where this was done in history. I have in mind bootstrap operations, in particular.

It is convenient to take bootstrap operations because they exclude the inductivist view of science *a priori*. Inductivism does, of course, permit and even encourages the use of one theory for the purpose of discovering another. But the inductivist views the use of a theory in discovery not as bootstrap – it is for him climbing a mountain (or in special instances climbing a ladder). The mountain climber (and the ladder climber) makes sure to the best of his ability that a step he is going to take will put his forward foot on solid grounds. Of course, the mountain climber can never make absolutely sure that his next step will be safe until he has put it behind him. Yet, his success in making any step rests on the stability of the place on which he has rested his foot in the previous step. In a bootstrap operation the opposite seems to be the case: the support cannot sustain one well enough, yet one uses it just long enough to make the next one. Literally, of course, the image is exaggerated to the point of impossibility – we cannot pull ourselves by our bootstraps alone – yet the allusion is to ever so many situations which are notoriously precarious yet which we voluntarily enter in the hope to exit very quickly from, and into a better position than ever before. An example may be walking fast in a swamp or a speculation in a bull market by a penniless customer who, however, has enough credit to speculate; the customer resells at a higher rate before he has to pay. An image may be that of treading water; but this only keeps our heads above water until rescue comes and is so not a sufficiently adequate image.

Examples of bootstrap operations in science abound, though historians of science have not yet studied them as much as they deserve. The simplest examples are indeed theories used to calculate results which conflict with the very same theories in order to eliminate the conflict. The most obvious examples are the steps leading to the personal equation; – the equation came to eliminate discrepancy between theory and observations (without the theory there would have been no discrepancy discovered) – or Roemer's theory of the speed of light which comes to rescue the Keplerian theory of Jupiter's moons; or Bradley's theory of abberation which rescued Newtonian celestial mechanics. The bootstrap operation is in the ract that these theories – Kepler's or Newton's – could not be developed without the use of some prior optical theories, which they now helped to replace! Better optics leads to better astronomy and vice versa. A still simpler example would be the use of the perfect gas law to develop the kinetic theory of gases, without which Van der Waals could not possibly have developed his replacement of the perfect gas law. Another example, somewhat less familiar, would be the use of very classical optical theory and radiation theory to develop astrospectroscopy to the point it could be used as an aid to the development of quantum theory. The clearest and most startling example, however, is something like Newton's use of Kepler's third law to establish the constants peculiar to the solar system (masses of planets) so as to be able to calculate the irregularities, i.e., the deviations from Kepler's laws (all three laws, that is) which the system exhibits. Or, to take another and more troublesome example, on occasion we use the old (sometimes relativized) quantum theory to identify spectral lines which we measure exactly, and compare the results of the measurement with the new quantum theory; but when there is a discrepancy we may, and at times do, alter the identification of the line under study. If these kinds of practice do not feel like a bootstrap operation, if all this does not give the logically minded reader a slight shudder, then that's the end to the problem: he can just as well ignore it or delight in his own solution to it as the case may be.

What is felt to be needed here, quite intuitively, is both a clarification of the logic of the situation, and an added measure of constraint. We may feel some fear that Newton contradicts himself when using Kepler while planning to calculate the deviations from Kepler. This is no catastrophe to begin with, but a challenge all the same. It is now agreed by all and sundry that Newton's calculus was not in order, and that it has meanwhile been successfully put to order by his nineteenth century successors. Perhaps the same holds for his astronomy. This has not been shown to be the case, but a suspicion is lurking all the same. And even if there is no inconsistency, there may still be too much arbitrariness in the whole enterprise. If we can alter almost any part of our system, then any change seems *ad hoc*. If we can change the identification of any spectral line, then what good is spectroscopic test? Even when no change is required, the attractive taste may have gone. And, of course, we have a purely-logical argument to shake confidence in the situation. We have tried to blunt Duhem's thesis by confirming our auxiliary hypotheses; but, Duhem would retort, in order to do this, the auxiliary hypotheses must be refutable. Are they?

Here we see most clearly the need for further constraint. Yes, we sometimes overthrow our auxiliary hypotheses, sometimes we rescue them with the aid of further auxiliary hypotheses. Are these not cycles on cycles and fleas on fleas? This must make us feel quite uncomfortable.

Take the auxiliary hypotheses which go into any spectroscopic calculation (except very few of the very easiest). Once you realize that a quantum spectroscopist can alter his identification of a given spectral line if he fails to calculate its characteristics (mainly wavelength and intensity, but also spread), you wonder how much trust you put in his results on occasions when they do fit. It seems as if he cannot lose but only win. It seems as if his ability to win does not reflect on the trustworthiness or otherwise of his specific theory. It seems as if the quantum spectroscopist has too much leeway. Too much for our taste, that is.

IV. THE NEED FOR CONSTRAINTS IS QUITE REAL

Strangely, this idea – it is hard to refute so much that goes on in science – is precisely the claim presented by Poincaré and Duhem. The spectroscopic empirical support of quantum theory which is made available by the success of the calculation (the data are much easier to obtain than the calculation) is invalidated by the fact that there is no empirical undermining made available by the failure of the calculation. Hence there is no corroboration in science; its theory is accepted *a priori*, not empirically; as a mathematical framework, not as the truth or putative truth about Nature. It is in the frontiers of science where there are so many continuing auxiliary hypotheses, both empirical and computational, that Duhem's thesis begins to look real.

Yet we are still uneasy; we may be uneasy because we want science to fix our beliefs, because we have inductivist predilections, of which Duhem may wish to chastise us for our own good. But we may be uneasy, feeling that there is more liberty prescribed by Poincaré and by Duhem then a scientist notices in his daily researches. We may feel not the need for more empirical support, but for more stringent rules of the game. (This, of course, will make us more sympathetic with Popper.)

Even this need, the need for more constraints, is not new; it has been felt already, and by Duhem of all people. He did so when responding to Poincaré's famous claim that it will always be more advisable to reform our auxiliary hypotheses in order to take account of new experiences rather than to reform our central hypotheses, the hard core of our science, such as Euclidean geometry of Newtonian dynamics. In his criticism of Poincaré, he said this. Tinkering with all the working hypotheses and auxiliary hypotheses, necessary for the saving of our central axioms, is usually advisable; but a moment may arrive when the tinkering becomes so cumbersome that we may prefer to upset our fundamental axioms which we had guarded against refutation for so long. Even Euclidean geometry and Newtonian mechanics may have to go. And yet, Duhem was bitterly opposed to Einstein until his dying day. He had the intuition that all the minor alterations would pave the way to the major alteration so as to prevent any revolution – that the major alteration is decomposable to myriads of minor alterations each made independently until the last one of these would effect the revolution in a *coup de grace*. This, of course, is Duhem's continuity theory. It is neither historically true, nor intellectually clear enough. Already Michael Polanyi has suggested in a manner which seems to conflict with that of Duhem, that there are changes in the history of science of varying magnitudes. Polanyi also claimed that the leadership of the scientific community decress these changes; I do not think Duhem would have endorsed this view. Thomas S. Kuhn has repeated Polanyi's ideas in a more popular presentation. He suggested that there is both continuity and revolution, in a presentation which is much more detailed then Duhem's; and he also expressed Polanyi's authoritarian view which ascribes to the leadership the right and duty to declare a revolution. Kuhn's system, then, is at least overdetermined since it accepts Polanyi's major points and much of Duhem's continuity theory. At times I think Kuhn endorses all of Duhem's theory, considering a major revolution to be just a minor step after many accumulated small changes; at times when I read Kuhn about leading scientists' sleepless nights I see him more of a Polanyiate and less of a Duhemian. Polanyi himself says that a scientist is constrained by his feelings, by his ineffable personal knowledge; and perhaps Kuhn shares this irrational theory. It is hard to say. It much depends on a detailed exegesis of Kuhn's view on the mechanism of normal scientific research: thus far he has said less about it then one might think after a cursory examination of his work. And his later expositions are often explicit changes from his earlier ones, often implicit ones, and his view seems now more clouded than when he first explained it. I wish to state, then, merely as a general impressionistic view, that reading Duhem, Polanyi, or Kuhn, one has the impression that science is much more constrained than they describe it to be: they do claim that it is rather constrained, but they do not say how and they seem to mistify the reader. Indeed, it seems that Polanyi explicitly enough criticizes his opponents (chiefly the inductivists and the Popperians) on these lines: I agree with you that science needs constraints, and I understand that your ideas of verifiability and/or of refutability in science is constraining enough; yet there is so much in science which is unverifiable and irrefutable. It is still subject to constraint, but not to yours; rather it is subject to the scientist's own good sense and his own feeling of when he gives himself too much leeway.

The only writer who has explicitly argued, however, against conventionalism and against Polanyi, charging them with being much too lax, is Adolf Grünbaum. What Grünbaum has said, in a detailed example, is that it is not as easy to blame a working hypothesis or an auxiliary hypothesis for a refuted prediction – that *ad hoc* amendation of a refuted system, though in (logical) principle is admittedly always possible, is in practice much less easy to achieve than one may think when one accepts Duhem's argument. Nor does Grünbaum agree with Polanyi's alternative. When a refutation occurs, says Polanyi, a scientist may ignore it and take the responsibility without knowing why; we may trust a scientist to find his way sooner or later, even if he does not quite know as yet why he is so unperturbed by a given seeming refutation, says Polanyi. Polanyi even offers a historical example or two – but I will not explain here why I consider them invalid. Enough it is to note that, as Grünbaum notices, this theory is dangerous as both authoritarian and irrationalistic but that all this does not invalidate it.

To return to Duhem's argument, Grünbaum's observation that it is not easy to apply it is correct, but it may be not damaging in the least, nor does it tell us enough about the technical constraints which science does undertake. For, saying that any move in the advancement of science requires imagination and is therefore difficult, does not refute Duhem; indeed it is much in accord with Duhem, perhaps it even accords with Kuhn's view (it is too vague for me to decide this matter). But I do not wish to register any disagreement with Grünbaum, of whose views I do not claim expert knowledge (he is notoriously taxing on his reader). No doubt, what I am going to say he has at least in part anticipated, in part he may wish to differ from. But I cannot say.

V. SCIENCE CONSTRAINS ITSELF BY AUXILIARY HYPOTHESES

The constraints which I think science undertakes are often matters of calibration. I say undertakes because on occasion science may repeal or postpone this undertaking, as we shall see. Indeed, any part of the scientific endeavor is a matter of undertaking alone, and so one may, and sometimes (*pace* Popper) one does, feel free to alter one's procedure at will. One may, moreover, simply disregard a refutation, even a very obvious and strong one, with no good reason or even with no reason at all, if one shows awareness of what one is doing and of what this amounts to. What exactly this amounts to we shall soon try to explore.

The constraint undertaken – and sometimes repealed – then, is related to calibration, both the successfully undertaken and the one skipped. The unsuccessful calibration, we remember, disqualifies the experiment before it is performed. Now, if the result of the experiment refutes a theory which one wishes to declare still unrefuted, one is always at liberty to recheck the calibration. This means that one tries to put the blame for a false prediction on some auxiliary hypothesis and then to show this by refuting it empirically – which is often easier said than done. Theoretical scientists throughout the history of modern science have questioned the results of observation and required rechecking – sometimes with detailed specifications, sometimes not – claiming that the results must accord with their views. In almost all such cases, rechecking was undertaken soon enough, either corroborating the theoretician's *a priori* accusation, or forcing him to make a different move, or give up his position. If he does choose to make a different move which is again a challenge to the observer, some observer may take up his challenge and the story might develop accordingly; or the community of science may view him as a bore and a write-off.

Polanyi suggests that here the authority of the leadership of the community of science comes into operation – when it declares that a theoretician has gone far enough and now ought to desist: if he does not, he may be penalized and lose his status. In my view, however, it is a matter not of leadership but of democracy. As long as there are enough people interested in the theoretician's struggles, I think, they may cooperate with him by checking what he suggests ought to be checked. It is very doubtful whether Planck could succeed in his venture without such close cooperation with experimenters. And he was both a high heretic and the secretary of the physical society to which he belonged. And he did not always accept the verdict of his experimental colleagues, of course, though he was not as stubborn as Newton was, for example. In particular, when a theoretician is his own experimentor, as Oersted or as Faraday was, a much higher latitude was allowed (by himself, and at his own risk).

Oersted, it is true, was somewhat penalized for his outlandish views, but he was also compensated (he failed to be appointed for the chair he felt was due to him, but as a consolation he received a stipend to build a private laboratory). Both Oersted and Faraday habitually blamed their instruments as too insensitive to detect effects which they were sure were present. This, to return to our problem, looks much too arbitrary: no matter what the missing effect is, we may always blame its elusiveness on the insensitivity of our instruments. This, surely, is easy enough! What is there to stop us from making such claims? Why, then, are such claims so seldom made? What was so special about Oersted or Faraday?

Take a modern effect: non-conservation of parity. What Lee and Yang felt was that the law of conservation of parity was rescued with great ease – too great – when a particle whose generation sometimes violated the law was declared to be two different particles (the so called tau-theta paradox). In other words, Lee and Yang did not begin with an arbitrary assumption but with observing an assumption made by others to be too arbitrary. They countered it with another arbitrary hypothesis: parity conserves only approximately. They made a hypothesis about the conditions more favorable and those less favorable for the conservation of parity, and were thus able to recommend an experimental set-up where the non-conservation of parity could be observed. And they were successful.

It is not clear what would have happened if the two particles which Lee and Yang declared identical had shown a difference other than the known one, or if the experiment which they devised had ended in failure rather than in success. It is this kind of situation which, I think, ought to be examined more carefully.

What I wish to suggest, first, is that when a calibration is corroborated, its overruling has to be corroborated as well. Here is the requirement for corroboration based on prior corroboration of a contrary hypothesis in a manner hardly open to correlation with credence or credibility. Rather, the corroborated claim takes priority until its contrary is corroborated. It is simpler, we say, to take the corroborated thesis (as true) then to reject it in favour of its contrary, until the contrary be corroborated. But here simplicity may be synonymous with preferability. To make this rather ordinary and fuzzy presentation more cogent, let me put it thus. If one puts one's view to empirical examination and another performs the examination and comes up with a result not congenial to one, if one can dismiss the uncongenial result by the mere guess that the other's performance was not good enough, then *a priori* the other's performance is useless because one's excuse amounts to saying, there is no need for any empirical examination of my views. That is to say, if one is *a priori* willing to discard as inaccurate any empirical evidence, then it is simpler not to start with the experiment in the first place.

To make simplicity signify more, we can say, in some circumstances the introduction of some contrary hypothesis introduces simplicity from some new direction and thus it may tip the balance. For example, it is simpler to say that we have not observed deviation from the law of conservation of parity because the law holds than because our instruments are insensitive. But it is also simpler to say that two seemingly identical mesons are identical indeed, which entails that the law does not hold. For another example, it was simpler to declare the deviations from Gilbert's theory of terrestrial magnetism to be due to local disturbances than to shift the magnetic poels from Earth's geographic poles. But when evidence culminated the picture changed. After Ampère's hypothesis concerning magnetism as due to currents, Earth's magnetism was viewed as the result of currents due to Earth's charge and rotation. Consequently, the picture changed again and the magnetic poles found themselves immediately replaced in their original place at Earth's geographic poles; and the observed magnetic poles had to be viewed as mere deviations from the true ones. This was a very big *ad hoc* adjustment; again we see how sometimes such adjustments thrust themselves upon us. Now these had to be corroborated. How? One cannot observe deviations of observed poles from the poles and so the theory as it stands thus far cannot possibly be either refuted or corroborated. In order to render it testable one needs further hypotheses! The history of the theory of Earth's magnetism is very complex and not sufficiently studied; at least I am ignorant of it. Let me merely mention on that it includes another case of a revival of an old hypothesis – Hansteen's revival of Halley's hypothesis of two pairs of poles. Finally, it has been decreed that the Earth's magnetism is not due to the diurnal rotation of the charged Earth but due to the rotation of Earth's charged core which is due to, but not coincidental with, Earth's diurnal motion. The terrific corroboration of this hypothesis came with the rise and corroboration of the hypothesis that Earth has altered its magnetic polarity a few times in its history, whereas, of course, nobody assumes that Earth has changed its diurnal rotation significantly ever since it came into orbit. (The corroborations of the changes of Earth's magnetic polarity came from the analysis of the ages of some magnetic rocks. If our theory of radioactivity will change drastically these corroborations may all wither away.)

It is difficult for me to decide how evident it is that all the developments of the kind discussed in the previous paragraph are bootstrap operations, that it is inconceivable that one hypothesis would be corroborated without taking another for granted and vice verca – always taking the most corroborated for granted to upset the other completely and replace it too. But I shall leave it as it stands for the time being.

VI. REVOLUTIONS OCCUR WHEN BOOTSTRAP OPERATIONS FAIL

To further clinch matters, let me draw attention to the uncalibrated part of our instruments. There are assumptions which no one questions and so no one tests. Some calibrations, obviously, depend on the geographic location of the instruments; some only on altitude or on lattitude; some not on location at all. How do we decide such matters? Most clearly, with the aid of our whole theoretical cum empirical background knowledge. Hence, Duhem is quite right; when a result comes unexpected we may always suspect that it is due to the location of the experiment. Do we? Usually it would be deemed just mad.

When we say it seems mad, we usually mean to say, you are unlikely to succeed in pursuing this line of thought. Philosophers of the common-sense persuasion from Locke to Duhem have suggested – implicitly, to be sure – that what is seemingly mad will never succeed. As Bohr has hinted, when the seemingly mad succeeds it looks more ingenious than when the plausible succeeds. The seemingly mad may conflict with deep-seated metaphysical theories which are shared by all or most well-corroborated physical theories. But the seemingly maddest may succeed: the odds are smaller and the stakes are higher: the game is fair.

It was Mario Bunge who, in his *Scientific Research*, recommended to aim always at the smallest upheaval with the hope that it fail and so lead us to a bigger one. I find this a magnificent modification of Popper's proposal: Everyone says that corroboration is a Good Thing and Popper says that a refutation is a Good Thing as it leads to upheavals. Now Bunge says, when you rebel against a minor part of the Received Opinion, offer a minor correction and hope your correction will not do, thus leading to the search for a more fundamental correction. This, also, is a magnificent modification of Duhem's proposal (though Bunge strongly rejects Duhem's instrumentalistics views): try the easiest and smallest modification, but hope to fail at once, not hope to succeed and look for the next small modification. So, whereas Popper says, it is the desideratum of a good hypothesis that it be corroborated before it be refuted, Bunge says, if you have a good hypothesis you may suffer to corroborate it for a while; but if you invent a poor one, hope to refute it at once! I find Bunge's idea better than Popper's. In particular, Bunge claims that all this makes sense because an idea which receives its grounding within science does so not only, or even not chiefly, by empirical means, but chiefly by the way it interlocks with all the rest of the given body of science.

But now, it is clear, when the new modification, or rather correction, goes far enough from the old view, then we may need to go back and find intermediary steps in order to correlate the old and the new views. We will then need approximate theories between the old and the new, and each approximation will have to be corroborated before it can be used. For example, between Newton's and Kepler's theory, we have the theory of interaction, in accord with Newton's law, only between the sun and each planet alone, and we have to juggle even that theory since the sun interacting with Jupiter is the same as the one interacting with Saturn. But we check the result empirically and in corroborating it we calibrate our system of constants. There are many other examples, from the history of the determination of atomic weights, from the history of genetics, and others. There is no doubt that in the history of measuring atomic weights bootstrap operations were made all along the way, and that many re-futations for every atomic measurement were abundant – and correctly ignored – at each stage of development prior to the development of mass spectrography which overruled many auxiliary hypotheses as redundant. No sooner was one mass determined more accurately than the others, and it was used as the new anchor both to increase the precision of the mea-sures of other atoms and to attack new compounds. The story of the determination of chief constants of quantum theory, to wit, h, k, e, m, and c, bears a great similarity. So, I suggest, this is a general feature of the development of physics:

In every bootstrap operation the assumption used need not be accepted as true – they may be good approximations – but each of their empirical applications needs corroboration in order to make it harder to ignore, by those who wish to ignore, the results of the future experiments. We may say, then, that corroboration is what gives a claim its factual status; but not that the claim for factual status cannot be overthrown.

One final and striking instance – not for bootstrap operations but for what counts as factual – is this. When levels of celestial long wave radiation were on occasion observed, depending on geographical locations and times of day, there was no difficulty in assuming that the variations were due to terrestial transmissions of radio messages rather than to cosmic variations. Of course, in some cases these were corroborated when precision was so increased as to enable one to dicipher the messages rather than some cosmic variations. Of course, clearly, there are celestial sources of radiowaves. There was no problem here to decide which is which since every move was highly corroborated – things were fairly much on the surface – and the high degree of actual corroboration is what gives more facticity to a theory; what is more fact-like and what is more theory-like varies according to experimentation.

And so we can also offer a slightly new variant of a very widely held theory – I think due to David Hume (in his *Dialogues*) – of what we consider fact-like. One must take some precaution here, since the word 'fact' may be stretched without severly violating common usage. For example, Laplace drove home his opinion that Newton's mechanics is a verified generalization by calling gravitation's (alleged) absolute obedience to the inverse square of the distance a fact. Clearly, in some sense Newton's gravitational theory, even if true, is not fact-like. In this essay the word 'fact' is used in the narrow sense established by William Gilbert (1600) and Robert Boyle (1662), namely that which is may be reported by an eye-witness. There is no doubt that much which is reported by eye-witness is rejected a priori, e. g. testimonies concerning witchcraft, or a posteriori, e. g. outright lies, and so a criterion is needed. The criterion is this: a testimony is accepted as true eye-witness testimony if it is more likely to have been made due to observation than due to other causes. If we have an exhaustive list of causes for a person's having made a statement, and the probability with which each cause would make him utter the statement, then, by Bayes' theorem, we may find out whether his having observed it is the most probable cause of his assertion. But considerations of this kind exist even if our information about the causes is not complete.

In particular, one must notice, we take here an *a priori* stand about what causes are excluded – e.g. witchcraft – what causes may be operative under what conditions – e.g. lies – and what truthfully may be reported as observed yet be an illusion, or the result of defects or of inexact instruments of observations. Here let me draw attention to one classical difficulty, which has been noted by many writers, including Bacon and Popper: if we use theory to judge the truth of a testimony, then we shall end up dogmatists, by simply having to deny the truth of all testimony which refutes our theories. This led Bacon to the taboo on all theory. It led Boyle, (and many others including Popper) to the dictum: when in conflict, prefer observations to theory.

This rule, Boyle's rule, we saw, is empirically refuted. Also, it is easy to replace Boyle's rule: we can assume the theory to be true, and we can assume the theory to be false, and replace it by diverse alternatives. We can judge the situation from each viewpoint in a different manner, but choose the viewpoint from which we judge on different grounds altogether, for example from the viewpoint of high testability, or of interest. In particular, we may deliberately pursue more than one line of research. This means that neither theory alone nor facts (or alleged facts) alone decide matters, but we balance them against each other in different manners – using different theories and hence viewing the same record differently in each case – and choosing the most congenial balance or balances. This, too, is a bootstrap operation: neither fact nor theory decide our next step but the balance and interrelation between them. This would run contrary to what most authors say. They say, when you use theory to trim facts and facts to slightly modify theory, then you will always succeed. Grünbaum says, this is an exaggeration – and he is right. Yet we do trim both theory and fact to fit each other! Duhem says, hence complete refutation is impossible. Whewell, Popper, and others say, we should not trim the facts! I say, we do so regularly, but feel unhappy about it and find it necessary to recheck both facts and theories, trying to make the trim look either more congenial or more impossible; in the latter case we are heading for a revolution.

Science, it is well-known, is in a constant mess. The reason is, I feel, that while we check the details of both theory and fact we stumble upon more interesting and significant problems, and often leave in the middle our attending to older problems and tasks. To see, in particular, what kind of problem is most likely to be left behind, we have to have a closer look at corroboration, since what is left behind, is the job of fitting corroborated fact-like hypotheses with corroborated theory-like hypotheses which do not fit together comfortably by themselves. The bad fitting is only possible because the way the two are corroborated may be widely different.

VII. CONCLUSION

Thus far I have treated corroboration without going into detail except to say it is a failed refutation. Popper has written a few memoirs on the specific qualities of corroboration. I deem his detailed theory of corroboration unsatisfactory on the ground that in technology standards of corroboration (as expressed in regulations of, and imposed by, various government agencies concerned with public safety) are different in different times and places and much depend on institutionalized scientific theory. In science corroboration plays different roles and so it is even less amenable to general treatment. For example, when corroborating a refuted theory (e.g. the old quantum theory) a researcher may feel he is progressing in the direction of his search for an as yet unrefuted hypothesis, yet by Popper's detailed theory no refuted theory can ever be corroborated. Another difference of function is between corroborating fact-like hypotheses and theory-like ones. In both cases we still see Popper's idea of a corroborable hypothesis being refutable and thus improbable. But the idea of refutability as content, I feel, suffers here: though high content is always high improbability, the converse does not always hold.

A theory-like hypothesis is rich in content in its very generality; a fact-like hypothesis seems very *ad hoc* when judged as theory-like; this is why it is taken seriously only *ad hoc*, only to justify a stated observation report!

An *ad hoc* hypothesis is less contentful, and so, according to Popper, less testable. Yet, its very arbitrariness makes it improbable that it will be asserted, taken seriously, etc. This accords with Hume's criterion: *whatever is improbable that it be asserted without being an observation report, but probable as a result of observation, is taken as an observation report.* (Hume, within the tradition, meant taking a statement to be an observation report to be the same as believing that statement to be true; this, we saw, is an error.) The very *ad hoc* character of the observation report which makes it not improbable enough to qualify as a theoretical hypothesis, is what makes it, when reported, qualify as an observation, or as fact-like hypothesis; without this quality a report made on the witness stand is rejected as eyewitness testimony, though it may be endorsed as expert witness testimony.

Examples abound; let me mention a few striking ones: the alleged fact that only iron has magnetic qualities, that there is a symmetry between negative and positive electricities, that all planets rotate in the same direction. Each of these examples led to far reaching theories, and each of them has been refuted – but not easily and not at once.

Often, to conclude, the *ad hoc* nature of a fact-like statement is rooted in the theoretical background against which it is couched; given a different theoretical background and it fully falls into place, as the expression goes. If an observation report is at once a corollary of our scientific theory, then it is unproblematic. If it conflicts with our scientific theory, either we reject the theory or try to find an excuse for not rejecting it. When, however, a small theory which well integrates in our theoretical background is attacked by a well corroborated fact-like theory and all its defences are refuted, then a revolution may be under way. Such events may be rare, but they are the more intersting ones. At times we alter our whole theoretical outlook around a rather fact-like theory which then gets refuted. *We then look silly from any viewpoint except that which takes the process to be a bootstrap operation!*

APPENDIX: PRECISION IN THEORY AND IN MEASUREMENT

An intuitive idea concerning degrees of precision is widely accepted, and it is that we *in*crease precision of theories by paying attention to ever *de*creasing orders of magnitude of measurements which we incorporate in these theories. We increase precision of measuring or of predicting measurement of length, for instance, if we pay attention not only to centimeters but also to millimeters, microns, ångströms, and so on. And our theories are precise to centimeters, then to millimeters, and so on respectively. The idea is that increased precision is the process of capturing a point within nested intervals, and that this is reflected both in experimental and in theoretical progress. An intuitive analogy may be drawn with series of photos made by a camera which approaches the moon's surface. A corollary from this intuitive idea which has been stated by a number of writers is that observation reports and theories are not over-thrown even when superseded by more accurate observations because within the limits of their own precision-range they are still valid. When a law in physics is stated, it is stated usually without such qualifications; but this is a mere matter of convenience, we are told. The idea is that a tacit proviso is understood and the law should be understood to be not the statement "such and such is the case" but the statement "within this and that limit of accuracy such and such is the case"; in such a formula-tion, the law, once verified, remains unshaken regardless of further pro-gress in the field.

That this idea is popular amongst physicists in this or that variant is an empirical fact. It is also a fact that many physicists, e.g. R. Schlegel, in *Completeness in Science*, consider Heisenberg's principle to be a law limiting the degree of accuracy which will ever be achieved; consequently, some part of physics – containing both theory and observation-reports – is not only here to stay but also never to be superseded. It sounds strange that after the Einsteinian revolution and all that, scientists may view any part of science as final, but I think I may claim here to be making an observation of a fact, strange as the fact may be.

The most obvious criticism – valid or not, this remains to be seen – of the view of increased precision of theories as decreased order of magnitude of imprecision of measurements (observed or predicted) is this: Evey result of measurement can be amplified. The obviousness of the previous statement makes one puzzle whether it is relevant to our present discus-sion, and, on the assumption that it is, how could anyone overlook it. The puzzle will disappear once it is noted that the controversy is not concerning the amplifiability of small effects but concerning its relevance.

Amplifiability was used by Schrödinger to argue that quantum theory cannot be a complete description of nature because a solution to a quan-tum equation is often a weighted sum of possibilities of an outcome of a quantum process and the outcome may be amplified to kill a cat so that the solution would be amplified to a weighted sum of the possibility that in the end the cat will be dead and the possibility that it will remain alive, but only one of these two will be realized. The point of Schrödinger's argument is that the statistical interpretation of Born may render the weighted possibilities into relative frequency in the small but not when one member of a random sequence is isolated and amplified. Whether Schroedinger's criticism is valid or not is hard to say; the answer to it is that, upon repeating Schroedinger's thought-experiment, we get a relative frequency of dead cats but no prediction concerning any individual member of the sequence, just as in life-insurance. (We assume that probabilities are weighted possibilities.)

This point was taken up by Reichenbach in his *Philosophical Foundations of Quantum Mechanics* to dismiss the idea that possibly the world is indeterministic in the small but deterministic in the large. Doubtless this criticism is valid and, indeed, no one has ever seriously advocated the thesis Reichenbach criticizes. The question whether Schrödinger's criticism is valid, then, depends very much on whether indeterminism may be viewed as a complete view of nature in the sense that only a part of the future is predetermined by the laws of nature and the function of science is to be able to predict in principle the whole of that part and nothing else. This idea is that the nested intervals converge not to a point but to an elementary interval which is the objective limit of precision in physics. What exactly is this interval is hard, if not impossible, to say, as Einstein, Podolsky, and Rosen have ventured to argue.

So much for the quantum aspect of the ideas of increased precision of theories as decreased order of magnitude of imprecision of measurement and the criticism of it from amplification. The criticism, however, is much more general, and can be discussed independently of quantum mechanics, since, obviously, observed magnitude of any order can be amplified.

That this is so is obvious from the fact that we do increase the order of precision of observation without altering ourselves at all as instruments of measurement. All increased precision of observation beyond a rapidly achievable limit is achieved by amplification. The amplification which was historically important was optical magnification, whether of the optical instruments in question or of the scale on which they were mounted, etc. But the limit of magnification is that of resolution power, and resolution power is a wider concept of a degree of precision available than magnification; for example it is an essential factor in spectroscopy which can hardly be called optical magnification. A little reflection will show that any increased precision claimed in any field of micro-measurement includes a claim concerning resolution-power and amplification of separated parts within the limits of the claimed resolution-power. It is a known fact that all micro-observations are beyond the reach of the untrained whatever his eyesight, etc. This shows how very theoretically loaded such observations are, and the kind of theory involved concerns the objects whose 'images' are separated and separately amplified. Without theoretical instruction we cannot identify the reading of an instrument as an amplified micro-effect.

There are amplifications that have nothing to do with optics. A crucial experiment between Galileo's theory of gravity ($a = g = $ const) and Newton's ($a = $ const $1/r^2$) within eighteenth century conditions, is possible by amplifying any phenomenon of gravitation by sheer repetition; a pendulum clock adjusted on one altitude will be unadjusted *according to Newton but not according to Galileo* once it is shifted to a different altitude or geographical latitude. And however small the maladjustment might be, it would be detectible over a sufficiently long period (if the instruments' resolution-power permits) – as indeed it was (and to Newton's advantage of course).

Thus, the amplification does occur under conditions specified by a theory. Theory makes us identify (a) a micro-effect amplified by unusual instruments (pendulum clocks) with (b) macro-effects allegedly magnified by unusual natural (lunar orbit) conditions. The suggestion, therefore, that within a given limit of accuracy a theory still holds even when superseded is false, since the qualification concerns not so much the limits of accuracy but their dependence on certain conditions which may vary naturally or artificially. Classical statistical mechanics holds within limits of accuracy which depend on temperatures and which are thus very wide almost everywhere – the notable exception being those parts of the earth's crust which are occupied by laboratories. Newtonian mechanics holds in the same way where the limits of accuracy depend on relative speed and strength of gravitational fields.

Even when the effect in question is naturally small in our environment, we may fill our environment with amplifiers and thus render the old theory into a less and less useful instrument of measurement and prediction in daily life; or change our environment by travelling to parts where the conditions for the imprecision of the once accepted theory become increasingly common.

Professor Phillip Morrison has commented on the above saying that precision may apply to totally different kinds of widening of our field of observation, such as rendering ultraviolet light visible. Not only is there the idea of precision in the mode of rendering ultraviolet light visible but also the degree of precision of a theory may be reassessed after such conversion. It is no doubt the case that the theory of increased precision via nested intervals does not easily or fully apply there. My point previously was merely that where the theory does apply it applies only under strict conditions to be described by a superseding theory and he has extended it to cases where the theory should apply but fails to.

The current ideas on precision, then, have to be replaced; precision is only deceptively obvious, and to handle it more adequately we have to relativize it and make it more dependent on specific conditions.

Popper has claimed that degrees of precision are monotonic functions of degrees of falsifiability. He has also relativized his idea of degrees of falsifiability, relative to given fields of potential falsifiers. It follows that he has relativized degrees of precision. I do not know, however, how satisfactory this implicit relativization is, and I doubt that it is entirely so; as I shall argue in Chapter IX below, both his theory of degrees of falsifiability and his view of degrees of precision as monotonic functions of degrees of falsifiability may be questioned. In particular, since the degree of precision of a given theory can be adequately described only with the aid of a superseding theory, the degree of precision of the superseding theory itself can be said to be higher only when a crucial experiment between the two can be designed, either by devising new means of separation and amplification or by constructing or discovering those conditions under which, the new theory tells us, the old theory becomes highly imprecise.

The above observation may be obscured by the fact that observations are often stated within limits of accuracy in reports which refer to current theory but not to preceding theory. The limits of accuracy in such cases are sometimes the limits of the accuracy available to the observer. These facts indeed sometimes run counter to the above observation. The question is whether highly precise observations of this kind are of any value. The answer to this question, I think, cannot be given *a priori* on philosophical grounds. No sets of observation absolutely fit the theory at hand, and when the fit is considered *close enough and when not* is relative. In particular, a new theory with a better fit may be triumphant for a

while and then people may worry about the fit not being perfect and look
for a still better theory, etc. So the same data may support the same theory
in the old situation and condemn it in the new. There exist historical
instances to this case, such as from classical chemistry, from modern
spectroscopy, and from general relativity.

Thus, when an observer highly increases the degree of accuracy of
observation while referring only to one current theory, his observations
may be useless if the fit is judged good enough, or useful if he calls the fit
into question – quite in accord with the above observation of the relativity
of degrees of accuracy. For instance, when the atomic weight of oxygen
was deemed close enough to 16 there was neither any point in increasing
the accuracy of the measurement of its deviation from 16 to further
decimals nor even any point in isolating its isotopes to find the more
precise atomic weight of the predominant isotope. Things changed drastically, of course, when 16 was not good enough any more because
nuclear physics should yield the exact deviation from this weight (when
the mass of a proton is taken to be 1).

The above example may indicate that precision is not a value in itself
but a tool for furthering the ends of science from day to day and, more
generally, it may be useful in constructing interesting explanations and
in testing them. The sentiment expressed here is broadly the same as that
expressed by Popper, but the details may differ. It seems that Popper
equates explanation and testability, if not in his *Logik der Forschung* then
at least in an appendix to the *Logic of Scientific Discovery* where he
presents explanation (E) as monotonously dependent on confirmation
(C) which, again, is monotonously dependent on testability. Were testability, explanation, and precision all monotonously interdependent,
precision would be much simpler.

CHAPTER 8

TOWARDS A THEORY OF *AD HOC* HYPOTHESES

That *ad hoc* hypotheses are both repugnant and useful is a known fact. The joke of the biologist about teleology, that like a mistress, one wants to have it but not be seen with it, is more characteristic of *ad hoc* hypotheses. And, indeed, for mechanists, every teleological hypothesis is terribly *ad hoc*; Spinoza called teleology the shelter of ignorance because it was *ad hoc*. When Newton said his gravity was not occult he argued from the fact that it was not *ad hoc* but a powerful explanation. Copernicus was indignant about Ptolemy's epicycles but had some himself, of course. The reduction of purpose to cause is *ad hoc* all too often. And Newton's optics was *ad hoc*, as William Whewell argued at great length. Now Copernicus' complaint is not in itself unreasonable: Ptolemy's epicycles were old hands and gained the legitimacy of regular customers, whereas his own were stop-gaps. This idea can be generalized.

It is clear that one epicycle in Copernicus' system had a special privileged position, namely that of the moon – as Sir Francis Bacon observed, not knowing whether this was quite kosher. Now, clearly, whether it is so, and more specifically, if yes why, is easier to feel than to express.

Let us assume that what is rational is what has a general rationale rather than has no better cause than a whim. Now the *ad hoc* is neither due to a whim nor due to a general rationale. And so it can be either irrational, an excuse to stick to a rationale which turns out to be less general than originally assumed, or non-rational, either in being an exception or in being a *sui generis* or anything in between, e.g. a case of an unstudied territory.

The general rationale is at times a scientific theory, at times (and more often) a metaphysics. Let us take them separately beginning with scientific theory.

I. 'AD HOC' HYPOTHESES WHICH BECOME FACTUAL EVIDENCE

An *ad hoc* amendment to a refuted hypothesis may be a qualifying clause excluding the refutation with no further ado. This is just too high-handed and so utterly arbitrary and so inexcusable. It may be an auxiliary hypothesis explaining the irrelevance of the case. Consider the hypothesis, 'all birds have feathers' and its alleged refutation, 'bats have no feathers'. The exclusion of the bats may be the auxiliary hypothesis that bats are not birds. This may be construed as the auxiliary hypothesis that 'not all birds have feathers but all flying vertebrates are birds', is refuted by flying bats. Still, it looks rather arbitrary to pin the fault on the latter hypothesis. It really looks as if now the dual theory, 'all flying vertebrates are birds and have feathers', degenerates into a verbal device; all feathered vertebrates, whether flying or not, we name birds. This is evidently highly unsatisfactory since it explains nothing and introduces a verbal convention, i.e. an arbitrary element, for no other purpose than to show stubborness.

Now suppose we say all and only birds have feathers, all birds lay eggs, have a beak, two legs, and two wings each, etc. Then, first and foremost, birdness or the name bird can drop out, and the hypothesis now says all feathered animals lay eggs and have a beak, two legs and two wings each. Flight now has dropped out of the picture on account of there being flying non-birds. The hypothesis looks decent enough and it need not refer to flight, after all. Yet the bare omission of flight, the very move made, is suspect of arbitrariness. Felicitous as our hypothesis is, we may still feel it would have been more felicitous had it covered flight as well.

Now this uneasy feeling will be easily allayed, as the reader knew all along, the moment bats are discovered to be mammals proper and on the hypothesis that no mammal has a beak and wings or lays eggs. Also, of course, the existence of chickens and ostriches which do not fly make flight more problematic and so its exclusion *pro tem* more understandable i.e. *ad hoc* but not so arbitrary as to be helpless.

The picture can get complicated again. The discovery of the duck-billed platypus who has a beak and lays eggs yet is a mammal may raise afresh the problem, is the bat not a bird. Fortunately, the theory behind taxonomy makes it quite satisfactory to omit flight altogether and center on philogenetic characteristics which make the bat definitely a member of a different family altogether.

We can go further. Adolf Grünbaum has contributed to the debate on *ad hoc* hypotheses a very important observation which runs contrary to almost all that philosophers said about the topic, from Copernicus and Bacon through Poincaré and Duhem to Popper: there are instances in history showing that it is not always easy to generate *ad hoc* hypotheses. Also, as we saw, the *ad hoc* hypothesis needs some empirical backing to make it preferable to its possible competitors, and this may not be easy to come by. Let us take an example. Combustion, according to phlogistonism, always reduces the weight of the fuel: the ashes are lighter than the original fuel. This is not so in the case of metals. An *ad hoc* explanation came in the form of Archimedes' hypothesis, and was repeated even by a writer of standing like Peter Shaw although it is immediately clear that the order of magnitude of the Archimedes effect in the air is much too small and so cannot effect the case one way or another. Macquer had the hypothesis that during combustion of metals a secondary process takes place which adds more weight than the weight lost through the primary process. But he could not identify it, let alone empirically back it.

These examples prove Grünbaum's case quite conclusively, of course, except that historians of science may well exclude these examples *ad hoc*. Yet, as long as a theory, in our example phlosistonism, is engaging enough, we may put aside the troublesome cases in the hope that they will be covered one day by an *ad hoc* hypothesis that will be empirically vindicated. After all, we can imagine a universe where Macquer's hypothesis comes true and is empirically supported.

Examples may help push this point further. There are examples like Macquer's, and even more extravagant ones which proved correct. Consider the known fact that metals cannot be electrified. William Gilbert's identification of electricity and magnetism with matter and form respectively is clearly refuted by it, yet he seemed scarcely to be bothered by his own division of things to electric and non-electric, as well as to magnetic and non-magnetic. When, over a century later, Stephan Gray declared metals electric all the same yet whose electricity escapes in a secondary process, no one declared his hypothesis *ad hoc*, even though it is evidently more extravagant than Macquer's, simply because he gave it the full empirical backing by electrifying metals while stopping the secondary process by preventing the escape of the electricity, by the use of what was previously known as electric matter and which he presented as insulators.

The difficulty, then, which Grünbaum indicates, is rooted in the fact that a good *ad hoc* hypothesis, though *ad hoc*, is at times empirically testable with relative ease, leading to its establishment or refutation. When established it gains a fact-like status.

Another example. To save Kirchhoff's spectroscopy from refutation a few elements were postulated and their presence in the sun's atmosphere was blamed for puzzling spectral lines. Now one of these, helium, was later established on earth, all the others were later declared chimerical. Kirchhoff's law was in part rescued by other means, in part qualified and modified in a quite *ad hoc* manner and remained so until the advent of the new quantum theory, at least.

It would be quite *ad hoc*, I confess, to justify those who use a given method successfully and to condemn those who do exactly the same unsuccessfully.

The postulations of the existence of certain elementary particles, particularly the magneton and the neutrino, have the same characteristics. Indeed, the neutrino was more *ad hoc* than the magneton, since the neutrino came to save a theory from refutation, whereas the other came to clinch a theory. Yet the one was found and elevated from the status of a very *ad hoc* hypothesis to the status of an attested fact, whereas the other remains still somewhat *ad hoc*. And, indeed, Dirac's theory which encompasses it is somewhat suspect; but other theories with doubtful *ad hoc* amendments exist all around, and the *ad hoc* amendments are well tolerated. The most powerful example is the inapplicability of Mendelism to human skin colors which is rescued by the mere say so of geneticists that over twenty genetic factors go into the making of human skin color. We still wait to see this say so put into detailed specifications; they may then easily be established or refuted.

To conclude, I observe that we differentiate *ad hoc* between *ad hoc* amendments. We have those which we approve of, either because they later earned the status of facts or because they rescue theories we favor. And we condemn others. We forget that the method, purpose, and initial legitimacy are the same, and the differentiation is only *post hoc* – posterior to further investigation. The procedure is rational in both cases, or else posterior investigations are not necessary. Those, then, who condemn *ad hoc* amendments as *a prioristic* because they come – undoubtedly – to rescue theories or hypotheses from empirical refutations, those same people, then, act arbitrarily or are defenders of a much deeper (methodological) *a priorism*.

Grünbaum's profound observation, then, has an interesting corollary. A scientist who wants an *ad hoc* hypothesis but cannot create one on the spur of the moment can leave the task *pro tem* unattended and pretend or hope that one is coming fairly soon that will be empirically well substantiated. This is a high risk a scientist takes, of course. The question is, is the risk reasonable, and how does it grow in time?

II. THE CONVENTIONAL ELEMENT IN SCIENCE

This brings me to the following observation about the tentativity of all scientific activity: almost every scientific hypothesis is backed by promises of *ad hoc* hypotheses to rescue it from existing refutations. Let me first explain what this amounts to.

We all know that scientific hypotheses are tentative in the sense that they can be overthrown by empirical evidence any day now. There are a few writers who do oppose this thesis, among them Pierre Duhem, Ernst Cassirer, Michael Polanyi, Thomas S. Kuhn, and Imre Lakatos. The thesis is usually known as Popper's, but it really is due to Boyle, Faraday, and Einstein; we may label it as Boyle's (skeptical) thesis. Now even Duhem *et al.* agree that facts clash with theory; their point is only that usually theory is rescued *ad hoc* and so it is naive to believe that what overthrows a theory is its clash with the facts rather than our demand that under certain conditions the clash should overthrow the theory. The question is, under which conditions?

Here I must draw attention to Popper's specific, and superior, version of Boyle's rule. When Boyle postulated the rule he considered it as quite unproblematic on the grounds that fact is certain and theory is not, so that if we are certain of the truth of a factual report which contradicts a theory, then, *eo ipso*, we are certain of the falsity of the theory; *ergo*, we have already rejected it. Now Popper denies the certitude of factual reports and even enjoins us to try and refute them. So Boyle's argument will not hold for him. So he needs an added factor, an agreement to reject a refuted theory, and a reason for the agreement: the desire to avoid dogmatism. The convention Popper proposes is to endorse every observation report which has passed tests of agreed severity as true and thereby to have rejected any theory which conflicts with it – until the time when the report is ousted by another report.

Much ink has been spilled over the question, is not Popper a conventionalist because of this? Surely in a sense he is, and trivially so. But conventionalists proper are not all those who propose any old convention, but those who propose to stick to theories as conventions and rescue them *ad hoc* against refuting facts. Here, clearly, Popper is anti-conventionalist. But why endorse a convention to defend reports rather than theories? The answer can be found in the previous section: the claim that an *ad hoc* hypothesis has been turned an observation releases it of its *ad hoc* nature to a great extent: we prefer *ad hoc*ness on the factual rather than the theoretical side: the *hic* of the *ad hoc* is the observed fact, whether the one causing trouble and leading to modification or not.

This discussion seems subtle, but I think it can solve an irritating problem, and one which Klappholz and I addressed (*Economica*, 1959). The problem we were studying was from economic theory. It is well known that the profit-maximization hypothesis was tested in diverse ways, the most obvious of which is the direct questioning of businessmen: do you maximize profits? Now, some economists take this to be a reasonable and proper method, and, as the answer is not universally in the affirmative they consider the theory to be refuted by simple observations. Others may want the theory untestable and untested. Yet the majority, we think, want tests, though not of this type. Why?

Milton Friedman has said that since the profit motive is as abstract as Galileo's vacuum, to refute it, as to refute Galileo's vacuum, is besides the point. In other words, we know from the start that the profit motive is not exclusive and not always overriding and we ignore this fact as we ignore friction and the like. In other words, before a test is conducted we may be able to say – more or less (we may change our minds, etc.) – whether its outcome would constitute a refutation or not. And of course, the more we are willing to accept refutations (*a priori*) the higher the content of our theory, e.g. if we do not ignore friction (air friction or other) but add a viscosity factor to our theory.

So far the view advanced by Klappholz and myself is in line with Popper's idea of a convention of accepting Boyle's rule, but it suggests a greater specificity: rather than endorse Boyle's rule *a priori* and generally, I can address my opponent in a debate with the following request: do tell me, if I conduct a certain investigation, is there a chance that you will change your opinion? The very meaning of your opinion depends on that answer.

This is no contradiction to Popper's view, because when you refuse to accept a refutation we may construe your view to mean that it is not contradicted by certain reports. In order to contradict Popper we need examples where clearly a theory does mean to exclude certain reports, these are made, yet the theorizer is quite recalcitrant. Is there such a case? Does such a case conform to the view of Duhem *et al.*?

Yes, such cases do exist. No, they do not conform to this view. The classical example is Dirac's case of refusing to admit the truth of reports which conflicted with his theory. The reports, however, were checked and refuted. This is easy to analyze. Dirac's rejection of the reports was on the mere ground of their conflict with his theory. Everyone would say that this is quite arbitrary and high-handed. Perhaps also dogmatic. But let us look closer: had the reports persisted and Dirac persevered he would thereby perhaps qualify as a dogmatist (perhaps not, depending on what he would have to say for himself). But as the reports swayed his way we can say that his arbitrary and high-handed position was vindicated by the facts. The identification of arbitrariness in some sense with *ad hoc* will, again, lead us to say that tentative misconduct is not only condoned but at times rewarded. This point, as a methodological point, was first made by Sigmund Freud.

This, however, has nothing to do with the view of Duhem *et al.* On the very contrary: the view of Duhem *et al.* presents no problem here, whereas we are cognisant of a problem: was Dirac right? And solve it by saying, yes because he was doing it only for a brief while.

The point which I use *contra* Popper, and which I am ascribing to Freud, was generalized by Popper himself: whenever we propose a theory to explain the given facts, it usually leads us to think that the facts are slightly different. For example, Newton tells us not of Keplerian ellipses but of perturbed ones, not of Galilean constant gravity but of almost such. Hence, at once the theory is refuted. Yet we give the theory a chance by rechecking the facts.

The question then rises forcefully, how much tolerance should there be? How long should we allow a flagrant and unsettled contradiction between theory and fact?

III. REDUCING THE CONVENTIONS

To this question Paul Feyerabend has a marvelous response: whatever answer you give this question will itself be a theory – a theory of science, that is – which is itself subject to the same questioning. This looks like infinite regress, but Feyerabend has a surprise in store: rather than a meta-theory leading to infinite regress, we can have a competing theory that does not: when a theory is defended *ad hoc*, then rather than examine its degree of *ad hoc* defence, we compare it with a competitor and try to prefer the one which is less *ad hoc*. For a classical example the secular behavior of Mercury was in bad shape from a Newtonian viewpoint and generations of mathematicians and astronomers were trying to see what can be done about it. They chiselled out the difficulties by calculations, and they offered *ad hoc* hypotheses to explain them away – the body alpha being the best known of these. Yet the *ad hoc* hypotheses were rejected with ease and nothing remained to be done but to keep chiselling the differences bit by bit by honest calculations of detailed existing perturbations. Then Einstein came to the scene, the difference was almost covered, yet he said the small remainder cannot be handled by a Newtonian theory, only by general relativity! (See above, Chapter 2, Appendix, § 8.)

This story is also not in accord with Duhem's and his followers' tolerance for *ad hoc* hypotheses, though it looks so. It is not that *ad hoc* hypotheses were tolerated till Einstein no longer required their services. Rather, they were not tolerated, and the contradiction was tolerated until resolved or until Einstein said it can't be resolved and so we must oust Newton.

And who is Einstein to tell us this? It is not his authority but that of the crucial experiments between the two. For clearly, now Newton is so much at a disadvantage no one hopes to remedy all the contradictions his theory holds with factual reports: the hope of reconciling them is gone.

Do we, nonetheless, have instances of *ad hoc* amendments to resolve contradictions? Indeed we have myriads of them. The writings of both Poincaré and Duhem are full of such historical examples. The example of Macquer's phlogistonist hypothesis, incidentally, is not good enough, because that hypothesis was barely testable at the time. But the classical example in recent discussions, especially of Grünbaum and Popper, is very good. The Lorentz-Fitzgerald contraction does take place and does rescue current theories and is, indeed, taken over by later theories, though no longer *ad hoc*. Apply this to Macquer and you could see that he could confirm his hypothesis and say, indeed when metal rusts and thus gives out phlogiston it also absorbs air – as indeed we know it does, since rust is metalic oxide! Yet with Macquer, even when confirmed it would remain *ad hoc*, but in Lavoisier's theory it was nothing of the sort, as absorbing oxygen was identical with combustion, calcification, etc.

What this will indicate is that rather than resolve the issue we would like to prevent it from rising, or rather push it away. Contrary to the traditional condoning of *ad hoc* hypotheses, and in line with Popper's and Grünbaum's approaches, we see, once an *ad hoc* hypothesis is introduced we are unhappy about it and try to eliminate it. Moreover, we make it as hard to introduce them as possible. Rather, we say, we note the need for an *ad hoc* hypothesis and tolerate it *pro tem* – until the debt either becomes too big and bankrupcy is declared, or else the *ad hoc* hypotheses come with the factual evidence that supports them.

But this is the way we like things to be, not the way things always are. I do not mean that whenever an *ad hoc* hypothesis is needed, instead of keeping the token someone makes an *ad hoc* hypothesis. This is quite in order since some people ignore it and others test it, and thus either offer an empirical backing to it to render it a fact or refute it. Rather, we sometimes take seriously an *ad hoc* hypothesis which we cannot refute or back by experiment.

The classical example, we remember, is the neutrino. Without it, beta decay refutes the strict conservation of evergy. Yet it was not tested until the days of non-parity, over a generation after Pauli has proposed it. Why?

The answer turns out – not surprisingly if we remember our reluctance to concede *ad hoc* hypotheses – quite involved. The Bohr-Kramers-Slater hypothesis of statistical conservation of energy was refuted at that time. This is one classic example of a strict Popperian refutation – one theory, one experiment, clash, and the rejection of a theory. This made it hard to accept beta-decay as a refutation of strict conservation: as energy conserves, in fact, its conservation must be strict or statistical. Of course, there is a way out: conservation can be at times strict, at times not. But why? Under what conditions is it strict? No one knew. (This supports Feyerabend's idea that the decision to permit this or that degree of *ad hoc* is theory-laden, and undermines his theory as we have here a decision to exclude an *ad hoc* hypothesis based on pure ignorance.) And, assuming strict conservation, the existence of the neutrino was least *ad hoc*: it fitted most of the existing framework. The discovery of the neutrino, incidentally, modified our view of it almost entirely: it is still chargeless, of course, as a matter of known fact, and it is still a particle as a matter of general principle; all else has changed!

Let me recapitulate. We saw, with Grünbaum, that *ad hoc* hypotheses are hard to invent and need empirical testing and confirmation. We saw that we can agree that an *ad hoc* hypothesis is needed either while we provide none or while we provide an outline of it, until we find a good one which gets empirical backing or until the hope for finding one is given up. The question, how much patience we have before giving up is at times decided by a new theory, at times by circumstances. Yet the circumstances may produce a stalemate, especially when no one cares about the need for improvement. This, to return to my initial point, was Copernicus' complaint: not that Ptolemy used *ad hoc* hypotheses, but that the failed to rectify them after so long a period. Was Copernicus in the right? Perhaps not; I think his real complaint lay elsewhere, namely in his heliocentrism; and the real question is, was his complaint launched from a heliocentric viewpoint or was it one logically prior to heliocentrism and one of which the latter relieves us?

IV. METAPHYSICS AND 'AD HOC' HYPOTHESES

In my view judging Ptolemy from a geocentric viewpoint and judging him from a heliocentric viewpoint we obviously come to quite different conclusions. And, no doubt, Copernicus' harshness to Ptolemy is in part at least rooted in the latter's heliocentrism. Now this is quite alright, except that it is question-begging in the sense that a geocentrist need not be overly impressed with it. That is to say, everyone was always unhappy about the epicycles and/or eccentricities. The question was, what else could be done? And the geocentrists, we know, did all they thought they could within geocentrism. And so within their framework they could not find themselves half as culpable as Copernicus wanted to present them; and why should they have endorsed his heliocentrism in any kind of self-incrimination?

The answer is, to my surprise, at least, seemingly Feyerabendian: it was from the viewpoint of the new theory that the trouble they were working on turned out to be an unsurmountable scandal. Hence there is no difference between the Ptolemaic and the Newtonian efforts to array theory and fact so as to eliminate all contradiction and all *ad hoc* theory.

I do not wish to ascribe this Feyerabendian view to Feyerabend, both because he never expressed it and because it is not satisfactory. For, whereas the Newtonians were working in part on *ad hoc* hypotheses, in part on *non-ad hoc* elaborate calculations, the Ptolemains were mixing the *ad hoc* and the elaborate calculations. The initial theory, the geo-centric theory, was thus doubly defended against facts, both by *ad hoc* hypotheses and by promises of better calculations; there was no chance of ever collecting such a doubly protected promise, and so Copernicus had the right to lose his patience.

More congenially, we can say, then, what Copernicus offered is not so much a complaint from the heliocentric viewpoint, as a complaint from a neutral viewpoint which may make it more desirable even for a geo-centrist to look for an alternative, and then, by accident, the man who threw stones at your glass windows just happens to stand round the corner selling glass. You may resent him and refuse his services; this will not, however, restore your glass.

With this I can now try and reexamine Feyerabend's story of the secular motions of Mercury. Contrary to Feyerabend I will say this. Ad-mittedly, astronomers and mathematicians existed who tried to prove consistency between Newton and Mercury with no *ad hoc* hypotheses. Admittedly they went a long way towards achieving that goal and ad-mittedly they had all reasonable hope to reduce the difference between observed and calculated paths to within the limits of observation error. But once Einstein came in, then without admitting the truth of his theory one could see the hope to be very questionable! Not just because his theory agreed with facts within the limits of observation without further ado and thus was superior. No. But the logical fact that certain differences conform to one mathematical regularity made it desperately hopeless that it should also conform to another, so different.

The similarity between Newton and Einstein, between any old and new theory, is nowadays quite commonplace: the old is, of course, an approximation and a special case of the new. But the difference between old and facts looks most different from the old viewpoint than from the new! This is what makes us abandon the old.

Am I giving the same facts as Feyerabend in a different wrapping or am I giving a new theory? I do not know but let me show the important difference between our views if I can.

In Feyerabend's case the secular motion of Mercury was problematic for Newton but not for Einstein. In Copernicus' case epicycles were problematic for both the geocentrists and the heliocentrists. In both cases, I feel, the very introduction of a new theory altered the situation from a neutral viewpoint, whereas according to Feyerabend the new Einsteinian theory offered a new viewpoint from which to judge the old situation. To clinch matters, let us try a case where the problem for the old view radically differs from the problem for the new view.

My choice is Ampère's hypothesis of minute currents which gives matter its magnetic properties. The old theory had the problem, why only the few magnetic materials showed magnetism? When diamagnetism was discovered, the question became more pressing. The problem got most pressing when Faraday discovered that the opposite of diamagnetism is paramagnetism, to be distinguished from the very strong ordinary magnetism which is known as ferro-magnetism. If molecular currents are universal, why should there be such a division? With the Rutherford-Bohr theory of the electron in the atom's shell, it became clear that all matter should be of the same kind of magnetism. The Bohr magneton theory then, indeed, made all matter diamagnetic. Ampère's hypothesis designed to explain ferro-magnetism now turned out to explain diamagnetism alone. As to paramagnetism, it remained to be later explained in part by the spin of the electron.

Nobody, I think, protested. The new viewpoint of quantum theory reversed the direction of Ampère's currents, shifted the problematic from diamagnetism to paramagnetism, and everyone accepted the switch as a matter of course. The reason, of course, is that the case is highly localized and is treated within the context as we find the context with no intention of turning the context to fit our local problem.

Is this fair? Is this proper? First let me say it accords very well with the Poincaré-Duhem approach: the general situation prescribes our attitude much more than a mere single hypothesis, much less a single fact. And so, here, perhaps, we can look for the strongest support for the view of

Duhem *et al.* that therefore a fact cannot by itself overthrow a hypothesis.

Let me also stress, so as to avoid confusion, that Duhem would certainly not endorse the Cassirer-Kuhn-Lakatos view, since he both declared facts as imposing themselves on us, and logic to be always similarly or more obligatory. What Duhem said was, once theory conflicts with facts we must modify some of our ideas or others, but it matters little which, and the smaller the modification usually the better. Duhem's followers, however, sometimes say this: since we can make this that or the other modification, the very acceptance of a fact is not the same as the overthrow of the theory it conflicts with. At least Cassirer said so outright and perhaps also Lakatos.

Duhem would not agree. Conflict he saw as logical, and thus as immutable. But the theory conflicting with the fact he saw as a huge apparatus modifiable in innumerably many ways.

Can we see here, then, a conflict which nobody bothered about? Are there such conflicts? Polanyi offered such examples, but his examples are ones which men of science never admitted, such as Miller's 1926 repetition of Michelson's experiment with different results (which are deemed erroneous by the world).

But there is little difficulty in illustrating many facts which clashed with theory and yet were left to rest for a while: the theories were not rejected, the facts were not rejected, there were not even good *ad hoc* hypotheses to reconcile them, and fields were declared to be "in a mess". Is this what Cassirer, or perhaps Kuhn or Lakatos, could have in mind? I do not think so. On the contrary, Cassirer says, in the name of simplicity we usually, though not always, do oust a refuted hypothesis. And Kuhn says, after the *ad hoc* modification becomes too much, we make a revolution: he is not in favor of too much mess either. What Lakatos would have said I do not know. He died too young.

I do not know then what is meant by the claim that we do not always reject a theory on the basis of conflicting evidence. I observe that this view is now gaining much popularity in the field of the philosophy of science. I do feel that the field is now in a mess.

V. WHAT IS A MESS?

To conclude, let me say this. If we know what the mess is, as Feyerabend indicates, then we are already on the way to clean it up (or, as he says, we already have). A question well put, to put it otherwise, is half a solution. The mess, then, is a poorly articulated question, perhaps even a fully unarticulated one.

Since this is put in a suspicious form of a proverb, or of folkwisdom, let me adapt it, quite *ad hoc* (in a very different sense of *ad hoc*) to the accepted framework (quite metaphysical, though in the guise of methodological) of the accepted theory of explanation.

It was William Whewell who first described scientific systems as hierarchies of explanation. Yet he stressed that the explanations are far from being stratified. (To the extent that systems were known before, methodologists, expecially Sir Francis Bacon, took it for granted that they were stratified.) Today's explicans, said Whewell, is tomorrow's explicandum, but together with newly discovered facts. Hence, the explicandum is part theory part fact. Examples abound: the new magnetic theory of elementary magnetic dipoles (Poisson) explains the old theories (Gilbert to Coulomb) plus facts such as the facts of magnetization by random processes.

There is a problem here: how do we choose a new fact to go with the new theory to be explicanda for a still newer theory? Whewell does not say. He takes it for granted that intuitively we see a connexion and want to explain it. But then, how come we always have a fact handy to go with a given theory in such a fashion?

Things look much better from a Popperian viewpoint. Popper starts – he is in error to insist on this starting point, but this is another matter – from any given testable theory, and says we should try to refute it. He says, if we cannot refute it, we feel stale and then try to explain it by a better theory, i.e. by a more testable theory. If we do refute it we want to explain it plus its refutation by a new theory.

Here we have a hypothesis about the history of science: we try to find the most universal theories as well as to refute them. Hence, we try to put as many theories together as explicanda, plus their refutations. We have no theory, in Popper's view, as to which pairs of given theories are best suited as candidates of explicanda for the new theory of the near future. For this, of course, we need a metaphysics. Also, then, Popper does not say which theory and which facts may be paired together. But clearly, he has an obvious feeling that a refuted theory is a good explicandum when coupled with its refutation.

Thus, we can say, an optimum case is, old explicans and its refutation and ... = new explicandum.

And now comes the crucial point. As stated the new explicandum is inconsistent, of course, and so can be explained only by a contradiction and by any contradiction.

Examples abound, but the simplest is Kepler's Law plus the irregularities that, Newton says, were puzzling astronomers in his time.

Clearly there is an answer to my crucial point and it is very forceful and convincing and true. It is also very simple: the new explicandum is not the old explicans verbatim plus its refutation, but the old explicans in some modification or other plus the refutation – and this takes care precisely of the inconsistency in question.

Precisely. Except that the question is, which modification is chosen and how. Here comes my point: a well formulated explicandum, or a well modified one, is half the explicans. Of course, we may try to modify the explicandum differently so as to attain different versions of both explicandum and explicans, thus leading to crucial experiments galore.

This explains a very important and puzzling fact. Every research physicist knows that new insights to a problem may be attained by the mere transformation of an equation, more easily a set of equations, from one wording to another. Now we all know that rewording itself means no alteration of content, and yet rewording in mathematics may require much work and ingenuity. Is it worth it? Why?

The answer is that different wordings 'suggest' different kinds of prima facie modifications and 'suggest' or 'lead to' different kinds of explicans.

What this shows is that we often do not know what is our explicandum, that the explicandum often comes in a mess, and that cleaning it up is half the job. Often the job is done in one stroke and the explicans offers the explicandum.

The point of the present chapter was this, often we offer *ad hoc* hypotheses as excuses *pro tem*. Further, often we offer token *ad hoc* hypotheses, obviously lame and half-articulated ones, but still acceptable merely because they are understood to be excuses *pro tem* anyway. Still further, we may offer no *ad hoc* hypothesis at all, but a mere token for it. Now the most *ad hoc* hypothesis is the *ad hoc* modification of a given theory to avoid a given refutation. Often we use a token rather than the *ad hoc* modification. One way or another, once the refuted theory in modifica-

tion plus its refutation are explained by a new testable theory, the modification plus its refutation are explained by a new testable theory, the
modification is *ad hoc* no longer.

In reality, I feel, we are never free of all *ad hoc* measures. But we keep
replacing them, thus keeping our credit high.

APPENDIX: THE TRADITIONAL 'AD HOC' USE OF INSTRUMENTALISM

In his classic 'Three Views Concerning Human Knowledge' Popper tells
the story of the battle between realism and instrumentalism as that between the forces of light – of science, of Copernicanism, of Galileo – and
the forces of darkeness – of religion, of the Church of Rome, of St. Robert
Cardinal Bellarmine. And he ends the story with the observation that
with the widespread acceptance of Niels Bohr's instrumentalism amongst
physicists the Church of Rome won the war without firing a single shot.

I accept the beginning but not the end of this story. Instrumentalism
is the denial of any informative force from scientific theory. It may apply
to (a) science at large – by enemies of science and by friends of science –
and it may apply (b) to a given defective or defunct or superceded scientific theory as a consolation prize by its scientific opponents. One might
go to a position between (a) and (b), and declare (c) all or some of today's
science but not all science and certainly not tomorrow's science to be a
mere instrument. Or, one can (d) use the instrumentalist philosophy as a
supplement to today's unsatisfactory realist philosophy.

Consider first, then, (a) any comprehensive version of instrumentalism.
Of the enemies of science I shall not speak here (see, however, appendix
to Chapter 10). The friends of science who are both positivists and instrumentalists, i.e., those who say both that only science should be heeded
to, and that no scientific theory has any informative content, they are
forced to see the world as a string of unconnected events. Hume found
this distasteful and was driven by it to desert philosophy in favour of a
game of backgammon. Mach and Wittgenstein were driven by it to
mysticism.

But instrumentalism was more often used (b) to apply to your theory
at the same time as realistic claims were made for mine. Thus, the Cartesians presented Descartes' theory realistically and Newton's as a mere

instrument; the Newtonians, especially Helmholtz, viewed forces acting at a distance as real but fields of force as mere instruments. And Maxwell claimed fields to be real but Lorenz's advanced potentials a mere instrument.

Now, somehow, viewing our opponents' theories as a consolation prize may be all right, as in the above examples, and quite objectionable, such as Bellarmine's. Now Bellarmine was only continuing an older trend, that was started by Ptolemy and strongly rejected by Copernicus and his followers. Indeed, the arch-instrumentalist and the paradigm of instrumentalism are Ptolemy and his system in all its transformations from his days to the days of Galileo. He, doubtlessly, viewed his own theory as a mere instrument of prediction in a mood of self-deprecation and while ascribing a realistic or philosophical statue to a theory he liked better. He had, in brief, two desiderata, and two doctrines; and each of the two doctrines answered only one desideratum: simplicity (which is a condition for truth, since nature allegedly is simple) and explanatory-cum-predictive value; or, in a more traditional jargon, agreement with nature and agreement with appearances. He would have preferred a theory which agrees with both. He even accepted that in some recondite manner any theory which agrees with nature also agrees with appearances; yet agreement with appearances must be itself apparent. And so, Ptolemaic instrumentalism was (c) an *ad hoc*, a stop-gap, the best of a bad job for the time being.

Almost the same attitude is expressed in modern times by Laplace regarding probability. It is a stop-gap for the time being, not a theory which agrees with nature: God does not play dice, as Einstein restated it. Even those who do not accept Laplace's and Einstein's determinism, even those who think that determinism is false and its adherents backward-looking, even they do not condemn Laplace and Einstein. Hence the fault of the Ptolemaic attitude lies elsewhere. When Copernicus found it shameful, he meant that the stop-gap was tolerated for too long; when Galileo found it appalling, he found it apologetic and shifty. And this, the apologetic attitude, the view that, unless and until we can restore Aristotelean physics, all current physics is a mere instrument – this is the classical version of instrumentalism. It was shared by Galileo's arch enemies, Bellarmine and Duhem; it is shared by quite a few physicists today; but it was never anywhere near crossing the mind of Niels Bohr, of course, and so I cannot see him as an instrumentalist.

This is no apologia for Bohr. Popper has spotted some instrumentalist ingredients in Bohr's views as expressed after Einstein, Podolsky, and Rosen had criticized his view of quantum theory. These ingredients are errors, and should be eliminated; but instrumentalist traces, even when presented apologetically, are not instrumentalism and not any capitulation to the Church of Rome.

There is, however, a difference between, say, instrumentalism which reflects (b) a somewhat tolerant attitude towards an objectionable theory, or the instrumentalism of Helmholtz which reflects (c) the demand for certain improvements of a theory before it passes from the status of a mere instrument to the status of a realistic scientific theory, and (d) Bohr's instrumentalism that comes to exonerate quantum theory from the realistic assault on it by Einstein, Podolsky, and Rosen.

This is not to say that Bohr is an out-and-out anti-realist or instrumentalist. For, the vacillation between realism and instrumentalism, between nature and convention, between any two given poles, is often very reasonable on the following assumption. Suppose you have two extreme positions, each of which is untenable; you may intuitively feel that the truth is the synthesis of the best elements of both, and feel that you approximate it sufficiently closely if you alternate between the two extremes, taking the best of each. When W. Bragg said in the 1928 Glasgow meeting of the British Association that he accepted the wave theory on odd days of the week and the particle theory on even days [p. 19 of the proceedings of the Glasgow meeting], he presented things in a terribly *ad hoc* way, meaning, of course, that he does not mind applying any of these two theories *ad hoc*, relative to the circumstances at hand. The vigilant philosopher may object to Bragg's arbitrariness: juggling is a stop-gap which may impede progress. True; but the juggling may also aid progress. We just do not know.

There is no better evidence that contemporary physicists are not as instrumentalist as they profess to be, than in their attitude towards applied mathematics which, almost by definition, is the home of theories which are supposed to be mere instruments, to have no depth but mere predictive value, such as classical dynamics, especially the continuum theory. The distinction between elasticity and quantum theory is made from the start from a realistic viewpoint by the same people who defend quantum duality from the apologetic instrumentalist viewpoint – thus exhibiting the desire to find the happy synthesis. Requiring them to stop doing physics until they straighten out their theory of science seems to me to be an excessive demand; trying to purge them of the vestiges of the brand of instrumentalism and/or the brand of realism they happen to know amounts to the same. I see no use for such a crusade.

It is here that one can place Bohr's view, I feel. It is, no doubt, not an instrumentalism plain and simple: Bohr never denied the existence of sub-atomic particles. Already in his original paper on complementarity he vacillates between the two poles. There are inductivist elements in presentation – nature imposes on us conflicting images – and there is a Kantian element or an instrumentalist element – whatever is in principle beyond our horizon should be left outside science. Though his reply to Einstein, Podolsky and Rosen is a severe modification of his earlier view it is not so much of a change as to justify trying to make him out a fully fledged instrumentalist.

CHAPTER 9

THE NATURE OF SCIENTIFIC PROBLEMS
AND THEIR ROOTS IN METAPHYSICS

According to Popper's philosophy the perfect division of labor in research would soon stop scientific progress. His view explains why in the history of science many investigators have concentrated on a handful of problems. The problem arises: How did investigators coordinate their choice of scientific problems? By what criteria did the bulk of investigators of a given period decide which problem was fundamental or important?

There exist a variety of such criteria, but one criterion stands out as the most important. Those scientific problems were chosen which were related to metaphysical problems of the period; those scientific results were sought which could throw light on topical metaphysical issues.

My aim is to present this as a historical thesis. I do not contend that scientific interest devoid of metaphysical interest is in any sense illegitimate or inferior. Investigators may wish to study a small part of the universe without bothering to study the universe as a whole, without even bothering to ask how their partial picture integrates with man's picture of the universe as a whole. Yet I contend, firstly, that very frequently problems, theories, and experiments which are traditionally regarded as important are highly relevant to the metaphysics of their time; and secondly, that my first contention provides a solution to the question of how the choice of scientific problems is coordinated.

This is all I wish to assert in the present chapter. I shall discuss problems of demarcation of science, of pseudo-science, and of metaphysics, mainly to dispel some vulgar errors concerning metaphysics (namely the identification of it with pseudo-science) and its role in the scientific tradition. I shall argue that metaphysics can progress – not so much in order to defend metaphysics as to expound my view of metaphysics as a coordinating agent in the field of scientific research.

I. SCIENTIFIC RESEARCH CENTERS AROUND A FEW PROBLEMS

Since there are more scientific problems to be studied than researchers to study them, a complete avoidance of overlap between projects is quite possible. The more the number of existing problems exceeds the number of researchers, the more one would expect the actual case to tend naturally toward the ideal of complete absence of duplication. But the facts are quite otherwise. Here are two historical examples where numerous obvious problems have been ignored. Diffusion is a phenomenon with instances widespread in physical nature: river water rapidly mixes with the oceans' waters, smoke with the atmosphere, salt with soup. Until the late eighteenth century no one paid any attention to this phenomenon and the scores of problems it raises. Priestley seems to be the first who studied it; Dalton concentrated on it for a while. Yet though Dalton's study received great publicity, only a handful of thinkers worked on diffusion before the celebrated studies of Maxwell rendered it an integral part of physics. My second example is elasticity, which was left almost entirely unstudied between the days of Hooke and of Young but was studied more and more seriously in the nineteenth century, only to be relegated in the twentieth century to the borders of applied mathematics and technology.

Whether concentration of intellectual power on a few problems is advantageous or a waste has hardly been studied because of misconceptions about science. Popper's theory of science answers this question unambiguously: perfect division of scientific research work will quickly bring scientific progress to an end. This theory makes the 'friendly-hostile cooperation' between individuals crucial for progress. Some offer new ideas, some offer criticisms of these ideas, some offer alternatives to these ideas; if they all worked on different problems there could be no cooperation. Robinson Crusoe would be unable to sustain the development of science, because of his limited capacity to criticize himself and thus to get out of the routine of his way of thinking.

The existence of a variety of problems to be solved, and the fact that newcomers to science have a great variety of reasons which draw them to science, would by itself render science almost Crusonian. But by some process which has not yet been studied or even noticed, the more a person's interest develops, the nearer it approaches the interest of other students of the same field. Somehow interests coordinate themselves. And my problem now is what is this means of coordination (though I shall not discuss here the way by which individuals learn to apply it).

Undoubtedly, there exists a variety of coordinating factors. New

economic and political needs, new mathematical or experimental techniques, offer new avenues which are somitimes explored. Yet, by and
large, there are minor and often secondary factors – secondary, because
developments of techniques and of their fields of application often follow
interests. By and large, widespread scientific interests may be shown to be
connected with some metaphysical problem of the day. It is my contention that whatever the starting point of a person's interest in a science, the
more that person's interest develops the closer it approaches the general
interest, the interest which dominates the tradition in that science, and
that this general interest springs from, and flows back to, metaphysics.

Most philosophers and historians of science would vehemently oppose
this view. Descartes, as is well known, developed a philosophical theory
in which metaphysics provides the framework for science. His ideas were
greatly improved by Kant, but this was the last significant effort in this
direction; for good reasons or bad Kant's idea has been universally rejected. In this chapter I wish to rehabilitate metaphysics as a framework
for science, but within the framework of Popper's critical philosophy,
with certain modifications, of course.

My view is this. Metaphysical theories are views about the nature of
things (such as Faraday's theory of the universe as a field of forces). Scientific theories and facts can be interpreted from different metaphysical
viewpoints. For example, Newton's theory of gravitation as action at a
distance was interpreted by Faraday as an approximation to a (future)
gravitational field theory. An interpretation may develop into a scientific
theory (such as Einstein's gravitational field theory) and the new scientific
theory may be difficult to interpret from a competing metaphysical viewpoint. Metaphysical doctrines are not normally as critizeable as are
scientific theories; there is usually no refutation, and hence no crucial
experiment, in metaphysics. But something like a crucial experiment may
occur in the following process. Two different metaphysical views offer
two different interpretations of a body of known fact. Each of these interpretations is developed into a scientific theory, and one of the two scientific theories is defeated in a crucial experiment. The metaphysics behind
the defeated scientific theory loses its interpretative power and is then
abandoned. This is how some scientific problems are relevant to metaphysics; and as a rule it is the class of scientific problems that exhibit this
relevance which is chosen to be studied.

II. THE ANTI-METAPHYSICAL TRADITION IS OUTDATED

My own interest in physics originates from a very early interest in metaphysics; the present essay may be no more than a projection of my own case history into the history of science at large. In my undergraduate days I used to resent the hostility toward metaphysics displayed by my physics teachers; my present view is in a sense an inversion of theirs. They derided all metaphysics as the physics of the past; I extol some metaphysics as the physics of the future. But I wish to be fair to their view, and perhaps the best means to arrive at a fair attitude to a doctrine is to try to see it in its historical perspective.

Francis Bacon's anti-Aristotelian-metaphysics, which was the first fanfare of the modern positivists, was very valuable. In launching an attack on Aristotelian metaphysics, he overenthusiastically took it to be an attack on all metaphysics. This was an exaggeration, and a very understandable and effective one at a time when Aristotelian metaphysics reigned supreme. Then came the victory of Copernicanism and of the Galilean-Cartesian metaphysics. This development admittedly altered the situation. From then onward Bacon's exaggerated idea might have been profitably cut down to size by studying the difference between Aristotle's bad metaphysics and Descartes's good metaphysics. Yet this is debatable, since at that time there was still a need to encourage experimentation rather than speculation. Moreover, throughout the seventeenth and eighteenth centuries metaphysics was closely linked with religion; and religion had to be banned from scientific discussions for very obvious social and political reasons. Since the early nineteenth century both of these factors have become negligible, but other factors have taken their place; fortunately for the positivist knight-errants, there was the task of slaying such awful metaphysical dragons as the Hegelians and the existentialists. Unlike Aristotelianism, positivism has not been useless during its period of obsolescence. It is still fighting bad metaphysics, under the somewhat absurd guise of fighting metaphysics as such.

In addition to being an overzealous criticism of irrationalist metaphysics, positivism has also served the rationalist metaphysician. Metaphysics can easily degenerate into pseudo-science by providing a framework for *ad hoc* explanations instead of scientific ones. The Baconian-positivist attack on metaphysics as *ad hoc* or pseudo-scientific helped the good metaphysician by putting him on guard against irrational practices.

It is unfortunate that the merits of positivism are so often exaggerated, since positivism is conducive to ignorance. I have met physicists who know about only one metaphysician – Hegel – and only one detail concerning him – that he said when a doctrine of his turned out not to accord with facts "so much the worse for the facts." Rarely has anyone paid more dearly for a silly joke.

It is not my purpose here to disprove positivism but I feel I have to stress that in this essay I am speaking of good metaphysics while intentionally ignoring bad metaphysics, after having acknowledged the partial justice of the positivist attack on it. Every field of human activity ought to be judged by its very best, and it is time to notice that examples of bad metaphysics do not show that all metaphysics is bad. One can show that all metaphysics is bad, but only after abandoning the ordinary or traditional meaning of the word 'metaphysics.' This word is used by Hegelians and by positivists to signify the theory of the cosmos as a whole, of the very mystery or essence of the universe. In his *Tractatus* Wittgenstein accepted Newton's metaphysics as a framework for physics, but he did not call it 'metaphysics'; he considered 'the mystical' alone to be the subject matter of metaphysics. The positivists, the Hegelians, and the mystics, rightly claim that the mystical is unexpressible. This is a point which Russell rightly considered (in his *Mysticism and Logic*) trivially true. Metaphysics in the sense of a theory of the mystical is hence impossible. My own use of the word 'metaphysics' here is in its traditional and much narrower sense. Metaphysical doctrines are to be found, first and foremost, in Aristotle's *Metaphysics*, especially in Book Alpha: all is water; atoms and the void; matter and form; etc. There are a variety of sets of first principles of physics. Do these belong to scientific physics? Are they entailed by scientific theories? Are they useful for scientific research? I think they do not belong to scientific physics (though in principle they might). Metaphysical ideas belong to scientific research as crucially important regulative ideas; and scientific physics belongs to the rational debate concerning metaphysical ideas. Some of the greatest single experiments in the history of modern physics are experiments related to metaphysics. I suggest that their relevance to metaphysics contributes to their uncontested high status. And yet, I contend, the metaphysical theories related to these experiments were not parts of science. This raises the problem of what kind of relation between a given theory and observable facts renders that theory scientific.

III. A HISTORICAL NOTE ON SCIENCE AND METAPHYSICS

The term 'speculative metaphysics' and the term 'speculations,' when used as synonyms for 'metaphysics' (by Boscovitch, Faraday, and others), indicate the view that metaphysical doctrines are products of the imagination, in contrast with scientific theories which are – allegedly – products of inductive inference from facts. It was indeed this view which led to the tradition of divorcing science from metaphysics. The first modern positivist, Francis Bacon, presented the two methods, of induction and of speculation, as irreconcilably opposed to one another. The proper inductive investigation, he proclaimed, can be conducted only in the absence of all preconceived notions. Those whose minds are full of speculations are entirely unfit for proper scientific experiment and observation, much less for theorizing inductively: they are biased in favor of their speculations, and this bias makes them ready to observe only those facts which verify their speculations and unwilling to observe those facts which refute them. Consequently, they achieve not the truth but the reinforcement of their own preconceived opinions, and their biases thus become prejudices and superstitions.

Bacon's violent opposition to metaphysics was less violent than the ultra-modern one. His opposition to metaphysics was merely an opposition to its method; it was not an opposition to the abstract character of metaphysics but to the leaping to metaphysical conclusions. By developing science properly, by starting with observation and then slowly developing theories by gradually increasing the abstractness of knowledge, by ascending the inductive ladder properly without skipping any step, Bacon held, we shall end up with the most fundamental theory, namely, with scientific metaphysics. This metaphysics will be scientific because it will have been achieved, not by the speculative method, but by the inductive method.

Scientific metaphysics was later defended by Descartes and by Kant, each of whom considered his own metaphysics to be a body of certain, and hence scientific, knowledge. Their idea of certitude differed from Bacon's; it was based on *a priori* reasoning rather than on inductive inference. Consequently they viewed metaphysics as the beginning, not the end, of scientific inquiry. Buth both in viewing science as certain, and in taking it for granted that metaphysics must be scientific or perish, they barely differed from Bacon. It was William Whewell, the disciple of both Bacon and Kant, who first defended unscientific metaphysics from a scientific point of view.

In Whewell's view scientific doctrines do not emerge inductively from facts; they are first imagined and then verified empirically. And he considered his own (Newtonian-Kantian) metaphysics *a priori* valid, namely, demonstrable independently of empirical evidence. In accepting Kant's *apriorism* he rejected Bacon's view that all preconceived ideas are verifiable by virtue of their being prejudices, contending that much as people had sought to verify Newton's optics, much as they were prejudiced in its favor, they ultimately rejected it. His problem was how to explain why assent to Newton's mechanics was justifiable and assent to Newton's optics unjustifiable. He wished to find out the proper canon of verification and show that Newton's theory of gravity, but not Newton's optics, had conformed to it.

In brief (and in a slightly improved version), Whewell's canon can be put thus: proper verification is the result of severe tests. The procedure of severe testing is this: First try to explain known facts and state your explanatory theory as explicitly as possible. Then try to deduce in a rigorous manner from the theory a new prediction of observable facts. Then, and only then, decide by observation whether this prediction is true or false. If the prediction is false then the theory is obviously false too; if the prediction is true then the theory obviously explains the new facts without adjustment ('adjustment' being a suitable alteration or addition). In the latter case, Whewell declares, the theory is verified. Newton's theory of gravitation had been severely tested, and consequently the result of the tests could either refute it or be explained by it without any adjustment. In contradistinction, Newton's optics never stood the risk of a test and hence never explained a single new fact. Many new facts were alleged to be explicable by Newton's optics. Even Laplace had endorsed this allegation. Yet upon a simple and clear examination, which Whewell executed in a most masterly fashion, each of these new facts turned out to be explicable not by the original theory but by the adjusted theory.

Both Bacon and Whewell were interested in the problem of the demarcation of science. But their interests stemmed from different roots. Bacon considered Aristotelianism, which was then the academic metaphysics, to be the chief impediment to the advancement of learning. Whewell viewed Newton's metaphysics, which was by then the academic metaphysics, as demonstrable. His problem was not metaphysics but the overthrow of the allegedly verified Newtonian optics. Thus, while Bacon demarcated science mainly from metaphysics, Whewell demarcated science mainly from pseudo-science.

Since, according to Whewell, science begins by the invention of explanatory hypotheses, he was all for every possible source of inspiration. And he viewed all (reasonable) metaphysics as such a possible source. He gave a striking example for this. Kepler had developed his scientific hypotheses, Whewell maintained, in an attempt to carry out Plato's metaphysical program as outlined in his *Timeus*. This idea of Whewell's was so revolutionary that this great philosopher is now almost entirely forgotten because Mill and his followers condemned him as an intuitionist. (This charge is, of course, quite untrue. Whewell relied not only on intuition but also on Kantian transcendental arguments and on empirical tests.)

Initially, Popper's interest in the problem of demarcation was similar to Whewell's, though his examples were different; it was Marxism and Freudianism which he viewed as pseudo-scientific. His demarcation of science may be contrasted with Whewell's thus: Whewell demands that a scientific theory be testable and emerge triumphant from the tests, while Popper merely demands testability. Neither of them is hostile to metaphysics, and both contend that metaphysics is sometimes important as a source of scientific inspiration. A remnant of positivist prejudice may perhaps be detected in Popper's lumping together (like Bacon and unlike Whewell) of a few kinds of nonscientific theories, including metaphysics, pseudo-science, and superstition, under the one label 'metaphysical.' Though I dislike this label, I do not think it matters beyond leaving some ambiguity concerning the difference between metaphysics and pseudo-science.

IV. PSEUDO-SCIENCE IS NOT THE SAME AS NON-SCIENCE

Popper's idea (pseudo-science is untestable) is a marvel of simplicity. It explains why no matter how bad a pseudo-scientific doctrine is, its proponent may regularly win debates. It resolves the conflict involved when we feel obliged, against our own better judgment, to take a theory seriously because its proponents seem to be entirely undefeatable. It amounts to a proposal not to embark on the game before fixing its rules, before deciding in advance what kind of argument, if any, would be capable of defeating the proponent of a theory, and determining not to try to defeat him if he turns out too evasive to be vincible. As Whewell has pointed out, no kind of argument will defeat the proponent of any theory if he is allowed to adjust even minor details of his theory in an *ad hoc* fashion. On this Whewell and Popper are agreed. Yet wonderful as Whewell's ideas about pseudo-science are, by demanding too much from science he threw out the baby with the bath-water.

According to Whewell, scientific theories must also have withstood test. Consequently, he viewed as pseudo-scientific those theories which falsely claim to have withstood test. This leaves unclassified those theories which are testable but have been obviously refuted. As Whewell considered these to be neither scientific nor metaphysical, he confusedly implied that they are pseudo-scientific, especially when they are submitted to recurrent readjustment and retest. According to Popper such theories are scientific, for he only demands testability; according to Whewell they could not be considered scientific, and so he held them in contempt. He knew that Newton's optics had been falsely held to have been verified. Yet he did not see that as long as verification was considered a hallmark of respectability, the immense respect for Newton gave these false claims an immense appeal. But if the requirement of Whewell and his predecessors of a respectable scientific theory is too stringent, is not Popper's requirement of a respectable scientific theory, namely, a high degree of refutability, a trifle too lax?

Traditionally, a variety of characteristics have been attributed to science. Popper accepts some of these attributes, such as high explanatory power, high informative content, abstractness, generality, precision, and simplicity; he rejects others, such as obviousness and verifiability. He seems to have claimed in his *Logic of Scientific Discovery* that the characteristics in the first group are all reducible to one, to testability. This is his justification for requiring only this one characteristic of a theory before labelling it 'scientific.' I have little doubt that Popper will fully agree that the spurious simplicity of some monistic doctrines (such as Marxism or mechanism) rather than their spurious explanatory power has deluded some people into regarding them as scientific. Simplicity, however, is traditionally viewed (since Leibniz) as the paucity of assumptions relative to the amount of factual information they explain, so that there is no need to differentiate between simplicity and high explanatory power for the purpose of demarcation. And Popper would say the same concerning explanatory power, which, in his opinion, increases with refutability. For my part, I consider that the various characteristics of science are less often dependent on each other than Popper suggests. But I still side with Popper in viewing spurious refutability, rather than, say, spurious simplicity, as the chief characteristic of pseudo-science, and for two reasons. First, whatever else may characterize a scientific theory, the very acceptance of the proposal that scientific theories are agenda to be tested renders Popper's proposal to check whether a doctrine in question is testable or only spuriously testable a matter of supreme practical importance. Second, the claim of pseudo-science is the claim for empirical character. And empirical character is nothing else but empirical refutability, as I shall soon explain. Thus, Popper's demarcation between science and pseudo-science does not require any amendment even on the assumption that he has erred in correlating the various characteristics of science. As to his characterization of science as such, it requires a reformulation if, as I think, his way of correlating the various characteristics of science is in error. I think we have to characterize scientific theories not only by their refutability, but also by their simplicity, high explanatory power, etc. This has an immediate bearing on the problem of selection of scientific problems and of scientific theories which is the topic of the present chapter. According to Popper we always look for the most easily refutable theory. In my opinion this is not the case.

V. POPPER'S THEORY OF SCIENCE

Popper's arguments for his claim that empirical character is empirical refutability are very compelling. Logically, observation reports can contradict theories but not entail them in any way. Philosophically, Popper's view is the doctrine of learning from experience as a special case of learning from mistakes, of the critical method. Socially, it presents students of nature as human rather than as unerring supermen. Historically, it opens wide vistas of new studies of the history of science uncharted by the modern science textbook. Popper's greatest contribution to the philosophy of science seems to me to be rooted in the simple idea that since empirical character is empirical refutability, scientific research is a special case of Socratic dialogue. But I deny that the empirical character of science is all that makes science what it is.

It is not difficult to find empirical developments, i.e., empirical refutations, outside the field of science. Thales's metaphysical doctrine ("all is water") was refuted empirically when water was first decomposed; Moebius (as I. Lakatos would say) may have refuted empirically the mathematical theory "all surfaces have two sides"; Faraday refuted empirically some spiritualistic superstitions; Marx's prophecy about the geographical location of the socialist revolution has been refuted by his Russian followers; and this amounts to the refutation of his materialism since it entails the valuelessness of imaginative ideas; the very important philosophical doctrine about the universality of common sense (which even Duhem still advocated) is empirically refutable by comparative studies. Necessarily, either such cases should be viewed as scientific or Popper's proposal should be considered inadequate. My choice is the latter: I propose to use Popper's convention as a convention concerning the empirical character of science, not concerning empirical science as such. There is no difficulty in admitting that daily experience, as well as some developments of mathematics (or metaphysics, or any other field of intellectual or practical development), manifest a certain empirical character, even though they do not belong to empirical science. Empirical science manifests its empirical character more systematically than mathematics, and it manifests other characteristics as well, which are lacking in mathematics.

But what about the claim that theories manifesting empirical character, i.e., refutable theories, also necessarily manifest the other characteristics of science, i.e., they have informative content, explanatory power, simplicity, abstractness, generality, and precision? I simply reject this claim. As I have said earlier, I interpret a great deal of Popper's discussion in his classical work to be an attempt to support this claim. I consider the value of that part of his discussion as a valid criticism of his opponents and as stimulating heuristic material, but as very far from being a finished product.

To maintain my thesis I must contradict Popper here. He would say that research is conducted toward the finding and the testing of highly testable hypotheses, whereas I say that it is very often conducted toward the finding and the testing of metaphysically relevant hypotheses. And as a rule, I shall later show, research tends to begin with hypotheses which have a low degree of testability or are not testable at all. Consequently investigators often have to use great ingenuity to test a barely testable hypothesis, and even first improve a hypothesis to the point of rendering it testable to some degree. If the aim of science were merely producing testable hypotheses and then testing them such procedures would be irrational. But the aim of science, or rather the aims of science, are different.

The aim of science is to attempt to comprehend the world rationally, as we all agree (including the positivists who should disagree). But this is too vague. What is the rational method and what is comprehension? Rationality, said Popper, is manifest in empirical tests. He later generalized this: the rational method is the critical method. Is metaphysics rationally debatable? Yes. I shall argue that the study of a hypothesis of a low degree of testability is often conducted with a view to criticizing some metaphysical theory upon which it may have some bearing. So much for rationality. As to comprehension, Popper views it as deductive explanation, and he has suggested that explanatory power goes with refutability. I deny that explanation is the only method of comprehension. As I shall show later, the attempt to coordinate our various explanations within one metaphysical framework is not explanation, yet it is, in some weaker sense, an attempt at comprehension. Moreover, I deny that explanatory power is always dependent on refutability. Already in the last section of his great book Popper has noted that some theoretical systems may have some explanatory power and yet be untestable. I have already mentioned examples of refuted theories of little or no explanatory power.

Degrees of testability are, I think, of little practical importance. All that matters is that we may test in at least one way an interesting theory. According to Popper, there are two factors contributing to the degree of testability of a theory, the number of possible events which may refute that theory, and the probability of each potential refutation. To my mind the possibility of observing the next refuting event is all that matters, not the number of possible refutations. As it is the number of all excluded possibilities which is the content of the hypothesis, content is not the same as practical testability. *Ad hoc* explanations have some empirical content yet are untestable. Explanatory power is not content, and not even truth-content (i.e., that part of a theory's content which is true), but I should say (in agreement with Leibniz's idea as I understand it), known-truth-content (i.e., the overlap of a theory's content with the class of true observation-reports). And high explanatory power is not the sole characteristic of a satisfactory explanation. As I have learned from Popper himself, a satisfactory explanation must be independently testable. Thus, Weyl's theory which unifies Maxwell's and Einstein's has a high explanatory power and a high degree of testability, but no known independent testability, and thus it is not considered scientific. Simplicity depends not only on explanatory power and the paucity of parameters, as Popper mentions in his early work, but also on depth, as he now says. Nor does abstractness go together with universality: Boyle's law is more general but less abstract than the theory of consumers' demand, and the Heitler-London theory is more abstract but less general than Schroedinger's theory.

The result is pluralism: we may admire one theory for its boldness, another for its explanatory power, another for its elegance; and yet another, I suggest, for the light it throws on some topical metaphysical issues.

There seem to be very good reasons for Popper's correlation of a higher degree of testability with a higher degree of explanatory power, etc., and these reasons are of heuristic value. One reason of Popper's is this: If one theory explains another theory, it is obviously not less refutable than the other. If one theory explains another theory as a first approximation, then it is more precise, and a higher degree of precision goes together with a higher degree of testability. This is so because a more precise theory excludes more (logically) possible states of affairs, thereby possessing both a higher informative content and a better (*a priori*) chance of being refuted, or a lesser *a priori* probability. These arguments are valuable but insufficient and partly incorrect.

In his classical paper 'The Nature of Philosophical Problems and Their Roots in Science' Popper has given an admirable account of Pythagoras's metaphysics and the history of its refutation. When I read this excellent essay I decided to study under Popper; so the title of the present chapter adverts to his, partly for sentimental reasons. Yet, perhaps because my prejudice in favor of metaphysics came first, I was unhappy about his taking Pythagorean metaphysics to be scientific. Since his reason was that this metaphysics was refuted, I was bound to examine his refutability criterion for the demarcation of science. I now propose his empirical refutability criterion to be the criterion of empirical character, not of empirical science as such. Empirical science is the set of highly informative and simple explanations which exhibit independent empirical character – satisfactory explanations, for short. I owe this idea to Popper himself: in his lecture courses Popper presents science rather in this way than in the way he does in his classical *Logic of Scientific Discovery*.

VI. SUPERSTITION, PSEUDO-SCIENCE, AND METAPHYSICS USE INSTANCES IN DIFFERENT WAYS

Bacon justified his lumping together metaphysics with superstition and pseudo-science by saying that the method of them all is that marshalling verifying or confirming examples or instances and persistently ignoring counter-examples or refuting instances. This is much too coarse a characterization; to refine it we must first notice a few of the different roles that instances may play in intellectual activity.

The role of an instance may be solely presentational: we understand an abstract idea better when we are told how to apply it to concrete cases. So long as the purpose of an instance is elucidatory, an author is at liberty to choose his instances so as to avoid a discussion concerning their truth of falsity, and the more obvious the instance the better. One should either take a presentational instance for granted or use another in its stead. The moment an instance is sufficiently significant to be not easily dispensable, it has additional roles.

The most important role of instances is their role as refuting instances. This is the crux of Popper's solution of the problem of induction: learning from experience is learning from a refuting instance. The refuting instance then becomes a problematic instance, namely, an instance which ought to be explained by a new theory. The last important role of instances is that of showing how high is the explanatory power of a proposed theory. Perhaps one may consider the instances explicable by a theory as problematic for those who wish to propose an equally good alternative to it. This would explain why usually previously refuting and/or problematic instances are presented as explained instances of a theory though that theory explains many other instances as well. So much for instances in science.

A common, though by now highly suspect, role is played by instances which Bacon has called 'clandestine.' A clandestine instance hints at a possible truth. For instance, a miraculous recovery may be due to unknown causes or due to the excellence of the doctor in whose charge the patient was at the time. If we accept an instance as clandestine, we need not at once accept the theory it points to (in our example, that Dr. X is excellent), but we are well advised to investigate the matter seriously. And the more clandestine instances there are that suggest a particular theory, the more seriously we should take the theory.

The most obvious characteristic of the superstitious is their serious approach to clandestine instances; the root of this lies in their want of a critical attitude. Not all errors are superstitions, only those concerning which we cannot conceive that we may be critical towards them.

In this sense of 'superstitious,' medieval empirical research was largely superstitious. The taking up of clandestine instances, hints which Mother Nature has mercifully thrown in our way, was quite routine procedure then. In modern times, mainly under Galileo's and Boyle's impact, this has been outlawed.

This immediately raises the question of the difference between a problematic instance, which requires explanation, and a clandestine instance, which should be ignored. The chief difference between them, I think, is that of attitude both toward theory and toward fact. When we have a problematic instance we first try to explain it and leave the question of the truth or falsity of our explanation to be discussed in a critical fashion afterward; whereas following a clandestine instance we hope to find the truth even though we may not fully understand it or fully formulate it to begin with – even though, that is to say, we are not capable of subjecting it to rational discussion straight away. And the same applies to facts. The fact constituting a clandestine instance, being a wondrous hint, should be taken seriously at once; whereas a problematic one should be capable of critical examination, and hence it must be repeatable.

Yet a critical attitude is but a necessary condition. While it is true that unrepeatable facts are useless, too many repeatable ones are left unstudied. Footprints in the sand are as repeatable as one could wish, yet science says precious little about them. In my view the ignored phenomena are those which our metaphysical frameworks are too poor to interpret (in the sense discussed below). They are too problematic. The same applies to theories, like elasticity theory, which are too difficult to incorporate within the existing metaphysical frameworks, and hence are not scientifically interesting.

Next comes the confirming or verifying instance. Whenever someone marshals instance after instance, challenging you to examine their truth, it is on the tacit assumption that if his instances are true his theory is also true. If you admit his instances and yet reject his theory he will marshal more instances. If you prove impervious to all his instances he will proclaim you unreasonable.

Confirming instances play the same role today as clandestine instances played in the Middle Ages. They play the role of clandestine instances for the uncritical audience and explained instances for the less uncritical audience. They are usually unsatisfactorily explained instances, yet the poor explanations are overlooked by audiences who are impressed because they are striking clandestine instances. For my own part I prefer to view all confirming instances as explained ones. For presenting an unsatisfactory explanation is still an attempt to explain, an attempt at a rational procedure; marshalling clandestine instances is plainly irrational.

To take an example. If someone throws a child into the river, Adler would interpret this act as one of self-assertion. And he would say the same if someone else rescues the child from the river. Thus, says Popper, opposite modes of behavior toward others are both somehow covered by Adler's doctrine. Hence it is no explanation. Adler's doctrine plus one of a given set of additional hypotheses, selected to suit each of the different cases, will indeed explain each action. But then all these explanations are *ad hoc*. The feeling is conveyed that many cases have been strikingly explained by one single hypothesis because Adler has claimed that these instances are indications of self-assertion, clandestine instances for his theory.

An example to this effect which has greatly impressed me is Freud's story of a married woman who unthinkingly signed her maiden name. Freud interpreted this as an unconscious expression of suppressed discontent with her husband and, indeed, he triumphantly added, a few months after her pen slipped in that ominous fashion the poor lady was divorced. This is pseudo-science at its peak; it is a glaring case of a clandestine instance thinly masked as explained instance. Since some married women divorce their husbands without having accidentally used their maiden names, and since other married women use their maiden names by mistake without ever asking for a divorce, clearly in this special case Freud erroneously claimed that the error and the divorce were explained by his theory of slips of the pen. Yet it does appear as if this theory spectacularly explains the unexpected relation between a slip of a pen and a divorce.

The mark of pseudo-science is the use of confirming instances. The practitioner of pseudo-science, unlike the superstitious, is not surprised by criticism. On the contrary, he is often painfully aware of the existence of critics; he is only too ready to meet his critics and argue with them. He will claim in the argument that every relevant case is an instance of his theory, that his critics' challenge can easily be met, that the critics do not see the immense explanatory power of his view simply through being so hostile toward it. When his explanations are scrutinized, however, it will be seen that the critic's facts are explicable not by the theory itself, but by the theory plus some additional hypothesis. Usually the additional hypothesis is so trite and plausible that one hardly notices its having been added, and those who make a fuss about it are prone to be successfully dismissed as mere pedants. Yet the great ease with which the pseudo-scientist so impressively explains all phenomena rests on these trivial (and usually acceptable) additions, not on the original theory.

Popper has accepted the claim of the pseudo-scientist that he can interpret all phenomena. He has stressed (in his "Personal Report") that since pseudo-science can interpret any conceivable (relevant) phenomenon it is not refutable by any conceivable phenomena, and hence it is untestable or unscientific. This is very neat, and quite important, yet perhaps it ought to be more explicitly stressed that though pseudo-scientific doctrines have high interpretative power, they have low explanatory power. This characteristic pseudo-science shares with metaphysics.

When Thales said that all was water, he provides a few instances for his doctrine, instances which led Aristotle to hint, and Bacon to assert, that he had based his metaphysical doctrine on facts by using the inductive method (to wit, that his metaphysical doctrine was to some degree scientific). Thales used the freezing and the evaporation of water as examples of his doctrines. He also claimed that solid deposits left in kettles by boiled water, and solid deposits in river-mouths, were instances of water turning into solids.

It is difficult for me to say what would be Thales's answer to such questions as why can we not turn a whole bulk of water into a piece of chalk. Quite possibly Thales, being the first metaphysician, was partly superstitious and partly (in some sense) a pseudo-scientist, and also (as Aristotle states) partly a mythologist; I do not know. Yet I suspect he was really none of these. I imagine he was asked such questions and in reply simply confessed his ignorance. Descartes's answer, and Newton's and Faraday's (whose doctrines I shall soon discuss), however, are clear and straightforward: we are not unaware of the lacunae in our doctrines, they would say, and we shall try and find some scientific theories to deal with your question in due course.

There is a similarity and a difference between pseudo-science and metaphysics. Freud's theory of the slips of the pen, like Descartes's and Newton's and Faraday's (if not also Thales's) metaphysics, sketch possible explanations. Metaphysics may be viewed as a research program, and the false claims of pseudo-science as the result of confusing a program with the finished product.

One corollary of this is that metaphysics can degenerate into pseudo-science. This corollary seems to me to be true, and exemplified by Aristotle's metaphysics, which becomes appallingly *ad hoc* when applied to phenomena, as in his *De Caelo*. I find the following corollary more interesting: it may be possible to elevate a pseudo-scientific theory to the rank of metaphysics. The first step in this direction is to strip it of its pretentiousness by making its logic clear. Expurgated, Freud's theory may be viewed as an interesting metaphysics of psychology. I therefore consider Popper's verdict on Freud and Adler much too harsh.

As instances of a metaphysical doctrine are not clandestine or even confirming, what kind of instances are they? Thales's instances, I think, served two purposes: one presentational, and one to show that his doctrine, be it true or false, is not as fantastic as it sounds. Newton's metaphysics, which asserts that the universe consists of atoms with their associated conservative central forces, was instantiated by his theory of gravity. This instance served a more significant role than a merely presentational one. It illustrated the potentiality of his metaphysics and thus constitutes a challenge to construct instances of that metaphysics which are satisfactory explanations of all known physical phenomena. I shall call such instances 'conforming instances.'

Since Newton's metaphysics does not specify what central force causes gravitation, Newton's theory of gravity does not follow from Newton's metaphysics; it is not an explained instance. Otherwise Newton's metaphysics would be refuted by the refutation of his theory of gravity, which it was not. Newton's metaphysics does not follow from his theory of gravity: the one asserts that all phenomena are governed by central forces, whereas the other is confined to fewer phenomena. Generally, a metaphysical doctrine neither entails nor follows from any of its conforming instances. Nor does it follow from the set of all its conforming instances unless it may be assumed that the set is exhaustive. Since such assumptions are testable, the metaphysical doctrines in question would follow from scientific theories, and thus they could legitimately claim scientific status.

This is the ideal case. To my knowledge it has never been achieved. The doctrine that arrived closest to this ideal was Newton's metaphysics as it appeared around 1800. Yet the ideal had an immense driving force. The debate about metaphysical doctrines often concerns their status, and this often leads to the development of scientific instances conforming to them, or to the discussion of whether such developments are possible. Thus the desire to render a metaphysics scientific leads to viewing it as a scientific research program whose satisfactoriness is open to critical discussion. To illustrate this I shall discuss in the next section the possible unsatisfactoriness of such research programs, and in the following section their possible satisfactoriness.

VII. METAPHYSICAL DOCTRINES ARE OFTEN INSUFFICIENT FRAMEWORKS FOR SCIENCE

The methodology of this and the next section is a generalized Cartesian methodology, and the generalization I am offering is possible only within Popper's framework. Descartes's metaphysics (which was an improvement on Galileo's), was a clockwork view of the universe. It explained almost nothing; it was not intended to explain anything. Descartes claimed that any scientific hypothesis which he could endorse must be one which conformed to his metaphysics. He added that explanatory hypotheses conforming to his metaphysics could always be found. Boyle made the same claim concerning his own semi-Cartesian metaphysics, and so did Newton concerning his own metaphysics (in his preface and the *Scholium Generale* to *Principia*). But this repeated claim of the metaphysician is often false. It may be argued that his doctrine allows insufficient room for explanation, that it provides too narrow a framework. When this is felt to be the case, the demand for a new metaphysical framework arises. Metaphysics stagnates in scienceless (or uncritical) cultures; it is progressive in scientific ones. It progresses then because existing metaphysical doctrines are felt to be constricting frameworks, and thus unsatisfactory.

Thales's doctrine aimed at explaining (physical and chemical) diversity and change by assuming an underlying and unchanging unity. Any such approach runs the danger of being too successful and thus self-defeating. For the assumption of an unchanging unity leads to regarding observable facts as illusory. This was the magnificent discovery of Parmenides, and it was this discovery that made him deny the existence of diversity. We have no right to despise him for having preferred his own logic to common sense, for having proclaimed appearances to be illusory; rather we should admire his dazzling logical acumen. But for him we might not have had Leucippus and Democritus. The greatest novelty of their atomic doctrine is that it expressly allowed for both unity – of the atomic character of matter – and diversity – of the atoms' shapes, sizes, and spatial order. To put it in quasi-ancient idiom, atomist metaphysics is a program to explain the many not by the one but by the few; it is thus more accommodating than the metaphysics of Thales.

(Of Parmenides's other great logical discovery, of the nonexistence of the void, I cannot speak here beyond saying that it was the cornerstone of the theory of space developed by Leibniz, Faraday, and Einstein. Further details of the story of Parmenides and Democritus, as well as the story of the downfall of the Pythagorean program, the demonstration of its narrowness, and its rectification by Plato, as well as the relation between Plato's program and Euclid's geometry, have been admirably presented by Popper in the paper already mentioned. The great role played by both Democritus's and Plato's programs in the seventeenth century have been beautifully told by Alexandre Koyré. The relation between Leibniz's program and Einstein's scientific theory of space is discussed in Einstein's exciting preface to Max Jammer's *Concepts of Space*.)

My next example of an unsatisfactory metaphysics is Cartesian meta-physics, which contained the thesis that all (non-inertial) motion was due to push. The example for this was the suction pump whose (pull) action had been scientifically explained as due to atmospheric pressure (push). Lifting a jar seems to be pulling it upwards, but in fact it is pushing the jar upwards by the handles. Now this last example was implicitly criti-cized by Newton. If the jar is strong, or contains light material, it can be lifted by its handles; otherwise, pushing its handles upward hard enough will only constitute lifting the handles while leaving the jar itself on the floor. We must admit, then, that lifting it by its handles not only involves pushing the handles upward but also pulling the jar itself upward with the aid of the attractive forces which keep the jar and its handles con-nected – by the forces of cohesion. This example justifies Newton's claim (preface to *Principia*) that his program was in the first place more accom-modating than Descartes's, and so could be more fruitfully adopted even if ultimately we should return to Descartes's program.

But Newton's metaphysical program, too, was so naive, that one may wonder how it was accepted for so long. Assuming, with Coulomb, that electric forces act solely between electric charges and that gross or ordin-ary matter is subject to the forces of gravity and cohesion only, why then does the charge remain on the charged body and pull it along when mov-ing toward, or away from, another charge? This question (which was raised in 1800 with the discovery of electrochemistry) clearly indicates that gross matter is in some sense electrical. Yet so strongly impressed were people with Newtonianism that twenty years after Faraday had produced wonderful scientific theories which incorporate the supposition that gross matter has some electric characteristics, these theories were almost unani-mously ignored (the exceptions were Kelvin and two other, rather minor, physicists). Those statements in the *Encyclopaedia Britannica* of the eight-een forties and fifties which appear to allude to Faraday's theories are certainly contemptuous and derisive.

It is not accidental that Boltzmann explained in 1885 (in a letter to *Nature*, p. 413) the general opposition to Maxwell by the general adher-ence to Boscovitch's, not to Newton's, program. Boscovitch had modified Newton's program to permit one material particle to dispose a variety of forces. This he did because he had discovered that otherwise the program would not accommodate any explanation of the phenomena of elastic collisions. But his program became popular only after Faraday imposed his view that gross matter had electric properties.

Incidentally, the indifference to Maxwell which worried Boltzmann shows that even Boscovitchian metaphysics may be highly dangerous; but it is a truism that any idea may become dogma.

So far I have only spoken of the requirement that metaphysical doctrines be sufficiently wide frameworks to accommodate possible future scientific theories. In the next section I shall speak of the requirement that metaphysics be inspiring and lead to the development of scientific theories.

VIII. THE ROLE OF INTERPRETATIONS IN PHYSICS

A statement of fact or a scientific hypothesis restated in terms of a new metaphysical doctrine is a new interpretation. New interpretations are only too often unsatisfactory explanations of the original statements – for example, interpreting a motive as a sex motive, or survival as due to a high degree of fitness, or change of a person's pattern of behavior is due to physical change in the brain. But the logic of interpretations is made clearer, and so is their possible usefulness, when we take examples from physics.

The handles of a jar stick to the jar. To repeat this in Newtonian terms, the particles of the jar are attracted (by some central forces, that is) to those of the handles. Is this restatement a circular or a satisfactory explanation? We do not know. How small are these particles? What is the magnitude of these forces? One may try an estimate on the basis of known facts – perhaps the force needed to tear the handles off. Or perhaps it is easier to measure the force of cohesion by observing a drop of water hanging on to a solid surface, where cohesion counteracts gravity. Or perhaps it is still easier to observe a drop of water in a tube, where the balance of cohesion and gravity is perfect, and where the weight of the drop and the area of the contact surface are more easily calculable. It is easy to develop the first step of Laplace's theory of capillarity by thinking along this line: restate the connection between the inner diameter of the glass tube and the height of the water-column (or mercury surface) in it in Newtonian terms. As Newtonian forces are central, it would follow at once that the narrower the glass tube the higher the water column (and the lower the mercury level) in it. And the relative curvature of the fluid surfaces will be equally easily explicable. This is a particularly fortunate interpretation.

Another example: Newtonianism forces us to view light as either particles or waves in an elastic medium. Each of these interpretations leads to obvious questions which may be given testable answers. Faraday's metaphysics, to take another example, which views the universe as one field of forces, invites the view that light consists of vibrations of the lines of force in empty space. Faraday himself considered light to be waves of the magnetic field of force. For decades he tried to test this hypothesis and failed. And Faraday's interpretation of the electric current as the collapse of an electric field is another example of his failure. Tyndall rightly declared that Faraday's theory of the current was unsatisfactory. But he was too eager to reject it offhand (being a dogmatic Boscovitchian); by further specification Poynting soon rendered it highly satisfactory.

Interpretations apply not only to facts but also to theories. Faraday accepted Coulomb's Newtonian theory of electrostatic forces, but reinterpreted it in his field conception. His interpretation seemed unsatisfactory, and he was painfully aware of this. He succeeded in rendering it satisfactory by looking for curved lines of electric forces, which his interpretation of Coulomb's theory, though not Coulomb's theory itself, allowed for. He thus found that electric lines of force curve in the presence of dielectrices, i.e., materials like glass or sulphur. It is no accident that Coulomb denied the possibility of dielectricity: he was a Newtonian. Nor is it accidental, I think, that Cavendish failed to publish his own discovery of dielectricity: he wished to work on it further and reincorporate it within Newton's metaphysics, and he died before accomplishing this formidable task. That this task could be performed with but a slight deviation from Newton's program Faraday knew, and he outlined ways of doing it, without however being able to do so himself for want of mathematical technique. The technique had been provided by Poisson, and Faraday said as much, but he was too neurotic about mathematical symbols to write them down on paper. Shortly afterwards Liouville was in a quandary because Poisson had, on his death-bed, asked him to make Poisson's own work the topic of a prize essay, and Liouville felt understandably apprehensive in view of Faraday's discovery which seemed to him not to fit Poisson's Newtonianism too comfortably. Kelvin, who was a young lad then, related all this in a letter which he wrote to his father from Paris, and he added a description of how relieved Liouville was to hear that Kelvin could interpret Faraday's discovery in an almost Newtonian fashion by using Poisson's own method. This was Kelvin's first published paper.

But there was no escape from Faraday's inspirations. Kelvin's theory of the dielectric assumed gross matter to possess electrical properties; his theory was not Newtonian but Boscovitchian. It soon transpired that Boscovitch's program needed modification. Gauss and Weber tried, and the attempt continued until 1905. By then it was clear that the program had to be given up; it looked as if Faraday's program had won out at last. Yet this program too was abandoned very soon after. It was deterministic and determinism had to be abandoned.

IX. THE HISTORY OF SCIENCE AS THE HISTORY OF ITS METAPHYSICAL FRAMEWORKS

The world is full of well known yet unstudied phenomena, of often heard but seldom debated theories. Historians of science all agree that some theories – Copernicus's, Maxwell's – and some experiments – Oersted's, Michelson's – are of supreme scientific importance.

That Oersted's experiment was of metaphysical significance is obvious in view of the supreme prestige Newton's metaphysics enjoyed at the time. The greatest problem in physics between 1820 and 1905 was, could there be a (satisfactory) Newtonian (or semi-Newtonian) explanation of Oersted's experiment? Study of this problem led to Newtonianism losing its interpretative power. It soon transpired that the only unrefuted satisfactory explanation of Oersted's experiment was Maxwell's, and it became an urgent task for Newtonians to interpret fields in accord with Newton's metaphysics, which means – since for Newton forces are attached to matter and since Maxwell's equations are not invariant to Galileo's transformations – with the aid of the assumption that space is full. A scientific version of this assumption was refuted by Michelson and Morley. In 1904 Kelvin still hoped that another Newtonian or Boscovitchian interpretation of electrodynamics could be found; but though a few shared his hope no one did anything about it, especially since his misgivings about Maxwell's theory were not shared by others. Undoubtedly, Maxwell's theory was so significant because it was a satisfactory explanation which conformed to Faraday's metaphysics. Undoubtedly Planck's theory became so important in 1905 when Einstein showed its conflict with Maxwell's theory because it seemed a major breakaway from Faraday's program.

I do not know why the significant events in the history of science should be metaphysically significant, but I have so far found it almost always to be the case. I suggest the theory that significance with respect to (pure) science is usually significance with respect to science's metaphysical frameworks. It is understandable that if metaphysical frameworks are research projects they should be taken very seriously, but why should all (pure) research projects be geared to a few metaphysical doctrines? Indeed, I think most research projects are not intended, at least not consciously intended, to be relevant to the dispute between the few competing metaphysical doctrines of the day. Yet those projects viewed later as significant show a capacity to throw light on current metaphysical issue. I can see no other explanation of the situation but that it is essentially metaphysical interest which gives (purely scientific) significance to this part of science rather than to that; hence, most (pure) scientists are more interested in metaphysics than they seem to be.

There are many studies which are not directly related to metaphysics. Take the continuum theory; it is the study of properties of matter, especially elasticity, on the assumption that matter is continuous. This study belongs to applied mathematics or technology rather than to pure science because it is based on a metaphysically unacceptable assumption. Its value for pure physics becomes apparent only when it is shown to throw light on an important scientific problem related to metaphysics. Indeed, since the Newtonian interpretation of the wave theory of light is the theory of the elastic ether, the rise of the wave theory caused immense efforts to be made to create any theory of elasticity whatever which might be used as a tool to render the ether theory scientific. Prior to that, the effort to develop a theory of elasticity were strictly in the Newtonian mode. We see how a significant plan of scientific research was first directly and then indirectly metaphysically relevant, and later it lost all relevance and with it all significance. Present day aerodynamics interests only few non-aeronauts, but it will interest more of them if it will reveal some bearing on existing metaphysical issues.

But what about scientific work unrelated to metaphysics? Let us take two examples. Jenner's study, his attempt to refute some village superstition, was highly idiosyncratic. Possibly it was connected with Bacon's idea that superstitions are dangerous to science, and yet as hardly anyone except Jenner undertook such researches, his work may well be viewed as idiosyncratic. The device of vaccination, which resulted from his study, was for long chiefly of practical value. The mechanistic interpretation of vaccination is identical with the theory of antibodies. It is thus a metaphysical theory. In the popular literature it is often presented pseudo-scientifically. Biochemists have used it as a program and found scientific instances which conform to it; they are still searching for others. This story shows how one idea entered into the mainstream of science because it fitted a metaphysical framework.

My second example is the discovery of the asteroids. It is insignificant. It refuted Hegel's doctoral dissertation, but this was of no value in any case. It refuted Kepler's metaphysics, but nobody had ever taken notice of this metaphysics. It agrees with Bode's law, but this law is related to no metaphysics. The discovery is insignificant because it has no direct or indirect relevance to topical metaphysics. It may, however, become significant, if asteroids are going to play some role in a future cosmogony.

APPENDIX: WHAT IS A NATURAL LAW?

Let me first elucidate the central issues concerning natural laws, and if what I am going to say is going to sound trivial, it is because I intend, I do not know with degree of success, to sound as trivial as I can – since I fear I may lose sight of the wood for the tress.

When Aristotle presented the two definitions of man, one as a rational animal and one as a featherless biped, he was hoping that everyone would find the one definition congenial, and the other problematic – chiefly because both are true. The uncongenial definition would not be uncongenial were it false, he felt. This is why Antisthenes, who opposed Aristotle's theory of definitions and who considered Aristotle a pompous professor anyway, threw in his face a plucked chicken, which is a featherless biped, to illustrate the falsity of the definition of man as a featherless biped and thus to dispel the uncongenial air about it. (Notice that the plucked chicken refutes 'man is a featherless biped' if and only if the word 'is' in it is symmetrical as in a definition.)

A few centuries later Sir Karl Popper, who has been called Antisthenes Redivivus, by the way, wrote a paper on the same topic, proposing the metaphysical hypothesis that for every universal statement which is not a law of nature there exists somewhere in the universe an instance contrary to it. Popper himself withdrew this hypothesis as implausible – let us take note of this fact – and thence came to accept the contradictory hypothesis that there exists some true universal statement which is not a law of nature. (See his *Logic of Scientific Discovery*, Appendix *x.) Let me state at once that I share the feeling of quite a few who find this hypothesis very problematic as well; and at the very least it certainly is also a metaphysical hypothesis. I wish to stress, first and foremost, that we have here two contradictory hypotheses, formed in a language which is of necessity acceptable and comprehensible to all with the exception of the word 'natural law' or its equivalent, and that both hypotheses are metaphysical. Hence, a logical positivist would have to declare the word 'natural law' to be meaningless, or else define it so that one of our contradictories will be a tautology and the other a contradiction.

The two contradictory hypotheses, let me reformulate, are 'all true universal statements are natural laws', and 'some true universal statements are not natural laws'. A logical positivist, such as David Hume, may be interpreted to have said, a natural law is any true universal statement, thus rendering our two hypotheses a tautology and a contradiction respectively. Also, Hume may be interpreted as having said, a natural law is any statement about causes, we never observe causes but only constant conjunctions, and therefore causation is a meaningless concept. Consequent on this reading, both of our metaphysical hypotheses become meaningless. I wish to stress that there are here two readings of Hume, *identical in content*, and different in wording. The one says, we have no concept of causations; we have a concept of constant conjunction which is confused with that of causation. The other reading of the same texts is, we have no concept of causation *except that* of constant conjunction.

The literature on natural laws may be presented, at the first approximation (to exclude statistics and such), as the question, is there a deeper connection between one couple of events constantly in conjunction then between any other couple of events constantly in conjunction? We do have an ordinary concept of deeper connexion: we all feel that a couple of businessmen are seen often together either because they are similar – say they have similar interests – and so frequent the same establishments, or because they cooperate and collude. We will all consider their being together accidental in the first case and due to some (efficient) cause in the second. Similarly, we all agree that many phenomena looked very natural to non-scientists and to scientists of the past, yet are mere accidents. Many phenomena which are temperature dependent, for example, are so seemingly natural simply because prior to our present century no man could observe temperatures below 100 degrees K or above 1000 degrees K and live to tell the tale.

But this argument carries little force: these intuitive cases do not take us very far. There are many universal statements concerning the properties of matter which were alleged to be true but which experiments in high or low temperatures have refuted empirically and thus we know to be false, or to be true only after they are properly qualified, perhaps, to ordinary temperatures. There is little trouble here as yet. But consider a universal statement which remains unrefuted even on the present widened domain of experienced temperatures. Suppose we were not able to refute it ever; will it thereby qualify as a law of nature? This question is the watershed of philosophy, dividing positivism from all else.

The idea of essential definition, like that of man as a rational animal, is that the universe is law-abiding, like a righteous person; that in the fibre of the universe there are certain constraints which prevent the universe from entering certain states. Of course we would like the constraints to exclude some logically possible states, and so the theory of essential definitions cannot be accepted in its Aristotelian form. We want essential definitions, however, not as tautologies yet not as any old true universal statement (such as man is a featherless biped).

Suppose this is so; suppose God has forbidden the universe from undertaking or exhibiting some logically possible events. Now the universe does not undertake or exhibit all possible events – this is a corollary of the trite claim that some events are exclusive of each other for one given space-time region. The question is, then, the following. Consider *any* universal statement which is *not* a statement of a constraint; the universe is permitted but not forced to exhibit at least once in its four dimensional menifold an instance contrary to it. Will it? If yes, then this is a strange meta-natural-law, saying, all permitted possibilities are tried by the universe at least once. (Is this so of necessity or by accident?) If not, then some true universal statement is true by accident, not due to any constraint imposed on the universe (say by its Creator).

I have used the word 'constraint' now, rather than 'cause' or 'law of nature', because this word cannot be rejected by positivists without their getting into some difficulty in economics, for example, or even in mathematics. Now, in these fields constraints are represented by certain universal statements taken to be true. But in these fields *not* all universal statements taken to be true are regarded as constraints. Hence, we may now contemplate the nature of constraints and see, after offering some criteria of constraints, whether all true universal statements are statements of constraint.

Here we can easily consider the form of the subjunctive conditional as peculiarly agreeable to the formulation of constraint-like statements; similarly, we can consider those conditionals which are clearly contrary-to-fact as obviously non-constraint-like statements. (See C. I. Lewis, *An Analysis of Knowledge and Valuation, VII, 6.*) Suppose a statement $(X)[P(X) \supset Q(X)]$ is true by accident, then we could imagine that if I somehow succeeded to be a P without being a Q, then the statement would thereby become false. If, on the contrary, it is not an accident but a constraint on all X's which are P's that they cannot be non-Q's, then I would not manage to be a non-Q P: if I were a P I would also thereby be a Q willy nilly. And so, the subjunctive conditional naturally seems to serve as the form of natural constraint. The contrary-to-fact conditional is one which may easily be designed to be true by accident. Take any description of an accident A, say 'today is Tuesday'; the universal statement $(X) [\sim A \supset P(X)]$ is true by the mere virtue of the accident that A is true.

Of course, many contrary-to-fact conditionals are erroneously viewed as false, many are considered to be true not as accidents but as constraints. The same may be said about subjunctive conditionals and more. Thus, though if I were to jump from my roof, etc. I would fall, etc., the law of gravity expressed thus may be an old and refuted version, such as Galileo's. Yet there was the feeling, particularly strongly expressed by Nelson Goodman, that if we could offer a criterion to distinguish these, from others which may be restated as subjunctive, then we would come closer to finding which universal statements are true laws of nature and hence true, and which are not. Here we come (again?) to the historical fact that natural necessity, which relates to the fibre of the universe and thus to metaphysics, concerns philosophers interested not so much in the universe as in the validation of science – perhaps as knowledge and thus as the truth; perhaps as merely probable.

Popper's experiment in offering criteria for natural necessity is regrettable not so much on account of its repeated failure. After all, experiments like that sometimes succeed only after a few minor amendations, and though W. A. Suchting's strictures are eminently valid they may call for only minor amendation. And so Popper conceeded (addendum to the 1968 edition of his work) that Suchting's strictures were correct, made small amendments, and declared he lost interest. Meanwhile Suchting has refuted the latest amendment as well. In any case, we may note, Popper's experiment in offering criteria for natural necessity is regrettable because it is performed (as he tells us in his Addendum) in order to prove the obvious to the heathens – an aim I consider unworthy of a scholar of his stature – namely, that we may know what is natural necessity in the abstract without thereby knowing whether a given true universal statement is accidental or not. I think it is obvious that this is so; we know what is a tautology or a logical necessity, yet we have no decision procedure for it; why should we be in possession of stronger tools regarding the decidability of natural necessity then regarding logical necessity?

I think Popper is in error in ascribing to Goodman the hope of finding such a decision procedure. It seems to me that what Goodman wishes to say – he is, I admit, vague enough to permit many readings – is that he wants a decision procedure not for truth and for natural necessity but for plausibility and for putative natural necessity.

Nelson Goodman and Michael Polanyi seem strange bed fellows, yet they agree on one point; certain generalizations from experience are so implausible that they are rejected as useless – not as false – without any examination of the facts, and regardless of their possible truths. Goodman's examples are (a) allegedly contrary-to-fact conditionals, (b) generalizations to arbitrarily limited space-time domains, and (c) those formulated by the use of arbitrarily deformed concepts like 'Grue'. Polanyi gives the example of a paper in *Nature* of empirical data for the hypothesis, all gestation periods counted in days are multiples of Π, where Π is the ratio between the circumference of a circle and its diameter. He offers other examples, like Velikowski's theory, and even his own theory of adsorption of 1914, then rejected off-hand as implausible yet now accepted. Polanyi does not complain about his own ill fate, nor does he offer a criterion for plausibility: he thinks there can be none. Goodman, I think, looks for a criterion, though he uses the term 'projectibility' rather than 'plausibility'.

I view the problem of natural necessity as metaphysical. I therefore think that any attempt to offer an explication of it, be it Popper's or Goodman's, must either be inadequate or amount to a metaphysics proper. Whatever our definition of natural necessity may be, say under conditions C the statement L will express a natural necessity. We may ask, is the fulfillment of C is ever necessary? Is it necessary for L to be a natural necessity that C be true? Etc. Assuming all this to be true for moment, then it seems more reasonable to relinquish the explication of the concept of natural necessity in abstracto in favour of the concept of natural necessity relative to a given metaphysics. At the very least I propose it is easier to offer a relativized concept of natural necessity – relative to a given metaphysics – than an absolute one. I have offered such a concept above: I have suggested that a universal statement is a putative law relative to a metaphysics if and only if it conforms to that metaphysics. I have offered a vague idea – not at all formal – of what conformity to a metaphysics is, and I have illustrated it with examples from the history of science, chiefly of physics. If space permits I shall briefly repeat my slender outline. Before that let me say that in my opinion a hypothesis is plausible or projectible in Polanyi's or Goodman's senses if and only if it conforms to the metaphysics relative to which it is judged. This leaves open the question, when is a metaphysical hypothesis plausible, and when is any other kind of hypothesis plausible. I have tried to handle some of these questions above, but not all; in any case, let me say briefly when a hypothesis conforms to a metaphysics. Though my presentation is sketchy, it seems to me very intuitive.

A metaphysical hypothesis is implausible in any case except when it belongs to a fully fledged metaphysics. (This is why both contradictory hypotheses with which we began the present discussion were unsatisfactory.) A fully fledged metaphysics is a theory of the world capable of becoming physics, or scientific. Thus, for example, Democritean atomism and other doctrines of what things are made of, look as if they can become scientific (and some of them did become scientific). Atomism as a metaphysical doctrine does not follow from a given scientific atomic theory and it does not entail any. But if we have a series of scientific theories explaining every known phenomenon, a complete scientific theory, each part of which conforms to atomism, then metaphysical atomism would follow from that complete series of scientific theories. We can then say that atomism was so enriched as to become scientific. Clearly, the law of definite proportion, reciprocal proportions, and the like, are all atomic or conform to atomism, whereas the continuum hypothesis, and thus elasticity, does not. The point can more easily be illustrated with Newtonian metaphysics which views the universe as an aggregate of impenetrable atoms governed by a finite set of central forces. When a Newtonian wishes to explain a phenomenon he looks for a central force; a law which is not that of a central force may look suspect to him, for example laws of diffusion, or it may look to him like a law which will soon find its proper Newtonian expression, like laws of chemical affinity or of electromagnetics.

All this reinforces the point that we may have an intuitive idea of a natural necessity, yet without knowing what is a true instance of it. For, and this I contend is a historical fact, what looks a natural necessity one generation may look an accidental universality at the next – I explain this by reference to our changing metaphysical framework. This is particularly obvious in the case of competing (metaphysical) schools. Indeed, whereas metaphysics guides us in what we may consider *a priori* a plausible putative natural law, science guides us in criticizing our metaphysics and replacing it by a better one. Science, thus, indirectly alters our idea of natural necessity. It did so in the past, and, I contend, it may do so again.

CHAPTER 10

QUESTIONS OF SCIENCE AND METAPHYSICS

The idea of science propounded here is a combination of two views. First, the Cartesian or rationalistic view, or the deductivist view, of science as subordinate to metaphysics – recently revived by Meyerson, Burtt, Koyré, and others. Second, the critical view or the hypothetico-deductive view of science as Socratic dialogue par excellence – a view we owe to Sir Karl Popper. In this chapter I wish to relate both views to a new branch of philosophy, the logic of questions, or erotetic logic. I shall briefly mention what has thus far been done in the field, describe my dissatisfaction with it as not a true dialectic or a logic of Socratic dialogue, and, finally, attempt to link the more advanced (dialectical) part of the logic of questions with my view of the role of metaphysics in science. Briefly, the Socratic method is a method of critical cross-examination, i.e., of trying to find errors in a given answer by eliciting more answers to ancillary questions. This requires the ability to produce answers and to recognize a given statement as either an answer or not an answer to a given question. The analysis of answers was taken up by various authors, including C. L. Hamblin, David Harrah, Nuel Belnap, and Lennart Åquist. The question, what does one do when one has no answer to a given question? has been studied by Sylvain Bromberger. If I understand him correctly – which he doubts – his view does not differ from mine: metaphysical theories help us devise answers to some sorts of questions and so act as selectors.

I. HOW DO WE SELECT QUESTIONS?

The question of the selection of what to say, whether here and now, or generally, has seldom been discussed. We all agree that a speaker should not be capricious in the choice of what he says. At most we allow his choice to reveal his own standpoint, which may, to some extent, be the outcome of arbitrary choice. The lack of arbitrariness or caprice is what makes the correspondence theory of truth, the objectivist theory of truth, most attractive. Truth is seen as independent of space and time; sometimes we even say, truth is independent of space, time, and circumstances. But there is one particular circumstance which signifies, and which is (rightly) most often stressed by the enemies of the correspondence theory of truth – pragmatists, relativists, and subjectivists. It is true *to me*, says the subjectivist. I decree the truth – for me and my followers. The circumstance which the objectivist by implication declares irrelevant is the speaker. The truth-value of what I say is independent of me or of my convenience or of my survival as the fittest. Do not believe me on my say so, and do not believe me even in order to save my life. Believe me only if you think I am saying the truth.

The requirement to avoid arbitrariness in our pronouncement is seldom discussed, yet it is best manifest in the objectivist theory of truth. The reason for the lack of mention of the requirement to avoid arbitrariness is, indeed, that it is covered by the requirement to speak the truth. The traditional view among men of science is that since scientists speak the truth they need not worry about avoidance of caprice: there is no caprice in truth. Much has been said about the fact that scientists and philosophers easily get on each other's nerves. The reason, I think, is that scientists take their own objectivity for granted, whereas philosophers often find this highly problematic.

Philosophers may find this problematic on any of three counts. First, they may reject the objectivist theory of truth. In this case I am as impatient with them as the scientists are. Second, they may be skeptics who doubt that anybody knows the truth. I am a skeptic myself and have encountered much impatience from scientific colleagues. Third, they may suggest that as a scientist selects certain truths from a larger body of truths his selection may be arbitrary. Many sophisticated scientists are willing to pay attention to this objection, though they usually consider it quite answerable. I agree with them, but find the answer not so easy to formulate. In this chapter I shall try to present not the answer but an answer, one possible objectivist criterion of selection. The criterion of selection will not be a criterion of selection of truths but of putative truths. For, as I say, I wish to start from a skeptical premiss, denying that we know truth from falsehood (except for tautologies, of course).

Our problem, how to speak while avoiding caprice as much as possible? itself stems from our ignorance. If we had knowledge, not only would we know how to avoid caprice, but we would even find speech redundant.

Those who know, share the silence of the community of knowers. We talk, we exchange ideas, we have dialogues, at least in order to even out the differences which are idiosyncratic – as a necessary condition for the search for the truth. So, we talk, for one thing, because we do not know the truth. Pretending to know the truth may easily make us replace the unknown truth by the sense of the meeting, namely what is widely accepted as true. For, though there is no point telling you what you know already, I may tell you what you want to believe anyway, so as to reinforce your desire. The ironing out of idiosyncracies is only a necessary condition for truth, not a sufficient one; it may also serve self-deception. And so, (1) we do not know the truth, and (2) we may mistake popularity for truth. Finally, (3), our selection of truths or of putative truths may follow our own personal predilection or the predilections of the group to which we talk. Deception by selection of truths is an ancient sophisticated substitute for lying. These, then, are three objections to the idea that the requirement to speak the truth covers the requirement to avoid caprice.

I shall take these three objections in this order, dispose first of the first two, concerning knowledge of truth, and concerning consensus, in order to concentrate on selectivity, criteria of selectivity, rules for the selection of putative truths, or rules concerning what constitutes significant putative truths.

The objectivist theory or the correspondence theory of truth has frequently been taken to extremes, with that resultant sectarianism and fanaticism which extremism so often entails. In order to be objective, it has traditionally been claimed, one must insure or secure or guarantee that one is being objective; in brief, prove one's correctness. This is an incredible idea: allegedly, objectivity requires proof of objectivity and proof of objectivity requires proof of truth. No doubt, proof of truth is to be taken as a sufficient condition of proof of objectivity which is to be taken as a sufficient condition of objectivity. Also, the making of sufficient conditions necessary is always adding a safety margin. But the safety margin may be too wide; it may require from us more than is at all possible. Turning proof from a sufficient to a necessary condition for truth rests on the doctrine that one is always able, in principle at least, to offer some proof. This exaggerated zeal for truth leads to the conclusion that as a responsible citizen I speak the truth – and hence, my opponents are all knaves and/or fools. This complex of views can be ignored; it may optimistically be declared a thing of the past. Admittedly, it still has circulation, but so have so many defunct currencies.

The idea of guaranteeing objectivity by proof is often now replaced by the idea of aiming at objectivity by inter-subjective inspection. The inspection begins from the speaker's notion of why he chooses to say what he says, examines his appeal to whatever relevant factor he wishes to appeal to, examines the truth of his evidence, but then leaves it at that, and lets each member of his public to be judged unto himself. Let us take for granted now that we can aim at objectivity by intersubjectivity. We then require not that we speak the truth but that we put forward a peculiar sort of examinable putative truth, that is, that we put forward some peculiar *bona fide* claims. The claims have to be peculiar in that they may interest our public enough to perhaps try and examine them. This, I think, almost fully disposes of my first objection. I admit I do not know the truth, but I shall try to speak truthfully and offer my assertions for your critical assessment.

Now to dispose of my second objection concerning unanimity as truth surrogate. It is hardly possible to please any large audience which has not been filtered through a monolithic ideology. Even on the supposition, which I believe to be true, that any philosophical audience will share certain presuppositions, it is hard to get at the sense of the meeting. First, as R. G. Collingwood noted in his *Essay on Metaphysics* (1944), some presuppositions are not expressible; or rather, as I would like to correct him, some of *today's* presuppositions are not *yet* expressible – they have not yet found their proper expression. The inexpressible (or not yet expressed) presuppositions constitute, Collingwood asserts, a point of view or a viewpoint. They constitute, he adds, a metaphysics. With modifications as stated, I go along with him.

Even if some of the presuppositions we all share can be articulated in sufficient numbers to constitute a public lecture, there will be quite a few who will think it nothing more than a string of platitudes. The rest of this lecture may be a case in point. Now, some audiences love to be reassured, some to be challenged. Therefore, how to avoid pleasing by appeal to prejudices, is hardly ever a serious problem.

It is hard enough to interest many members of even a fairly homogenous audience in one *question*, let alone to get them to agree on the *answer* to it.

The fact that some intellectual groups are nearly unanimous about what is a question worth pursuing at the moment is something of a miracle. The question I now wish to raise interest in, is just this: Are there any inter-subjective criteria for ranking the significance and/or interest of a given question or set of questions? There is a growing conern in the profession with the question of questions. Philosophy seems now to be on the verge of developing a theory of problems. Arthur and Mary Prior have already christened it 'erotetic logic' (in their paper on Bishop Whately's logic of questions, *Philosophical Review*, 1955). Before I go on I wish to summarize what follows in one paragraph.

II. WE SELECT QUESTIONS WITHIN GIVEN METAPHYSICAL FRAMEWORKS

The question, which question is worth pursuing? is a crucial question on which there is hardly any literature – but which is asked and answered regularly. The canons employed in this procedure can be made explicit and improved upon. These canons are part of what we take to be the viewpoint of the community to which the questioner belongs. That viewpoint includes, to repeat Collingwood's assertion, the metaphysics of the community. Not that the individual is bound by his society's viewpoint – indeed, he may invent his own. But for a new metaphysics to signify, it must capture an audience of investigators. It does that by raising questions that interest a public of students willing to spend time attacking them. The metaphysics, then, is operationally – but only operationally – equivalent to the cluster of questions it gives rise to, to use Bromberger's phrase. This is why Faraday, the inventor of a new viewpoint in physics, spoke of 'that duty to science which consists in the enunciation of problems to be solved...' So, I suggest, that question is best and most worthy of pursuing, which is most likely to alter our viewpoint, our metaphysics, our whole view of the universe. Science has made a habit of altering our viewpoint, and there is method in this madness. Moreover, this method is contrary to current methodological views, particularly those of K. R. Popper.

III. THE LITERATURE ON QUESTIONS

The interrogative literature is scant. Such literature as there is certainly includes the Pseudo-Aristotelian *Problemata,* but ignorance prevents me surveying, still less discussing, the history of the interrogative literature. I wish only to draw attention to a strange lacuna. Learned academies of all sorts make a practice of offering questions to the public, often against promised high remuneration, but the cases are not studied; even famous cases are shrouded in mystery. There is the case of Robert Boyle telling his fellow members of The Royal Society of London which questions to pursue in their empirical studies. Sometimes he had no success. An example is, what increment of the product of pressure and volume of a given gas is due to what increment of its temperature? So many times Boyle asked people to study this question, so many times he expressed his disappointment in the neglect which this question suffered; and nothing happened for over a century. Why?

Boyle also raised questions which were studied. This must have encouraged him, or else he would not have written what amounts to the first scientific questionnaire, *The Natural History of a Country.* It was published first as a paper, and then as a short book which saw several editions. Its questions were for travellers, concerning flora, fauna, climates, and customs. We cannot understand Captain Cook, or Captain Bligh, without knowing the tremendous influence of this volume. It was first noted by J. F. Fulton, the distinguished Boyle bibliophile. Its influence spread far and wide. Let me mention one example. Albert Chamisso, who marks the end of the Age of Reason, and who has a foot in romanticism, a refugee from the French Revolution, wrote a fantastic autobiography, *Peter Schlemiehl,* about the miseries of a fellow who foolishly sold his shadow to the devil. When Peter loses hope of ever again finding happiness he becomes a traveller. He – Peter as well as Chamisso – utilizes his travels and naturally writes books; on flora, fauna, climates, and customs, until he – Peter – is tired and passes away in a poorhouse. Chamisso doubtless never read any of Boyle's works. They simply entered the scientific lore.

We do not know the content of this lore, and can only partly reconstruct it. It is possible, for example, that the *Notes and Queries of the Royal Anthropological Institute* are linked with Boyle's *Natural History of a Country,* but I, for one, do not know the connections or the significance. We know, for another instance, that Newton's *Queries* in his *Opticks* dominated much of eighteenth century science and were imitated, say, by

Joseph Priestley in his great book on different airs. The total effect is not yet known or studied – except very superficially. We know that in the eighteenth century a prize was offered for the discovery of any connection between electricity and magnetism. We also know that the first such discovery was made half a century later by Hans Christian Oersted. Did he collect the prize? Was the prize cancelled by then? Why? I do not know. There are many books which mention the prize. None tells us whether Oersted collected the prize or if not, why not. The absence of any progress whatever could have led the academic institution involved to believe that the problem was insoluble. We know from contemporary evidence that the first response to Oersted from Paris was of utter incredulity: the news was taken as a hoax.

I hope this indicates how little is known about the history of questions and how significant it seems nonetheless. We can see this even in our own lifetime. There is consensus on, say, what are the current leading problems in physics, and every physicist can recite the semi-official list. Where a disagreement about ranking questions rages abroad, there may be debate about given answers to certain questions, but it is hardly ever about the ranking of the significance of these questions.

IV. THE LITERATURE ON THE LOGIC OF QUESTIONS

Coming to the literature on the logic of questions. It has not been surveyed or studied as yet. It has two parts, preliminary and contemporary. Peirce and Collingwood did the preliminary work, which is not technical enough for modern taste, but is highly problem-oriented (in particular, how do we choose questions to study?) and dialectical. The contemporary writers have contributed much to the technical side of it. With the exception of Åqvist, however, they are not dialectical, and, with the exception of Bromberger, they are not problem-oriented. Let me expand on this claim before I go into the more technical part of erotetic logic.

To begin with Peirce and Collingwood, it is not my concern here to ascribe priority, although, it is true, the idea that a true question must have a true answer was repeatedly expressed by Peirce long before those modern writers who make so much song-and-dance about it. Nor do I want to labor the point that the idea that questions hang on presuppositions is Collingwood's. The point I wish to make is that these two writers have written down certain wise maxims which are in a sense well-known, in a sense still ignored.

When Peirce recommends that before examining the truth-value of an answer to a question we examine whether the answer fits the question; when Collingwood asserts that an examiner of a serious question collects first the previous answers to it and studies their shortcomings; when Peirce and Collingwood see a question as both an expression of dissatisfaction and an attempt to prompt an answer; when so engaged, they are in the no-man's-land of what is obviously the wisdom of the ages and also un-examined mythology. What present-day erotetic logicians (the logicians of questions) do with this mythology is to elevate part of it to the rank of logic by putting it in modern hieroglyphics. (I am using the word 'hiero-glyphic' in the original sense of sacred writing (carving) rather than the vulgar sense of indecipherable writing.) I approve of the exercise, but cannot conceive of it as settling any issue whatever. On the contrary, it only presses our questions harder. Why is the mythology, what the myth expresses, so significant yet so obvious? Is it so obvious? David Harrah, for example proves that every statement is an answer to some question: the statement that p answers the question, is it the case that p? The ques-tion p?, he says, is the invitation to choose between asserting p and assert-ing non-p. He then identifies dialectics with questions-and-answers. This hardly does justice to the role of questions in dialectics proper. Questions should not merely elicit answers, but also constitute cross examination and criticism of answers to previous questions aimed at the exposure of a putative truth as a falsehood. The same injustice, surprisingly enough, can be found in the work of Collingwood, his own critical attitude not-withstanding. I shall go further and argue soon that dialectics criticizes questions, or, if you will, the presuppositions they rest on.

The example of a presupposition to a question which is used in the literature may well illustrate this. It is Collingwood's example, and almost everyone else's: Have you stopped beating your wife? Everybody knows, says Nuel Belnap, (*An Analysis of Questions: Preliminary Report*, 1963, p. 125) that this question is unfair – it rests on two presuppositions which we hope are not both true: (a) you have a wife (b) you beat her regularly. Now this is very baffling. This question, have you stopped beating your wife?, is unfair not because of its presuppositions but because of insinua-tion. The nasty attorney for the other side, let us say, asks an insinuating question in a run-of-the-mill courtroom drama. Only a fool would lose his cool. The best response is to take the question literally: No, I have not ceased beating my wife. The rules of procedure have to protect fairness. The attorney for your side has the right to cross-examine you next. Why, my dear fellow, he asks all puzzled and bewildered, why have you not stopped beating your wife? Because, you proceed in an equally cool manner, I have not started as yet. And I cannot very well stop what I have not started, can I? And this is a simple case of dialectic – even in common courtroom drama. If all courtroom dialectic were of this sort it would be much duller than it is. Yet erotetic logic has not even come that far.

The error which the erotetic logicians commit when declaring 'have you stopped beating your wife?' unfair is of a great significance. As Åqvist has shown (*A New Approach to the Logical Theory of Interrogatives*, 1965, p. 75), we regard the question as unfair only if we think its presuppositions to be false. It is certainly fair when the presuppositions are presumed to be true. When we do not know whether the presuppositions are true, then (as Harrah and Belnap themselves notice) we run a risk; if risk entails unfairness and if unfairness is to be excluded, many questions will be censured and this will quite frustrate our quest for knowledge! Hence Belnap's blanket opposition to the 'have you stopped beating your wife?' type of question is seriously objectionable.

Another example is that of Cooper Harold Langford, of Lewis and Langford fame, and though it appeared in the *Journal of Symbolic Logic* (1939), it has not been followed up – to the best of my knowledge – even by those who have pursued at some length the topic, which is the existential import of questions. Is your brother older than you? asks an attorney in court. You do not have a brother. What is the true answer in this case? Boolean, says Langford, is unlike ordinary English here: Boolean permits both 'yes' and 'no' as true and proper answers, whereas ordinary English admits no true answer at all. Of course, Boolean also permits as a true answer 'both yes and no', but we need not go into this fascinating and often misunderstood type of answer, beyond observing the well known fact that to say both yes and no is to say, I have no brother, and vice versa. Now, on the basis of this, a witness may well confuse one party in court – say by choosing to answer, yes, and thus insinuate that he does have a brother. It will be the task of the attorney for that party to ask the witness, have you got a brother in the first place? It is doubtless easy to report courtroom cases in accord with Langford's analysis. Hence, legal English is more Boolean than ordinary English where, in response to a question which may be misleading, the rules of polite conversation all too often require that one explain the answer so as to prevent misunderstanding. There is no doubt, in any case, that the rules of polite question-and-answer sometimes radically differ from the rules of cross-examination, scientific or legal.

Suppose, further, not only Boolean algebra, but also Russell's theory of descriptions. Take again the question, is your brother older than you? and replace it by, the person who is your brother, is he older than you? This question, entailing, as it does, the false presupposition that you do have a brother, is unanswerable. In the language of erotetic logicians, it is to be criticized rather than answered, or merely pseudo-answered. In court the attorney to the other side must object, the court must sustain the objection, the attorney questioning you must withdraw his question and break it into two: first, do you have a brother? If yes, he proceeds to the second half, is he older than you? If not, rules of relevance prevent him from asking the second half. Again, this is so standard a court procedure, that one even sees it regularly in courtroom melodrama. The case which, as erotetic logicians say, involves a question not to be answered but to be criticized, then, is common enough and indeed involves a break down in questioning because a tacit presupposition has been exposed and may have to be given up. This, in my opinion, is true progress in any kind of investigation – legal, scientific, or metaphysical – especially when the presupposition does turn out to be false as suspected. The rules of polite question-and-answer for the same kind of question are much more involved, and I shall not discuss them here.

Peirce and Collingwood saw as the central problem of erotetic logic the following one: which problem should we invest efforts in studying? The question has wide ramifications. I shall not discuss Abraham Wald's study in his decision theory of the cost of additional information and its possible influence on an improvement in the decision. I shall rather discuss Bromberger's remarks which go beyond the field studied by Wald, though with much less tangible results. First, however, let me show one case where it is of great significance to ask the question, which question a researcher is advised to study, which not?

Michael Polanyi tells (in his *Personal Knowledge*) of an experiment performed by a leading British physicist which should have opened up hosts of new and exciting questions, yet which was not taken up by anyone. Naturally, Polanyi was puzzled and asked a few physicists, why did they ignore the result? They shrugged their shoulders. It turned out that they were right: the experimental claim was based on an error. So much for Polanyi's story, the story of an experiment which ought to have drawn the attention of the leading thinkers in the field but did not. Polanyi uses this as an example of personal knowledge, of knowledge which is valid, useful, yet not given to articulation. He thinks it points to the idea that to become a scientist you must breathe science and feel science with your whole being in order to know which scientific question to pursue and which to ignore. If you go by the obvious you may easily go astray. This is why the minor scientist follows the major scientist or the senior scientist. This is why science is the activity of the community of scientists.

In brief, Polanyi believes that the agenda of scientific investigation, the priority of questions to be studied, is determined authoritatively by the doyens of science who, *eo ipso*, are always right. It is perceptive of Polanyi to notice, what has been noted all too seldom, namely that the choice of questions is often an important matter. This is, indeed, how Young Turks become established – by overthrowing an old agenda and putting forward a new agenda. Not necessarily a new view, a new answer, but merely a new question, and with it, to return to Collingwood, a new set of presuppositions. But Polanyi is unable to admit that Young Turks sometimes take over a scientific tradition because in his view the doyens of a tradition are always right. What shall we say to that?

Polanyi's case of an unpursued seemingly important experiment, is narrated with much too much mystification. It seems to me that the case is amenable to a much simpler and much more reasonable analysis. The procedure involved is rather simple and was invented and implemented by Robert Boyle: do not call an experimental result false, rather call it unrepeatable, and not even that unless it is decreed by others to be unrepeatable or unless you have yourself failed to reproduce it. Usually an unexpected result is first corroborated by experts in the field. Young upstarts ambitious enough to strike the iron while it is hot may take the risk of assuming the reported experimental result to be true and forge their way. If they were right, they get great rewards for being second only to the trailblazer, and if wrong all they lost was a little time. If they wish, they may, more cautiously on occasion, repeat the experiment and only then proceed with the questions it gives rise to.

It is amazing how much established procedure there is in each established field; how changes, especially of outlook, effect these procedures, yet how little literature there is on all this. After all, an experiment is a question to nature; and by the time an experiment is designed, the question to nature is well fixed. (Otherwise we say that the experiment is ill-designed and expect no enlightenment from the process of carrying it out so prematurely.) The process leading to an experiment is a lengthy process of choosing a question cluster (to use Bromberger's idiom) and slowly nailing down one narrow question. This procedure, so important, is still hardly examined.

V. THE INSTRUMENTALIST VIEW ON THE CHOICE OF QUESTIONS

What, then, determines which question will be pursued by scientists? Not surprisingly, perhaps, the only widespread answer offered to this question is technological. I often ask scientists orally, or try to divine from their written work the answer they would give to the question, how do you decide which question to pursue next? If I elicit any answer at all it is, I choose any question which I have the tools to pursue. The tools may be experimental, such as high energy accelerators, mathematical, such as recent solutions to non-linear equations of a given type, or scientific, such as a general theory of superconductivity. The answer that techniques determine agenda has been expounded at length only by the celebrated instrumentalist philosopher Pierre Duhem, in his magnum opus *The Aim and Structure of Physical Theory*. (See also Millikan's autobiography.)

Duhem's answer has truth in it, but it is a false answer, and one which is empirically refuted. Undeniably, there exist in any given field of investigation some standard techniques. No doubt new techniques invite new investigations. The discovery of radioisotopic organically assimilable elements has opened a vast field of research by enabling us, with the use of tracing techniques, to study diverse questions of assimilation and of metabolism and more. In some cases the rendering of a new technique useful for research is most ingenious, for instance Fraunhofer's discovery of the method of comparing the diffraction indices of diverse materials by the use of solar absorption spectral lines (which, subsequently, were named after him). A more obvious instance is Einstein's use of the absolute differential calculus in his development of his theory of gravity.

True, then, as Duhem's view is of many cases, it is most unsatisfactory even in those cases. For, the response these cases should evoke is not, thank goodness techniques evolve which enable us to tackle new problems! Rather, it should provoke us to ask, how come techniques evolve and how come they are applicable beyond their original intended domain of application? It is no accident, as I have tried to show elsewhere (*Journal of the History of Ideas*, 1969), that the absolute differential caluclus was there when Einstein needed it; rather, the problems which led to the development of that calculus, as well as to Einstein's theory of gravity, have a common ancestry – in the philosophy of Leibniz. It is not always that the questions asked by the inventor of a technique relate directly to the questions asked by the one who applies it. The questions answered by radiophysics differ greatly, for example, from the questions asked by the biochemist. Attempts to relate the two, however, constitute very interesting parts of physics and of physical chemistry.

Sometimes, of course, behind a technical question there is a theoretical question meant to be solved by it. When J. J. Thomson invented the primeval television tube he was interested not in the tube and in what it could bring about, whether television or mass-spectroscopy, but merely a theoretical question concerning the mass and charge of the smallest particle. This example, and its like – and there are many of them – utterly refute Duhem's theory of choice of questions. In these cases it is not that new techniques allow the study of new questions, but that the interest in new questions provide incentives for the development of new techniques. New techniques do not evolve, as Duhem's theory tends to suggest, they are developed with given interests in mind. How do these interests get chosen?

No doubt, existing techniques do constitute a factor determining choice of problems, only not the sole factor. Sometimes they do constitute the sole factor, and that is when the new technique is so powerful that its employment is satisfactory for almost everyone. Lakatos has observed that the techniques of modern formal logic, of the combined sources of the Frege-Russell-Whitehead stock, and of the more formalist Hilbertian stock, has offered so much challenge that all problems to which these techniques are not relevant were forgotten for a generation or two. This is understandable, but also it is a certain loss – not merely of leaving certain avenues unexplored, but also of leaving the lone explorers of these avenues isolated and forgotten.

VI. COLLINGWOOD'S PECULIARITY

Such a lone explorer was R. G. Collingwood. One reason why his explorations of the logic of questions and answers were neglected is that they were unrelated to orthodox formal logic and seemed – at least to him – to be in conflict with formal logic. He explicitly suggested, in open conflict with all other logicians, that the meaning of a statement is not a constant proposition but a variable which depends on the question it comes to answer. Even the truth-value of a statement, he was bold enough to assert, can vary from question to question. Nowadays we can go further. The meaning of a question, David Harrah now contends, itself depends on our total background-knowledge. The idea of dependence on total background-knowledge is Popper's. In his memorable review of the works of Harrah and Belnap, in the *Australasian Journal of Philosophy* of 1964, Hamblin, himself a former pupil of Popper's, suggests that the whole of Harrah's use of information theory is thereby swayed from the classical Shannonian use closer to that of Kemeny and Carnap, and nearly arrives at Popper's position.

Now, if Collingwood is right, then, to some measure, the meaning of an answer depends on the question it answers; and if there is the smallest measure of truth in Harrah's claim that the meaning of a question depends on background-knowledge, then, not only information theory, but also logic, has to alter quite radically. What is amazing, though, is that even so, much of the logic of questions can be developed within the old system – contrary to Collingwood's expectations. (Even much of the relativisation of answers to different bodies of background-knowledge has been developed quite classically. See Åqvist, *op. cit.*, final section.)

The question remains very simple: is there a compelling reason to accept Collingwood's claim (with some modifications)? The authoritative assessment of Collingwood, unfortunately, is that of Alan Donagan, as expressed in his comprehensive *The Later Philosophy of R. G. Collingwood* (1962). Donagan is doubtless right in his criticism of Collingwood. The criticism is very similar to what goes on between present-day erotetic logicians. No matter what Harrah or Belnap think of Åqvist's criticisms, these are often devastating, no one in his senses will consider them the touchstone for erotetic logic. This is perhaps the chief difference between a loner and a team: a criticism which in a team is a stimulant for further work permits us to ignore the loner in good conscience. But our real reason for ignoring the loner is that we do not share his preoccupations, that we are left cold by his questions and quests.

Collingwood's questions and quests, however, refused to lie dormant for long. They kept cropping up, and even in diverse places. Whenever an attempt was made to integrate our background-knowledge (to use Bunge's idiom), relating questions or interests, Collingwood's questions came close to the surface. I shall not dwell on these topics, each being sufficient for a separate study. Let me only mention two conspicuous items. The one is Chomsky's theory of ambiguity: every sentence is ambiguous and may be read differently depending upon where we put the emphasis, and the resolution of the ambiguity depends on context. The second, or perhaps even the same, idea comes from Gestalt psychology, just as a picture may look different depending on the observer's frame of mind, the viewpoint from which he approaches the picture, so a sentence may have different meaning depending on the question preceding it.

These are most important and glaring instances. But chiefly we have, philosophically, discovered that research is the pursuit of questions, some fruitful, others not. This is another topic which Collingwood thought important and tried to interest people in, to no avail. (Even Donagan, in 1962, failed to sift the grain from the chaff.)

Briefly, Collingwood believed that science is certitude, and so he believed in induction. He also believed that science consists in putting questions to nature. The combination of these two ideas is what he has called the Baconian method – quite incorrectly, since it is Whewell's, not Bacon's, methodology which Collingwood was following. So Collingwood rightly asked nature good questions and put his hypotheses to test, but wrongly went on to verify his answers. Hence, his colleagues concluded, and Donagan concurs (p. 200), Collingwood's method is defective. Hence, concludes Donagan, Collingwood's view of questions in science is erroneous. It is particularly here that Popper's theory offers a tremendous boost to Collingwood.

VII. THE LOGIC OF MULTIPLE-CHOICE-QUESTIONS

C. S. Peirce once said, science offers tentative hypotheses, indeed questionable hypotheses, sometimes even 'almost incredibly wild' ones, which are thus almost-questions: questions are doubts, and assertions are beliefs, he said. Not so, say Popper, Bunge, Bromberger, Lakatos, and others. We may accept a hypothesis for the purpose of finding an object of belief, and we may accept a hypothesis for the sake of finding an object of study and examination. Should I invest the next few years of my life in question x? or, will it soon turn out to be a worthless question? Or, still worse, will it turn out to be worthless only years later?

Popper said once, we need to fall in love with a problem. I am very dissatisfied with this. One must fall in love with a problem, to be sure, but one may try to make it the right one! Indeed, Popper's theory of degree of falsifiability says more. It says look for as highly testable a hypothesis as possible. This, surely, relates to the choice of problems. How exactly is not yet clear – it is being studied. The logic of questions keeps pressing and now people are working on a technique to embed it. Thus, techniques suggest questions but also questions recommend developing techniques. In 1955 the Priors said that this cannot be done within formal logic. Yet Hamblin started the trend, I think, with the use of simple ideas from logic and information theory. Roughly, Hamblin equated (already in his doctoral dissertation, University of London, 1957) a question with a set of alternatives plus the instruction to choose one: e.g. 'What is the color of my horse?' means, choose between 'the color of my horse is white, is brown, etc.' Now the disjunction may be complete and so a tautology – at least until challenged in accord with Arthur Pap's posthumous paper in *Mind* about the synthetic *a priori* character of the complete list of colors. We may, however, exclude green on the basis of the background knowledge that no horse is green. The disjunction *sans* green, then, presupposes 'no horse is green', and from this Harrah concluded that questions presuppose information. If the information be false, Belnap and Åqvist add, we do not answer the question, but criticize it, correct it, rectify it, or pseudo-answer it. This, of course, is more interesting and more dialectical. It is, indeed, very significant that for the first time use is made of the truism, a question well-put, is a question half-answered. But this is only a beginning. Erotetic logicians can handle some yes-no questions, though this takes some effort. Also, they can handle questions of the kind, which of a given set of alternatives is true. This kind of question includes who, which, when, etc. questions, which Carnap has called W-questions. If the list of people is not complete, who questions may be cumbersome, or unanswereable as yet, but still not seriously problematic.

A little reflection will make this quite plausible. When a customer asks an airline computer which flight to book from a to b? the computer gives no anser to the question. Rather, the computer can translate the question into a multiple-choice question, and the customer then decides the answer. The multiple-choice list may be incomplete on various counts, or complete (relative to given background-knowledge!), as the case may be, with both question and answer subject to certain constraints (explicit or embedded in the given background-knowledge).

What this indicates is, first and foremost, that contemporary erotetics is a logic of questions, hardly of the choice of an answer or of a decision. This may explain the fact that, contrary to Collingwood's expectation, so much of it could be developed within traditional formal logic.

This, however, is not to say that all is smooth within the logic of questions: as I have hinted, the situation is more of work in progress than of a job completed. Even when we restrict the study to seemingly trivial cases, we may get a number of interesting results. Peirce has translated 'Is this the way to the city?' into 'This is the way to the city, eh?', which is the same as the request to determine the truth-value of the statement 'This is the way to the city'. This sounds perfectly straightforward and quite unproblematic, until one tries to embed it in a formal system with requests and knowing what constitutes an answer to a question (i.e. turning it into multiple-choice) and presuppositions. It soon transpires that some questions are risky, even yes-no questions (have you stopped asking risky questions?), as we have seen already.

Thus, much of present-day erotetic logic renders questions into multiple-choice questions, their transformations, and the limits of their satisfactoriness. But we have a completely different kind of question, which Bromberger has labelled p-predicaments, with p for perplexity or puzzlement. When we ask, why?, we may know whether a particular statement would count as an answer. Bromberger has also introduced the term b-questions – b for beyond our capacity – for example questions which were asked by classical physicists but could not be answered prior to the rise of quantum theory. I shall discuss the difference below.

How do we go about such p and b questions? How do we explain? i.e. how do we answer *why* questions? Popper says, we cannot answer this question at all – there is no method of discovering answers. But even he admits that there exist partial methods, though he is reluctant to discuss these. To my surprise I have found that my own study of the existing partial methods (particularly of Faraday's field-theoretical method of raising new kinds of questions and of looking for new kinds of solutions to them), have led me to dissent from a few details of Popper's view of the methods of science. Let me sketch here briefly a theory of partial methods, and contrast it with Popper's theory of explanations. But before that I wish to discuss the theory of explanation in general.

VIII. BROMBERGER ON WHY-QUESTIONS

In his famous paper on 'Why-Questions', in *Mind and Cosmos* (1966), Bromberger criticized Hempel's deductive model of explanation by showing, with the aid of many instances, that not all deductions are explanations. His aim was to supplement Hempel's model with the aid of the logic of questions and answer. Before following him through I want to discuss his criticism because it has puzzled me for some time. Surely Hempel knows that not all deduction is explanation. Hempel knows that any statement is logically equivalent to its double-negation but the two do not explain each other. Bromberger has laboured to construct many non-explanatory instances with a universal and an existential premiss and an existential conclusion, even a conclusion which does not follow from the existential premiss alone. I dare say Hempel knows this too. Moreover, there is really no difference, as far as explanation is concerned, or as far as deduction is concerned, between the case of a universal and an existential premiss with an existential conclusion, and the case of a universal premiss with a conditional conclusion composed of the two existential ones in the proper order. The question really is, do Bromberger's examples constitute a criticism of Hempel's model? What is Bromberger saying Hempel has overlooked? Did Hempel ever claim that all deduction is explanation? Is Bromberger's criticism also relevant to any other model of explanation, from Aristotle's to Descartes', to Newton's, to Whewell's, or to Popper's?

For Aristotle, explanation is deduction from a definition or from a statement of an essential quality. When we explain Socrates' mortality by means of 'Man is a featherless biped' our explanation is defective; but when we use 'Man is a rational animal' together with 'All animals are mortal', etc., our explanation is satisfactory provided our deduction is valid.

In spite of all savage attacks on Aristotle, his metaphysics, his vagueness, and what-have-you (attacks which are often just, in my opinion), Aristotle wins every time. Take the following: All metal responding properly to a touch-stone is gold; Tom's coin responds properly to a touch-stone; therefore Tom's coin is gold. This is a valid inference, the premisses are eminently reasonable; the application of the inference for ages and ages has not been criticized by any historian of science, of technology, or of economics. Yet anyone who calls this inference explanatory will be met with an amused smile. Take, per contrast, all metal which is yellow and whose average specific gravity is d_{Au} is gold, Tom's coin etc., therefore Tom's coin is gold.

This comes closer home. Go deeper: All metal with specific gravity d_{Au} is gold, etc., and you are still closer home. Still closer is, All and only atoms with the nuclear charge Z_{Au} are gold; atoms with nuclear charge Z_{Au} have the nuclear mass N_{Au}; the number of atoms in a cubic centimeter of solid metal under normal circumstances is, etc. etc.... therefore Tom's coin is gold. In other words, we do recognize certain deductions as explanations, classical or modern, an inaccurate modern (without isotopes) and an accurate (with isotopes) modern, and we grade them as to their degree of satisfactoriness. Incorrect as certain classical explanations surely are, they nonetheless are explanations proper, though less satisfactory than other explanations we have; whereas deductions from accidental statements referring to touchstones are not viewed as explanations, even when endorsed. Why then did Hempel, as well as others before and after him, refrain from dividing those deductions which are explanations from those which are not?

That some deductions are not explanations is a fact with a history too well-known to be forgotten. Soon after Aristotle offered his theory of explanation as deduction from essences, a new idea of deduction was used not as explanation but to save the phenomena. It is commonly well-known that Ptolemy did not consider his deductions explanatory but purely mathematical. All these distinctions, between explanatory theory, or philosophical theory, and mathematical theory or purely descriptive and/or predictive theory, never died. The Copernican tradition was opposed to the mathematical tradition and in favor of a philosophical tradition. Galileo, and later on Descartes, make it clear that in physics the philosophical is the geometrical. What is the geometrical is less clear – it is, I suppose, Cartesian metaphysics or some such system. Equally clear – to Galileo, Descartes, and others – was the fact that theory had to have deductive force to be explanatory. Thus, though Descartes' astronomical theory did comply with his metaphysical principles, it was not satisfactory in so far as it was not deductive. Newton's theory of gravity seemed to be the reverse: deductive it surely was, and to a high degree; yet it did not comply with Descartes' metaphysical theory. Perhaps it might have been considered a mathematical theory, purely descriptive, designed merely to save the phenomena. Orthodox Cartesians did, indeed, tend to consider Newton's theory purely mathematical and not at all philosophical. Even when he was an orthodox Cartesian, Newton resented this. His principles were mathematical principles of natural philosophy, even though as causal explanations they were not yet satisfactory. He confessed he could not deduce his theory of gravity from a satisfactory causal theory.

Causal explanation now becomes a mystery. In Aristotle, causal explanation is deduction from definitions of essences. In Descartes, causal explanation is deduction from a hypothesis conforming to Cartesian metaphysics. (I have discussed the theory of a hypothesis conforming to metaphysics in Chapter 9 above.) When Newton was a Cartesian he agreed; even then the status of his own theory of gravity was obscure; when he ceased to be a Cartesian the status of his own theory became even less clear. Meanwhile two things happened which led to the elimination of causality from causal explanation altogether. The first was Newton's answer to the charge of assuming occult qualities, the second was Hume's attack on causality.

The attack on occult qualities was a confused critique of the moderns, especially the Cartesians. They charged the Aristotelians with two charges as if they were one. The first charge is that Aristotelian explanations in terms of occult qualities are circular or *ad hoc*. Moliere's example is the paradigm: opium puts you to sleep since it has *vis dormativa*. The second attack, to use Hosper's terminology, was that Aristotelians explained the known phenomena in terms of unknown or hidden (etymologically, occult = hidden) essences. Somehow, it was felt, if the explanatory principles were better known, better comprehended, than the explained phenomena, then the circularity too will vanish. There is much force to this idea. Yet the demand to explain by the known only, should not be confused with the refusal to allow for circular explanation: at best the one covers the other; they are certainly not identical. The requirement to use only known explanations may be sufficient for the exclusion of circularity: it certainly is not necessary.

Newton did not clarify matters; rather, and not for the only time, he exploited the confusion of his critics in order to repel the attack: his theory of gravity, he said, was not of an occult quality, since it was not circular. His disciple Roger Cotes even declared gravity to be an essential quality, thus making Newton conform to the Aristotelian-Cartesian theory of explanation, though not to Cartesian metaphysics. Kant later declared Newtonian gravity self-evident.

I shall not dwell on Hume's attack on all cause, all essence, all substance. Let me merely say that those who took some heed of his criticism, yet refused to go all the way with Kant, quite naturally found in Newton's idea of non-circularity the only traditional element still available for epistemologists.

IX. THE NEED FOR A METAPHYSICAL THEORY OF CAUSALITY

Thus, the idea that causal explanation is a non-circular deduction has won the day. It is common to William Whewell, Sir Karl Popper, and Carl G. Hempel, among other thinkers. But what constitutes non-circularity or non-ad-hocness? For Whewell, non-circularity amounts to empirical verification, much as for Newton; for Popper non-circularity amounts to refutability; for Hempel it amounts to confirmation. Therefore, one might interpret their views as the claim that explanations are all and only deductions which are verified, refutable, or confirmed, respectively; but one may interpret them differently. In this case the theories of explanation offer only necessary but not sufficient conditions; hence we cannot deduce from them that any theory is explanatory; hence they are not explanatory; hence they are not satisfactory.

Assume, then, that these theories offer necessary and sufficient conditions for a theory to be an explanation. It may be impossible to offer a counter-example to Whewell's theory if we insist that verification is impossible. But if we use Whewell's conditions for verification instead – i.e., that a theory stand up to an independent severe test – then, surely, there can be a counter-example to Whewell's theory. Indeed, my example of the touchstone is a counter-example to all three theories, Whewell's, Popper's, and Hempel's. The touchstone hypothesis should count as explanatory by all three, yet I dare say they will all agree that intuitively this is defective.

What the counter-example shows is that causality or causal explanation cannot be fully captured by sheer methodology. I have explained, in my paper in the first volume of the *Philosophical Forum*, how the methodological idea of causal explanation broadens the classical concept (e.g. to include statistical theory); as we see now, it broadens matters too much. The metaphysical distinction between essential and accidental, the link between causal and essential, is still intuitively felt when touchstone theories are denied the status of causality.

So much for causal versus accidental theories. We also have, we remember, philosophical versus mathematical, where philosophical corresponds to causal but where mathematical does not even correspond to accidental but more to the fictitious. All three models of explnation – Whewell's, Popper's, and Hempel's – cover equally well both electromagnetics and the theory of elasticity. Indeed, formally the resemblance between the two theories is astounding even to this day. Somehow, we know, electromagnetic field theory was meant as an explanation proper of electromagnetic phenomena; and with good reasons, we feel. Somehow, we know, the theory of elasticity was meant not as an explanation but as a mathematical tool for description of the phenomena of elasticity; and for good reason, we feel. Hence, all three theories of explanation are either inadequate or false.

So much for my exegesis on Bromberger's critique of Hempel. Why only of Hempel, and why he does not elaborate, I do not know. I have found the elaboration necessary. I do not yet know what Whewell, Popper, or Hempel, would say about these criticisms; I dare say there are a few answers to them. I feel, somehow, these answers are not very good. The very idea of Whewell and his followers, including Wittgenstein of the *Tractatus*, is to reduce causality from metaphysics to methodology. Popper even expresses this idea – whenever possible, and especially in the case of causality, reduce metaphysics to methodology – openly in his classical *Logic of Scientific Discovery*. I find this idea interesting, important, and useful. But on causality, for example, it breaks down.

A few more words about the modern instrumentalism of Duhem and his followers. They deny that any theory is explanatory, and claim only deductive or descriptive force for all theories. This abolishes the distinction between the theories of elasticity and electromagnetism. This, equally, makes nonsense of the dissatisfaction commonly expressed about the eightfold way theory which, admirable as it surely is, is regrettably more descriptive than explanatory. The dissatisfaction proves that intuitively even scientists who endorse instrumentalism mean to do so merely in order to reject essentialism. They simply respond too violently to a past mistake.

X. COLLINGWOOD IN A NEW GARB

Bromberger's modification of the deductive theory of explanation is, a deduction is an explanation if and only if the premiss answers the question, why is the conclusion true. For example, if we ask why is the sky blue, the theory of dispersion fluctuations answers our question and so explains why the sky is blue. But to say that all skies with atmospheres are blue, our sky has an atmosphere therefore therefore our sky is blue, is proper deduction but it is an explanation to a smaller or narrower extent. And to say that Tom's coin is golden because it responds properly to a touchstone will not do at all, because to say so is not to answer any why-question (it would answer a why-question if it talked about our belief that Tom's coin is golden).

This is very convincing. We cannot answer, why does a tuning fork vibrate? by, because it is made of a continuous medium. Here the continuum theory of elasticity is evidently non-explanatory. Even to say, the tuning fork vibrates in a series of harmonics each subject to force proportional to the width of the vibration, even this is recognized as not an explanation but a highly sophisticated restatement of the question. Perhaps, when we accept the Newtonian schema of explanation we deny that decomposing a vibration into series (Fourier analysis) is explanatory, and also that the acceleration of the parts of the vibrating fork is proportional to the width of the vibration; but we shall, then, accept the force causing this acceleration as explanatory. And here, when Bromberger says explanations are deductive answers to why questions, he enables us to distinguish between essential and accidental theories, as well as between philosophical and mathematical or fictional; when we philosophically accept Newtonian forces we accept a part of elasticity theory as explanatory, and the anti-atomic part of it as mathematical or fictitious; otherwise we declare the whole of it fictitious.

So far, then, so good. But Bromberger has only raised again our old question, whereas he seems to think that he has answered it.

Things are, really, very difficult and baffling. We do not even quite know what is a question. We seldom consider the question, what is the distinction between x and y? a question proper – except in exams. (Actually, it is ungrammatical because of the misuse of the definite article; discussed at length in Chapter 2 above). Popper declared all what is...? questions bad as they conform to methodological essentialism, tabooed since Galileo and Bacon (both of them epistemological, but anit-methodological, essentialists). Yet what is...? questions can easily be converted into why-questions, and so not outlawed, unless we can answer the question, what is a question? But this question is taboo as well, and so we may convert all what is...? questions into why-questions. However, why-questions are very puzzling. Yes-or-no questions, as well as w-questions, to use Carnap's term (i.e. questions of the kind, who, where, which, how), are treated by erotetic logicians as multiple-choice problems. These, though quite bothersome in so many aspects, are fundamentally less problematic than questions which are not multiple-choice questions. How, what, and why-questions can be multiple-choice but need not be. When we have a question which is not a multiple-choice question we may have a definite and clear ability to recognize an answer to it, yet be utterly unable to conceive or to discover any answer to it. This is what Bromberger calls a p-predicament. The existence of p-predicaments is a great puzzle: how is it that I feel so easily and clearly that x is or is not an answer to y the moment it is given to me, yet a moment earlier x was as remote from me as the mystery of the universe?

How is a p-predicament possible? I do not have an answer to this question, but I do have a hint of an answer, a matrix which I hope will be filled and thus provide an answer – true or false – to the question, which I have posed in Chapter 9 above. My hint of an answer is not substantially different from that of Bromberger. It may be introduced by noticing the fact that Bromberger's description of the p-predicament is not complete. There is little doubt that we can conceive of an historical instance of a person in a p-predicament concerning a given question, who died without having any answer to it, and who would have recognized, had he lived longer, an answer by one of his successors as an answer to his question; yet who would not have recognized an answer to the same question given by a much later successor. I shall not even bother to outline an example – it is too easy to do.

And so, my correction or supplementation of Bromberger's description of a p-predicament is as follows. A person in a p-predicament is puzzled over a question to which he has no answer whatever, who feels that presented with a putative answer he will immediately perceive that it is or is not an answer to his question; but whose feeling is justified by the facts up to a point and not necessarily to the full. I dare say my supplementation to Bromberger's description is rather trivial because no one ever said that the list of answers to every question is finite; indeed, some questions, we know, avail themselves of infinitely many answers, and it has not been claimed by anyone that given an infinite list of answers anyone can decide, at once or at any given length of time, whether a given answer is included in that list or not.

Hence, the fact that a person is able to decide, in some cases, that a given answer is an answer to his question, must be relativized to a general matrix of answers that the person has, to some general presupposition, articulated or not, that a person has regarding any possible proposition which in the future may pose as an answer to his question. This general presupposition, when unarticulated, is a person's point of view, and in any case it is his metaphysics regarding the question at hand.

And so, at last, I have managed to close the circle, and relate the father of the logic of questions and answers, R. G. Collingwood, to his latest successors. It is not a mere sentimental point, but there is a logic of the situation which drives Bromberger to discussions akin to those of Collingwood. It is possible to show that articulating metaphysical viewpoints, i.e. matrices of answers to given questions, turn them from mere puzzles and p-predicaments to manageable w-questions, to sort of multiple-choice questions. I have argued my view on this point elsewhere at length, and so has Bromberger, whose view is not too different from mine. I shall not elaborate; I shall give a quick and simple instance.

The problems solved within Newtonian metaphysics are all why questions turned into which questions, p-predicaments turned into multiple-choice. Newtonian metaphysics views the universe as a conglomerate of diverse kinds of particles each kind exerting a kind of central force on its own kind. So the question of celestial mechanics becomes, what is the central force acting between heavenly bodies, and the answer is, one proportional to the inverse square of the distance. The question what makes Boyle's law true is answered, a central force proportional to the inverse of the distance. What makes a tuning fork vibrate? A force proportional to the distance. Electricity? The inverse square of the distance, again. And so, we have infinitely many functions of the distance, and each question is translatable to the question, which function? A why question is thus turned into a which-question.

This is an over-simplification. In some cases, though the above picture was deemed true, it was not deemed useful enough, and more ingenious methods of turning a why-question into a which were used. Laplace's theory of the tides and his theory of capillarity are such instances. But I shall not dwell on this here.

What is more exciting about all this is the idea of not answering but of criticizing a question – which we have met with Harrah and Belnap. A why-question, a p-predicament, cannot be easily criticized. Why is the sky blue? can only be criticized by the contention, the sky is not always blue. To which the response should be, why is the sky bleu under conditions x? This, no doubt, can be criticized again and lead to the response, why is the sky blue under conditions y? Such dialectic really cannot take us away from the level of observed phenomena, no matter how sophisticated it may otherwise become. Not so when we transform a why-question into a which-question.

When we change why is the sky blue? into which central force renders the sky blue? we can criticize the question. Today we are all content that the sky is blue not thanks to any central force but thanks to fluctuations of distribution of atoms of atmosphere and thanks to non-central interactions between them and photons. We have transcended Newtonian mechanics by showing that it handles certain why-questions poorly. To use the idea of Åqvist in the language of Harrah and Belnap, we correct a false question here, and thereby, I should add, transcend one viewpoint, one set of presuppositions and go more deeply to a newer viewpoint, to a newer set of questions.

There is much more to say on this topic. Since the list of answers to a given question is, in Newtonianism, quite infinite, it is, in a sense, impossible to correct Newtonianism. Einstein was aware of this fact. He noticed that one can always correct Newton's theory of gravity by choosing a function of the distance close to the inverse square. Yet he dismissed such attempts as too arbitrary. And so, my answer is but a matrix, merely an outline of an answer, not a full answer. It is under certain constraints which we must specify more fully, that we can correct a question and replace it by a deeper one, that we can transcend a viewpoint and attempt to construct a deeper one. I shall not discuss this issue any further. Let me merely conclude that to criticize a question is only a first step forwards replacing it with a better one.

When we translate a why-question into a which-question we do so from a given, often developed viewpoint. When we transcend the question thus put, when we transcend the viewpoint from which it was put, the which-question fizzles out and we are turned back to our why-question, to our p-predicament. Or, alternatively, a b-question, one beyond our horizon, enters it as a new p-predicament. We may try to solve it directly, we may try to develop a new viewpoint in order to handel it; we may, more dialectically, try to do both with the aid of the partial success of each. This comes close to the theory of Bunge which I reviewed in *Synthese*, 1969: we try to integrate our general view of the world, he says, into a coherent picture. This, I say, we do by imposing a viewpoint on our background knowledge. Not successfully, of course, but with partial success.

Hence, when we ask a why x? question, we want a theory from which x follows, but not necessarily any theory. We want one which conforms to present day metaphysics, or one which may help us develop a new metaphysics, as the case may be. (These two cases parallel somewhat Kuhn's normal science and revolutionary science). Hence, what explanation is causal depends on current, or on tomorrow's, metaphysics, depending on the situation at hand. And we choose questions which our metaphysics turns from why-questions to which-questions, sooner or later arriving at a stumbling block which may force us to alter our viewpoint or metaphysics altogether. Still better, and closer to reality, we may operate with more than one set of presuppositions, turn a why-question into two different and competing which-questions, and see which is preferable. This is a sort of crucial experiment between sets of presuppositions of metaphysical points of view – a crucial experiment not like the one in science which selects one answer from a set of answers to a given question, but which selects one translation to a which-question from a set of such translations to a given why-question. But all this is very far from being put into a formal language, or even into a fairly rigorous presentation, and it is still all up in the air.

This, then, is my view of the dialectic interaction between science and metaphysics, presented in the light of the latest developments in the logic of questions and answers.

It is in this context, or a similar one, that we can see the force of Robert S. Cohen's introduction of why-not-questions. For, formally, there is no difference between why- and why-not-questions: why-not-x becomes why-y when y is defined as not-x. But his questions, such as why there and not here (i.e. western but not Chinese science), why this and not that, make sense from a given specific viewpoint, from which one may wish, as Cohen says, to develop a logic of comparative analysis. For more details the reader is invited to examine R. S. Cohen's 'The Problem of 19(k)' in the *Journal of Chinese Philosophy* **1** (1973), 103.

APPENDIX: THE ANTI-SCIENTIFIC METAPHYSICIAN

The contemporary instrumentalist tradition stems from a frankly, and even initially, anti-scientific tradition.[1] It is the traditional view not that the world is flat (Berkeley, Hume)[2], but that the world of science is flat whereas the world of metaphysics is deep. This tradition, I suppose, goes back to Hegel. T. M. Knox, the English translator of Hegel's *Philosophy of Right*, ascribes this view to Hegel, and contrasts "The categories of 'essence'" with "mathematical and empirical science or... formal logic" and even "those philosophies which adhere to scientific method instead of abandoning it in favour of reason and the philosophic method..." I tend to agree, and find adumbration of this view in the Preface to the *Phenomenology*[3], but I confess I find no clear statement of it anywhere in that work, in the *Logic*, the *Encyclopedia*, or in the *Philosophy of Right*. I suppose it is what Hegel would have liked to say; but it took about a century to learn to say it as crisply as contemporary Hegelians have learned to say it, judging from Bradley's aptly named *Appearance and Reality* of 1893.

Sir William Hamilton commended logic as the science of essence and condemned mathematics as merely formal (i.e., descriptive, or purely instrumental). George Boole's earliest publication is couched in the form of a retort to Hamilton: if logic is part of mathematics then his view is absurd.[4]

Subsequently the anti-scientific tradition jettisoned not only mathematics but also formal logic – as Hegel had recommended in the first place, according to T. M. Knox. Let me quote, though, some contemporary authors who are sufficiently influential and who hold this view, such as Croce and Sartre.[5] Let me begin with Croce.

> Science... cannot be anything but... philosoph. If natural *sciences* be spoken of, apart from philosophy, we must observe that they are not perfect... The so-called natural sciences... are surrounded by limitations... They calculate, measure... Even geometry... rests altogether on hypotheses... What of properly naturalistic they contain, is abstraction and caprice... The concepts of natural science are, without doubt more useful; but...

And here is what Sartre says:

> ... the further research of the scientist will reveal it as purely a thing – i.e. stripped of all instrumentality. But this is because the scientist is concerned only with establishing purely external relations. Moreover the result of this scientific research is that the thing itself, deprived of all instrumentality, finally disappears into absolute exteriority.

> ... the concept of objectivity, which aimed at replacing the in-itself of dogmatic truth by a pure relation of reciprocal agreement between representations, is selfdestructive if pushed to the limit.

But science has more extreme opponents. I shudder to quote Heidegger; even his interpreters are often too unclear. It *seems* to me that Paul Ricoeur claims that Heidegger's *Holzwege* advocates the instrumentalist view of science;[6] but I cannot say.

The following quotation from Maurice Merleau-Ponty[7] is both conventionalist-instrumentalist and apriorist:

> Husserl says in the first volume of the *Logical Investigations* that the physicists proceed by "*idealisierende Fiktionen cum fundamento in re*" – that is, by idealizing fictions which are nevertheless founded in the facts. Let it be, he says, the law of Newton. Basically it makes no assertion about the existence of gravitation masses. It is another one of those idealizing fictions by which one purely conceives of what a gravitating mass would be. Then one determines what properties it would have, on the supposition that it exists.

According to Husserl, Newton's law says nothing at all about existence. It refers only to what would belong to a gravitational mass as such.

The method actually used by physicists, therefore, is not the chimerical induction of Mill... It is rather the *reading of the essence*... That which gives it its probable value... is rather the intrinsic clarity... ideas shed on the phenomena we seek to understand...

I think the main question to ask about these people is what do they think sticks and stones and the cow and the moon are made of, or how and why their machines work. It seems obvious that they do not care, and so their views on these issues are vague, confused and outdated (see Sartre's discussion of contemporary physics, *loc. cit.*). These views, however, *not* the deep metaphysical doctrines which their advocates propound, are what these advocates offer instead of sience – unlike Bellarmine or Duhem who accept Aristotelian physics; and unlike philosophers who, with Hume and Mach, think the world is flat; and unlike Bohr and his followers whose discussions still reflect efforts to create newer and deeper pictures of the world. Their instrumentalism is simply the best excuse they have for their ignorance of and disregard for science.

NOTES

[1] Sir Karl Popper, 'Three Views Concerning Human Knowledge', reprinted in his *Conjectures and Refutations*, London 1963.
[2] Popper, 'A Note on Berkeley as a Precursor to Mach', *op. cit.* See also p. 211 above.
[3] G. F. W. Hegel (transl. by T. M. Knox), *Philosophy of Right*, OUP, 1952, 1967, pp. vii–viii and *The Phenomenology of Mind*, OUP, 1967, p. 88.
[4] George Boole, *Mathematical Analysis of Logic*, Cambridge 1847, opening.
[5] Benedetto Croce, *Aesthetics as Science of Expression and General Linguistics*, New York 1953, p. 30. J.-P. Sartre, *Being and Nothingness*, special abridged edition, 2nd paperback edition, New York, 1965, pp. 176, 283.
[6] M. S. Frings (ed.), *Heidegger and the Quest for Truth*, Chicago 1968, p. 68.
[7] Maurice Merleau-Ponty, *The Primacy of Perception and Other Essays*, Evanston 1962, p. 69.

CHAPTER 11

THE CONFUSION BETWEEN PHYSICS AND METAPHYSICS IN THE STANDARD HISTORIES OF SCIENCES

Let us compare histories of science to a series of portraits. You have throughout the history of art beautified portraits, whether painted or photographed and then retouched. The wrinkles are ironed out and the person's expression in the portrait is friendlier than in real life. The person has in common with his portrait only the outline, the general contour, and even that not very accurately. Some people like portraits of their idols beautified in this way. I am not one of them. I like to see the wrinkles as signs of the hardships of life; of the sweat and the sleepless nights. This is my personal predilection. I therefore dislike much history of science that is being written today because it is a beautified portrait. Indeed, it is worse: it is, to shift the metaphor, a success story. Its theme is: Mr. So-and-So had this bright discovery and Mr. So-and-So this brilliant idea; and this is how we have arrived at the peak of this wonderful mountain called contemporary science. This is not to my liking. I like to see the wrinkles in a portrait; the setbacks or the road to success; in the history of science I want to see the discrepancies, I want to see the quarrels, personal quarrels as well as, and even more so, intellectual quarrels. I want to know, for instance, why Volta and Galvani quarrelled so that Galvan lost his job, but I want more particularly to know about their disagreement about animal electricity, and to understand it. I need not tell you that most histories of science are entirely reticent about this matter. On the other hand, I also love the general or broad outline of the history of science, or what E. A. Burtt calls the metaphysical foundations of science. In my view science is to be valued not because it gives us utility, but because it helps us to develop our metaphysical theories of the universe. So I am in a quandary. I want to see the contour alone and I want not to lose sight of the wrinkles, of the detailed discrepancies and difficulties, of the hardship.

To put this in a more intellectual form, I will now discuss an historical example: the views of Faraday, whom his friend Dumas called a great philosopher and a metaphysician.

Faraday thought that forces do not act at a distance for reasons that I shall soon explain. He therefore had to declare Newton's theory false (though a good approximation, of course). His paper on the conservation of force, which is one of his latest and most philosophical works, boldly presents a very unorthodox view. He begins with the admission, even with the stress on the fact, that Newton's theory was empirically verified better than any other theory, and as strongly as anybody could ever expect, by the discoveries of the new planets. Yet Faraday rejected that theory. He declared that forces can *not* vary with the variation of distances according to the inverse square law or any other way because forces cannot vary. If forces conserve then Newton's theory, taken literally, is false. The root of the error was Newton's theory of action at a distance which, Faraday reminds us, Newton himself rejected. There is no such thing as action at a distance, Faraday claims, because force, contrary to Newton's theory of force, is not a property of (or a relation between) material particles; force, rather than matter, is the primary entity of which the world is constructed.

This is Faraday's position. It is reminiscent, as Dumas has pointed out already, of ancient Greek metaphysics. Thales says, "all is water", and Faraday says "all is force". Faraday comes as a great philosopher, as a metaphysician, and overrules the best and the most well-verified theory of his time because it conflicts with his metaphysics. It seems quite possible that this is the kind of situation which causes hostility towards, and impatience with, metaphysicians and metaphysics. If so, then those who try to dismiss metaphysics in the name of science are merely retaliating, and they have a point. If historians of science speak against metaphysics as a result of all this, then they also have a point. The metaphysician arrogantly presents his speculations, and then comes forward and overrules, if need be, the best scientific theory, instead of humbly accepting the situation as an indication that he should withdraw his metaphysical speculations.

I have now translated my problem from the metaphor to the theory of science. I want to see the details of science, and hence I do not want to see the metaphysics overruling it; but I do not want to see metaphysics deprecated. This conflict is my problem. And this is why my metaphor is inadequate. In a portrait you may see the contour as well as the wrinkles, and the painter may have, at most, a problem of balancing the two. But the existence of a conflict between the metaphysical contour and scientific details raises a more difficult problem. If E. A. Burtt has given us only the metaphysical contour but not the scientific details, we may feel that the details may be added to his picture. But this is not always possible, as I have shown in my example. We may, of course, centre our attention on the scientific details which do fit in well with Faraday's metaphysics, and ignore those which do not. This, however, will not help us to solve the problem. It is not only an ostrich policy, it also makes Faraday look like a prophet. Faraday, it is often said, prophesied, or divined, Maxwell's electromagnetic theory of light and Einstein's general theory of relativity. Now divination is, again, a success story and since most historians of science love success stories they often indulge in the history of divination. They ask who was first to divine that light was electromagnetic; was it Faraday, was it Ampère, before him, or was it even Father Beccaria still earlier? I find the study of the history of the divination of the future developments of science distasteful and even embarrassing. To me it seems clear that Faraday was not interested in divining the future development of science when he developed a crude idea of gravitational fields; he was simply concerned with an immediate problem. His prescription or suggestion was to overcome his problem by devising such a gravitational theory which agrees with his metaphysics, and explains the known facts at least as well as Newton's gravitational theory.

This provides my solution, for what it is worth, to my problem of how to see the contour and the details at the same time. The standard approach is both to denounce the metaphysics of scientists and to commend parts of it as divination of future scientific developments. Burtt's alternative is to view metaphysics as the foundation of science. My alternative is to view some metaphysics as the possible foundation of future science; to view it as often conflicting with existing scientific theories and as incentives to alterations which would remove the conflict. In my view, then, the interaction between physics and metaphysics is by way of metaphysics prescribing programmes for future scientific development.

As you know, programmes are neither true nor false but commendable or condemnable. Now, if the metaphysics is true, the programme it prescribes is obviously commendable. The converse, it seems, is not universally true: Faraday's metaphysics is false, yet the programme based on it was fruitful. The greatness of contemporary physics, in my personal opinion, is that it gave rise to a better metaphysics than Faraday's. And yet Faraday's metaphysics, being better than Newtonian metaphysics, could give rise to better scientific theories. Faraday's theory was not a system of scientific divination, nor was it a detailed scientific theory; it was a programme for possible future scientific theories. Metaphysics often is, so to speak, the contour being filled up later on with details – sometimes more successfully, sometimes less successfully (where success is, of course, not material but intellectual), but usually with great effort and along with controversy and trial and error.

The confusion between physics and metaphysics in the standard history of science is objectionable not only because it turns some metaphysics into the divination of future scientific developments while leading to the dismissal of other, less successful, metaphysics. It is also objectionable on the ground that the metaphysics thus considered is trivial and uninteresting. It is my opinion that although the standard historians of science rightly caution us against metaphysics they usually report the history of metaphysics rather than the history of physics, and indeed, only that part of the history of metaphysics which is a dull success story, not the one which is an exciting story of conflict and immense efforts to render a metaphysics scientific. I will give an example from Poincaré. I want to give an example from Poincaré to avoid giving the impression that I criticise the standard history of science because it is not Popperian. As a follower of Popper, I look for the wrinkles on the understanding that the significant history of science is mostly wrinkles, mostly refutations, mostly criticisms; but here I am criticizing people not for ignoring Popper but for ignoring Poincaré.

The example from Poincaré is so obvious, and *prima facie* so strong, that those who reject it have to explain why they do so. Of course, 'obvious' does not mean true, but anybody who dissents from an obvious contention should answer it. To my knowledge Poincaré's contention was never answered, never in fact taken up, although it is in one of the most well-known classics, his *Science and Hypothesis,* and in one of its most crucial chapters. Poincaré discusses the possible formulations of the law of conservation of energy and its status in each formulation. The most common formulation is: energy equals kinetic energy plus potential energy, and the energy of a closed system is constant. Now, do we exactly know what kinetic energy is? Let us assume we do. But do we know what

potential energy is? Poincaré says, no; we are vague about it and for a
good reason. If you look closely at the formula you see that by 'potential'
we quite often mean gravitational potential, the Laplacian potential. But
then, we should have to say that this formula is false the sum of kinetic
and gravitational potential energy in the system is not constant, since part
of it may be converted into heat energy. We have missed a factor there;
we forgot the existence of heat energy. We therefore amend the formula:
energy is now the sum of three factors. Do we know what heat energy is?
Let us suppose that we do. We then find out that we have forgotten yet
another factor; electromagnetic energy. We add this as well, and so our
formula will contain more and more factors.

Now, says Poincaré, you have your choice between two alternatives.
The one alternative is this: you lay down each item of the possible forms
of energy, and describe them in sufficient clarity and detail, so that you
have a refutable hypothesis. The hypothesis is refutable with the discovery of a new form of energy which has not been listed so far. I hardly
need tell you, says Poincaré, that the moment the refutable hypothesis is
refuted you should discard it. The other alternative is not to lay down the
list of all possible forms of energy: either you present the list as incomplete or unfinished, or else you leave the formula in its original version,
with only two forms of energy – kinetic and potential, and confess that
the meaning of the word 'potential energy' is ambiguous. In this case,
says Poincaré, no fact can invalidate the law. For my part, I would say
'refute' rather than 'invalidate' because I believe in truth rather than in
validity as the aim of science; but I shall not discuss this point here. To
return to Poincaré, then, he said, the law of conservation of energy in its
ambiguous version is irrefutable by experience; hence it is a tautology. To
be more precise, Poincaré doesn't say 'tautology'; he says 'a kind of
tautology'; which is kind of choice. The phrase 'a kind of tautology' may
mean 'a quasi-tautology' and may mean 'a tautology of a certain characteristic'. According to Poincaré's theory, the second alternative should be
the correct one, but according to his discussion the first alternative should
be the correct one. Indeed, here there is a very interesting split in Poincaré's philosophy, but I shall not discuss it here, because whether the law
is a tautology or a near-tautology surely is trivial. The chief point is this:
the law of conservation of energy in its vague version – with the potential
energies unspecified – the irrefutable version, is indeed trivial. To this

Poincaré would agree whether he would ultimately decide that the irrefutable version is a tautology or a near-tautology.

The confusion in the standard histories of science of the trivial version of the law with its more substantial versions, is consequent on the attempt to beautify the picture of the history of science, because the substantial versions include conjectures and refutations; correction of the law of energy in each stage. The historian of science who does not like the discrepancies and criticisms resorts to generalities and speaks of conservation of energy in the ambiguous formulation just because he tries to forget that the history of science is full of refutations. I will give you one example.

Poincaré originated a hypothesis according to which in uranium, or uranium salts, electromagnetic rays are captured and then emitted after a while. Becquerel refuted this hypothesis. As a result of this very famous refutation of Poincaré's hypothesis, we all know, it was found necessary to add nuclear energy to the list of forms of energy. I ask you, what history of physics mentions this refuted hypothesis of Poincaré, or similar cases of important refuted hypotheses? I have found very few which do, and it is those books which I admire; the rest I think do a disservice, because they present that version of the law of conservation of energy which is trivial, useless, and valueless.

But I do not wish to imply that the scientific and the trivialized ways of looking at the law of conservation of energy discussed by Poincaré are the only important ones. I agree with him that the trivial metaphysical version and the more informative, specific, and scientific, versions are distinct. But there is a third distinct case, where the law may play a crucial role in an interesting metaphysical dispute. To make this case distinct we must select a metaphysical system which clashes with another metaphysical system, or with existing scientific theories, over the interpretation of this law. In Poincaré's example the contour and the details are in perfect agreement with each other, so that the contour is of little or no value in comparison with the detail. So I shall now discuss my example of a metaphysical conflict between Newtonianism and Faradayism and see what function the law of energy conservation plays there. But I have to warn you again against confusing metaphysics with clairvoyance. There was a game, invented by Tyndall, I think, of discussing who was the clairvoyant who foresaw the law of conservation of energy for the first time.

First he said it was Mayer and then he said it was Séguin and then he or others went back and back and back. As you know, the law of conservation of energy is ancient. All of a sudden somebody discovers that it is ancient, and, perhaps because this spoils the game, they dismiss him saying "but the ancient law is speculative; we ask about its verified version", thus forgetting that they are speaking about divination. Verified, indeed! The law was never verified, and as Poincaré knew, it could not be verified. Rather, I have argued, in its many scientific versions it was repeatedly refuted. But the sad fact is not that this kind of confused talk about the history of the law of conservation of energy still goes on – nobody really minds whether Séguin or Mayer divined it – but that this kind of confused talk stands in the way of discussions which can lead to really interesting results, as I want to show now, by briefly discussing the role, or rather the different roles, which the same law plays in Newtonian metaphysics, and in Faraday's mataphysics. ·

First, within Newtonian metaphysics each potential energy, as is well known, can turn into kinetic energy and *vice versa*, so that possibly one potential energy can first become kinetic energy and then another potential energy. But, a potential energy cannot turn directly into another potential energy. This is so for the following reason. When gravitational potential turns directly into, say, electric potential then, to Newtonians, this can occur only as a consequence of gross matter, or heavy matter, turning into electric matter. But as within Newtonianism the law of conservation of each kind of matter is accepted, the process is impossible. When we speak of heavy matter in the Newtonian sense, we already have in mind the idea that heavy matter conserves, that the quantity of heavy matter in any closed system remains constant. Thus, in Newtonianism a potential cannot change directly into another potential.

Comment from the floor: I'm sorry I don't agree with you. If you take two massive bodies which are attached together by a spring and move them apart, as a result of the attraction of gravity they will come close together, and therefore originally you had no motion and if the spring were under no tension, no potential energy.

Agassi: I am grateful for this interruption, because it deals with the example which Kant and Boscovitch have discussed and which led them to invent a new metaphysics. Newtonian metaphysics rules out this example as the following corollary to Newton's theory of gravitation shows. If without motion some gravitational potential disappears, or turns into something else, than some heavy matter has also disappeared, or turned into something else, which is impossible. But then, ask Kant and Boscovitch, how can ordinary heavy matter be elastic? If you assume that two heavy atoms possess both gravitational energy and elastic energy you already admit that they interact not strictly according to Newton's law of gravity simply on account of Newton's law of addition of forces. Since the billiard balls are elastic, we must assume that their atoms are both elastic and heavy. To exclude the possibility of elastic and gravitational forces acting simultaneously Kant and Boscovitch suggested that the gravitational force acts when particles are placed at a great distance from each other and the elastic force acts when the same particles are placed at a small distance. If so, then Newton's law of gravity does not hold in the small-range. This is the first modification of Newtonianism designed to allow one and the same particle to possess many forces. Without this modification Newtonian metaphysics allots one kind of matter to one kind of force (because of the law of addition of forces), thus leading to the invention of a multitude of matters – gross, electric, magnetic, caloric, etc. Newton himself, incidentally, was not a Newtonian, but a Cartesian, a point which Faraday emphasized in his debates with the Newtonians.

It is a strange fact that in spite of the criticism launched by Kant and Boscovitch, unmodified Newtonianism remained popular. It remained popular even after the criticism was levelled from a different angle in a stronger fashion in the early 19th century. If gross matter interacts only with gross matter and electric matter only with electric matter, then the two kinds of matter are two universes apart. If so, why does an electric charge remain on the Coulomb test body? Why does it not jump? More precisely, why does it jump only under some specific conditions in the form of a spark? If, on the other hand, we assume that there is a small factor of interaction between gross matter and electric matter, then every gross particle will have both electric force and gravitational force and by the law of addition of forces these two should unite, with the consequence that two gross particles do not interact precisely according to the inverse square law, at least in the presence of an electric particle. Thus, the second alternative leads to the modification of Newton's theory of gravity, of his inverse square law. The only way out of this dilemma is to invent a new

metaphysics. And the new metaphysics of Kant and Boscovitch may provide a way out of the difficulty because it disallows this unification of
forces: it says that there is one force acting at one distance-range and another force acting at another distance-range, as the Boscovitch diagram
makes clear. But this solution is not satisfactory because we know empirically that gravitational and electrical forces have the same distance-
range.

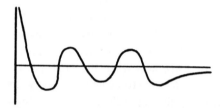

Fig. 1. *Boscovitch's diagram* presenting the force of interaction between two material
particles as a function of the distance between them. On the extreme right the function
coincides with the inverse square function.

Oersted wrote his doctorate on the Kant-Boscovitch model, just before
the discovery of electrochemistry; he tried to explain this phenomenon as
soon as he learned about it. He said to himself, I suppose, the Kant-
Boscovich model is not satisfactory because the distance does not play
any important role in electrochemistry. But, he presumably said, what is
important to electrochemistry is a certain chemican setting, and the contact between the various parts of the pile. So, instead of saying different
forces act in different distance-ranges, he said that different forces act in
different set-ups. Now this theory has a far-reaching consequence, because, according to it if a set-up is changed then the force acting within
it is changed. In other words, what we call force is just a manifestation of
something deeper, of an underlying reality, which does not change. The
same force appears in different manifestations when in different settings;
but it is always the same force, the primordial force. This is the law of
conversion or conservation of force.

Holding this view Oersted tried to make electric force convert into
magnetic force and it was no accident that for about twenty years he was
working almost alone on the topic, and he was entirely alone in his
persistence, because nobody else believed his metaphysics. He introduced

the most violent electric discharge as a part of his setting because he knew that it causes other transmutations, of the electric force into heat and light forces, and he tought, by extrapolation, that if the current is very strong a part of the electric force converts into magnetic force as well. This is why he worked with bigger and bigger batteries although his experiment can be successfully conducted even with the weakest battery he could construct. But he got no result for a long time because following Newton and Kant he thought that the force he was looking for was central. When he ultimately made the discovery of electromagnetism, he corrected his errors of detail, but considered the discovery to be a strong support of his metaphysics, especially since he thought that non-central froces may fit his own metaphysics but not Newton's or Kant's.

Faraday accepted Oersted's metaphysics from the start, and he later accepted as a matter of unshakeable faith Oersted's law of conversion of forces; every two forces convert into each other at fixed ratios. Hence, he concluded, electricity can convert directly not only into magnetism, but also into gravity. We all know that gravity and electricity interchange via kinetic energy; this did not constitute any problem. What Faraday wanted was to discover a case of a change of gravitational forces directly into electric. And this, as I have said before, means to a Newtonian that Faraday wanted to change gross matter into electric matter. Hence, the Law of Conservation of Matter within Newtonianism conflicts with Faraday's progam. It is for this reason that Faraday gave up matter altogether. He realized that the Law of Conservation of Force and the Law of Conservation of Matter contradict each other.

The Law of Conservation of Energy occurs both in the Newtonian and in the Faradayan metaphysical systems. But the two systems clash even with respect to this law, and it is this clash which I wanted to tell you about. I think that it is this clash which makes the trivial Law of Conservation of Energy so very interesting, because the clash is not trivial and the two clashing attitudes are interesting and deviate interestingly in the way energy is viewed.

I have contrasted first the scientific and the metaphysical versions of the Law of Conservation of Energy. Secondly, I have contrasted the status of the metaphysical law in isolation, and its status within a metaphysical conflict. I shall now briefly mention the law in its unscientific commonsense version, and in isolation. I should like to quote the story which

Benjamin Franklin relates in his autobiography. When he worked in a London printing house his co-workers were surprised that he was the strongest in spite of the fact that he was a teetotaler and they drank strong beer; they were so ignorant that he had to explain to them that there can be no more strength in the beer than in the rye from which the beer was produced. This is the Law of Conservation of Energy in isolation, and it is, indeed, trivial: everybody knows it one way or another: if you don't put fuel into the machine, including the human body, it will not work. This, I contend, is the triviality which is sold to us by many historians under the cover of history of science, along with scholarly and heavily documented discussion about how much of this triviality was known to Séguin or to Mayer. In order to understand why the conservation of energy was to important in the last century, I contend, we must consider the metaphysical systems, metaphysical conflicts, metaphysical difficulties, of the time, which were deeply connected with the scientific problems of that time.

APPENDIX: REPLY TO COMMENTATORS

The critical comments on the foregoing are very interesting and valuable. I regret that for lack of time it is practically impossible to do them justice. I wish to discuss only one point concerning which I strongly dissent from my commentators. It seems that a change of attitude is taking place amongst some historians of science: the antimetaphysical attitude is becoming less and less universal. Now by 'metaphysics' we mean various things, but at least one thing we did always mean, and that is the doctrine of the substance of universe. This is what Aristotle understood metaphysics (or first philosophy) to be concerned with, as well as Bacon, Descartes, and Boyle, and Meyerson. There exist paradigm cases of metaphysics: "all is water", "all is matter and form"; and, to take a less paradigm cases, "all is force", which is Faraday's metaphysics. It is, perhaps, congenial to some people to forget the mistake of antimetaphysics and to reintroduce metaphysics either via the back door or via a new name, as Gerald Holton has suggested. There is nothing in a name. But we made a mistake when we tried to overthrow metaphysics, though the mistake was made for a good reason. We should remember and record our mistakes, especially those made for good reasons, so as not to repeat them. It is not the name that I am concerned about, but the statements of metaphysics and what is done with them in science. Kuhn and Hanson have raised the question of the scientific status of such metaphysical theories as those I have discussed. Are they empirically refutable? Are they within the framework of science or are they separate? Are they separable? I do not really quite know. But one thing I do know is that metaphysics does present a great danger of overruling and overriding all empirical science. This we have seen in the example of Faraday's attitude towards Newton's theory: it is the best scientific theory, but it does not tally with my metaphysics so I prefer to reject it all the same. This is very dangerous, and we should remember the risk because it is this that made us antimetaphysical to begin with. If we are going to become pro-metaphysical again, as I am advocating now and as my commentators are advocating as well, we should not forget the risk. It is easy to adopt metaphysics and throw science overboard, as Hegel said true philosophy should. It is easy to advocate science and be hostile towards metaphysics. It is less easy, or less obviously possible, to advocate both simultaneously. In justice to the anti-metaphysical attitude we should remember that it has this rationale. If we reintroduce metaphysics by another name we may easily confuse all this, and forget the rationale for the antimetaphysical attitude as well as the fact that it was an error nonetheless.

CHAPTER 12

THE CONFUSION BETWEEN SCIENCE AND TECHNOLOGY IN THE STANDARD PHILOSOPHIES OF SCIENCE

The distinction between pure and applied science seems too trivial to draw, since applied science, as the name implies, aims at practical ends, whereas pure science does not. There is an overlap, to be sure, which is known as fundamental research and which is pure science in the short run but applied in the long run; that is to say, fundamental research is the search for certain laws of nature with an eye to using these laws. Still, this overlap shows that though the distinction is not exclusive it is clear enough. The distinction between applied science and technology is a different matter altogether. All philosophers of science equate them, whereas it is clear that technology includes, at the very least, applied science, invention, implementation of the results of both applied science and invention, and the maintenance of the existing apparatus, especially in the face of unexpected changes, disasters, and so forth. The distinction between applied science and invention, to my knowledge, was made by only one writer, the most important writer on technology, perhaps; I am referring to H. S. Hatfield and his *The Inventor and His World*. Hatfield does not draw the distinction explicitly, but he uses it clearly and systematically enough. Applied science, according to his view, is an exercise in deduction, whereas invention is finding a needle in a haystack.

My own concern with all this comes from my studies in the theory of confirmation. Contrary to most, if not all, writers in the field, I hold that confirmation plays no significant role in science, pure or applied.[1] The contrary impression seems to me to stem from the fact that both invention and the implementation of novelties – from applied science or from invention – require confirmation. The standards of confirmation are legal, and they are set by patent offices in the case of invention and by institutions in charge of public safety and of commercial practices in the case of implementation. The excessive demand for confirmation is the tool by which the complacent postpones the implementation of novelties.

I shall return to the distinction between applied science and invention later on and also touch on the other points of technology. Here let me say that the literature in the philosophy of science is oblivious of all this, usually confusing pure science with applied science and both with technology. The inductivist philosophers of science insist that pure science concerns probable beliefs and that technology concerns decisions, both relating to the same probable hypotheses. They confuse belief and decision. Instrumentalist philosophers do not believe in the probability of pure science but in its usefulness: not only do they equate pure with applied science and applied science with technology, but they even consider all three as identical. Both leading groups of philosophers of science appeal to the fact that science and technology are so very enormously successful. To go to the root of their confusion[2] let us examine views on success. We shall later be able to show how different scientific success is from practical success.

I

Living in a robust progressive society, we are surrounded by the tendency to evaluate people by the measure of their success, more specifically by the measure of their success in attaining high economic and social standing, which is usually known – rather inadequately, but let this ride – as material success. Obviously, success philosophy is rather vulgar, and material success philosophy particularly vulgar and even stupid. I have no intention of speaking against success, for without some success there is no action; nor do I wish to speak against material success.

But to value success is not to say that success is all that matters, or that material success is all that we value, with the exception of saints, perhaps. Many works of fiction comprise plausible counter-examples to the identification of success with material success, usually by describing convincingly heroes who are wretched though materially successful (e.g., Bergman's *Wild Strawberries*). A less convincing yet true counter-example is Thomas Alva Edison. He had been very poor and miserable throughout childhood and adolescence, and he consequently loved both money and fame – even to an exaggeration; and he was no saint, being really a difficult man and not always honest, and being prejudiced against intellectuals, women, and Jews; yet he always insisted in words and stubborn deeds that the challenge of invention was to him in itself much more of a remuneration than all the benefit it accrued him, financially or socially.

He even brought upon himself material ruin so as to put pressure upon himself to wrok hard at his inventions in order to make money which he then quickly disposed of again, and so on, until he was too old to continue.

This example shows how complex and unconvincing may be the case of any person's personal success. And so let us ignore personal success from now on, material or otherwise. When we speak of success impersonally, we speak of the successful execution of a task regardless of its personal significance to the one who carries it out. Can we evaluate a person by the measure of the success of his performance? Can we measure the man by the measure of his achievement? Obviously, this, too, is rather vulgar. The latest biography of Edison, for instance, by Matthew Josephson, has been written partly, if not mainly, in order to break away from the vulgar success philosophy which permeated previous biographies, thus making them unpalatable to the more sophisticated.[3] At the very least, we all must agree, if we do measure people by their success, then we should take into account the fact that success or its absence depends partly on luck, good or bad; we must adjust the evaluation of the person, then, by considering the factors which were beyond his control yet affected the outcome of his activities. Judging people by their luck is poor judgment. How they can stand up against bad luck and how they can make use of their good luck contribute to their personal makeup much more than how lucky they were.

II

My colleagues in the field of the philosophy of science love to speak of the greatness of science as being identical with the success, the great achievements, of science. Consequently, when we speak of the adventure of science, we are hypocritical: we all admire Captain Scott, since though he failed he was very brave; but Joseph Priestley was seldom accorded similar consideration. Not Priestley, but his opponent, Lavoisier, is admired – and not for his sense of adventure either, but for his alleged success.[4] When you stop to think about it, you may find that success is rather bewildering and calling for much explanation and re-examination.

Somehow the assuredness and faith some of us display in the success of science seem rather smug. This impression is usually dismissed by the claim that the faith in science is amply justified. But when the faith in science is justified, science may become much less of an adventure and, hence, not much of an achievement. The justification of science makes it stupid rather than prudent not to apply it. Moreover, the justification is a principle of induction, and the principle of induction is usually justified by our successful reliance on science in practical affairs. This is circular, but never mind; the principle of induction may indeed be a method for success; it may be the golden goose which lays golden eggs regularly, or a computer which makes its followers rich. If so, then we may compliment the scientists and their followers no more than we may compliment the owner of a golden goose or of a computer. If there exists an inductive algorism which can be fed into a computer and which is most successful, then science must be least exciting and devoid of all adventure. This ideal of science, especially of applied science, as based on a precise algorism, can be found already in Laplace's *Philosophical Essay on Probabilities*, where he expresses his hope to see even judges on the bench replaced by computers.[5] Perhaps the judges of his day were so inhuman that replacing them by machines would have been progress from the negative to the zero degree of humanity; still, the very thought is chilling.

Those, however, who think that viewing science as based on an alogrism debases science usually stress that in science we need both luck and intuition. Both words, 'luck' and 'intuition,' are indicative of our ignorance, perhaps of our essential ignorance. For, obviously, if this ignorance were temporary, science based on algorism would be attainable. Though they are both mysterious, luck and intuition seem to be opposite poles: whereas we admire those who contribute to scientific progress when we view them as inspired, we somewhat deprecate them when we say they were lucky. Understandably, Pasteur and Edison alike felt deprecated when their success was viewed as due mainly to luck. They both repeated Lagrange's mystical formula about luck coming only to the prepared mind. One might push the argument and say that even having a prepared mind depends on luck. Indeed, both Oersted and Einstein humbly viewed their own inspiration and talent as mere luck. And yet pushing the argument so far is scholasticism: if a typical scientist were lucky enough to have one of two wishes granted, these being for talent and for material success, he would doubtlessly grasp his luck and wish for talent, as King Solomon is alleged to have done, hoping to have the chance to use it.

Hence, it is neither talent nor luck but, rather, adventure, namely, the bold use of talent and the grasping of luck when it comes one's way, that we wish. It is all too easy *not* to use either talent or luck by the mere lack of a sufficient degree of courage and perseverance. As Edison said so aptly, "It has been so in all my inventions. The first step is an intuition – and it comes with a burst; *then*, difficulties arise.... I have the right principle, and am on the right track, but time, hard work, and some good luck are necessary too."[6] And again, "The trouble with other inventors is that they try a few things and quit. I *never quit* until I get what I want."[7] Of course, Edison was very lucky to be able to work so hard, so boldly, so cleverly: most people can do nothing to alter their predicament. But Edison's luck is so very different from the luck of the born princess or of the person who got rich at his first gamble; it was not the luck that brought success but the luck that made him able to strive for success.

III

Query: can we admire science as the product of the bold use of the investigator's imagination and of his good fortunes, and yet justify our trust in science? If we justify our trust by a criterion or a formula, the criterion or formula may turn into a Laplacian algorism which renders science mechanical, but if we have no criterion or formula, we may have no justification. Thus, in his *Logical Foundations of Probability*, Carnap tries to differ from Laplace but fails. He indorses the view – which he attributes to Einstein and Popper, but which might be more justly attributed to Brewster and Whewell, if not to Galileo and Kant – of the imagination as an essential ingredient in science.[8] And yet he provides the Laplacian formula for deciding which hypothesis to believe in and to act upon, and his formula provides the Laplacian algorism of projecting the past into the immediate future and thus generating mechanically the best possible hypothesis.[9]

One cannot argue cogently against the creation of any working algorism, that is to say, against any systematization. Descartes systematized geometry and enabled people to prove theorems mechanically rather than cleverly; solid-state physics is now systematizing one of the most known fields of clever invention, namely, metallurgy; and yet no one is any the worse for these systematizations. When Abraham Wald systematized some decision-procedures, his ideas were so powerful that they were kept for a time as war secrets.[10] Any partial systematization, such as Descartes's or Wald's, merely covers some ground and sends the adventurer further afield in search of new frontiers. But total systematization excludes all adventure. In any field, any alogrism is welcome; yet, were an algorism universal in that field, creative thinking in that field would be redundant altogether, once and for all. If the field in question is the generation of ideas in general, algorism in it is the end of creative thinking in all science and technology.

But it is possible to invent a criterion or formula for justifying our faith in and reliance on a scientific theory without taking away the spirit of adventure from science. A formula for trusting science which is not a means of generating mechanically the theory to be trusted is possible, though Carnap, at least, has tried to produce it and failed. Such a formula was invented by William Whewell[11] and reinvented with improvement and modification by Karl Popper;[12] my dissent from his philosophy is rooted in my rejection of the formula, but I must acknowledge its superiority. Let me present it first and discuss the question whether it is a measure of success later.

It runs as follows. A belief in a theory is justified to the degree to which it was corroborated by experience; that degree depends on the explanatory power of the theory in question, of the degree of testability of that theory, and of the degree of severity of the test which it has thus far passed successfully.

Let us not examine this formula in detail, so as to avoid a rather academic exercise. Let us first try to get the spirit, the general feel and approach behind it, and see how acceptable that is. Up to a point, I think, it is, but only up to a point. To begin with, let us consider the positive general characteristics of this formula. The formula takes full account of the challenge of research. Thus, it is in full accord with the above quotations from Edison, as well as with the following one: "I would construct a theory and work on its lines until I found it untenable, then it would be discarded and another theory evolved. This was the only possible way for me to work out the problem."[18]

Both Whewell and Popper stress the need for problems, for inspired solutions, for the usefulness of criticism in the development of new solutions, and for the success of theories which stood up to criticism and thus proved their mettle. The chief difference between Whewell and Popper, it is well known, is that Whewell believed in the finality of such success whereas Popper believes in its tentativity. Also, one need hardly say, in this division no one sides with Whewell today, even though most philosophers of science accept neither view. The majority indorse a view which is nearer to Whewell's than to Popper's in that they replace Whewell's finality not with Popper's tentativity but with probability. This notion is either meaningless or else it is a theory of probability in the classical sense, and thus it yields a Laplacian algorism.[14]

IV

So now we have a formula – the Whewell-Popper formula – which perhaps justifies our faith in science, and if so, without doing injustice to intuition and luck. Let me now show you that to judge the greatness of science by its success, when success is measured by the Whewell-Popper formula, does not accord closely with the admiration of science as an active adventure, which I am advocating.

Imagine an Einstein, racking his brain, developing his most inspired ideas, sifting them, elaborating them, deducing from them both explanations and tests; he then retires to his den, full of nervous anticipation. The world notices his work, considers it, concedes that it is possibly admirable, and sits patiently and waits. Then comes an Eddington, translates the test into action by designing all sorts of instruments, by mobilizing funds, by organizing workshops to prepare the instruments, by organizing an expedition, by supervising and participating in the experiment, by calculating the experimental results and comparing them to Einstein's results as calculated from Einstein's theory. All the while Einstein is supposed to be sitting back and waiting, and all the time we are supposed neither to admire nor to dismiss him; and then Eddington may give the green light, and we all burst in admiration for Einstein; or he does not, and we do not.

In such a case we want the scientist to intuit not only a brilliant idea but also certain truths. In such a case the scientist is more of a fortune-teller than I would like to be the case. It was said that Faraday had a nose for the truth; Edison was called the "wizard of Menlo Park"; maybe all this is true. If so, the truth is not very palatable, at least for those who find it less enjoyable to admire Faraday or Edison because he is a wizard than to admire them as intellectual adventurers.

Have we not gotten into an impasse? That we want success is self-evident: to say that we want success is to say no more than to say that we want to achieve something. If we did not want anything or if we had no expectation of achieving it, we would not act at all. And if we do not want the success of applied science to be a matter of wizarding or uncanny insight, we must have an algorism. So, it seems, if we do not want an algorism, we do want some wizarding! Considering this impasse, we may be more patient with my colleagues who view science as a success-algorism. Michael Polanyi[15] is surprised that most philosophers of science deny that applied science or technology contains an intuitive element when this claim is a standard criterion of all patent offices: a machine than can be created by anybody according to a publicly known alogrism will not be granted a patent; the claim for a patent must involve (by legal standards!) a claim for originality. Polanyi exaggerates the wissdom of the legal patenting apparatus (consider the Selden case, and see Edison's 'My 40 Years of Litigations'), yet by and large he is right.[16] Still, we may well understand the philosophers whom he criticizes: reluctant to acknowledge the wizardry of the inventor, they decide that invention has an algorism to it, and thus they imply – perhaps unwittingly – that inventors need not be original. They equally imply that scientists need not be original, even though we are not supposed to grant Ph.D.'s for dissertations which we do not judge original – this also by some legal standards. This does not disturb the inductive philosophers who insist on the idea of science as unoriginal.

The idea that science is no magic and hence it is based on some algorism is by no means new, by no means peripheral to the traditions of science, and by no means a merely unintended aspect of inductive philosophy. Its chief corollary from our viewpoint is no news either; it has been stressed again and again that the measure of a scientist's success is, provided he is a proper scientist, the measure of work he has invested in science and no more. This calls for equal admiration for every hard-working contributor – all contributions will be thankfully received – as was stressed by Bacon, by Boyle, and, more recently, by Malinowski as well as A. P. Usher, leading historian of technology.[17] Bacon also said that inventors should be honored by statues made of different material from gold to wood and in different sizes, depending on the greatness of their inventions. Similarly, the fact that we do not admire in equal measure every hard-working inventor and every anthropologist who has conducted field-work is well known. Usher dismisses this inequality as cults of personality and historical fiction; young anthropologists are told that every fieldwork is of equal importance, and young chemists and engineers are told similar stories. Yet most young engineers and anthropologists think that all this is a mere scientific myth. Professors reiterate the myth of equality of all contributions, and students reiterate the myth of great adventures. Who is right? What is the real criterion of success?

The instrumentalist philosophers who identity – with Duhem – pure science and applied science have the idea of simplicity as the criterion of success.[18] Mach viewed simplicity as success of mental economy, which is very different from material success.[19] Material success, no doubt, is often awarded to very complicated theories,[20] and to be honest we need economists to assess such material success. And, of course, we should choose materially successful economists to do this job. Moreover, all economists will agree that the archeology of extinct cultures, for instance, is of hardly any practical value, except as a highbrow means of entertain-ment, perhaps. Most philosophers of science, however, do not indorse this instrumentalist philosophy, preferring to it some inductive philosophy which justifies their claim that science is useful by extrapolating from past to future material success. They say, "when we make a decision, we base it on induction."

These people just do not understand the essentially adventurous nature of decisions. Neurotic people, it is well known, who do not possess the ability to decide, often say, "I cannot decide because I do not know enough."[21] They obviously deceive themselves: when we know enough we have no problem at all, and so we need not decide at all. When we say we must decide which route to take, we imply that we have no knowledge as to which route is the shorter, or the safer, or the better one by any accepted criteria; alternatively, if we have not accepted a criterion, we may be saying that we do not know which criterion is the correct one. So, one thing is quite clear: we just do not decide by induction.

Let us then examine the sort of decisions which pure scientists, applied scientists, and technologists face; the sort of questions they have to try to answer. This will give us, I hope, some insight to the sort of adventure they may face.

V

One reason why I admire Popper's philosophy of science is just its playing down of success. Roughly, but very seriously, let me put it thus. Everybody admires science for its ability to predict. Meaning, of course, its ability to predict *correctly* or, more precisely, sufficiently correctly to pass as the ability to predict *successfully*. (The criteria of success are vague and invite us to confuse science with technology.) Only Popper, to my knowledge, has stressed that the ability to predict with some degree of precision is what characterizes a scientific theory – regardless of the correctness of the prediction. Popper had predecessors, to be sure; Galileo, Boyle, Faraday, Whewell, Peirce, and Edison spoke favorably of mistaken predictions in science, and doubtless others did too. But they all insisted that finally it was the successful prediction that counted, that the errors were important only as stepping-stones to imminent success which really characterizes science. It was Popper's province to discover that this is not necessarily so.

According to Popper's theory, a scientist needs a good hypothesis. He may work for many years with no success, or with almost no success. Nothing in Planck's researches after 1900 equals his success in 1900, for instance. And so, according to Popper, the adventure of science may lead to no result; there is, in science, such a thing as utter failure. But, according to Popper, once success was achieved in that a good hypothesis has been found, and once the work has been invested to show the goodness of the hypothesis, then even if the hypothesis leads to predictions which are refuted we may value that hypothesis. The goodness of the hypothesis, according to Popper, may be measured by its explanatory power, as well as by its testability, which is the ability to deduce from it predictions which may be checked by experiment and observation, quite independently of whether the predictions are later found to be true or false. Hence, our appreciation of science need not follow the Whewell-Popper formula, since that formula speaks of theories which have yielded only correct predictions thus far.

To my regret, Popper has expressed recently a view quite different from the one I have just ascribed to him. Referring to discussions with myself, he has said a good hypothesis should also be corroborated before it be refuted.[22] He uses the Whewell-Popper formula as a measure of the goodness of a hypothesis or of our appreciation of it, thus demanding from scientists some measure of wizardry, though he does not go as far in applying the formula as others would: he does suggest that once a theory is refuted we then withdraw our belief in it, but he does not suggest that in such a case we also withdraw our appreciation of it. Nevertheless, I wish to ignore this aspect of Popper's writings, and if you do not like it, you may view the ideas I attribute to Popper as his ideas in my modification or in my distortion, as you wish.

My best reason for rejecting the use of the Whewell-Popper formual as a measure of success is that I follow Popper's theory of the process of trial and error in science as something essentially different from trial and error in many other situations, such as more everyday ones. Trial and error is progress by elimination. In small sets of alternatives, elimination by trial and error may insure success. Most everyday situations are like that, and indeed, no one in his senses would view researches in such situations as adventurous. When our television set breaks down, we take it for granted that there is a finite set of possible faults it may have. Usually we test them one by one and finally eliminate the fault; there is no adventure in the search. The classical argument against the theory of induction, as you well know, is that there exist infinitely many possible explanations of known facts. Lord Keynes has acknowledged this criticism and has postulated the principle of limited variety which he has found implicit in earlier inductivist writers.[23] Assuming, quite *ad hoc*, by the way, that only a finite number of hypotheses may explain known facts, he advocates the most unadventurous theory of science.

Consider now the adventurous method of trial and error in everyday life. This is known as looking for a needle in a haystack. The haystack is supposed to be infinite for all practical purposes. It is quite conceivable that one goes on eliminating more and more possibilities about where the needle may be and yet not progress to any appreciable extent toward the correct position of the needle. Looking for a needle in a haystack is adventurous; it demands intuition and luck.

According to Popper, the search for a good hypothesis is looking for a needle in a haystack. But, according to his philosophy, once a hypothesis is found which solves the problem at hand, that is, which explains the facts which puzzle us, and once it is found testable, then progress has been achieved. (He even claims ability to prove this,[24] but let us not press him on this point.) Even if a good hypothesis is refuted, things will never be the same again. The refutation of a previously unrefuted hypothesis is surely a new discovery we partly owe to its inventor. The task after the refutation of the new hypothesis is not the same as before: we now wish to explain, not the same set of facts which the refuted hypothesis has explained, nor the same set of facts plus the refuting facts, but the refuted hypothesis, as a special case and as a first approximation, plus the new fact.[25] What Einstein explained is not all the known astronomical facts but Newtonian astronomy plus the facts it failed to explain, just as Newton explained Kepler's theory and the deviations from it. This is a fundamental methodological point. The only way to evade it is to claim that pure empirical data exist; as you may know, such claims are becoming increasingly difficult to maintain.[26]

The metaphysical theory corresponding to this fundamental methodological point, of explaining not bare facts but previous theories and their refutations, is the theory of the approximation to the truth by levels of explanations. This theory may be false; it does, however, justify Popper's rejection of the claim that scientists must intuit the truth, and his claim that the intuition of a possibly true explanation is enough of a step forward for the time being. If so, then though finding a good solution is like finding a needle in a haystack, the eliminating of that solution as false is not like finding that the needle is not in a small portion of a haystack. Pictorially, scientific theories are not explorations of small portions of a given haystack; they are sieves through which the whole haystack is passed and whose meshes are used to build sieves of ever finer mesh. In this picture of science as progress through conjecture and refutations, there is no room for confirmation or corroboration.

VI

So much for pure science. What I wish to show next is that all this does not apply to applied science or to technological invention. In applied science there are two standard kinds of problems, deducibility and applicability. Given a theory and given a problem, the applied scientist asks himself, can I solve the problem while using that theory, and is my solution true? The first kind of question is essentially mathematical, and this explains why the terms 'applied science' and 'applied mathematics' are so

often used interchangeably by people who will never use the words
'science' and 'mathematics' interchangeably. Mathematics, pure or applied, has its own kind of adventure of which I shall not speak beyond
saying that it is the only kind of adventure an applied mathematician or
an applied scientist may encounter. One may be even more specific here
and add that in applied science, unlike pure science, the problem of deducibility is to find initial conditions which, together with given theories,
yield conditions specified by practical considerations. This is, indeed, how
Popper characterizes technology at large,[27] which is a rather narrow view
of technology.

The second question, namely, is the solution to a given problem which
he has deduced true, hardly allows itself of intellectual adventure, though
designing the test may be, and executing it may also involve physical adventure. That the testing of results of applied science does not belong to
applied science is rather obvious: even the applied scientists in industry,
not to say in the universities, wear white collars, not white lab coats or
blue overalls. They cannot and need not perform tests. That they cannot
is a point of great significance, which I wish to label as 'Hatfield's law.' It
is this: there is always a gap between applied science and the implementation of its conclusions, to be filled by invention.[28] The law is trivial in the
sense that applied science does not issue programs for computers which
implement its results; it is less trivial if we understand that applied science
does not issue programs even for skilled workers without gaps to be filled
by clever inventors.

To take a very simple instance[29] of the difference between invention
and applied science, let us take a case involving both. The inventor Edison wished to replace gas street lighting with electrical lighting. The crux
of the difficulty which disappointed other inventors was practical: the
amount of copper needed to conduct electricity along the streets of a city
would be too large to warrant investment. It then occurred to Edison
that this obstacle could be overcome if high tension and high resistance
were employed. This was a technological idea, which could be correct or
incorrect: Edison needed an applied physicist to tell him that, and his
applied physicist, Upton, applied Ohm's law to the problem and concluded that by using a tension of 100 volts the quantity of copper needed
could be cut into one one-hundredth of the originally calculated quantity
required. The conclusion surprised even Edison, though it followed

logically from a law he knew very well. Upton's job was finished on paper. Applied science is, thus, providing answers to given questions when these are implicit in a given theory. Posing the question and seeing the possible technological significance of the answer to it was one of Edison's inventions on the road to electric lighting.

Edison's opponents did not think the obstacle – the need for immense quantities of copper – could be overcome except by connecting the lights in series. Applying Ohm's law to this solution, they found it impracticable and showed that the light given by the many electric lamps needed to light a city would be too small to be of any use. Doubtlessly they were quite right in their applied science but rather wanting in imaginative invention. This shows how different are applied science and invention. That the two can overlap was shown by my example of the collaboration of Edison with Upton; that they are distinct is symbolized by the distinctness of these two individuals, their training, and dispositions.

One need not blame Edison's opponents, or any other people, for lack of imagination; for by doing so one may underrate the power of the imaginative inventor. Not only was Edison's idea so very new, it also required many other innovations to adjust to his new scheme. For instance, whereas dynamos in his days gave out constant currents, his idea required the invention of a usable dynamo with constant tension. It is conceivable that his idea of high-tension currents would not work if he could not invent such a dynamo. How much more understandable, then, it is that his opponents were thinking with existing dynamos in mind. And this is not all. When all means for high-current lighting were found, there was no filament which could stand the conditions imposed by Edison well enough to be practicable. For all he knew, there may not have been such a filament. He himself conceded that much at the stage when he worked on platinum filaments which could not be produced commercially prior to the commercial development of machines to transmute base metals into platinum. It is incredible how lucky Edison was and how much he dared gamble on his good fortune.

Invention depends on finding facts for which we have no clues, at least a sufficiently wide absence of clues to make the haystacks in which they lie practically infinite. Otherwise, a team of applied scientists would find it, to be sure. And so the failure of a technical inventor is as final as the failure of a failed wildcat prospector.

Not seeing the wildcat character of invention, not seeing it as the miracle of finding a needle in an infinite haystack, often leads to misunderstanding and to underestimation of inventors and their sense of adventure. The most definitive and thorough history of photography is probably that of the Gernsheims.[30] Yet their view of Daguerre's invention is incredibly naive; they claim[31] that Herschel was much greater than Daguerre, since after merely having heard about Daguerre's success Herschel repeated and improved in a few days the results that initially took years to develop. The truth is that there was no reason at all for anyone to think of the possibility of latent images to be developed. The problem of photography for decades had been how to fix an image which is visible at once (perhaps while turning it from negative to positive). No one thought of latent images to be developed, and everyone knew of visible images which soon disappeared. Thomas Young applied this knowledge of temporary images to photographing infrared interference – the first piece of photography of any value, and a typical case of applied science which is in no way an invention, as was Daguerre's thought of the theory of latent images to be first developed and then fixed. Daguerre had the thought under most peculiar circumstances, and the thought appeared surrealistic to him at the time and years after.[32] Once his idea proved successful, hosts of improvements were possible within a relatively short period.

It is because invention is a theoretical rather than a practical activity – though to a practical end, of course – that inventors may illustrate their inventions while using most crude machines; but it is because their claims for inventions must be corroborated that they must use some machines, unlike theoretical scientists and even applied scientists who may never bother about practice.

We see, then, that though corroboration is needed in technological invention, it is of no use in science, pure or applied. The question is whether corroboration has anything to do with belief, and I think I have shown that in technological invention corroboration has something to do with the nature of the problems at hand, which is always, how one can do successfully something specified; whereas in science, pure or applied, belief is not a matter of scientific method either. As Popper has stressed, in pure science we may try to refute the most corroborated view, and in applied science truth is of little importance. In applied science, that is, the question of the truth of the theories to be applied hardly even enters, though the question of the applicability of results from it its crucial and is answered by simple tests. Thus, we do not believe, but we still apply, Newtonian mechanics, though not to systems involving high velocities.

VII

Technology involves various factors, and I have thus far concentrated on only two of them, namely, applied science and invention. The implementation of the results both of an invention and its maintenance are important fields which I cannot discuss here beyond saying that in the field of implementation corroboration is most important, and the degree of corroboration required prior to implementation is socially and legally determined. It is doubtless that here the requirement for corroboration is a matter of public safety and that though no amount of corroboration insures success, the legal requirement for corroboration does eliminate some dangerous technical innovations prior to their implementation in the market. But if we demand much corroboration, there will be no implementation.

To show how complex the problem of implementation is, and how difficult, I should mention the classic trial of General Billy Mitchell, who thought he had corroborated his theory of the importance of air force bombers and the consequent diminishing importance of naval warfare. His sincerity is beyond question and is the main reason for the fame of his court-martial. Another reason for his fame is rather vulgar, namely, the truth of his prediction of an unannounced Japanese air attack. But, of course, he was mistaken in most of his views except about the power of air raids, and implementing all his proposals might have been a disaster.[33] Success philosophy is what makes the various books (as well as the movie) about him so very unsatisfactory. Without success philosophy the story of Billy Mitchell becomes much more interesting and enlightening, especially in relation to the requirement of the complacent – such as his superiors – for more corroboration before implementing any novelty. But I shall not go further into it. I wish only to say that it exemplifies one point of significance from our present viewpoint.

Those who believe that empirical evidence renders a theory credible view the implementation of that theory not as an adventure but, on the contrary, as the reduction of risk and thus success. In truth, the implementation of a novelty is risky even when no risks seem possible. Testing is performed so as to eliminate some risks – some 'bugs,' to use Edison's jargon – and some risky innovations. But sometimes the prevention of one risky implementation entails the taking of other risks, as Mitchell's story shows. This argument has been used in a recent U.S. Senate subcommittee investigation into the disaster of the atomic submarine 'Thresher,' which sailed in spite of known and eliminable risks. Whether the argument that too much testing and insisting on safety may lead to other risks was used properly or in order to shield comrades in arms I do not know, but such questions can be discussed rationally, and will be, perhaps, by future historians of technology.

As I have said before, success is not something to be proud of, but a puzzle to be explained. The discoveries of modern biology and modern physics about the hostile world in which we live should make our very survival a great puzzle. But, on the contrary, philosophers of science refuse even to see the obvious fact that luck and intuition are essential to success. Yet I must say this in their favor: When any phenomenon takes place regularly, we wish to explain it and thus systematize it. Discovery and invention look regular enough nowadays, and so they seem to require a systematization. That there is a flaw here I have tried to argue, but I cannot put my finger on it. The dichotomy between algorisms versus lucky intuition seems a fundamental fault in our way of thinking. Obviously, mixing algorism with intuition and luck will not destroy the dichotomy. One way out is to view systematization dynamically: some past successes which looked lucky can be explained by later systematization. This eases the pressure of the difficulty, but no more. Take any unsystematic but regular process: to use Chomsky's work,[34] take a child's intuitive learning to speak. Not only do we not know how to explain it, it seems as fundamentally mystifying as Faraday's or Edison's uncanny ability to discover and invent to order.

To conclude, the confusions in the field are rooted in a difficulty which is shared by all of us. Our dichotomy between algorisms and lucky inspirations is very unsatisfactory. Philosophy managed to progress in various directions in spite of this fundamental obstacle. But this obstacle may be a most serious impediment for the progress of the philosophy of technology. So, this field may be at a dead end for the time being, or, if it will get going, may lead to spectacular results which are bound to revolutionize all philosophy and much of our way of thinking in many departments.

APPENDIX: PLANNING FOR SUCCESS:
A REPLY TO PROFESSOR WISDOM

Professor J. O. Wisdom has kindly offered some critical observations on the foregoing discussion[35]. In his comments he agrees with me that applied science and technology are distinct, but for different reasons. Since 'reason' may be either 'criterion' or 'end,' let me make it clear that I use a simple factual criterion to distinguish the two, and to a definite philosophical end. Consider the following inference: Corroboration (or confirmation or verification, or factual support, or agreement with experience) is important in technology; technology is identical with science; *therefore*, corroboration is important in science. The inference is valid; its conclusion is false; and its first premise is true; hence, the identification of science and technology is mistaken. This suffices for me; for those who do not agree that the conclusion is false, and even insist that it must be true because the two premises are true, for them different arguments might be of use. If they see that the identification of science and technology is erroneous independently of the theory of corroboration, then they might be more agreeable to reform their ideas of corroboration too.

It is a simple matter of fact, I think, that applied scientists use high-powered theories, difficult mathematics, etc., to solve problems in a way not open to those less well versed in these theories and deductive techniques; technologists are not usually as theoretically high powered and do not work out patiently and carefully chains of deduction – they hire applied scientists for that – but they are high powered in different respects, in being more familiar with technological situations, facts, problems, tasks, etc., from the availability and specific merits and defects of raw materials, to storage, to marketability. My example of a technologist and his work was Edison and his building of an electric street-lighting system; this involved, *inter alia*, invention, applied science, organization of implementation, economics, and even corporation law. When it came to applied science, Edison used an applied scientist – Upton – to make the calculations. Whether the results of the calculation were correct was not Upton's concern but Edison's – as long as the deductions were valid, of course. Applied scientists do not usually wear lab coats (except perhaps for the sake of prestige) as do experimental scientists and technicians – including engineers and inventors. When the New York electric system was inaugurated, Edsion was there, tailcoat and all. Soon troubles ensued, and he quickly tried to straighten them out. In no time, he looked as if he were wearing the previous year's overalls. Excitement of this kind may take place in technology and in experimental science; excitement in theoretical science and in applied science is very different.

Since Wisdom kindly clarifies my views, I wish to take leave to clarify his clarification on two points: success and scientific success. I have no objection to success, but only to success philosophy, to the view of success as the measure of worth. The better the society and the more favorable the circumstances, the more likely, perhaps, that success and worth would go hand in hand; that is the most one could reasonably say. Anyway, our society is not that just; not as yet, at least.

As to scientific success, I dislike the idea that it can be generated by a formula or an algorism, and I dislike the idea that it can be assessed by a formula. I have given an example of a formula pertaining to performing both tasks – Laplace's – and I have given an example for a formula pertaining to performing only the second task but *not* the first task – the Whewell-Popper formula as generally understood. I am not clear whether this point of mine is correctly presented by Wisdom.

The formula for corroboration which is not an algorism is the Whewell-Popper formula: A theory is corroborated if, and only if, after it has been invented and shown to explain the puzzling phenomena it was put to serve, it has been tested in an attempt to refute it, and that attempt has thus far failed to refute it.

I gave an example of how the Whewell-Popper formula would be misused as a measure of Einstein's worth because it would make Einstein's worth depend on luck concerning matters entirely beyond his control, such as the Eddington experiment. Wisdom shows quite clearly – and in my view correctly – that this does not apply to other factors, which were very much within Einstein's control, such as knowledge of experimental results attained *prior* to his studies. True, Einstein was encouraged by his ability to explain the motion of the perihelion of Mercury; this is irrelevant to my criticism of the use of the Whewell-Popper formula as a measure of worth, as it merely purports to show that in some cases such a use of the formula is not as objectionable. Also Wisdom is mistaken here on a technicality. Both Whewell and Popper would not consider the Mercury perihelion motion a corroboration, but they would consider Eddington's experiment a corroboration.

Wisdom fears, incidentally, that if a theory is refuted at the first test it will not be tested further, and so experiments in agreement with it might not be discovered. Perhaps. Also, let me add, if a theory is not refuted at first test, then possibly it will not be tested further and refuted – sometimes even with premature implementation, leading to refutations descending on us as catastrophes. There are four logical possibilities here (early corroborations or refutations giving or not giving false impressions), and there exist historical examples conforming to each of them. So arguments here go neither way, except, perhaps, in some limited historical contexts. But, as we are concerned here neither with history nor with psychology, arguments go neither way. My view that qua corroboration, corroboration cannot play any role in science was drawn from Popper's philosophy. A critic may show that my reasoning was faulty or that Popper's philosophy is false; if he denies either possibility he is no critic of my view.

But Wisdom is right in claiming that we do wish to know both the facts which agree with a given theory and those which do not. This is because we wish to know how much a theory explains; and the more it explains some puzzles, the better. Now suppose that we go back two or three centuries and find there a theory which explains a phenomenon discovered one or two centuries ago by one who had no knowledge of that theory; and suppose, further, that the phenomenon is not problematic today. Then, the present discovery that one old fact agrees with a somewhat older theory is of no interest. Why? Popper would say, because no one intended to refute that theory by that experiment, and so it is no corroboration. Scholars have been puzzled by Popper's stress on intention, and not without reason. I think intentions as such need not matter, certainly not the sincerity of the attempt to refute, which regrettably Popper stresses so much. But intentions may enter when they are more objective, such as the desire to solve certain problems. When we wish to know how much a theory explains, and wish it to explain much, surely a corroborating experiment does that. When we wish to find, if a theory is false, that it is false, then corroborations will not do but refutations will. No one wants his partner to cheat him; but some people want to know if their partners cheat them, when, and how they cheat them. Some do not want such information even if it is true, and I tend to feel like them. But I do want to know where my theories went wrong if they are not the absolute truth. It all depends on what we are after.

And if we are after solutions to certain problems and we have a sure method of solving them, we shall, of course, indorse this method; even if the method is only partially successful (in any sense) we shall indorse it. To speak against success here is folly. But suppose all our problems were easily solved and life were deprived of all challenge. Paradoxically, success would become failure by excess. The literature on the poor rich is maudlin and vulgar, but there is a kernel of truth in it.

When we learn to operate a machine or an algorism we may feel challenged, but when we master it we cannot enjoy a life consisting chiefly of operating it, no matter how successful the operation is in terms of economic and social standings or in any other terms. This is a psychological truth concerning youngsters and adults, humans and even some other animals.

If we feel challenged when we learn to operate a machine or an algorism, but not when operating it, then such a progress is not mechanical or algoristic. If so, how is it that practically everyone learns to operate the algorism in the same way? We all learn our mother tongue to an incredible uniformity. Einstein's theory, being non-formal, may be open to myriads of understandings or misunderstandings. Yet most men of science understand it in one, two, or three standard interpretations. I think this is an interesting fact. It may have to do with the fact that Einstein's theory came primarily to answer some questions, and that somehow questions have a logic of their own which is capable of directing people again and again toward understanding neither by mere mechanics nor by mere luck and ingenuity which defy generalization. My difficulty is exactly this. Suppose that statements about intuition can be generalized like statements about other operations. Then intuition may feel different from memorizing by rote, but it is not: Then, we can have a sort of intellectual progress by rote, a sort of science-making machine. If intuition is so unique, if we have no science-making machine, how is it that science progresses so steadily? This question seems to be of some significance in the philosophy of technology: As Dr. Jarvie has pointed out, the American space program is based on an estimated rate of technological progress. This optimism strikes me as most incredible. Can we explain it?

NOTES

[1] See Chapter 2 above, especially the appendix.

[2] K. R. Popper, 'Three Views Concerning Human Knowledge', *Contemporary British Philosophy* (ed. by H. D. Lewis), London 1956, pp. 355–88; and *Conjectures and Refutations*, London and New York 1962, pp. 97–119. See also my 'Duhem versus Galileo', *British Journal for the Philosophy of Science* **8** (1957), 237–48.

[3] Matthew Josephson, *Edison* (paperbound ed., New York 1959), pp. ix–xi.

[4] See my *Towards an Historiography of Science*, Beiheft 2, *History and Theory*, The Hague 1963; facsimile reprint, Weslyan U. P., 1967, pp. 8, 17, 42–48, 84, 85, 86, 105, 106.

[5] Pierre Simon, Marquis de Laplace, 'Application of the Calculus of Probabilities to Moral Philosophy', *A Philosophical Essay on Probabilities* (paperbound ed., New York 1951), chapter X.

[6] Josephson, *Edison*, p. 198.

[7] *Ibid.*, p. 216.

[8] Rudolf Carnap, *Logical Foundations of Probability* (2d ed.), Chicago 1962, p. 193.

[9] 'A General Estimation Function' in *ibid.*, sec. 99 ff. See also his 'On Inductive Logic', *Philosophy of Science* **12** (1945), sec. 10 (reprinted in *Probability Confirmation, and Simplicity* (ed. by M. H. Foster and M. L. Martin), New York 1966). Cf. Popper, 'On Carnap's Version of Laplace's Rule of Succession', *Mind* **71** (1962), 69–73; and my 'Analogies as Generalizations', *Philosophy of Science* **31** (1964), 351–56.

[10] J. Wolfowitz, 'Abraham Wald', *Annals of Mathematical Statistics* **33** (1952), 9: "... was put on the restricted category and made available only to authorized recipients. Wald chafed greatly under this restriction." Cf. Statistical Research Group, Columbia University, *Selected Techniques of Statistical Analysis for Social and Industrial Research and Production Management*, New York 1947, pp. viii–ix.

[11] William Whewell, *The Philosophy of the Inductive Sciences*, London 1840, Vol. II, Part 2, Book XI, chap. i, secs. 6, 10; chap. v, secs. 7, 8 (about testing), 10 (about techniques of testing: predictions), 11 (about the stringency of the test: new prediction), 12, 13 (correlating simplicity and testability, so to speak); chap. vi, sec. 12 (non-*ad hoc*ness). See also his *History of the Inductive Sciences* 3d ed.; London 1843, Vol. I, Book VII, 'The Discovery of Neptune': "Thus to predict unknown facts found afterwards to be true, is, as I have said, a confirmation of a theory which in impressiveness and value goes beyond any known explanation of known facts." And such confirmation, he says, took place only a few times in the whole history of man.

[12] Popper, *The Logic of Scientific Discovery*, New York 1959, chap. x and New Appendix *ix.

[13] Josephson, *Edison*, p. 198.

[14] See Chapter 2 for a detailed discussion of this statement.

[15] Michael Polanyi, *Pure and Applied Science and Their Appropriate Forms of Organization*, Society for Freedom in Science, Occasional Pamphlet No. 14 Oxford, December 1953, p. 2. See also p. 9.

[16] For Polanyi, see *ibid.*, p. 6, on the ability of the competitive system to cope with technological novelties. See also his *Personal Knowledge*, paperbound ed.; New York 1964, p. 177 n. See n. 17 below for the exactly opposite view. For the Selden case, see William Greenleaf, *Monopoly on Wheels: Henry Ford and the Selden Automobile Patent*, Detroit 1961, and the review of that volume by John B. Rae in *Technology and Culture* **11** (1961), 289. For Edison's 'My 40 years of Litigations' *Literary Digest* (Sept. 13, 1913), see Josephson, *Edison*, pp. 354–60, and his Index article 'Patent Infringement.'

[17] For the Baconian tradition, see my *Historiography* (n. 4 above), pp. 4–6, 15, 60–66, 81. For the Malinowski tradition, see Ian C. Jarvie, *The Revolution in Anthropology* London and New York 1964, pp. 1–7, and his Index article 'Fieldwork'.

A. P. Usher, *A History of Mechanical Invention*, rev. ed., Cambridge, Mass., 1954; paperbound ed., Boston 1959, p. 68: "Cultural achievement is a social accomplishment based upon the accumulation of many small acts of insight of individuals. The massiveness of this social process was long ignored or misunderstood. ...A conspicuous result of this disposition to put a part for the whole was the frequency of bitter controversies over claims of various inventors to a particular invention.... . These disputes all rest on the false assumption that the achievement was so simple and specific that it could properly be identified with the work of a single person at a single moment." A clearer and sharper statement is quoted from seminar notes by Thomas A. Smith in his enthusiastic review of Usher's volume, in *Technology and Culture*, **11** (1961), 36. "The concept of heroes and men of genius," says Usher according to Smith's record, "are literary and cult devices that simplify the historical record." See also paperbound edition, p. vii. For my view of the partial justice of such wild claims, see my *Historiography* (n. 4 above), pp. 31–33.

[18] Pierre Duhem, *The Aim and Structure of Physical Theory*, paperbound ed.; Princeton, N.J. 1962, Part 1, chap. ii; Part 2, chap. iii, and Appendix 'The Value of Physical Theory'.

[19] E. Mach, 'The Economical Nature of Physical Research', *Popular Scientific Lectures*, La Salle, Ill., 1907, chap. xiii; his *The Science of Mechanics*, paperbound ed.; La Salle, Ill. 1960, chap. iv, sec. 4.

[20] Popper, 'Three Views' (see n. 2 above), sec. 5: 'Criticism of the Instrumentalist View'.

[21] Erik H. Erikson, 'Autonomy vs. Shame and Doubt', *Childhood and Society*, 2d ed.; New York 1963, chap. vii, sec. 2, p. 253; Allen Wheelis, *The Quest for Identity*, New York 1958, chap vi, esp. pp. 183, 199, 201, 250.

[22] Popper, *Conjectures*, chap. x, sec. 23, and note on p. 248.

[23] J. M. Keynes, *Treatise on Probability*, Cambridge 1921, chap. xxiii, esp. p. 271.

[24] Popper, *Conjectures*, p. 217: "I assert that we *know* what a good scientific theory should be like, and – even before it has been tested – what kind of theory would be better still, provided it passes certain crucial tests." And on p. 242, "The second requirement ensures that...." Meanwhile this claim – Popper's theory of verisimilitude – has been amply refuted.

[25] Popper, 'The Aim of Science', *Ratio* **1** (1957), 24–35. See also my 'Between Micro and Macro', *British Journal for the Philosophy of Science* **14** (1963).

[26] See my 'Sensationalism', above, for a detailed discussion of this statement.

[27] Popper, 'Naturgesetze und Wirklichkeit', in *Gesetz und Wirklichkeit* (ed. by S. Moser), Innsbruck 1949, pp. 43–60, esp. pp. 53ff. English translation in his *Objective Knowledge*, Clarendon Press, Oxford, 1972, pp. 341–361, 352–3.

[28] H. Stafford Hatfield, *The Inventor and His World*, 1933; Pelican ed., 1948, pp. 111, 133, 134, 151, *et passim*.

[29] Josephson, *Edison*, chap. x, esp. p. 194. Cf. Usher, *A History of Mechanical Invention*, pp. 72–77 and 401–6.

[30] Helmut Gernsheim, in collaboration with Alison Gernsheim, *The History of Photography from the Earliest Use of the Camera Obscura in the Eleventh Century up to 1914*, Oxford 1955.

[31] *Ibid.*, p. 81.

[32] Helmut and Alison Gernsheim, *L. J. M. Daguerre: A History of the Diorama and the*

Daguerreotype, London 1955; American ed., published under the title *L. M. J. Daguer-re, the World's First Photographer*, Cleveland, 1956.

[33] Ruth Mitchell, *My Brother Bill: The Life of General "Billy" Mitchell*, with an Introduction by Gerald W. Johnson, New York 1953, pp. 12–13: "There was a moment when he seemed to be entirely right. But events in Korea, where our separate airforce has accomplished nothing... prove... But even if he had been wholly wrong...." If the Korean war offers new evidence in one direction then, surely, the six-day war goes in the other direction.

[34] Noam Chomsky, 'Explanatory Models in Linguistics', in *Logic, Methodology, and Philosophy of Science* (ed. by E. Nagel, P. Suppes, and A. Tarski), Proceedings of the 1960 International Congress, Stanford, Calif. 1962, pp. 528–50.

[35] J. O. Wisdom's comments are printed in *Technology and Culture* 7, No. 3 (Summer 1966), 348–70. So is Professor Jarvie's paper referred to in the end of this chapter.

CHAPTER 13

POSITIVE EVIDENCE IN
SCIENCE AND TECHNOLOGY

ABSTRACT. If the problem of induction were soluble, it should be solved inductively: by observing how scientists observe, etc. The fact is that scientific research is successful, and the real question is, will it be so in future? If there is a formula of induction by which success is achieved, then by this formula we can say, as long as it will be used science will succeed. If there is no formula it looks as if future success in scientific research is most doubtful. Hence, a transcendental argument for induction goes, there is an inductive formula. Since, however, such a view of induction is rejected even by inductivists as naive, the argument collapses. Hence the question is, on what basis do we project the future success of science? My answer is that this future success is built into our social institutions and is partially institutionally safeguarded.

I wish to address myself, in this chapter, to the so-called scandal in philosophy. The expression, you remember, is due to Immanuel Kant. He said, the fact that idealism is unrefuted is a scandal in philosophy. Alfred North Whitehead said, the fact that the problem of induction has not yet been solved is a scandal in philosophy. For my part, I do not think we are such a scandalous group, and I do not feel that the problem is a scandal – except, perhaps, in the Greek sense of skandalos, a snare, a stumbling-block, a frustration. But even this, I contend, is a thing of the past. Philosophy need not be frustrated any more. We can face in our daily studies myriads of interesting and promising problems, even of some crazy quasi-empirical nature. I wish to begin by posing to you one. It is alleged, very poorly by the way, that when a new invention is empirically backed by positive evidence to some given degree, it is rational to employ it. We know from experience that there are bureaux of standards, food-and-drug administrations, even chambers of commerce, which impose empirical standards in civilized countries. Also, that the accepted empirical standards are different in different countries and operate sometimes with happy results, sometimes not. I suggest that philosophers may find all this terribly relevant and interesting for their work. That is to say, I recommend the empirical study of existing standards of adequate empirical study as a new avenue for philosophy; this is a direct continuation of the traditional preoccupation of philosophers with the problem of induction, and it need not be frustrating, much less scandalous.

I. KANT'S SCANDAL

Kant's scandal, that idealism is still unrefuted – idealism of the Berkeleyan type, of course – has been somehow forgotten, or replaced by Whitehead's scandal. Philosophers who boast of some rational disposition and of some interest in science, consider Berkeley's philosophy utterly passé. Indeed, philosophers who wish to discredit the Copenhagen school of the interpretation of quantum theory call it Berkeleyan, using the epithet as the one of dirtiest available in the current philosophical dictionary.

What has happened to idealism? How is it that it has ceased being a scandal? I have two answers. My first answer is that Kant's scandal was transformed into Whitehead's scandal. In other words, nowadays philosophers are preoccupied with the problem of induction and postpone engaging in the problem of idealism because they feel that the solution to the one is the same as to the other; or, that it is the most promising way to arrive at a solution to the other. Solve the problem of induction, they say, vindicate induction, and you will thereby justify physical science, thus imposing on the idealist who is still reasonable the physical world postulated by physical science. Induction, then, may expel idealism.

Alas, the truth is the very obvious obverse. The very preoccupation with induction, the very stress on the need to justify by observation, is what entrenches idealism, as Bishop Berkeley so well argued when he showed that the distinction between primary and secondary qualities, between objective and subjective ones, cannot be vindicated by any induction, so that all experience must be forever subjective.

I shall return to this later. Now I shall take it for granted, and so I shall have to reject the first answer to my question, how come we have lost interest in refuting Berkeleyan idealism. I shall offer now my second answer to this question: idealism has been satisfactorily refuted, and so we need not try to refute it any longer.

My own teacher, Sir Karl Popper, has said, idealism is irrefutable; irrefutability, however, is not a virtue but a vice. I disagree with both of these statements. But let us accept them for now as true. Let us agree that idealism is not refutable – that is to say, not empirically refutable. Let us also agree that empirical irrefutability is not a virtue but a vice. Hence, idealism is refuted – not empirically but philosophically. Let me elaborate.

Berkeley said his doctrine is verified since it is and must remain unrefuted; Popper says it is no good for the very same reason. Either we have here two criteria which are beyond rational debate; in which case we can choose the newer and forget the older. Or, we have the older criterion rejected by rational means and replaced by a better one. If so, and if this rational process leads to the rejection of idealism, we can say that idealism is empirically unrefuted and irrefutable – but philosophically refuted.

I am loath to study public opinion; but when talking to a public one has to develop some idea of what that public knows, thinks, and worries about. I have the distinct impression from the current literature that the reasonable philosophical public of today, in particular philosophers of science, are willing to view idealism as unproblematic by accepting Popper's idea that its empirical irrefutability is a severe defect (some even say meaninglessness!). Yet they do worry about induction which they view as highly problematic. But if induction leads to idealism, as Berkeley says, then there is a flaw here. Or shall we say a scandal? Frustration, rather. The situation is frustrating indeed; and for good reasons, I think.

II. WHITEHEAD'S SCANDAL

What is common to Kant and Whitehead is the deep feeling that some proof urgently needs to be given, can easily be produced, yet is not yet available.

The faint hint is that philosophers are lazy – scandalously lazy. This is why Kant's scandal is no scandal at all – it took so much to overthrow Berkeley's idealism (assuming this, at least), that we cannot honestly blame Kant's contemporaries or successors for not having succeeded to accomplish this task: the task was not as easy as they thought. With Whitehead, the feeling that the problem of induction is urgent and must be soluble of course, is what led him to echo Kant. Is this really so bad? Whitehead himself was not a man of poor intellectual abilities; why did he fail to solve the problem?

Wherever there is a task which seems at once both so easy and so difficult, it may be advisable to explain the conflicting appearances first. For, as long as we are not clear whether the task seems easy or seems difficult, we are not clear even about its most superficial aspects; we do not have a clear expectation; we are, in fact, confused. We may confuse two tasks, one (seemingly) easy, one not. Or we may confuse two different settings, one in which the task is (seemingly) easy, one not. Or we may confuse two different criteria of satisfactory executions of the task, by one criterion the task is (seemingly) easy, by one not. Can this apply to induction?

It seems to me clearly to be the case. I shall not enumerate all the possible distinctions, not even all those which are exemplified in history. I shall not even mention all important distinctions along the above lines; I shall only mention one. There are two versions of inductivism, of the views of what induction is or should be. The one is a radicalist inductivism, and the other is tempered inductivism – one which recognizes its own limitations. The radicalist version is one which denies all knowledge, and all foundation of knowledge, except the analytic *a priori* and the synthetic *a posteriori*. As Hume and Kant have proven, this version is plainly inconsistent as it rests on a principle of induction which is neither analytic *a priori* nor synthetic *a posteriori*. And so the principle of induction must count as meaningless or as transcendentally proven by the very success of science and hence count as synthetic *a priori*. Indeed, Wittgenstein in his radicalist *Tractatus* declared the very principle of induction meaningless, and Russell in his 'Limits of Empiricism' accepted a tempered version of inductivism by declaring the principle of induction synthetic *a priori*.

(Here we see the force of Berkeley's argument from radicalist inductivism. We may accept the distinction between primary and secondary qualities on some *a priori* grounds; this will deprive us of claims for radicalist inductivism. It is only radicalist inductivism, not necessarily any tempered one, which imposes idealism on us.)

Some transcendentalists leave no room at all for induction, some transcendentalists leave everything to induction except the principle of induction. The middle ground between these two extremes is a vast *terra incognita*. Mill assumed, *a priori*, a principle of simplicity of nature which Keynes stated, at least with greater precision, as the principle of limited variation. The principle cannot work because we do not know what variant is excluded from, what variant is included in, the small group of those laws which nature is permitted to adopt. But once we make the principle more specific, then hey, presto! induction becomes no problem at all. This idea, incidentally, has been discovered by Descartes, and his *Principia* contains an explicit principle of induction which he (quite rightly) treats as not problematic in the least.

A clear example for all this is organic chemistry or molecular biology:

given a few well-known general laws and a list of atoms which a molecule may contain, and the search for the composition and even structure of any given molecule, however big or complex, is merely a question of time. Within the accepted framework there is a finite number of given possible alternatives; we may assign each an *a priori* and an *a posteriori* probability; we can verify one of them; we can render one of them probable – especially by induction by elimination. Induction is simply no problem at all for philosophy once the setting is sufficiently, clearly, specified.

And so, quite possibly, Whitehead felt the scandal because he felt that the problem was on the one hand easily soluble, on the other insoluble; yet he failed to distinguish the different settings which rightly make the problem very easy and very difficult respectively. If this is so, then the chief question is, can we solve the easy problem and ignore the difficult one? I think this problem of strategy is very alive in the literature though it is, as usual, poorly articulated. What kind of induction, we may ask, is science engaged in, and can philosophers confine their discussion to that induction which in actual fact is employed by active science? Come to think of it, do scientists, in actual fact, employ any rules of induction? Most philosophers say, but of course. Not all, however.

III. THE FACTS ABOUT INDUCTION

There is no induction, says Popper; and he explains why. But very few hear him. The few who do notice Popper's philosophy in their writings about induction, notice him not because he says, there is no induction, but because he speaks of confirmation or corroboration in a manner which they find more interesting. Some, in particular: D. Stove, J. J. C. Smart, and Wesly Salmon, think he is not very serious when saying there is no induction since his theory of corroboration is nothing else than a theory of induction.

This has made me stop and think. What is it that these philosophers think so obvious that they are stunned by Popper's oversight? How come both Bar-Hillel and Isaac Levi say that it is not possible for the intelligent person that Popper is to say what he says and mean it literally? What is so obvious, they all say explicitly (and I do them the courtesy of taking them to mean what they say), is that science and technology obviously exist and flourish.

Arguments from the existence of science are transcendental arguments: if my theory were false then science would be impossible; but science exists. This is the logical form of Kant's argument which he christened 'transcendental' and this is the logical form of Russell's argument, and Sir Harold Jeffreys, and of lesser lights. The way to avoid transcendentalism is to take the existence and success of science and technology as a datum to be explained.

We may then read the complaint against Popper as the claim that he ignores the chief datum we have. To some extent I do agree. I do declare that the success of science and of technology are empirical data that many speak of as if they have been empirically observed, as if they are so obvious that no one can fail to observe them. I do declare that these empirical data – the success of science and technology – are begging to be observed, with some attention to detail. It is the very superficiality of the existing observations, of the observations which are taken uncritically by the many, that causes the trouble, the communication-breakdown. People conclude from the superficial observations that Popper has seen even less then they have. Let us articulate for them what they do not bother to articulate because they superficially declare to be so obvious, clear, and unproblematic.

What they suggest, I think, is first and foremost that we cannot view the success of science and technology, given its range and duration, as a mere accident. If no accident, then it follows certain rules. Call these rules induction and there you are. Certainly, Popper's is not allowed, by his own dicta, to oppose this meaning or that. So much for the current criticism of Popper's claim that induction simply does not exist.

I think I have this right. If not, I too, like Popper's, fail to see the obvious; I have a blind spot. I was willing to consider this seriously, but it is a depressing thought and a blind alley. As Bacon has said, as even Wittgenstein has said, no one is a good judge of his own blind spots. And so I shall assume, until corrected, that my presentation of the criticism of Popper's denial of the existence of induction is right. Let us return to the chief datum, scientific and technological success, and its explanation: the success is due not to mere luck but to some method. Ah! says Popper's. You believe in the sausage-machine of science, in the view that there is a sure science-making algorithm. You all do! Even great scientists often say, even Robert Oppenheimer said, give us more money and we shall buy better people and equipment and produce more and better science. But science is no sausage-making product; there is no science-producing algorithm; hence there is no guarantee that more money – or anything else – will bring about more or better science.

This, I think, is the crux of the disagreement. Popper's critics are not very articulate, and so this paragraph is my attempt at an articulation of their views and criticisms. We do have methods – spectroscopic methods, analytic methods, radio-isotope methods, breeding and separation methods, diagnostic methods and tracing methods; we have methods galore – and they are successful. We expect the successful methods to work in the future, and be successful, or to give way to better ones. These are the hard data! Of course, the methods, at least when employed in research, are not algorithms; they are not Baconian; they are not sausage-machines; but they are reliable and steady nonetheless! We cannot say what differentiates a steady sausage-machine from a steady non-sausage-machine. We admit that science and technology need as a constant ingredient the invention of hypotheses and other aspects of ingenuity. Popper is quite right in claiming that even the successful empirical find – real find not just production – of really fresh evidence, is imaginative. In brief, science has some method – not a total method as in Bacon. So let us call it not induction but some-induction. Braithwaite and Smart and Salmon accept some Popperian methods as some-induction. But even Carnap's induction is intended to be only some-induction, not a sausage-machine; Carnap does agree, he declares, with Einstein and Popper, about the need for the imagination in the production of hypotheses.

And so there is some method, and it leads to more than some success with more than occasional regularity. Such is, still, our datum. And so the philosopher is frustrated in his attempt to locate this obvious datum. He may easily identify his some-induction with what I have earlier called tempered inductivism. Since tempered inductivism is not very problematic, some-induction may, perhaps, be easily vindicated. Somehow this sounds otiose.

We are closing in on the scandal in philosophy – in the sense of frustration, of course. We still have as a datum some-induction. Some successful method. To ground this success in some synthetic *a priori* principle is to beg the question, to shift the explanation in a most frustrating manner. Why does scientific method have some success? Because it follows the principle of Induction. This is too *ad hoc* and highly problematic.

The old view of science as a sausage-machine, as a total algorithm, is erroneous but very interesting and even a very cogent solution to certain problems as they appear historically. The partial view is problematic. The view that science and technology are totally algorithmic whereas the fine arts are totally spontaneous, for example, is doubly false but very clear and satisfactory. Now that we know that both scientific and artistic activity is partly routine partly inspired we may ask, how are they to be demarcated?

I have not seen all this ever expressed like this, yet I do contend that this is at the back of many writers' minds. When they say that science is predictive, they speak of the material success of science and thus over-stress the importance of the problem of induction. But they are not thereby vulgar materialists; they clumsily try to demarcate science by its success. If so, their ill success can be removed by regaining balance and re-membering some failures too.

IV. SUCCESS AND RATIONALITY

The problem of induction, many say, is, can we project past experiences into the future? Will the future be like the past? Alas, no; I am unable to expect rejuvenation. That is a witticism. Certain past experiences we do project, and we mean, though seldom say, project successfully. Indeed. Especially success must be projected (since we have to succeed to keep alive) such as success in predictions concerning the nutritious value of bread. Particularly, for those who earn their bread by science, it is important to project scientific success into the future. Will science be as successful in the future as in the past? Are the cuts in the Federal Government's budget a temporary set-back or am I losing my job and career and social status as a scientist?

The problem of induction is an expression of the wish to be reassured not by a condescending parent or priest, not by blind faith, but by rational means. Can we justly predict and explain the success of science? This is the problem of induction. Popper says it is insoluble in a positive manner. All he can offer by way of a solution is his notorious *modus tollens*, or, to be precise, the rule of retransmissions of falsity from the conclusion to the premises. In the explanatory model we assume a hypothesis and initial conditions from which a prediction follows. If and when the initial conditions are confirmed by experience but the prediction is refuted by experience then, we may say, experience refutes the hypothesis in question. Assume that all that experience can offer is such refutation of scientific hypotheses, and, in particular, no confirmation and no guarantee. If that is all to it, then all scientists and technologists are leading precarious lives with no insurance for the future. As a hard datum, this is not so. Banks are ready to offer long term credit to scientists at least as easily as to businessmen. They know science is secure. Hence Whitehead's scandal.

But all this is a mere dramatization of a very well known prosaic point. Given the principle of induction we can explain the success of science; and we can make the following conditional prediction: those who will employ the principle in their research will be – certainly or in most likelihood – successful. This is the famous infinite regress, step number one.

But if science has been successful due to the employment of this principle and if it will be successful as long as scientists continue to employ it (and banks do give them credit on the assumption that they will), why then can we not ask scientists what principle they employ? Why can we not become scientists or employ scientists to find out this principle?

This, indeed, is so puzzling one might wonder if it is not a scandal. Hume's subtitle to his *Treatise* may read, an attempt to make induction a part of empirical psychology. 200 years passed. I think that Popper's new philosophy of science, that the studies in the theory of rationality, of Popper, of Bartley, and others, have given us something that we may indeed call a method which is not an algorithm or sausage-machine and which those more sensitive to success than to the history of ideas and to our heritage in the philosophy of science may indeed call induction. It is a partial method with the partial guarantees – not the absolute one – that most civilized citizens of civilized communities have come to expect. Painters, for example.

A painter produces regularly and his art should not be produced by a sausage-machine. Here is the crux of our situation, in art just as much as in science: the gulf between them is part Baconian – science is a sausage machine – part romantic – art is not produced regularly. Both science and art have their routine side and their inventive side. I think that even the dichotomy between the routine and the inventive is here overworked; to do an utterly routine job we have to mechanize it so much with conveyor-belts and what-have-you, that you may just as well dispense with the worker altogether and fully automate production. There are fully automated productions in art, science, and technology, of course. They are unproblematic and uninteresting. You can apply the problem of induction to them. Asimov has invented a robot who is bugged by metaphysical problems and invents Descartes' philosophy. We can more easily apply the problem of induction to a robot: will he calculate tomorrow as correctly as today? But we do not. We are concerned with our success, not that of the robots; also our success as robot-owners and bread eaters, but more so as scientific and artistic creators.

But the robot we employ and the bread we eat may betray us; or they may be useless in a total catastrophe. A total catastrophe of one kind and of another is declared possible by science, and so in the long run some total catastrophe is, we are told, more or less inevitable. When insuring ourselves against future catastrophes we ignore such eventualities. The grand axiom of insurance theory is, there is no insurance against epidemics; if one in a catastrophe is compensated, this is because others like him prosper and pay him through the insurance-system, be this an insurance company proper, a federal agency, or the village system of mothers babysitting for a sick mother. And here, we can see, the artist and the scientist integrate well in the system by various means: he teaches, he administers teaching and production – he produces almost routine works, he is one of one hundred who idle in the hope of producing a masterpiece and they all feed on the one masterpiece produced among them (for example, we all sell logic texts to the publisher, and he makes his profit from very few of them).

The artist and scientist, then, have bread-and-butter operations which signify within the system, and wind-falls to boot; even regularly though not algorithmically. An algorithm or a quasi-algorithm is possible within a system: given our present system, the number of chemical combinations in organic chemistry and molecular biology is finite and so, even in accord with the strictest Popperian canons, induction (probability or eliminative induction) is demonstrably possible.

Can we say the same about the framework itself? No. It cannot be justified, especially since we tend to improve it, i.e. admit it to be faulty. But, like democracy, its the best we have. Like democracy, also, we do not give it up when we find its defects until we can replace it.

Jeffreys and many in his wake have said, we do not give up our belief in a refuted hypothesis until it is replaced by a better and more viable one. This shocked and puzzled my younger self of a two decades ago or so. Does Jeffreys mean to advocate the belief in a refuted hypothesis? This is absurd! But I do think he is right on technology; we do not believe in the framework we use, but we go on using it until it be replaced. And when we try a replacement we try pilot plants. A pilot plant is designed for the achievement of positive evidence, though its test is performed as attempted refutation: we want to be right at least to the extent of having a better system to implement. But if we are not, we prefer to learn about it in the pilot plant not in the main plant. Hence pilot plants want corroborations in Popper's sense of the word. But the degree of corroboration required varies from country to country, from time to time, it may be improved by trial and error too.

Socially speaking, or economically, we allocate some of our intellectual resources to pure research which is problem oriented, critical, and seeks refutation; we allocate some of our intellectual resources for technology which is task oriented, critical, and seeks corroborations. We support the two groups from the common pool; we do not like our inventors, scientists, and artists, to depend on their really inventive imagination because then they starve in attics before they can perhaps burst into fame and riches.

This solves the problems of success and of induction, in an empirical fashion and sociologically. Psychologistic induction is now replaced with sociologistic one.

V. THE SOCIOLOGY OF KNOWLEDGE

Our ignorance of some possible risks does not permit us to take known risks lightly; at least in this respect, institutionally speaking, *it is not sufficient for us to criticize the exisitng institutionalized theory or even to offer an alternative to it. The existing institutionalized theory at least works, and the new one has to be carefully examined and corroborated on a small scale before it is institutionally accepted.*

Here we see that testing in pure science is rooted in the purpose of science, the scientific search for truth, but testing in applied science and in technology in general is rooted in the institution fostering competition; whatever is not prone to competition is taken on faith. Why does this ostrich policy succeed? This is a problem for pure science of a Darwinian stock.

All this may easily be tested as well as applied to cases of retention and replacement of frameworks which are intellectual and institutional at one and the same time. I shall offer only one case – which is not very interesting but which has enormous practical possibilities. I chose it in order to illustrate an additional point made in this chapter, namely, that research is mingled with very quasi-routine jobs that would hardly qualify as research.

Any project that can be performed on a budget of $10,000 can also be performed, with not much effort, on a budget of $5,000,000. The question is, what is the benefit of spending an excess of such a magnitude? Since, doubtlessly, such excess expenditures are commonplace, we cannot dismiss this question as merely hypothetical. It is easy to analyze the situation of excess expenditure in terms of bureaucracy – boards of directors, chairmen, fund-raisers, etc. etc. That the multitude and the bureaucrats are institutionally disposed towards the safe, which is, to use the language of the last Lord Chamberlain – the last British censor of stage and screen performances – one step behind the avant-garde and one step ahead of the crowed. Such analysis, however, is incomplete: were such expenditures total waste someone would have complained more effectively. The function of over-expenditure of research funds, I say, is educational: it is teaching the academic public who would learn only under the guise of research. If this be so, then, those in charge of research funds are able to institutionalize new ideas.

Let me stress that this analysis, though not so well known, is by no means original with me. Let me quote from *Sponsored Research Policy of Colleges and Universities,* A Report of the committee on Institutional Research Policy of the American Council on Education, Washington, D.C., 1954, Summary (pp. 76–77): "The large scale projects afforded by sponsored research have enabled young investigators to acquire research techniques which can be learned only by participation in organized team research." As all techniques can be learned only by participation, the justification added to the explanation is true but pretentious. "Some of the sponsored research laboratories in universities have been superb training grounds for research engineers, providing them with actual research environment." (See also there, pp. 44–45, and references there, for more details.)

Now, assuming all this to be true, then research grant committees are powerful educational influences, a kind of body deciding on curricula for top students who are so good they cannot afford posing as students. Now, doubtlessly, the committees which decide on grants have criteria – vague and inconsistent, but criteria – and these may be formulated, criticized, and improved upon.

So much for institutionalized modes of shaping new institutional theoretical frameworks; what I wish to stress, however, is that theories exist which do act as institutional frameworks. This, I think, is what Marx spoke about when he said that ideas become material when they take hold of the masses. To be precise, mass theories are traditions, whereas theories upheld by intellectual leaderships are institutions, but I shall not enter here the fine distinction between traditions and institutions proper.

Popper's philosophy comes to replace a philosophy which was institutional for three centuries, and very successful it was. Popper's theory can improve matters by making controversy more openly the order of the day, by cutting out dead wood and accretions. But it will be unreasonable to expect it to become the institutional theory prior to its corroboration on a small scale in some pilot plants, such as workshops of science and technology. This is why philosophers are so hestitant to consider Popper's opinions freely: they fear to deviate too much from the institutionalized opinion. If and when scientists and technologists institutionally endorse Popperism, the philosophical fashion will follow suit all too quickly. Hence, Popper is mistaken in his efforts to dissuade people from their old fashioned theories merely by an intellectual appeal. It is more useful to try and create pilot plants, to work out wind-tunnels for ground tests prior to test-flights leading to institutionalized flights of the imagination. But there are different ways of institutionalizing a philosophy. In particular, it is easy to declare a philosophy acceptable by the public but only after incarcerating and sterilizing it and destroying all its possible social implications. I can hear the strains of the approaching band-wagon, and can only express my hope that when it passes by, not all philosophers of science will be adding their weight to the procession.

APPENDIX: DUHEM'S INSTRUMENTALISM AND AUTONOMISM

Duhem argues against the theory that the aim of science is the discovery of theoretical truth: if this were so, he says, science would lose its autonomy and be subjugated to metaphysics. In his view, however, science is not autonomous but subjugated to technology. By autonomy he means two things. First, autonomy as the independence of value; this independence is observable, for example, by the fact that scientists' opinions are not divided by metaphysical opinion but are unanimous. Second, autonomy not as expressed by unanimity but as identical with it. But this second criterion is not really a criterion but a confusion of a criterion with a touchstone of it.

A1. *Instrumentalism*

Let me contrast autonomism with instrumentalism. The instrumentalism-autonomism contrast is often overlooked, as Popper has noticed, because a small word, 'only', may make all the difference between the two. Take aesthetics. That beauty may be used, say for political purposes, is doubtless; that beauty is none but the achievement of smooth political efficiency, that smooth policitical efficiency is the sole aesthetic criterion, is a thesis which may be true but is certainly disputed. Generally, '*x* is useful for *y*' is a matrix all too often applicable, whereas '*x* is merely useful for *y*' is a version of instrumentalism. Opposed to '*x* is only useful for *y*' is *x*-autonomism: '*x* serves its own purpose'. Ethics may be declared autonomous or in service of politics, and vice-versa. Ethics may stand in service of aesthetics and vice-versa, leading to aestheticism in ethics and to moralism in aesthetics respectively. Autonomism offers ethics its own purpose – the moral – aesthetics, its own purpose – the beautiful – politics, its own purpose – the just – science or perhaps metaphysics, its own purpose – the true. Funnily, there seems to be one field which, *eo ipso*, cannot be autonomous, i.e., technology. Or perhaps we may view control of our environment as the autonomous domain of technology, with the control of physical environment leading to physical technology, of social environment to social technology, etc. Be that as it may, we all agree that truth or beauty or propriety or justice are not the ends of technology, that control is.

Duhem declares truth to be the end of metaphysics, and let us accept this. Accepting also control as the end of technology, we get into a small difficulty. Science becomes almost inevitably non-autonomous, and instrumental either to technology or to metaphysics!

A.2. *Unanimity in Science*

Duhem says the search for truth is the domain of metaphysics, metaphysics is divided, and science is unanimous; therefore, science is not instrumental to metaphysics; hence it is autonomous; hence it is instrumental to technology. The paralogisms are all very interesting. There is no unanimity in technology in some sense of unanimity, and some unanimity in metaphysics in another sense. Unanimity is a curious phenomenon, hard to pin down or even define. Duhem himself was engaged in a bitter controversy on electricity, which began with Ampère and Faraday and ended with Ritz and Einstein; and he hated Einstein both scientifically and methodologically. He declared the Einsteinian revolution too violent for science. Here science is dominated not only by technology but even by methodology. Is methodology autonomous? Do we have unanimity in methodology? What is the aim of methodology and is this aim its own or borrowed? These questions can barely be properly handled. Why does Duhem make so much of unanimity?

Duhem's argument is staunch common-sense. Engineers are more in unanimity about engineering (no schools of engineering) than metaphysicians about metaphysics (metaphysical schools and dogmas abound); science is nearer to engineering in this respect; hence science is nearer to engineering than to metaphysics.

Therefore, Duhem's argument has to be examined on common-sense grounds. For this, first and foremost we must rectify Duhem's statement: not science is autonomous, but science is more akin to technology than to metaphysics, is his thesis. Second, unanimity is now not a virtue, as he claims, but merely an observed phenomenon (up to a point) which is used as a touchstone in the decision between the two masters – metaphysics and technology.

The common-sense criticism of Duhem, then, is this: in so far as there is unanimity in science, it may be viewed as the handmaid of technology, but when a dispute rages, even between the Ampère school of action-at-a-distance and the Faraday school of action-in-fields, science should be viewed as a handmaid to metaphysics.

A.3. *Conclusion*

This seems to be a compromise between Duhem's instrumentalism and traditional realism, between Duhem's view of science as a handmaid of technology and the traditional view of science as the search for truth. But, as explained before, in any attempt at compromise, instrumentalism capitulates. Nobody denies that science *also* serves technology: Duhem's thesis was that science *only* serves technology, and this thesis was denied. Now the compromise, science serves both technology and metaphysics, brings us back to the pre-Duhem in position – but with the added recognition of the significance of controversy in science; perhaps also of the fact that scientific controversy is rooted in metaphysics.

CHAPTER 14

POSITIVE EVIDENCE AS A SOCIAL INSTITUTION

The current literature in methodology and epistemology is almost exclusively devoted to one topic: positive evidence, or favourable evidence, or empirical support, or harmony between theory and experience, confirmation, etc. etc. The problem concerning evidence which engages the current literature most is, *how* does evidence back theory? The present chapter is devoted to the question, what is the good of such backing? The literature views evidence as the basis of rational belief in theories and as the justification of their practical application. Now consider criticism rather than justification to be the key role of rational activity. You will then consider negative evidence as important and tend to view positive evidence plainly as failure, as the undesired outcome of attempts at criticism or at refutation. Since this is not the whole story, the identification of rationality with criticism, as advocated by Popper in his *The Open Society and its Enemies*, must be rejected. Once we view only *internal* criticism as rational, once we view rationality as goal-directed and rational criticism only with respect to given ends, then the picture changes. The end of pure science is a theory which is both true and comprehensive or encompassing explanation. Consequently, though negative evidence of a comprehensive explanation is more important than positive evidence, positive evidence is important too – as one which renders theory more comprehensive, or more explanatory. So much for theoretical ends. For practical ends any positive evidence plays a significant role if and when it serves such ends – if, that is, it can be put into use. Its relation to theory need not, then, bother us at all in such contexts. Alternatively, when wishing to employ a theory, negative evidence may prevent disaster, and so we devise tests to procure it. Positive evidence procured in such pragmatic tests may be useful in itself, or it may be of no pragmatic value of any tangibility – except as evidence that the road is clear of certain obstacles. When a theory thus tested and backed is applied with disastrous results, then the propriety of the test procedure, i.e., the severity of the tests, is used as evidence of lack of irresponsibility on the part of the applier. Yet new application is always a risk, attempting it is never obligatory and always a challenge. After its success application becomes institutionalized and thus obligatory whithin the society which adopts it.

Once we take institutionalized theories as if *a priori* true, then many theories can be verified as if conclusively. Philosophers who observe verifications are considered naive, and naive they are; but they do observe verifications – in applied science and in technology; and even in historical research. When our theoretical framework entails a limited number of empirical possibilities we can perform eliminative induction and verify an hypothesis by eliminating all alternatives to it. When the verified possibility is then refuted as well, so does its *a priori* framework: this is how applied science can aid pure science. Pure science aids applied science by allowing its tentative ideas, when sufficiently corroborated by some socially determined standards of sufficiency, to become the *a priori* framework to be taken for granted. The standard of corroboration is not Popper's, since the corroborated theory remains accepted even after its refutation, whereas for Popper the degree of corroboration of a refuted theory is negative; nor is the standard of corroboration logical probability, need I say, since logical probability precludes the Popperian idea of positive evidence as being nothing short of failed negative evidence. Rather, a corroborated theory is one which is highly probable relative to recognized institutionalized theory and in the light of evidence procured under test conditions. Hence, though the test is scientific, corroboration is of mere technological value, but of no theoretical value. For pure science some theory of verisimilitude looks, at first blush at least, more promising.

I

The current literature seems to echo the following ideas which are traditional since the founding of the Royal Society.

(1) Enlightenment is rational belief, belief which can be justified objectively.

(2) Science alone is rational. In opposition to science we have over-belief and under-belief – credulity and undue skepticism.

(3) Scientists practice what they preach. Rational technology is the applied science.

Since point (2) declares science and only science to be rational the sense of point (1), point (3), concerning rational technology being scientific, is a corollary to points (1) and (2). All three points may be viewed as definitions, but then the traditional view may be reduced to the proposition that the definitions are not empty: the traditional theory will read, there exist known instances or paradigms of rationality in science and in technology as so defined, and no cases of rationality besides the ones which fit the definition.

Of all the criticisms of the traditional views I prefer the one from the broad – and shallow – outline of intellectual history. The argument (which I have discussed in Chapter 9) rests on the observation that general intellectual frameworks contain general metaphysical presuppositions shared by scientists and other reasonable men, yet which alter in time. Some inductivist philosophers have noticed this and have modified inductivism accordingly to be the claim that induction is valid only with respect to such frameworks. Though this view goes back to Descartes and to Kant, it seems that Frederick L. Will is the originator of this view in our century. The great advantage of this view, of course, is that it solves the problem of induction within deductive logic. The slight disadvantage of this view is that it reinforces the adherence to an existing framework up to the very last moment of its existence, and the switch to the new framework immediately after it replaces the old one. But who should demolish the old view and who should erect the new one? If what Will recommends is the rational thing to do, then all our big medals have gone to the brave irrationalists. And so Will unwittingly endorses a medieval theory of the proximity of the gates of hell and the gates of heaven; it is too romantic a theory to be a serious contender for a theory of rationality.

II

Popper has replaced the received standard of rationality by an alternative standard:

(1′) Learning from experience is learning from facts to locate our mistakes. There is no rational justification of belief, but only rational justification of *dis*belief.

(2′) Science is rational in the sense that it is critical, in the sense that it consists of refutable conjectures plus the empirical refutations of some of them.

(3′) Technology often employes rejected doctrines (Newton's, Maxwell's) and sheer rules of thumb (concerning safety margins and procedures not amenable as yet to any scientific explanation).

To this I wish to add the following:

(4) Empirical support to a theory is a necessary condition for its applicability, legally imposed by bureaux of standards, or by food and drug offices, or by trade and industry offices and their like. Empirical support is never a sufficient reason for application. Empirical support never forces us to apply a theory; it only permits us to apply a theory if we are ready to take the further risk. The theories we are normally forced to accept are those which are accepted by leading reasonable people in our society – as the law specifies in many cases of *bona fide* mistakes.

Popper has devoted much of his time and energy to the criticism of the current doctrines. Again and again he tries to appeal to the rationality of the multitude of philosophers by devising newer and easier refutations of their views – but to little or no avail. I consider it rather unreasonable to act so instead of reason out an obvious set of problems: Need one bother with the current doctrine? If so, why? Need one bother with it merely because it is current? Why? What are the best means available to one decided to demolish a philosophical fashion? It seems to me that these problems are not particularly interesting, though they are on occasion practically important; intellectually it seems preferable to pursue, in relative solitude if need be, certain problems raised by the best available ideas, but social and educational considerations may put on us some demands.

As a sort of compromise between the intellectual and the practical aims, when these compete for our time, we may choose the middle ground. We may persue certain problems of some interest and not little usefulness. In the present case I suggest the following. What, if any, is the role of positive evidence in pure science? How can one explain the prevalence of the fashionable predileliction for positive evidence? In particular, can the traditional theory of rationality (point (1) above) be construed as a first approximation to a better theory? Why is there a semblance of universal agreement in science? If science thrives on dissent and criticism, why is dissent played down and dissenters disparaged rather then encouraged? To what extent is research handicapped by dogma, superstition, and prejudice, be it Aristotelianism, Marxism, inductivism, instrumentalism, or Popperianism? Finally, what is the role of negative and of positive evidence in technology, in applied science, and in fundamental research?

There are two peculiar aspects to these Popperian problems: they have an unmistaken philosophical flavour, yet they lend themselves to both empirical investigation and (social) technological application. The empirical study of the actual role of evidence is almost entirely a virgin field, and as far as the role of evidence in the commonwealth of learning is concerned, inductivist preconceived notions have precluded all empirical investigation of it. Even well-known facts have been constantly disregarded. Boyle, Duhem, Popper, and others, have assumed that in legal practice evidence is free of all philosophical problems. This seems to be obviously false, as Clarence Darrow knew and as many expert-witnesses can report. Also, the problem of the permissibility of using background information when discussing positive evidence is common to both philosophy and jurisprudence, but the twain have not yet met. The application of probabilities to linking testimonies to specific cases is extremely rare: the law often demands of witnesses to bridge gaps precariously rather than have their testimony and use experts to make positive identifications. The role of public-opinion in the determination of *bona fide* error is a similar common problem. But institutions other than the courts may be in similar predicaments. The risk of not purchasing information and the cost of information has already been studied by Wald, but only theoretically. Apart from cost there is the element of risk – in both the purchase of information and the foregoing of such purchase – as evidenced from the serious problems in the inquest into the disasters of early commercial jet-flight and the inquest into the loss of the atomic submarine, the Thresher. Such problems also loomed large in the congressional investigation of the problem of insurance of commercial jet aviation since American anti-trust laws were claimed to be a serious obstacle to the insurance of business ventures involving large scale risks such as jet flight insurance prior to the inception of commercial jet-flight. Finally, the tragic case of thalidomide is still a challenge to students of positive evidence and its practical aspects: was it mere luck or were the standards of tests higher in the U.S. than in Germany and thus reduced the disaster there, or were the U.S standards the more rational? Was the U.S. ban on thalidomide an application of a standard or a mere hunch? All theories of positive evidence that I have ever read or heard about are too poor to answer my questions about thalidomide!

III

Let us begin with the role of positive evidence in pure science. Is positive evidence in pure science of any significance? Why is it considered extremely important? According to Popper, science consists of series of bold conjectures and the empirical refutations of some of them. If so, science could, in principle, be gloriously successful with no failure of any test, i.e., with no positive evidence. Somehow this does not ring true. Popper himself has modified his own view, declaring it desirable to have some of our conjectures corroborated before they be refuted. Empirical corroboration, he says, is a sign of progress, of having come an iota nearer to the truth. This claim cannot be taken seriously. First, if corroboration plays an indispensable role, more than such a passing remark about this role is necessary. Second, Popper's demarcation of science must be altered so as to make room for any substantial role corroboration is alleged to play in the development of science. Thus, Poincaré's refuted but never corroborated theory of radioactivity may be unscientific according to Popper's new view of science as conjectures and corroborations and refutations.

Popper's characterization of the empirical character of science as empirical refutability is very satisfying first as the explanation of what we view as pseudo-scientific, as pseudo-empirical, namely as empirically irrefutable; second, it is part of a modern version of critical realism ("I err, therefore the objective world exists"). His restriction of all empirical character to science, however, seems objectionable, since empirical character may occasionally accompany theories traditionally viewed as paradigms of nonscience, this including superstitions (e.g., spiritualistic claims empirically refuted by Faraday), and prescientific empirical technology, and folk-medicine, metaphysical doctrines (e.g. Thales' theory that all is water, refuted by Cavendish and Watt), and epistemological doctrines (e.g., the theory – once universally believed – of the finality or utter verification of scientific doctrines, especially Newton's, refuted by the Einsteinian overthrow of Newtonianism), large chunks of value theory (such as that desirability and undesirability always resolve into final verdicts making each item either desirable or not) and much common sense (especially concerning sex and concerning race, past and present). Popper himself has sometimes characterized science as the set of all refutable conjectures, sometimes as the set of all explanatory refutable conjectures. He seems to consider these two characterizations as coextensive, but they are not. The latter is preferable not only as being the narrower but also as characterizing science by its aim, namely, comprehension or explanation (preferably true). Viewing science chiefly as purposive or as goal-directed activity is much more in accord with the sociological approach favoured by Popper, namely the rationality principle, or situational logic, or the principle of rational reconstruction. The word 'rationality' in the sense of purposiveness should not be confused with 'rationality' in the sense of enlightenment. Assuming, then, the purpose of science to be enlightenment, i.e., true explanation, we may try to explain the behaviour of men of science as the search for the true explanation through the testing of the best explanations we have.

Now, assuming that scientific theory is refutable explanation, our esteem of a given theory may be high if it is highly refutable and/or if it is highly explanatory. Popper speaks mainly of a high degree of refutability. Refutability, however, is rather a practical requirement, and practically the ability to construct one test to a given theory suffices to keep us busy for a while – provided the theory is valuable; the more a theory explains the more it fits our purpose, and if we are ambitious we try to test it first. (Einstein was encouraged when he discovered the high explanatory power of his own theory of gravitation.) If the test fails, then the failed refutation increases the value of the tested theory in that it increases by one the number of observed facts which it explains. This is why the empirical support of an admittedly false theory which, however,* explains a surprising number of facts, may still be quite significant; e.g., the empirical support of Bohr's model of the atom. By Popper's theory of corroboration, as Bohr's theory was refuted it could never be corroborated. Hence, Popper's theory of corroboration is empirically refuted. A better theory of corroboration should take account of the desiderata in Bohr's case – not to refute a theory but to use it as a stepping stone for the construction of a better theory. Thus, specific cases may have specific ends on top of the general ones.

Ideally, then, in pure science we would like to invent wide explanations, repeatedly support the true ones and very quickly refute the false ones. But we can find empirical support of false ones, so that support is no criterion of truth; we may still try to eliminate false explanations, in the hope of progressing towards the truth, and the wider an explanation the more seriously we shall take it and hence the more anxious we shall be to examine it. We may, in the interim, try to find out as many positive and as many negative instances to a given theory – until we discover a more interesting and challenging theory.

IV

And now to applied science. To a very large extent applied science receives its goals ready-made from technology (though here operational research is an exception and not such a simple case). In technology, our aims are to devise the best ways of doing things. We wish to devise new ways, and verify our hopes and refute our fears concerning them. This cannot be done. Alternatively, we would like to verify our realizable hopes, and refute the vain ones; we would like to verify our fears which have grounds in fact, and refute the others. Even this is too ideal. Yet, it is already sufficient to show the difference between pure science and technology – or even between pure and applied science: in pure science positive evidence in favour of an error is the failure to eliminate it; for technology the same may be success – if the evidence in question happens to be useful. In pure science, being right in prediction for the wrong reasons is not very desirable: in applied science being right for any reason may do very well, if the case is useful for technology; hence, in applied science things are not as clear-cut as in pure science.

Existing techniques of research are embedded within given theoretical frameworks. Within such frameworks conclusive verification is often possible. If such a verification turns out to be inconclusive after all, the framework within which it occurs has thereby been transcended. In such a case, applied science or technology gives up the problem as temporarily insoluble, perhaps to be relegated to pure science or to fundamental research. Pure scientists may be disappointed when a phenomenon they are studying is shown to be explicable within the existing framework which they wish to transcend (e.g., Faraday's explanation of Chladni's figures within classical mechanics, contrary to Oersted's hopes). By contrast, applied scientists do not wish to transcend but to employ more and more the existing hitherto successful framework. And so they may be very happy to explain within the framework a hitherto unexplained phenomenon, (e.g., colloidal chemistry, much biochemistry, even genetics); the same holds for the technologists, especially when they explain successfully an undesired feature of a process, (e.g, borings in internal combustion engines due to rapid combustion, leading to the use of delaying additives), or a side-effect (in pharmacology often due to impurity), which they can thus hope to eliminate. If the explanation is verified, the undesired feature is often on the way to being removed. But such research is confined to the existing theoretical framework. When the limit of the existing framework is reached the pure scientist may become very interested since the improvement of the existing theoretical framework is his concern. Examples of this kind are rare, but they exemplify the possible contribution of applied science to pure science; the failure to pump water above eleven yards was an interesting scientific discovery; similarly the failure to create controlled nuclear fusion was an interesting discovery in plasma physics. Examples of locating technical failures for the purely technological purpose of eliminating them may occasionally be more technologically important then scientifically interesting; e.g., the discovery of a certain cause of the standard failure to grow bacterial culture (every now and then cultures were known to perish) which quickly was instrumental in creating penicillin. But whereas the successful elimination of a failure *may* be scientifically interesting, the failure to eliminate some failure should be of interest for pure science if and when it is a scientific refutation. That is, when we have scientific reason to eliminate a failure, yet fail to eliminate it, then the technologist calls in the scientist.

So far reference was made to existing techniques and to their improvements within the existing theoretical frameworks or even on the limits of such frameworks. Yet the existing theoretical framework is something not sufficiently clear and definite. Moreover, the existing framework is irrelevant for too many technically important questions which technological research may pursue by trial and error or rules of thumb. Once a solution to such a question is backed by facts it may be incorporated within the existing framework and thus vastly improved within a surprisingly-short time. But even then the incorporation is done very vaguely and into a rather vaguely defined framework. Consequently, the risk involved in such implementation of an innovation is highly problematic. The problems are varied; they may be specific or general, social or epistemological. These problems are unstudied because of the prevalent mythology which views anyone alert to them as a Luddite or at least a suspect.

V

I now come to the problem of the fashion favouring positive evidence. I shall link it first to a much deeper problem. Primitive people personally depend on society but economically are quite autonomous. In the West, economic autonomy has been ousted by division of labour, but personal autonomy is increasing. Here are conflicting tendencies, already discussed by Russell and others long ago: economic interdependence calls for more conventional coordination, personal independence for less. How are these reconciled? Even though technology reduces our working-time it increases our need to coordinate. How is this need gratified? I suggest that the coordination is best achieved by an accepted common theoretical framework which we all agree to follow socially while considering ourselves utterly free to disbelieve it personally: we call this conventionalims, frank and open hypocracy, etc.; but we cannot shake it off. The framework is institutionalized and adhered to not because it is true but in order to coordinate wherever coordination is needed, whether in the theatre, in railroad time-tables, or in industry. We may try to reform the institutionalized theoretical framework, especially when this framework permits too much stupidity and immorality. But the western idea of personal freedom does not go so far as to permit people to act exclusively on their own best judgment or on the best scientific opinion of the day. The claim for an individually determined rationality, for subjective probability of the de Finetti and Savage school, is an epistemological fiction: it was invented in the eighteenth century as an ideal of a totally rational society – or, to be precise, not society at all but a bunch of rational individuals with clear-cut interests. In real circumstances, institutionalized public opinion must be considered as well, even when deemed mistaken. It is no accident that current epistemology shies away from its own social implications: these are rather ludicrous. In only one point is there deviation from all this: somewhere the leading philosopher Carnap agrees with Popper the heretic that we conventionally accept corroborations and stop testing a theory: hence, induction is much a matter of convention!

Let us examine the risks taken in the stopping of a test prematurely – logically all stopping of a test is premature – and the implementation of an innovation based on the corroborated theory. These risks are socially, not personally, determined; they differ in different Western countries. Again, remember thalidomide. Psychologically, a test-pilot may feel confident enough to fly a newly designed airplane, but the law and the underwriters are there to prevent him from so doing until certain ground-tests are accomplished. If the ground-tests are not successful, there will be no test-flights; if the ground-tests are successful they still do not insure safe flights, and therefore test-flights are still necessary prior to determining a plane's airworthiness, and that is why test-pilots take risks when they fly on test-missions.

Thus, the high degree of permitted risk is regulated institutionally. So must be the low degree of permitted risk, or else the spirit of adventure may decline. An underconfident test-pilot, or even an underconfident aircraft company, may decline taking risks; social institutions exist to insure that the underconfident will be replaced by the more adventurous. One method of securing this is institutionalized competition, whether between various military powers, various entrepreneurs, technologists, medical research-workers, or test-pilots, who covet prizes, medals, etc. Failure to undertake a risk may involve military inferiority, losing markets or new investments, etc. A competitor may consider his loss or gain in making the decision to pursue a venture or not, depending on whether the venture is going to be successful or not. Another method is that of insurance, and insurance often depends on such considerations and estimates too. These considerations and estimates are made within a theoretical framework and the detailed theory of them is due to Abraham Wald. The framework is normally a composite set of assumptions, party institutionalized, partly not. *It is usually very reasonable to work on the institutionalized theory because failure in such cases enables one to fall back on one's society as a victim of uncontrollable circumstances.* (A simple example: relying on a false railway timetable is more excusable an error than relying on one's own false judgment on the same matter even if one's own judgment is usually better than the railways' own timetable!) *This makes the adherence to institutionalized theory a form of insurance.* (One who does only what is expected from one in one's station is safer than one who shows initiative – even where initiative is recommended!) Those not able to draw this kind of insurance benefit, be they derelicts, outcasts, or spies, or even simply highly independent people (free lance or academics), are better off when they get rid of as much of their theoretical background as soon as they can. (This is what Popper ignores when he invests in efforts to convince his opponents.)

Hence, the following of institutionalized theoretical frameworks is always a form of insurance (particularly when the framework includes insurance methods of all sorts). Thus the framework operates as pooling risks. This hangs one person's and one institution's safety and stability on the safety and stability of the whole society. This is particularly useful for reform: we make a new and weak institution depend on old established ones. The question is, doubtlessly, does not this process of minimizing individual risks lead to the increased risk of destruction of the whole society?

Since the calculated risk of any venture is considered only within a theoretical framework, the risk of holding on to that framework is incalculable. Nor is the risk calculable of changing it for a better one rather than patching it up *ad hoc*. Those who dreamt that scientific advancement on the whole decreases the risks of mankind as a whole were forced to reconsider their views after science created the new possibilities of global destruction (by population explosion, radioactive fallout, etc.) These risks are known to us, but there may be risks not known to us, as there were for our forefathers. Global risks, known or unknown, are not insurable, since insurance is merely the pooling or sharing of risks and the diversification of resources. In a catastrophy there are not enough resources to compensate everyone. Progress permits further diversifications of resources, but it may lead to new global risks. The question, which of these two will win in the end (e.g., will space colonization precede nuclear destruction?) is handled by science fiction. It belongs to science fiction since it essentially transcends our knowledge.

Our ignorance of some possible risks does not permit us to take known risks lightly, and, at least in this respect, *it is not sufficient for us to criticize the existing institutionalized theory or even to offer an alternative to it. The existing institutionalized theory at least works, and the new one has to be carefully examined and verified on a small scale before it is institutionally accepted.*

Here we see that testing in pure science is rooted in the purpose of science, the scientific search for truth, but testing in applied science and in technology in general is rooted in the institution fostering competition: whatever is not prone to competition is taken on faith. Why does this ostrich policy succeed? This is a problem for pure science of a Darwinian stock.

All this may easily be tested as well as applied to cases of retention *and* replacement of frameworks which are intellectual *and* institutional at one and the same time. I shall offer only one case – which is not much interesting but which is of enormous practical possibilities.

Popper's philosophy comes to replace a philosophy which was institutional for three centuries, and very successful it was. Popper's theory can improve matters by making controversy more openly the order of the day, by cutting out dead wood and accretions. But it will be unreasonable to expect it to become the institutional theory prior to its verification on a small scale in some pilot plants, such as workshops of science and technology. This is why philosophers are so hesitant to consider Popper's opinions freely: they fear to deviate too much from the institutionalized opinion. If and when scientists and technologists institutionally endorse Popperism, the philosophical fashion will follow suit all too quickly. Hence, Popper is mistaken in his efforts to dissuade people from their old fashioned theories merely by an intellectual appeal. It is more useful to try and create pilot plants, to work out wind-tunnels for ground tests prior to testflights leading to institutionalized flights of the imagination. But there are different ways of institutionalizing a philosophy. In particular, it is easy to declare a philosophy acceptable by the public but only after incarcerating and sterilizing it and destroying all its possible social implications. I can hear the strains of the approaching bandwagon, and can only express my hope that when it passes by, not all philosophers of science will be adding their weight to the procession.

APPENDIX: THE LOGIC OF TECHNOLOGICAL DEVELOPMENT

There is psychological evidence against the suggestion that we can assess the credibility of a theory and then adjust our credence in it so as to make its credence equal its credibility. There is also psychological evidence against the suggestion that credence is an additive measure in any way. There is also historical evidence that much of scientific progress heavily depended on some metaphysical faith in is some new incredible ideas. And so the psychologistic theories of scientific assurance must be rejected. In many cases where tests are obviously too risky, even permission to test is conditioned by the prior procuring of positive evidence. The most obvious instance of this is the use of laboratory animals in medicine. This, however, is not to justify any of these procedures, rather these procedures are legal (stipulated) codes of justification, themselves unjustified (and unjustifiable), and perhaps even harmful.

This involvement plays a crucial role also in ventures into the unknown, and of insurance in technological innovation. The specific rules governing ventures are related both to involvement and to the use of positive evidence in insurance practice (e.g., the insurance of a test-flight is void if certain procedures of ground tests are waived). The rules, thus, insure stability by localizing and minimizing risks of new ventures. To proceed further one would have to fall back on a theory of social stability and planned social change, and, more specifically, of the balance between stability and planned change (including desirable and permissible rates of change, etc.).

Popper has developed a theory of social progress akin to his theory of scientific progress – it is the outcome of negative evidence. This does not yet explain social stability. For this we need add his theory of social institutions as interpersonal means of coordination and as accepted by convention. And here positive evidence may be the conventional, defective as it may be, means of checking institutional change: the higher the standards of required positive evidence leading to change, the more 'static' the society in question. But a 'static' society is not necessarily stable in the sense of durable. (Bellarmine applied the highest standards to Galileo's proposals, thus keeping the Church 'static', but leading to its loss of hegemony in the world of learning.)

A theory of social stability and of assurance in general can in this manner be incorporated into Popper's philosophical framework, then, with little or no alteration. Within this theory a theory of the role of positive evidence in implementing technological (physical, biological, social) innovations or reforms can be developed on lines adumbrated here (and in some of my other papers on the topic). Briefly, the guiding line is this: *whereas assurance is no guarantee of success, it is a prerequisite, a social condition for permission for trial.* The current inductive confirmation theory presents the acceptance of confirmation or scientific assurance – and hence also the actions based on confirmed theories – as obligatory for all rational beings. By distinction, the theory adumbrated here is that positive evidence should be viewed as a permit, in some cases a permit necessary by stipulation: for many types of novelty, positive evidence prior to implementation is required by laws; hence the case is open both to empirical study and to criticism and improvement. Whereas the obligation which the current confirmation theory posits is psychological (pertaining to the individual's rationality) and somewhat elusive (to say the least), the license of the present theory is conventional and hence can be examined by being applied to social cases, e.g., with food and drug regulations, public health regulations, binding testing procedures (in medicine and certain industries) or patent laws. This, then, opens up the possibility of comparative sociology of assurance or of positive evidence. Assurance in the psychological sense, then, need not abide by rules of propriety; but socially, propriety of assurance depends on severe test not resulting with negative evidence. Propriety, then, is no guarantee of success, but a mark of accepted degress of responsibility; this is in accord with common sense.

The principle, "there is no insurance against epidemics", when applied to the insurance provided to social institutions (rather than to an individual or even to a subsidiary institution, as usual) amounts to the rule "don't put all yours eggs in one basket", which is the idea of interdependence of social institutions, which is a version of generalized functionalism. The *guarantees* which institutions – insurance companies or governments alike – can offer, *are valid only within the social framework:* they do not guarantee the viability of the social framework itself. The social framework, thus, plays a similar role in technology as the current scientific framework in scientific research. For all we know both may collapse; but, also, both are susceptible to improvement through criticism. (This is a combination of Kantian and Popperian principles.)

The distinct role which positive evidence plays, then, is in the field of social implementation of technological innovations and alterations (physical, biological, and social). A particular case of this is that of social changes relating to the implementation and improvement of the institutions designed to aid the advancement of learning; and these include both current research techniques and current scientific ideas. If so, Popper's proposals for the social reforms of research conventions will not be widely implemented prior to tests leading to positive evidence sufficient by current standards. This may be achieved by instituting pilot-plants and their likes.

The usphot of all this is that a sharp dichotomy is introduced between science whose chief role is to provide testable explanations and eliminate the worst of these, and technology whose chief role is to introduce practical proposals, eliminate the worst of these, and volunteer implementation of the uneliminated. Whereas standards of criticism in science may be raised as much as it is within our reach, one can easily overdo the standards of assurance necessary prior to implementation. This may lead to stagnation or sluggishness which, in a competitive world both military and commercial, may be suicidal. It is the competitiveness, built into the social system, which both guarantees that positive evidence should be in accord with the Whewell-Popper theory and that not 'too' much positive evidence be required prior to implementation. These two guarantees are evidently in conflict which may be studied, regulated, perhaps even minimized, but only after critical procedures are institutionally reformed, as mentioned above.

CHAPTER 15

IMPERFECT KNOWLEDGE

Let us assume that claims for knowledge are often made which do not amount to claims for perfect or demonstrable knowledge. Let us further assume that when such claims are suspect it is possible to examine them and, subsequently, to declare them sometimes just, sometimes not (roughly in the manner followed in law-courts). Let us call such claims for knowledge claims for *imperfect* knowledge, and assume that these are sometimes just. The question is, what does imperfect knowledge amount to? How do we demarcate imperfect knowledge from perfect knowledge and from mere conjecture? In what follows I shall criticize the popular theory that imperfect knowledge equals a high degree of rational belief. My own view will be (paragraph 13 below) that imperfect knowledge differs from perfect knowledge only in that it makes allowance for acts of God, so-called.

I. EQUATING IMPERFECT KNOWLEDGE WITH SCIENCE
IS QUESTIONABLE

1. The traditional opinion concerning imperfect knowledge, if it exists, equates such knowledge with the body of highly probable hypotheses. This opinion rests on the assumption that scientists try to render theories probable in the light of experience. This latter assumption has been under the ceaseless barrage of devastating arguments from the pen of Sir Karl Popper during the last three decades or so. His alternative assumption is that scientists try, not to render theories probable, but to refute them; events which are erroneously considered as raising the probability of a given hypothesis are, he adds, failed attempts to refute that hypothesis. If one holds to the traditional opinion about imperfect knowledge while replacing the traditional assumption concerning the activities of scientists with Popper's alternative assumption, one comes up with he amended traditional opinion: imperfect knowledge is the body of hypotheses we have unsuccessfully tried our best and cleverest to refute.

2. Does Popper equate imperfect knowledge with the body of severely tested and still unrefuted hypotheses? Almost all of Popper's commentators say, yes, of course, in print as well as in conversation. I have found no sufficiently clear-cut answer to this in Popper's own written or spoken material. Moreover, the traditional opinion equates science with the body of highly probable hypotheses. Does Popper wish to equate, instead, science with the body of unrefuted though severely tested hypotheses? For years I answered this question in the negative. Today I know I had been reading my own wishes into his writings. Re-examining his works I find no sufficiently clear-cut answer to this question in his written works.[1] Dismissing the question of the authorship of this opinion until its rightful owner stakes a claim, the question of the truth of this opinion remains and deserves study quite apart from the fact that its popularity is on the increase: is it true that

science = imperfect knowledge = the body of well tested yet unrefuted hypotheses?

3. What is common to the old-fashioned equation of imperfect knowledge with the body of probable hypotheses, and the newly fashioned equation of imperfect knowledge with the body of hypotheses which thus far have stood up to severe tests, is that they both enjoy the respectability of being backed by a philosophy of science. Perhaps one may put it like this: the possibility of splitting traditional philosophy of science into two parts, one of which has been reformed and one, basic, which has remained intact, is what makes the opinions adumbrated here so very important. We do have a basic equation

imperfect knowledge = scientifically attested (rational) belief,

and a reformed one,

probable hypothesis
(the old-fashioned view)

scientifically attested (rational) belief

well-tested-but-as-yet-
unrefuted hypothesis
(the newly fashioned view)

Popper does assert that scientifically attested (rational) belief *may be* equated with the body of well-tested-but-as-yet-unrefuted hypotheses, though he is (regrettably systematically) ambiguous as to whether he advocates such an equation. In Chapter 7 I have criticized this equation to my satisfaction; here I wish to criticize the more fundamental, and as yet unchallenged equation,

imperfect knowledge = scientific knowledge.

As an alternative to this equation I propose the following hypothesis. Imperfect knowledge is that contention which would be perfect knowledge if certain conventional and unquestioned contentions were unquestionable. Moreover, all claims for imperfect knowledge are so understood, and therefore they become null and void the moment conventional and unquestioned claims upon which they hinge turn out to be untrue. Furthermore, the criterion given here can easily be expanded to become partially graded: a sociological (semi-institutional) criterion for the conventionality of an opinion can be given. This will render the present view highly refutable, and it may be refuted, but not so easily that I can say how. It may also be used with ease to engender examples which refute the popular equation of imperfect knowledge with scientific knowledge.

When rejecting the equation, imperfect knowledge = science, as I intend to do here, as well as the equation, science = the body of well tested though unrefuted hypotheses, as I have done elsewhere, I reraise the questions of demarcation of both imperfect knowledge and of science. In the present chapter I only refute the equation of imperfect knowledge with science and try to demarcate imperfect knowledge.

4. Let us consider perfect knowledge first. In his *Preliminary Discourse to Natural Philosophy* of 1831, Sir John Herschel has a small but interesting presentational difficulty, which he solves in a manner still acceptable today. He wishes to explain his claim that scientific knowledge is perfect knowledge, and is at pains to distinguish it from common and imperfect knowledge. I understand Herschel to say that perfect knowledge is perfect certainty, to be sharply distinguished from common and imperfect certainty which is the merest feeling of certainty. However strong the feeling of certainty is, it may accompany theories which are objectively open to doubt; scientific knowledge, he says, is perfect knowledge which permits

no doubt, and which should be endorsed with centainty because it is demonstrable. He calls this *mathematical* certainty, to distinguish it from the mere certainty of one's feelings, which he calls *psychological* certainty. Scientific knowledge, then, according to Herschel, is perfect knowledge, which is mathematical certainty, which is demonstrability. Scientific knowledge is nowadays considered imperfect, but the rest of Herschel's analysis is still generally accepted. Now we may feel certain regarding imperfect knowledge or even regarding the merest conjecture. And thus, not only perfect knowledge but also imperfect knowledge may be demarcated not psychologically but more objectively. The question is, what is objective imperfect knowledge, whether scientific or otherwise?

It may perhaps be gratuitous to note that were imperfect knowledge identified with the feeling of certitude, than all *bona fide* claims for imperfect knowledge would be acceptable. Yet this should lead to a slight modification of our initial premises. We have initially assumed that some claims for imperfect knowledge are not acceptable. We now have to strengthen this: some claims for imperfect knowledge are *bona fide* yet not acceptable.

5. Numerous philosophers have recently reaffirmed the view that *ordinarily* claims for knowledge, as well as for certainty, are not claims for perfect knowledge, yet they are in some (weak) sense objective. This makes claims for perfect knowledge made by mathematicians, some theologians, some philosophers, and even some scientists, not ordinary but, one might say, extraordinary. In any case, since claims for perfect knowledge are very clear, we may ignore them and center on the question, what are the objective claims for imperfect knowledge? Can we have objective certainty yet only imperfect knowledge? What, in such cases, is the objective grounds for claims for imperfect knowledge? Can we not adhere to the traditional opinion which identifies knowledge with objectivity and thus imperfect knowledge with partial objectivity?

6. Herschel endorses the opinion, which is already endorsed by Dr. Watts almost a century earlier, and whose beginnings can be traced to Bacon's preface to his collected works of two centuries earlier, namely the opinion that there are objective degrees of rational belief. According to this opinion, we ought to believe a theory to the extent that it is supported by

evidence (empirical evidence, usually), and suspend judgment concerning its truth to the extent that it has not been so supported. In other words, a theory is objective to the degree that it has objective empirical backing and should be held to the degree that it is backed. Unfortunately, this does not help us to understand what is ordinarily meant by claims to (imperfect) knowledge: the concept of degrees of imperfection of knowledge and the concept of partial proof are not clear, nor is the identification of these with the concept of partial belief. Admittedly, the concept of partial belief is somewhat clearer. Yet, historically, it was the clearer concept which was clarified first. A criterion for the degree of our belief in a given theory which can apply to ordinary circumstances was proposed by William Hyde Wollaston and reported by Michael Faraday ('On Mental Education'): the nearer to certain one is that a theory is true, the more one will be willing to bet on it even against high odds. Thus, the assigning of high probability is the measure of high confidence in the truth of a proposition, and partial belief or a degree of belief is partial confidence or a degree of confidence. Admittedly even regarding rational belief this clarification is not sufficient, since it only clarifies the concept of belief, rational or irrational. This, no doubt, was a difficulty felt by various writers, and fairly early in the day.

7. Faraday reports that when Wollaston told him of his criterion, namely that the degree of one's belief in a proposition may be reflected in his betting quotient on the truth of that proposition, Faraday was not very happy with it: he said, one is foolish to bet. Wollaston answered him, it is not an actual betting but a hypothetical betting that he had in mind. To this one might add a stronger argument of a more modern stock: people do gamble all their lives, not only when they invest in the stock exchange but when they make any investment whatsoever, in the broadest sense of investment: in a significant sense all action involves gambling.

This can be reformulated thus: fair betting quotients are necessary, not sufficient, conditions for an acceptable bet. When we elicit what one thinks is the fair betting quotient we do not force or even recommend a gamble, but merely decide one necessary condition for it.

This is, I think, quite acceptable to all parties. However, this only shows that we may elicit on some occasions what even a nonbettor may think a fair betting quotient is. It does not show that we can always elicit a fair betting quotient, it does not show that a fair betting quotient indicates one's degree of belief, it does not show that one's degree of belief is one's rational degree of belief (one may be under the influence of, say, an optimistic mood in the neighborhood), and it does not show that rational degree of belief equals current scientific opinion equals imperfect knowledge.

There is an enormous analytic literature on all this. The chief question the literature rests on is, how can we measure the degree to which one may believe a proposition given one's present knowledge of facts. The other questions are only lightly touched upon. Keynes, for example, only implies – however clearly – most of the identities listed in the above paragraph. Even if one assumes that all arguments in the literature are valid, one can still wonder whether rational belief equals imperfect knowledge.

II. EQUATING IMPERFECT KNOWLEDGE WITH RATIONAL BELIEF IS AN ERROR

8. Here I wish to refute mainly the equation of rational degrees of belief with imperfect knowledge, and incidentally current or accepted scientific opinion with imperfect knowledge. What is needed for this refutation is one lemma, concerning the betting quotient of certitude or perfect knowledge: *we bet*, if we do at all, *against any odds for a certainty and never against it*. That is to say, *no odds are fair against certainty*. It is therefore no risk to bet on a certainty; it is no sport to do so: gentlemen do not bet on a certainty.

Assume this lemma (the italicized bits in the last paragraph) to be true. Consider now the case of imperfect knowledge in this light. No matter how high is the probability that one's information is true, unless it equals unity it may be possible to calculate the odds for a fair betting on it. Moreover, however high the probability may be that one's information is true, unless it equals unity, the refutation of that information is not the refutation of the claim that it was probable: refuting one's expectation does not amount to the finding of the injustice of one's claim for high probability: only the frequent refutation of such expectations amounts to this. Indeed, we have ample (though *relatively* rare) realistic examples of forecasts which are highly probable yet turn out to be unrealized. (A good forecaster who forecasts with 90% accuracy may predict that 90% of the members of a given class A will have a property B. He may be right even when a high number of A's turn out to be non-B's. For this may be one of his 10% erroneous forecasts.)

Conversely, however often an expectation comes out correctly, this is no proof of its probability being the highest. This is a point which Russell has emphasized in his criticism of Reichenbach: only infinite sequences provide the accurate assessment of any probability. Thus, whether philosophically or on the basis of common sense, when we identify certainty with demonstrability with the highest probability, or with the probability one, we can be sure that we never are certain: we know with probability one that we can never assign utter certainty to any forecast. Therefore, the maxim, a gentleman never bets on a certainty, is not applicable ever.This conclusion is false: we have imperfect knowledge of cases in which it is ungentlemanly to bet.

We may rescue the maxim by saying, a very high probability makes the fair odds so stiff that there is no chance of the odds offered in reality to be fair. This rescue operation, however, will be rejected by most experts as not getting the point of the maxim. Of course a gentleman may be offered unfair odds and sometimes honestly agree to bet on them; what he cannot honestly do is bet on a certainty – under any circumstances. (If he bets on certainty under expediency, he views the expediency as one which allows dishonesty – love or war.)

10. To clinch matters, we should show that the certainty on which a gentleman may not bet is not a case of perfect knowledge of probability one, but of ordinary imperfect knowledge. This seems impossible because only perfect knowledge has probability one. And the last sentence implies the equation of imperfect knowledge with, at most, a very high probability. Hence, we cannot show what we should be able to show if our case were correct. Hence our case is incorrect.

But this argument confuses necessary and sufficient conditions for knowledge: perfect and imperfect knowledge do not share sufficient conditions: yet this is what the previous argument utilizes. Perfect and imperfect knowledge share a necessary condition which is fairly obvious. We cannot claim any knowledge of a refuted proposition.[2] We sometimes do claim perfect knowledge of a false proposition, as well as imperfect knowledge of a false proposition: there exist ample instances of either in history. But after a proposition is refuted, either the claim is withdrawn or the claimant is declared unreasonable.

This similarity, however, should not be exaggerated. Though in both cases withdrawal of the claim is mandatory, the assessment of the claim as it had been made in the past is different. It is the case that we do declare all claims for perfect knowledge of refuted propositions to have been unjustified even before the discovery of their refutations; yet this does not usually hold for claims for imperfect knowledge! [3]

11. Consider the claim that a certain statistical hypothesis is known to be true. Regardless of how one claims to have attained such knowledge, ordinarily one's claim will be dismissed if the ordinarily observed frequency is repeatedly different. This was noticed by Hans Reichenbach and Rudolf Carnap, who, however, chose to put it differently. In their opinion it is rational to gamble with truth according to some fair betting criterion, where the fair betting criterion is properly defined so that the computed fair betting quotient is always rational. Carnap adds that when an entrepreneur acts according to this criterion of rationality, then he acts rationally even on the occasion on which he subsequently loses, *because* in the long run this rational method will make the entrepreneur as well-off as possible. If all this is applicable, then there is a simple test for statistical imperfect knowledge: in the long run he who possesses more of it, and more properly utilizes it, must be better off than his neighbor. Hence, he who claims to possess more of it and use it more properly yet does not get better off is making a false claim.

12. The regrettable fact with the above criterion, however, is that even if it were applicable, the above test for a claim for knowledge will not work because seldom are claims for knowledge made concerning elements of random series. (This criticism is Popper's.) On the contrary, when referring to elements of random series one speaks not of elements of the series but of (imperfect) knowledge of the relative frequency of the series in question. Thus, a capable gambler in an honest house has an imperfect knowledge of the relative frequency and of the fair betting quotients of the gambles that are going on, not of any outcome of any given gamble. Of a given gamble he has an imperfect knowledge of the likelihood of its outcome,

not of its outcome itself.[4] Moreover, most ordinary claims for imperfect knowledge concern not easily discernible random sequences but either unrepeatable events or elements of unknown random series. Consider the question of the success of a certain marriage or of a certain business venture, or even of an investment in the stock exchange. Furthermore, very often claims for imperfect knowledge of unique cases are tested, and it is a simple empirical business to describe the way these are tested. Thus, Reichenbach's or Carnap's system is seldom helpful in practice, yet imperfect knowledge is often a matter of practical significance. Moreover, se wee that imperfect knowledge is not the same as likelihood, though we may have an imperfect knowledge of a likelihood.

13. Sir Francis Bacon has made a suggestion which has nowadays become very popular, namely that we consider unjust claims for knowledge as including false promises. Thus, the alchemist makes promises to deliver a recipe for making gold; which he cannot deliver. This is very obvious in all claims for perfect knowledge: when a proposition claimed to be known with certainty turns out to be false the claim must be declared unjust. When we wish to extend this to imperfect knowledge we must be cautious – not only because of the difference between perfect and imperfect knowledge, but also because of the nature of promises under ordinary circumstances. Ordinarily, the interference of an act of God exempts one from the charge of having failed to deliver what one has promised: *all promises are nullified by acts of God which render their execution impossible. A claim for perfect knowledge is a claim for a share in the divine* (Plato, Bacon, Spinoza) *in that it promises no act of God* (short, perhaps, of a miracle) *to nullify it.*

   With this difference taken care of we can safely equate perfect and imperfect knowledge: *imperfect knowledge that x becomes perfect knowledge that barring an act of God x. Also, the same holds for the imperfect knowledge that x is probable or likely* (Shimony).

14. This opinion, finally, enables us to find the error in the argument, since perfect knowledge that x is probability one, imperfect knowledge that x must have a lower probability. Imperfect knowledge that x too is probability one, but not of the same proposition x. Hence, when we say I know that he will propose to her within a fortnight, meaning 'know' in

the sense of having imperfect knowledge (soft sense, so-called), then we mean, I know that *if* they are alive and well for another fortnight, if he is not soon inducted into the armed forces, and if she does not all of a sudden decide to visit her aunt in New Orleans, etc., etc., then, etc. Now, it seems cheating to declare imperfect knowledge of x to be the perfect knowledge of a quite different proposition – namely of if y then x. However, this exercise in clarifying claims for imperfect knowledge has a perfect precedent in the case of clarifying claims for perfect knowledge. Bertrand Russell's analysis of the claim for perfect knowledge of the Pythagoras theorem p is making it depend on Euclid's axioms e: so that in his view the perfect knowledge is not of p, as claimed, but of if e then p. I need not say this analysis is now common knowledge.[5]

III. IMPERFECT KNOWLEDGE-CLAIMS ARE QUALIFIED BY PUBLICLY ACCEPTED HYPOTHESES

15. According to the opinion presented here we have a test for imperfection of knowledge-claims and a (partial) grading of them. A perfect knowledge claim is utterly unqualified. We can make an explicit list of qualifications: the longer the list (or the higher its content) the less perfect the knowledge claimed. A qualification is on the list if its violation does make one withdraw the knowledge claim but without feeling unjustified in his (by now admittedly erroneous) claim.

Thus, the claim to know that Johnny is arriving in town on the evening train is less imperfect if one also thereby claims to know that the train will not be stopped by kidnappers than if one's claim is not construed unjustified under such strange circumstances. It is easy to observe that one and the same claim for knowledge may be sometimes unjustified, sometimes not. For instance, compare the claim made by a friend to the seemingly same claim made by a police inspector whose assignment is to guarantee Johnny's safe passage. Although kidnapping exempts the friend from the responsibility even though the claim has turned out to be false, the very same kidnapping does not exempt the police inspector from responsibility; hence the inspector's claim for knowledge is rendered unjust in the face of kidnapping. To put it differently, the friend but not the inspector may promise the arrival of Johnny without previously checking and excluding the possibility of a kidnap. The problem still remains, what is the criterion for the intervention of an act of God? Why is a kidnap an act of God for the friend?

16. We can put our theory formally first, without as yet offering a criterion for what counts as an act of God. When we speak of perfect knowledge, we either speak of a proposition whose probability is one, or of a proposition whose probability is one in the light of given evidence and the given evidence is known perfectly. This raises a problem for the formal theory of probability, because empirical evidence is never such that its probability is one, yet the literature takes it often as known perfectly. This is a paradox of the inductive theory of probability which I do not wish to solve. Rather let me now assume that we can compute the probability of a hypothesis on the basis of agreed or unquestioned hypotheses (just as in the axiom system of Sir Harold Jeffreys). Then, all claims for knowledge are claims that certain hypotheses have probability one. The perfect knowledge is a hypothesis which has the probability one even *a priori*. An inductive philosopher would also say, and I shall not bother to refute him again here, that we have perfect knowledge *a posteriori* of all propositions whose probability is knowledge fully based on given known (observed) empirical evidence. Otherwise the claim is of only imperfect knowledge, the degree of imperfection being perhaps proportional to the improbability of the conventionally accepted hypotheses which are used as (supporting) arguments in the claim for knowledge.

There may be a few objections to this formal presentation of the opinion advocated here. I shall not enumerate them. I am more interested in the question of empirical testability of my opinion than in the elegance of a formal variant of it. And, we can easily see, this leads us back to the problem of demarcating acts of God.

17. The opinion presented here, concerning the test for imperfect knowledge, may be empirically testable, provided that we do not shift the meaning of 'an act of God' in order to avoid refutation. It would be an improvement if we provide and utilize an empirically testable criterion of what people do usually consider as an act of God. Such a theory is provided with detail in the previous two chapters and elsewhere[6] but it may be briefly sketched here. As a first approximation we may say, whatever kind of event is considered as impossible by all people concerned is, when it

happens, an act of God. The rationale for this is obvious; we cannot blame the claiment for knowledge for a false claim that we share with him. But this approximation is untrue, and has interesting counter-examples both ways: we have events we think impossible which we will not allow to use as acts of God kinds of excuses, and we have events we think possible yet we do use as acts of God kinds of excuses.

When someone promises something which, our accepted views imply, is impossible, say a new invention, he thereby undertakes not to use our accepted views. This very instance indicates that whatever we all share with a claimant for knowledge may ordinarily be construed as a condition for his claim. (This case is not much different from the case of the inspector).

On the other hand, death cancels all appointments (though not all debts), earthquake may wipe out a wide range of business obligations – and these, then, must be construed as acts of God whose exclusions may be conditions for imperfect knowledge. Yet such events are considered all too possible. Hence acts of God are not only what our commonly shared views declare impossible.

18. A better approximation concerning acts of God is this: acts of God are events which we exclude by tacit but public agreement; partly on the ground of theories we all hold to be true, partly on theories which we agree to pretend to hold even though we do not believe them to be true. For the theories we all tacitly hold as true we offer no excuse; we do not even attempt to justify them. Not so for the theories we tacitly pretend to hold as true. These we not only do justify – we even insist that the justification be so simple and trivial that there is little room for controversy here. The reason for this is all too obvious: we may differ as to whether to adopt a fiction but hold the fiction that we are ready a adopt a fiction. This would make things quickly slip out of hand, especially since agreement is here tacit and so not amenable to subtle disquisition; and so every problematic case is to be excluded: whenever there is doubt as to whether a convention is accepted, we usually deny that it is tacitly accepted and demand its acceptance to rest on an explicit statement in a contract or some such.

19. This approximation, too, is probably false, and as yet it certainly is too sketchy and incomplete. It is more than sufficiently elaborate, however

to enable the reader to construct with little effort a variety of instances against the equation of imperfect knowledge with opinions endorsed by scientists. That the equation is false is, by the way, though novel to philosophers, no news to lawyers. (It is no accident that the doctors of the law were called hypocrites, and by one who did not think much of social convention of any kind.) It is an explicit point of English law that acting on an error and causing damage or even death, may be considered an accident under a certain condition. The condition is that the error is shared by the actor's community. Reference to the actor's honest belief, on the one hand, is no support to his defense; and criticism of his community's belief from current scientific belief, on the other hand, is no undermining to his defense. This ruling, obviously, is only possible if some false claims for imperfect knowledge are not culpable. That this, indeed, sometimes obtains has been stated earlier.

20. When we refute the identification of claims for imperfect knowledge with claims for scientific knowledge we refute the theory identifying imperfect knowledge with rational degree of belief with scientific opinion. This is so regardless of whether scientific opinion is declared to be the one highly probable, the one best stood up to severe tests, or the one which best fits our general scientific outlook at present.

What I insist on is that what is the presupposition for a claim for knowledge is one which we all take for true; or, rather, one which we all pretend to be true; still more precisely, we do so in a rather unproblematic manner. Abner Shimony suggests that I mention here the fact that there may be different orders of pretending that a proposition is true. The reason may be a matter of technicality, from the simplification of a calculation to the simplification of the application of an idea to a complex and/or problematic situation, and it may be a matter of social attitudes, such as a conservative or an aesthetic one.

21. Finally, the view presented here raises a few new problems. Here are two. First, how do we change rationally the institional framework within which we can claim imperfect knowledge? We have to base our proposed reforms on some imperfect knowledge; and so we seem to be trapped. The answer to this problem must rest on the claim that our institutional framework is built so as to be imperfect and allow inconsistencies so as

to allow us to use one part against the other. This is particularly the case when we make an institutional room for science: we neither endorse its conclusions automatically nor allow them to be declared utterly irrelevant (our hypocrisy is tempered by reformism). This leads us to our second question. How do they determine institutional knowledge in sicentific society, whether an inventor's world, its patent office, clubs, journals, and all; or the U.S. Space Center. I shall not go now into this fascinating discussion.

Let me conclude with one observation which brings back a large chunk of the old view which has been rejected early in this chapter. When examining the content of the presupposition to a claim for knowledge carefully, one always finds some elaborate opinion about the significance of tests and some reports about past tests, namely that they were performed honestly yet without leading to a refutation. And so, a theory of what hypothesis under what test may be 'projected' (to use Nelson Goodman's terminology), is a part of the publicly and tacitly accepted presupposition. But such a hypothesis cannot be generalized as it varies from one community to another, both concerning the 'law-likeness' (to use Popper's term) or 'projectability' of some observations, and as to the severity of the required tests.

Finally, in his last work John Austin has elaborated on Bacon's correlation of truth and promises, attempting to reduce the concept of truth to the concept of promise. The opposite direction seems much more promising – which is why truthful people are also honest people who do not break promises, even when they later do not make the relevant clauses in their promises come true.

NOTES

[1] I remember a number of spoken remarks from Popper which clearly indicate to me that in Popper's opinion today's science equals todays well tested but unrefuted hypotheses. I have severely criticized this opinion in my review of T. S. Kuhn's 'The Structure of Scientific Revolutions', *J. Hist. Philos.*, 1966. I do not think Popper will explicitly endorse this opinion.

[2] We may be still ignorant of the refutation. We may dogmatically deny the truth of some refuting evidence. We may hope it is false and be on the way to checking it. I shall ignore these and similar cases.The reader interested in them may consult Chapter 8 above.

[3] As I have learned from Imre Lakatos, the concept of perfect knowledge even in mathematics has undergone deep alterations. Consequently we have to narrow this. Only those who share our concept of a proof will admit they were in error when offering a proof of a false theorem though they should have known better. But even this may go

too far. Therefore, though I do think we always tend to frown about erroneous proofs, I do not like this.

[4] All theories of probability assume the likelihood of an outcome to be the measure of our imperfect knowledge of the outcome; this leads to a well known paradox: for every possible outcome we have a high degree of imperfect knowledge that it will not turn up, and so we have a high degree of imperfect knowledge that though one outcome will certainly show up, none will; which is obviously absurd. The absurd is to declare a perfect knowledge of a likelihood and any likelihood a measure of imperfect knowledge.

[5] Also, Russell's analysis is not contradicted by anyone, and in a sense deeper than the claim that p cannot possibly be known: the other analyses of the claim for the certainty of p in the literature on the foundations of mathematics are equivalent to Russell's in the sense that they can be extended in the same manner both within mathematics and in adjacent fields.

[6] See also my 'Conventions of Knowledge in Talmudic Law', in Bernard S. Jackson (ed.), *Studies in Jewish Legal History; In Honour of David Daube*, Jewish Chronicle Publications, London 1974 (also published as Special Issue of *J. Jewish Studies* **25** (1974), 16–34). See also next two chapters.

CHAPTER 16

CRITERIA FOR PLAUSIBLE ARGUMENTS

The most conspicuous datum in philosophy, perhaps the most important one as well, is the following:

(d)   Most philosophers consider the thesis of scepticism[1] highly implausible, yet they consider the traditional arguments in its defence highly plausible at first glance.

The rationale for the qualifier 'at first glance' in the datum is the following generally accepted thesis concerning plausibility,

(p)   A thesis defensible by plausible arguments is plausible, and vice versa.

Most philosophers would accept our plausibility thesis (p) as a matter of course, if not as analytic. Hence, our datum (d) without the qualifier 'at first glance' would allow us to charge most philosophers with inconsistency.

(c)   Most philosophers are inconsistent in viewing scepticism as both implausible and plausible.

The charge is answered, of course, by the qualifier which will force us or modify (c) by qualifying 'plausible' in it with 'seeming', and this qualifying 'inconsistent' likewise.

The outcome, then, would be:

(c′)   Most philosophers are seemingly inconsistent in viewing scepticism as both implausible and seemingly plausible.

Now, this raises a serious problem. Suppose a philosopher rejects a thesis as implausible. Suppose a colleague advances some plausible arguments in its defence. By our plausibility thesis (p) he will consequently have to change his mind. Suppose he is unwilling to do so. Suppose he is unable to do so since his intuitions are too strongly bent on his own views which are contrary to the thesis just proven plausible. Suppose he has suspicions that not all is as clear as he wished. Suppose even that he cannot convince himself on the strength of one or two plausible arguments that practically all his colleagues and illustrious predecessors were utterly in error. What then?

The problem is rather significant. On the one hand we may view as sheer intellectual timidity the refusal to alter one's view because one shares it with others. We may even view such an attitude as no less than intellectual dishonesty and ulterior motives. On the other hand, we consider as quite reasonable the readiness to pay heed to the best views around and the reluctance to rush to declare all one's colleagues in error on account of what looks to oneself to be a plausible argument. After all, what looks as a plausible argument is at times nothing but a sophism, or a slight and negligible error with unexpectedly great ramifications, or a seemingly unanswerable objection met by a new and unexpected answer which is obviously quite satisfactory by any standard. History contains a gallery of people who placed too much trust in what looked to them a convincing argument and who thus started on a road to cranky pseudo-science or pseudo-rationalism. And history is full of people who put too little trust in arguments which are convincing and which looked to them quite convincing but they were not at all sure.

Lest the reader think that the dilemma just presented is an abstract thought-experiment with little or no practical significance for sane thinkers, let me mention an example. The details of the example are given in my *Faraday as a Natural Philosopher* (Chicago University Press, 1971). Rightly or not, I have presented in that book the following story. In the mid-nineteenth century Newtonian mechanics was considered absolutely proven. Some thinkers even said it could never be explained by another theory; all thinkers agreed that it was absolutely true and definitely neither a mere approximation to, nor a special case of, another theory. At that time Faraday strove towards developing another theory of just these qualities. He advanced arguments which he found amply plausible. His contemporaries ignored his theory; even those who admitted some of his arguments in favour of his theory ignored it. In order to compensate him for this disregard of his theory, his comtemporaries rewarded him and praised him all the more for his numerous empirical discoveries (which he made as, and while searching for, plausible arguments for his theory). Consequently he wavered between viewing almost all the leading physicists among his contemporaries as dogmatic and viewing himself as a cranky pseudo-scientist. He found both horns of the dilemma most distasteful. The emotional burden of this dilemma was too much for him. He had a series of emotional breakdowns which finally led to utter senility.

Not all manifestations of this dilemma are equally dramatic, of course. As it will soon transpire, the dilemma cannot be very dramatic unless it concerns a very far-reaching thesis. Hence, Faraday's case is very far from the average, a very extreme case indeed. In principle, however, the dilemma is very common and was implicitly recognized by Francis Bacon who blamed some people for too much self-trust, others for too little self-trust, some for too much dependence on public opinion, others for too much willingness to oppose public opinion. What, however, is the just measure of self trust and of consideration for serious public opinion? At least Faraday tried hard to find the answer to this question, yet he failed. At times, it seems, Bacon and the whole tradition of the Enlightenment, as well as Faraday, recommended the utter disregard for all public opinion. But no one except perhaps Spinoza could live by this standard.

One may, perhaps, sympathize with the problem but declare it no longer pressing in view of the fact that we have now a long and stable tradition of scientific public opinion which one may safely rely upon without becoming a dogmatist. This, however, will not do. Unusual as Faraday's case may have been, it nonetheless refutes (should I say 'seemingly'?) the view that public opinion, when scientific, can be relied upon. It even refutes the weaker view, which is offered in the preface to the first edition of Popper's *Logik der Forschung*, of 1935, and which permits the reliance on scientific public opinion regarding its own problems. 'A scientist... can attack his problem straight away... For a structure of scientific doctrines is already in existence; and with it, a generally accepted problem-situation. ... The philosopher ... does not face an organized structure.... He cannot appeal to the fact that there is a generally accepted problem-situation.... Nevertheless... some... believe that philosophy can pose genuine problems.... And... all they can do is to begin afresh...'. This is refuted, regardless of whether Faraday is classified as a scientist or as a philosopher. He faced an organized structure of scientific doctrines, but not any generally accepted problem-situation: indeed, he tried to create one, or to alter the generally accepted one in a radical manner. He offered arguments which he considered plausible. Nonetheless he failed to arouse his contemporaries' interest. He had to begin afresh, as Popper puts it, and so he opted for the next generation; and there indeed he had better luck.

The gist of all this is to lend force to the problem at hand: when is it forbidden, when is it permissible, when is it imperative, for a member of a learned society, to declare a thesis plausibly defended? (The prohibition and the compulsion may be on the pain of expulsion from the society or of demotion within it; a point noted repeatedly and systematically by Michael Polanyi.) What are our criteria of plausibility? When may (should) we charge one (person or society) with an inconsistency of the type of (c), and when do we have to accept the plea that the inconsistency is only a seeming one of the type of (c')? When is it the case that scepticism is defended by only seemingly plausible arguments and under what conditions will scepticism win?

The problem, as my story of Faraday indicates, is not peculiar to philosophy alone. As many other stories, some of them quite well-known, clearly suggest, the same problem occurs in daily life, in political life, in technology, and even in science. (The story of Columbus is so interesting as it is a mixture of all these.) The problem when a thesis is plausibly defended, is all too common. It can be split two ways:

(PA)   When is an argument plausible, when is it only seemingly so (as yet)?

(PT)   When is a thesis sufficiently defended by admittedly plausible arguments to count as plausible?

To be more precise, the splitting should not be considered a separation, since under certain circumstances a thesis rendered plausible may lend plausibility to some of its more doubtful supporting arguments. This is a part of the bootstrap theory of rationality expounded below. Yet the present formulation of the problem, though it may be improved upon, will suffice to show the insufficiency of our initial plausibility thesis

(p)   A thesis defensible by plausible arguments is plausible, and vice versa.

Most philosophers of science will want to modify (p) quantitatively. Ignoring the problem (PA) as hardly troublesome, and taking all and only attested relevant facts to constitute plausible (inductive) arguments, they would present a rule for determining numerically the degree of plausible inductive support which given arguments afford a given thesis. They will then add the following rule:

(PIS)   A thesis more supportable by plausible inductive arguments than any competing thesis, is the one to be endorsed as scientifically plausible.

Even while taking the outlandish view that (PA) does not cause us any trouble, and even while taking the more outlandish view that we can determine quantitatively degrees of inductive support, even then we may still find (PIS) too vague, and for the following reasons. First, it speaks only of support, not of undermining, whereas, at least to begin with, many scientific theories (even well-supported ones) look seriously undermined by known evidence. This perhaps can be remedied by setting-off support against undermining (= negative support). But we know that the initial undermining may be taken lightly at first, and so it is quite different from the undermining which cannot as yet be removed despite certain efforts. Secondly, (PIS) does not say what makes two theses competitors. (The intuitive idea is that two theories are competing if and only if they are contraries, i.e. inconsistent unless referring to non-existent entities. The proviso, however, does not stay unproblematic and the range of contrariness can thus easily be expanded indefinitely – which is absurd. This problem is still unstudied, strange to notice.) Nor does (PIS) say what is the set of competing theses – all the known competitors or all the possible ones. If we take all the possible ones, then the support always spreads over infinitely many theories. (This is a standard topic in the literature on inductive logic: the problem of zero initial inductive support.) If, however, we take all the known competitors, then (PIS) may invite complacency at times of awareness that no known theory is plausibly enough defended, even though clearly such times are proper times to look for a new theory.

There is one further reason which, I think, invalidates (PIS) as well as all its traditional substitutes. It is this: even if it is at times valid, it must be qualified by additional criteria.

To show this let us follow the suggestion that we should take inadequacy before taking adequacy (a suggestion which Plato's early dialogues made a central part of our tradition). All existing theories of science consider only one kind of inadequacy, namely poor agreement between theory and fact. This is clearly not sufficient because, as the first reason above indicates, some such poor agreements are considered merely temporary setbacks, others not. Hence, there is need for some deeper criterion of inadequacy. Nor is the temporariness merely a matter of how long and how hard we have already attempted to overcome it. For it may indeed be the case that a theory is engaging us in other directions than the setback, or that the theory pleases us so much with its many agreements with facts that we are willing to turn a blind eye to the disagreements, as indicated in my second reason. Moreover, history clearly shows that other criteria are often deemed more significant.

When a theory is offered as a solution, it is seldom the case, contrary to Popper's above-quoted view, that we all agree what problems it comes to solve. It is, indeed, all too often the case that opponents expect from a theory that it should solve more problems before they deem it plausible. For example, Bohr's ability to explain only wavelengths, not enough wavelengths, and no intensities of spectral lines. This limitation is an uncontested historical fact. It was deemed by many scientists at the time to be a strong argument against the plausibility of Bohr's ideas. His paper on the correspondence principle came to combat that by suggesting new criteria and new avenues of research.

The moral from Bohr's case is very simple and obvious. Supposing each argument in favour of a thesis is deemed (at least for the sake of argument) perfectly plausible, we still have no consensus about how many such arguments make that thesis plausible. We have this divergence of opinion even in science, not to mention engineering, medicine, politics, everyday affairs, or philosophy. How shall we judge or choose between the diverse opinions? By what criterion?

This problem can be reformulated as the problem of desiderata for a plausible thesis:

(PD)   What are the desiderata reasonably expected from a thesis before it be, may be, should be deemed reasonable?

Now the question may be tackled both descriptively and prescriptively, allowing, of course, always for the fact that even the best of us fall short of their own standards, not to mention the possibility that our standards are improving, so that our predecessors, even the most illustrious scientists and philosophers and wise men amongst them, may fall short on two counts and not only one.

During the whole tradition of western thought there were two and only two answers to the problem of desiderata held consistently enough for any length of time to enable their holders to elaborate on them with sufficient detail and clarity to permit adequate comment on them without exegesis. The first is that of classical rationalism – Greek or of the Age of Reason – according to which a thesis can be deemed reasonable if and only if it is demonstrable. The second is that of all fideistic religious philosophy – medieval philosophy almost exclusively, some but not all modern religious philosophy, and some but not all philosophers who call themselves Marxists (rightly or not) – according to which a thesis is plausible if and only if it accords with the faith and then is supported or proven etc. by further arguments.

This brings us back to the very initial datum (d) from which we started. For what is common to both these answers to our problem of desiderata (PD) is that they include the following two assumptions. First, that (PD) is answerable. Second, that if scepticism is true, then (PD) is not answerable. Scepticism is thus rejected by all its opponents a priori on one transcendental argument: if scepticism is not rejected, then (PD) is not answerable; hence whether one is a sceptic or not becomes a sheer matter of taste; hence we need not endorse scepticism come what may.

It is, therefore, highly desirable for a sceptic like the present author to try and show how (PD) can at all be attacked by a sceptic. In other words, the problem the initial datum (d) poses to the sceptic concerns rational scepticism:

(RS)   is it at all possible for a sceptic to be rational?

Given (PD), the problem (RS) can be put more specifically thus:

(RD)   is it possible for a sceptic to decide reasonably about the reasonableness of the desiderata for a reasonable thesis?

Now (RD) is formulated in a manner most conducive to elicit a negative answer and so condemn scepticism. For, any decision about reasonableness which a sceptic may come up with cannot but be circular, and hence unreasonable, whereas (RD) is the search for a reasonable answer. Let us assume, for the moment, that indeed, scepticism is hopelessly circular; the non-sceptic can then reasonably admit the sceptics' charge of circularity, without being forced into scepticism thereby. The non-sceptic can reason as follows. The sceptics' criticism, admittedly, deserves study and ought to be demolished regardless of the effort required by such a task. Yet this does not mean that the sceptic wins, no matter how long the attempt to demolish his seemingly plausible argument has failed. For, to succumb to scepticism on this point is simply to replace one circularity by another, thus arriving at no gain and quite possibly, indeed obviously, at some great loss.

This last point, it seems, goes as deep into the anti-sceptical mode of argument as any. It should also be noted that not only classical rationalists but also classical fideists have asserted or implied it now and then. It is quite an interesting point, since it embraces a view shared by all non-sceptics of the necessary and sufficient conditions which render plausible a thesis plausibly defended. We have now presented the classical answer to our question (PT): when is a thesis sufficiently defended by plausible argument to count as plausible? And the classical answer is meant to be used in the attempt to vindicate the rationality of most philosophers (who, in accord with our initial datum (d), deny the plausibility of scepticism, though they admit it is plausibly defended), be they classical rationalists, or modified rationalists along (PIS) or some such principle, or fideists. What our discussion thus far clearly indicates are the following desiderata, the following necessary and sufficient conditions for the transmission of plausibility from arguments to the thesis they defend:

(1)   Circularity is bad and should be avoided, at least if at all possible.
(2)   A thesis should answer certain specifiable desiderata.
(3)   The arguments defending the thesis should cover these desiderata to an agreed extent.
(4)   The agreed extent, as well as the desiderata, should themselves be plausible.
(5)   We take as plausible the most plausible thesis we have, provided it is plausible enough by the other criteria. (Otherwise we have to keep searching.)

(The fourth desideratum, it may be noted, is relatively new. It is, for example, the central argument A. J. Ayer employs against the sceptics in his *The Problem of Knowledge* of 1956: The sceptics win all too easy a victory, but only after they make some quite unreasonable demands put on any thesis which may then be deemed plausible; the extravagant demands, he says, concern both the kind of desiderata and their extent. The fifth desideratum is really a variant of (PIS) and appears in the works of Keynes and Jeffreys on inductive logic early in the century.)

The thesis of the present chapter is that – surprisingly – the above list of five criteria is sceptical, not anti-sceptical. The argument in its defence would hinge on the interpretation of the plausibility of the desiderata which, according with (1), should not be circular, if possible. A bootstrap theory of rationality best answers the five desiderata listed above and is sceptical par excellence, where a bootstrap theory presents series of criteria of rationality, each an improvement on its predecessor by its predecessor's own lights. This, incidentally, will remove the circularity from the definition of plausibility since the plausibility referred to in the definiendum is more advanced than that referred to in the definiens.

This, however, is a mere nicety. There is a substantive and very well known important desideratum which a bootstrap theory of plausible argument should satisfy. It is a principle which has been invented by historians, but which has not yet been introduced into the discussion of the theory of rationality. It is the principle of the historical relativity of rationality:

(HRR)   What was rational enough yesterday need not be rational enough today.

Coupled with the classical, rationalistic or fideistic, theory of the desiderata for a plausible thesis (PD), the thesis of historical relativism of rationality (HRR) yields the thesis of historical relativity of truth:

(HRT)   What was true enough yesterday need not be true enough today.

This, incidentally, explains why so few historians of science endorse the thesis of historical relativism of rationality (even though this lands them repeatedly in trouble, as I have illustrated in my *Towards An Historiography of Science*, 1963, reprinted by Wesleyan University Press, 1967). Most of them endorse the classical theory of rationality as demonstrability or as scientific probability or such, but refuse to accept that truth is relative. They have a point here which is very attractive on both philosophical and commonsense grounds. But so is the thesis of historical relativism. This, then, may culminate with a major desideratum and a major thesis.

(Major Desideratum) We want a theory of plausible argument which embraces historical relativism of rationality (HRR) and rejects historical relativism of truth (HRT);

(Major Thesis) To achieve the major desideratum we need notice not only the improvement over the ages of our theories and of our stock of data, but also of our standards of rationality (preached or practiced).

The gratifying corollary of any bootstrap theory of rationality is this. We may, and indeed should, distinguish between a descriptive and a prescriptive theory of plausible argument; yet in so far as there is progress, what is merely prescriptive at one time may become fairly descriptive soon after. In that vein we may avoid charging traditional anti-sceptical philosophers with the inconsistency (c) but rather explain our main datum (d) by (c'), the claim that they seem inconsistent. The dilemma we mentioned above, between following one's own reason and following the mainstream, is likewise resolved in a manner which may stand a little elaboration before the close of this essay.

What seems to characterize the bootstrap theory is the interaction between the framework and its content. Whereas the fideist accepts the framework of faith as final, and whereas the classical rationalist accepts the framework (if any) of demonstration as final, in historical fact frameworks are accepted tentatively. (As Richard von Mises observes, even the Catholic Church could not escape all improvements.) The frameworks enable us to solve some problems locally and declare other problems locally insoluble, or practically insoluble. It is commonsense which tells us what can be taken for granted and what should be considered practically impossible – and the articulation of commonsense is a framework. When a practically insoluble problem is pressing enough, we may offer a solution to it which transcends the framework. The question then will be: can the new solution be the new, or the foundation of the new, framework? The more daring will take the plunge first and act on a framework which is hardly there and which may turn out to be a mirage. As Bunge suggests, it is commonsense to hope to be able to solve a problem within the given framework; it is ambitious to hope to fail in that and succeed otherwise; it is folly not to try to search for a more conventional solution (i.e. a solution within the given framework) first.

The choice between frameworks often follows a theory of rationality, whether more articulated or more intuited. There is, also, the interaction between the choice of a framework and the choice of a theory of rationality. It is no accident that avant-garde defenders of rudimentary frameworks from Galileo to Niels Bohr have also revised (and explicitly) our criteria of rationality.

Finally, the question of whether the old sceptics were rationalists or nihilists is likewise solved. The old sceptics did not have anything like a bootstrap theory of rationality. And so in a sense the rudiment of a theory here presented (largely due to Karl Popper, Mario Bunge, W. W. Bartley III, I. C. Jarvie and others) is not sceptical. Yet, by some lower standard of rationality which is more to the taste of the historically minded, they were eminently rational. Whether the new bootstrap operation theory is more akin to classical scepticism or to classical rationalism becomes less clearcut than it seemed before. All this is highly satisfactory and should now enable us to declare philosophers who insist on their attempts to answer a sceptic not very rational. With the further development and expression of the detail of the new theory of rationality (c′) will have to be replaced by (c) unless, and much more preferably so, our initial datum (d) is no longer true.

All this opens new and exciting vistas. The most gratifying corollary to all this is the solution to the following mystery. Many historically important debates have been left uncompleted; and no rational reconstruction could complete them in a satisfactory manner. Query: is there a logic to all this? Answer: Yes. A debate led by one set of desiderata may evolve a new set of desiderata which may then lead all participants to rethink and reformulate their answer – or even their questions.

NOTE

[1] By scepticism I mean the view that we 'cannot defend reason by reason', as Hume puts it (*Treatise*, I, iv, 2), or that we can never justify our theory either by proof or by probability.

APPENDIX:
THE STANDARD MISINTERPRETATION OF SKEPTICISM

Justificationists systematically misread skepticism as follows. Consider the following three propositions:

(1)   Justificationism: Rationality is justification (of one's views by some sort of proof).
(2)   Skepticism: Justification is impossible.
(3)   Rational Ethics: We should be rational (act on views we hold rationally).

The combination of all three leads, of course, to an impossible ethics. The conjunction of Justificationism and Skepticism, thus, leads to the denial of Rational Ethics; the conjunction of Skepticism with Rational Ethics leads to non-Justificationism – and the conjunction of Justificationism with Rational Ethics lead to non-Skepticism.

Cynicism = Justificationism and Skepticism
Non-Justificationism = Skepticism and Rational Ethics
Positivism = Rational Ethics and Justificationism.

Now almost all opponents to Skepticism, and quite a few adherents of it, understand the doctrine to be cynicism, i.e. the conjunction of Skepticism with Justificationism; the skeptical tradition, however, is to combine Skepticism with Rational Ethics. This tradition looks inconsistent, because Justificationism is tacitly attributed to it. The positivists' inability to imagine a non-Justificationist rationality is the cause of a certain narrow-minded rejection of skepticism. But not all rejection of skepticism is narrow-minded, and even where narrow-mindedness is present it is often less dogmatic (*pace* Sextus and *pace* Sir Karl Popper) than plainly naive.

CHAPTER 17

MODIFIED CONVENTIONALISM

> The Socratic Method is only an approximation-method,
> and belongs to the semi-fictions, the only kind that really
> come into question.
>
> (VAIHINGER)[1]

A philosopher wishes to include in a comprehensive view certain ingredients; this may be impossible; or, it may be possible, but beyond that philosopher's capacity. It is important, therefore, to distinguish between a philosopher's desiderata for a theory, and his theory's answering or not answering these desiderata. In different stages of his career, a philosopher may consider different desiderata and assume they do not conflict; but this assumption may be false.

In various places I have ventured to apply these general and rather obvious considerations to Popper's theory of science. Let me air one simple further example before going into the major example of the present essay. In a recent comment on criticisms by Bartley, Popper exempts himself from replying to these criticisms, on the ground that Bartley, though he had been a prized student of Popper's, chose to overlook quite a few aspects of Popper's philosophy (['*Remarks*'], 88). For example, his own problem-orientation since rather early in his philosophical career (96). No doubt, Popper has taken problem-orientation to be a desideratum for the theory of science fairly early in his philosophical career, and his quotation from an old text of his fully supports this claim. Popper, however, considers this as evidence that his own theory of science does answer this desideratum. We may say to this, in the absence of any theory of problems and of problem-solving, it is difficult to know how to judge this claim. We may say, alternatively, that every given task implies the problem, how best can one perform the given task? The task of finding a hypothesis, then, or of choosing it, of accepting or rejecting it, of refuting or confirming or verifying it, are all problems which render each and every known theory of science problem-oriented, if not practically all philosophy. Granting that not every task is a problem, I do not find

Popper's early philosophy of science – his theory of scientific character as refutability and his theory of learning from experience by refutation – to be particularly problem-oriented. In his lecture courses Popper himself has presented a more problem-oriented philosophy of science. Scientists explain puzzling phenomena and try to render their explanations less *ad hoc* or less arbitrary than these look at first. There is much in common between Popper's two theories of science, the oral and the written. For example, reducing arbitraries is done by corroborating, and attempts to corroborate are the same as attempts to falsify (the difference between corroborations and refutations is not in preparation, in the question addressed to nature, but in outcome, in nature's answer). But Popper claims that these two theories are identical. I have elsewhere (['*Nature*']) explained (at least) to my own satisfaction that this is not so; I shall have occasion later to mention a few counter-examples to this, as well as to show how one may present a more problem-oriented theory of science, which may indeed be viewed as a variant of Popper's.

In the meantime, I wish to return to Popper's desiderata and to the theories which possibly satisfy them. Let us begin with the question, how well his various desiderata harmonize. In particular, I shall examine his desideratum of refutability ([*Logic*]) and his desideratum of search for truth ([*Conjectures*], ['*Aim*']); they seem to agree with each other with ease; but the claim that they do is a hypothesis which may merit critical examination.

The ancient dichotomy between nature and convention, where nature stands for truth and convention for fiction, applies both to the philosophy of science and to social and political philosophy. A third view replaces fiction by *bona fide* error, and postulates degrees of approximation of convention to nature. This view is very close to one which has been proposed by Sir Karl Popper both in the philosophy of science and in social philosophy. He also declared institutions to be like scientific hypotheses and science to be a social phenomenon. This parallel between his two gradualist theories – of science and of society – may really be a version of modified conventionalism, though Popper himself favors a somewhat narrower version – of modified naturalism or modified essentialism.

I. THE PROBLEM

Let us begin with Popper's classic essay, 'Three Views Concerning

Human Knowledge' ([*Conjectures*]), where two traditional views, of science as ultimate and unalterable knowledge, and of science as mere utility, are labelled 'essentialism' and 'instrumentalism' respectively, and where Popper endorses the third view, of science as approximation to truth in stages, which can be labelled 'modified essentialism'. There is a lack of symmetry here, as there is little sense to talk about modified instrumentalism. This, as Popper stresses, is because the bold idea that science is merely instrumental for technology becomes rather trite with the slightest modification: no one denies that science is *also* an instrument. This is not a matter of scientific instrumentalism as such, but of any instrumentalism whatsoever, aesthetic, moral, or social. In general, '*x* is *also* instrumental to *y*' is all too often trivial, whereas '*x* is *only* instrumental to *y*' is radical and stimulating. It is the radical view, not the trivial, which is instrumentalism proper. And it is the slight change of the radical view which lands us in the trivial.

There is no point, therefore, in speaking of a modified instrumentalism – as each theory except instrumentalism itself is in a sense a modified instrumentalism; but there is a point in speaking of a modified essentialism as a theory which, like essentialism, aspires to the truth and, like instrumentalism, rejects essentialism, i.e., the idea that the last word on scientific questions is attainable, as rather naive.

This does not seem right; the feel is that, though it looks perfectly reasonable and unproblematic, we may well benefit from caution here. Let us first remember how different a feel one receives from Popper's *Logic of Scientific Discovery* of 1935 from the feel one receives from his 'Three Views' ([*Conjectures*]) of 20 years later. In his earlier work Popper contrasts not essentialism with instrumentalism, but inductivism with conventionalism. It is not easy to doubt – at least I cannot see how – that inductivism is a species of essentialism (in spite of some possible exotic exceptions in the long and varied history of inductivism, see my ['*Sensationalism*']). At least, it is not to be doubted that classical inductivists have repeatedly declared that the aim of science is the unearthing of the truth about the nature of things, that is to say, the true essences of things, and not merely convenient modes of expressing our past knowledge. Thus, Popper approvingly quotes an attack on conventionalism from the inductivist Joseph Black ([*Logic*], 82).

Consequently, still intuitively and superficially, one tends to endorse the two equations,

$$essentialism = inductivism;$$
$$instrumentalism = conventionalism.$$

Admittedly, both equations contain some injustice to the apriorists (see Chapter 18). But, at first blush, we may forget apriorism altogether and accept, however tentatively, the two equations above. And the intuitive superficial impression is now that Popper's view as expressed in his *Logic of Scientific Discovery* is that of a modified conventionalist, not of a modified inductivist.

It is difficult to nail down impressions, and problems need not develop out of the merest uneasy feelings which accompany them. But let us wait and see. The point where Popper sounds like a modified inductivist is very important: namely, the conclusion of his *Logic of Scientific Discovery*. And this well accords with his 'Three Views'. He says there, "ideas... are ... our only instruments for grasping" nature (280). "Science never pursues the illusory aim of making its answer final, or even probable", he says at the close of his earlier work. "Its advance is, rather, towards a finite yet attainable aim; that of ever discovering new, deeper, and more general problems, and of subjecting our ever tentative answers to ever renewed and ever more rigorous tests." And, finally, Popper speaks at the end of his work of his theory as that of movement in the *quasi*-inductive direction, contrasting it with theories – such as Spinoza's – of movement in the quasi-deductive direction; and he rejects movement in this latter direction, Spinoza's or any other, on general methodological grounds, on his desire to retain science as empirical as possible, which means for him to prefer those scientific theories amongst the available ones which are as testable as possible (277–8).

This is as far as it goes. The impression which the volume as a whole gives is that Popper is decidedly a modified conventionalist, hardly a modified inductivist – seemingly in clash with an impression he gives 20 years later. This, towards the end of his *Logic of Scientific Discovery* (Section 84) he suggests to drop the concept of truth altogether. In a later note he reverses this position, of course, and accepts truth as a regulative principle. This, however, seems to clash with the already quoted idea about the desideratum that the aim of science be 'finite yet attainable'.

Also, the quote about ideas being instruments for grasping nature seems not to agree with his present rejection of my own views (*[Conjectures]*, 248n): according to my view, he says, theories are the merest instruments of exploration; my view is, he suggests, a modified instrumentalism! More conspicuously, Popper rejects inductivism as a naturalistic view of science, and endorses instead, quite unhesitatingly, the conventionalist view of science as a game played in accordance with certain rule. Though he rejects the rules offered by *the* conventionalists and presents rules of his own, this clearly places him more as a modified conventionalist than as a modified inductivist.

Victor Kraft (*[Circle]*), (127 and notes 127 and 147) has even branded Popper as a conventionalist plain and simple. What part of Popper's early work particularly impressed Kraft as bluntly conventionalist was the idea of Popper, according to which we need a rule for accepting or rejecting factual statements or observation reports or eyewitness testimonies. The naturalist's view will force us to acknowledge all testimonies. This openly runs counter to the desideratum, sometimes expressed even by arch inductivists (and reported by Popper, *[Logic]*, 96), that testimonies be sometimes revisable. Nor can the discrepancy be resolved, since to the naturalist any rule is either imposed by nature and is thus not a matter of choice or makes us conventionalists who are willing to rest on arbitrariness.

This is not to endorse Kraft's view. On the contrary, I wish to ascribe to Popper the important invention of *modified* conventionalism, or softened conventionalism – or partial conventionalism, to present it in Popper's own way (*[Logic]*, Section 27) – though somehow this was overlooked by Kraft. We may admit the presence of convention and even of sheer arbitrariness in science, *provided* it be confined *and* reducible. And, contrary to Kraft's impression, the conventional element in testimonies is more easily reducible, particularly when the conventions offered by Popper are adopted, than the conventional element in theories, particularly if the convention offered by the conventionalists are adopted. Indeed, the new element in Popper's conventionalism which modified it so much, is the proposed convention to be wary of the inevitable presence of the conventional element in science and to prefer at any stage moves leading to its (partial, but increasing degree of) elimination.

The question, therefore, is, shall we equate conventionalism with instrumentalism and declare Popper a modified instrumentalist? Will this not come in conflict with his modified essentialism? Is not modified instrumentalism trite?

II. SCIENCE AND SOCIETY

When a philosopher develops his theories of diverse topics these may be unrelated. When they are related, the question, how well they fit together to form a system, may intrigue a reader. In our case the two chief philosophical concerns of Sir Karl Popper are the philosophy of science and social philosophy. I shall not discuss all the possible links between these two fields, nor even all those explicit in Popper's writings. I shall center on one and attempt to examine it somewhat critically.

The conventions of science which we have discussed, particularly the convention to examine and test scientific hypotheses, are social institutions. Popper attacks inductivism as naturalistic and thus psychological and thus question begging. In his *Logic of Scientific Discovery*, the view of science as an institution is merely implied – by contrasts. He stresses the contrast between psychologism and the view of science as intersubjectivity, as well as the contrast between the naturalistic view of science (be it psychologism or apriorism) and the view of science as based on agreement. "Agreement... puts... theory to the test" he says (106). "Coming to an agreement... is... to perform a purposeful action..." And, finally, he contrasts "a *justification* with a *decision* – a decision reached in accordance with a procedure governed by rules..." (109) – which contrast he clarifies by an analogy with a verdict by a jury. Perhaps the nearest he comes to institutionalism proper is where he stresses that language is a given for scientific inquiry (104).

Popper speaks explicitly of science as a social institution, or of the institutional aspect of science, espectially the conventions of testing as an institution, in his *Poverty of Historicism*, section 32. In a similar vein he speaks in his *The Open Society and Its Enemies* (ii, 218), of the "public character of scientific method", and of "the working of the various social institutions which have been designed to further scientific objectivity and criticism". Popper concludes (ii, 220), that "the individual scientist's impartiality is, so far as it exists, not the source but rather the result of this socially or institutionally organized objectivity of science."

(I wish here to express my great indebtedness to Sir Karl Popper's unpublished *Postcript*, to which I tend to ascribe a sociological view of science more distinct than in any of his published writings. But I check myself: though I have studied that work very closely, in manuscripts and in gallies, it is nearly two decades since I saw it last, an I do not trust my memory to such an extent on a matter of interpretation; in memory what is kept is usually interpretation, not the verbatim text, and this is liable to surreptitious change unless checked.)

Popper goes further. He attacks both psychologistic individualism and holistic or organicist collectivism – and both are versions of naturalism (*[Open Society]* i, 60; ii, 88 ff., 206 ff.).[2]

He contrasts this with what I shall call conventionalism, though he calls it legal or moral positivism, namely the theory according to which social rules and codes are merely arbitrary conventions (i, 68, 71; ii, 206). He offers instead his own view which I shall call modified conventionalism, though he calls it conventionalism or critical conventionalism (i, 60 ff; ii, 178 ff). According to this view, first, social institutions are conventions, including codes, rules, coordinations, etc.; second, they contain arbitrary elements as well as built-in errors; and third, conventions can be improved with the aid of criticism. On behalf of the third point Popper likens institutions to scientific hypotheses (i, 163; ii, 218). To conclude this picture, Popper does not claim that all social factors are comparable to scientific hypotheses. In his 'Towards a Rational Theory of Tradition' (*[Conjectures]*) he describes traditions as myths; and in his 'Back to the Presocratics' (in the same volume) he shows how mythology may transmute into science (or traditions into institutions) with the addition of a dash of definiteness and a dash of the critical attitude.

How far can we draw the comparison between science and society, between hypotheses and institutions? Surely, there is the (irreducible) dualism of facts and decisions: hypothesis is merely factual, whereas an institution has both a factual element and, more significantly, a conventional element.

This may be a restriction on the length to which Popper's analogy will go; and so I understand him to view matters. For my part, I would like to push the analogy all the way, for reasons to be outlined below. In my opinion a hypothesis which may be viewed as scientific is not merely one which answers certain desiderata and thus qualifies for the honorific status of science. On the whole, the idea of the honorific status of science, the implicit view that the task of the philosopher of science is to offer medals to scientists, appeals neither to Popper nor to his former students; but as for Bartley and myself at least, we feel that Popper still follows the tradition and demarcates science by a set of desiderata, which makes it difficult to dissociate Popper from those who give medals to scientists. In contradistinction, I prefer to follow Derek J. De Solla Price in considering hypotheses as scientific – for better and for worse – only if they are taken up by the world of science ([*Babylon*], 116, 124). Popper's view of science as the better, never the worse, hypotheses should at least be modified to say, the (better) hypothesis which answers certain desiderata is scientific only if, and to the extent that, the interpersonal conventions of critical examinations and testing apply to it. There are many crucial remarks in Popper's own works supporting this view. He does say in his *Logic of Scientific Discovery* (82–4) and elsewhere that scientific character is not a quality inherent in a proposition or a theory, but largely our attitude towards it, in particular our agreement not to protect it from refutations: when we decide to rescue a hypothesis from refutation it ceases to be scientific. The word 'we' in the previous sentence is somewhat ambiguous: it may be a euphemism for 'I', in which case the theory has a psychologistic tone: it may mean the scientists or the leading scientists, in wich case the theory sounds collectivistic; or it may mean the body of scientific tradition or institution, in which case the theory shows an institutionalist bias. There is no doubt, in view of Popper's staunchly institutionalist view of science, that a dash of the psychologistic and the full-blown institutionalist points of view must be included, and that the collectivist bias should not be added (but may appear – leading to what Price views as diseases of science). And so, I cannot see how Popper can escape the claim that for a hypothesis to be scientific it has to be critically entertained by the scientific community; a view which, I say again, I understand he wishes to reject.

There is another limitation on the extent to which we may draw the analogy between science and society. Science, according to Popper's views as expressed in his later period, aims at the truth as an ideal goal. Social institutions, inasmuch as they embed factual contentions and expectations, may share this characteristic. But whether social conventions *qua* decisions may aim at ideal goal analogous to the truth is an open question. This open question can, and perhaps should, be divided as to moral decisions and as to social conventions. Is there such a thing as the ideal moral code, the moral truth, final and unattainable? Is there such a thing as the ideal society, the kingdom of freedom, final and unattainable? Are these two ideals one? These questions about ideal or final ends are tantalizing. Some serious reviewers of Popper's *The Open Society and Its Enemies* have felt that it is most regrettable that Popper does not address himself to these questions of ideal or final ends. This may mean the desire to have these questions answered, coupled with the compliment to Popper as the man who is more likely than others to be able to satisfy this desire. It may also, however, be an implied criticism: Popper should have answered them because a philosopher should answer all questions (in principle), i.e., philosophers should build systems. This, of course, is an expression of the desire for an authority, but only after it has rationally proven itself to be acceptable – it is the ambivalence between the desire to remains in the closed society and the readiness to assume responsibility; Popper has discussed it at length in his *The Open Society and Its Enemies*. Perhaps, however, he should answer these questions about ideal or final ends because he has undertaken to discuss them, or accepted certain desiderata for his social philosophy which impose on him the obligation to answer these questions. However, the critics do not indicate which desiderata these may be.

Personally, I do not think Popper meant to answer these questions. At the same time we may want to extrapolate possible answers to them in order to better be able to criticize Popper's theory of science. Is Popper's theory of science criticizable? If not, should we endorse it, should we apply it and, if so, where; or should we strengthen it so as to render it criticizable? If it is criticizable, in what manner? This may hinge on the unanswered problems about ideal ends or final ends.

The dualism of facts and decisions is reflected in conventions or institutions: these are agreements of dual characters. We may agree about ends, and we may agree about means to execute them; in these cases the agreements are of decisions, different people come together to make similar decisions. But the question may be, what are the available means to choose from, or even what is the problem of choice. Here we have to form opinions; and again we have to do so in agreement in order to form a convention. It is obvious that we can criticize the factual element of a convention much in the way we can criticize any factual contention. Popper states ([*Logic*], 38, 83, 107) that there is a factual element in the conventions of science, namely the claim that following these may be fruitful in some definite sense of fruitfulness. Is this contention refutable, or is it only mildly criticizable? Or is it an article of faith? Our answer depends on our description of the conventions of science: the more informative and detailed any description is, the more likely it is to be refutable ([*Logic*] *passim*). For example, the conventions of science vary, but need not include an agreement on a *lingua franca*, be it Latin or English. The question of the advisability, given the fruitfulness and the dangers, of a *lingua franca*, is one eminently open to empirical investigations.

Somehow, to a reader familiar with Popper's exciting *Logic of Scientific Discovery*, the very mundane quality of my example may sound very strange and awkward. Philosophers of science seldom speak of a *lingua franca*, even when discussing the metaphysical tradition on the topic (from Leibniz to Frege, Russell, and Wittgenstein); because the topic sounds either too metaphysical for ordinary condiserations, or all too ordinary for even rather ordinary philosophers. Where, between the bluntly metaphysical and the bluntly empirical does the philosopher stand? What kinds of conventions does Popper discuss? Ordinary? Metaphysical? Say, ordinary; which institutions in Britain, in France, in the U.S., guarantee scientific criticisms? Which of these is more effective? Can they be improved? Are there direct incentives in these countries for criticisms? Does the Nobel Prize Committee consult a Popper-like handbook? For many years I puzzled as to whether Popper's statements concerning conventions are descriptive, in the way in which reports made by an anthropologist describe conventions in a given society (his or alien), or whether they are proposals, put in the way a member of a commonwealth (of learning) presents a proposed reform. Examinations of the relevant texts indicate to me that Popper is systematically ambiguous; I cannot say whether or not his view is descriptive or prescriptive. And so I cannot say whether or not the factual element in his view of the conventions of science is refutable, say, by historical examples.

Yet the problem becomes more difficult when we ask, should the aim of science be the search for the truth? And then, even when we agree about the ends, and the circumstances, and the most effective means, we can still raise further questions and ask whether the sole criterion for judging the organizations or the conventions of science is the alleged maximal

efficiency with which they allow us to approach these ends: we may have –
we do have – conflicting ends and competing institutions.

I think it is clear that Popper does assume that there are ideal ends, but
that he balks at the suggestion that there are ideal conventions to further
them. Perhaps the very intertwining, then, of science and society, or
rather of thought and action, absolves him from answering this question –
if it does not destroy this question itself: the science which will outline the
ideal society will be ideal knowledge itself (and the individual may be
ideal too), and thus will make the very need for institutions to safeguard
criticism and to test hypotheses (and individuals) quite unnecessary.

One could put it this way. Proper fallibilism is one which includes
human fallibility as an essential ingredient of any picture of mankind.
Therefore, any ideal which conflicts with this must be rejected: a properly
ideal humanity is not humanity. The ideal of final truths or of final ends,
or even of both, is not in conflict with fallibilism. The perfect method
cannot exist even as an ideal, because the ideal method ideally is the road
to perfection and therefore impossible.[3]

Thus, however vague and abstract our description of the conventions
of science, it cannot be the ideal or even part of it. Hence, if the conventions
are described too vaguely to be criticizable, there is no reason to assume
that we have captured the core of all reasonable descriptions of the conventions of science, the core which must be true. Rather, we may wish to
render the description more informative and thus more criticizable. Yet
we have to see to it that some intuitive demarcation is retained (after all
the increased informativeness), between the conventions of science in
general and, say, the rules and regulations of the Royal Society at its
foundation. This seems to me to be a strong desideratum. Those who
distrust intuition and want a reason for this desideratum, may find it in
the fact that the rules of the Royal Society were designed by conscientious
inductivists who hardly believed in rules and regulations in science (since
inductivism is naturalistic, of course), and in the following way. The conventions a philosopher of science describes, be he Poincaré, Duhem, or
Popper, are definitely not of the same type as the laws of the Royal
Society, or else the description is empirically refuted by historical evidence.
The conventions described by a philosopher, then, belong to a higher type
– like constitutions and canons.

I do not wish to decide here the criticizability and modes of criticizability of constitutions or canons. I only wish to draw attention to the fact that some constitutions and canons do allow for legal reform and even constitutional amendments, while some do not. And here, we can say, the constitution of science which Popper describes – or rather outlines – is the same as the constitution of the open society which he describes – or rather outlines – in the sense that both coincide with his doctrine of modified conventionalism. We cannot, therefore, deny that we observe the existence of a modified conventionalism in operation, much as Kant could not deny the fact of the operation in human beings of a moral sense and of moral enthusiasm. This is, indeed, how even ethics shifts from the domain of one's conscience to the domain of conventions ([*open Society*], Ch. 5): its canons of necessity pertain to society.

We see here the operation of modified conventionalism already noted in the previous section. The full force and novelty of this idea seems to me to stand out against the background of an almost universal rejection of the doctrine in the midst of a regular and daily adherence to it practice. In theory we have the old Greek tradition of polarizing nature and convention with its enormous impact on the whole Western tradition in social and legal philosophy, and in the philosophy of science, which still rests on polarizing nature and convention and on viewing nature as truth and convention as entirely arbitrary. This dichotomy, nature versus convention, still occurs both in social philosophy and in the philosophy of science, in the face of all the reforms which regularly occur in both society and science. This may be explained thus: attempts at a scientific philosophy start with canons of justification which impose the dichotomy. Attempts at a scientific social philosophy usually started from either a naturalist or a conventionalist philosophy of science, thus further imposing the same dichotomy on social philosophy (see my ['*Individualism*']). It seems to me fairly obvious that this last observation, though partly an interpretation of history and partly an application of Popper's ideas, is the transformation of an important philosophy into a new ideology, the idealogy of the New Renaissance. Its author is Popper's former student, W. W. Bartley, III. If my memory serves me right, there was a lecture, broadcast over the University of Illinois radio station in Urbana, Illinois, in 1963 and repeated there in 1964, in which Sir Karl expressed a similar appreciation of Bartley's broader view of non-justificationism. Unless Popper has retained his transcript or his tape, the lecture would seem to be lost.

To repeat, the broad idea offered here is both that justificationism is the moving spirit in the Western tradition (contrary to Popper's emphatic and repeated claim that it is the critical or skeptical approach), and that traditionally the dichotomy between nature and convention was the expression of justificationism. To repeat, both these components of Bartley's novel – and I do mean novel – philosophy are to be found in a way in Popper's writings, and even put with emphasis (but not sufficiently consistently). In particular, it was Popper who has shown that certain justificationist doctrines (though the term 'justificationist' is Bartley's) impose the dichotomy on us; that the dichotomy, nature *versus* convention, is the same as the true *versus* the false and arbitrary ([*Open Society*], i, 61; [*Conjectures*], 18); that for conventionalists the main thrust of their philosophy was their dissociation from naturalism; although they themselves often balked at the total arbitrariness they found themselves endorsing.

In moral philosophy, one of the constant and heroic aspects of Russell's studies, throughout his long career, was his confession that he could not jettison the dichotomy, but found each of its components highly objectionable, the one as too authoritarian, the other as too nihilist. Extreme conventionalism is advocated by quite a few legal philosophers, especially Hans Kelsen and his followers, all of whom find it necessary to qualify it somehow so as to escape total arbitrariness. Yet in legal philosophy it is H. L. A. Hart who first clearly broke away from the dichotomy, playing ethics against convention, and speaking of moralizing our laws in stages (thus employing a higher level convention to improve conventions). In the latest edition of the *Encyclopedia of the Social Sciences*, in the article on legislation, Benjamin Akzin still wonders at the degree of self-deception which legal philosophy engages in when justifying the law either on the basis of nature or on the basis of convention, but in the face of ever present reforms.

So much for the persistence of the dichotomy in social philosophy. In the philosophy of science the same phenomenon occurs. In his *The Value of Science*, Henri Poincaré finds in simplicity an anchor in non-arbitrariness. This has been criticized by Popper ([*Logic*], Section 46) who argues that for the conventionalist the preference for simplicity is itself merely a convention. "It is curious that conventionalists themselves have overlooked the conventional character of their own fundamental concept", he adds. "That they must have overlooked it is clear, for otherwise they would have noticed that their appeal to simplicity could never save them from arbitrariness, once they have chosen the way of arbitrary convention."

This criticism which Popper launches in his *Logic of Scientific Discovery* clinches the charge that, contrary to accepted desiderata, conventionalism presents science as utterly arbitrary. This criticism is echoed in Popper's *Open Society* (i, 237), where he draws the parallel between the philosophy of society and of science, and where, in the name of the dualism of facts and decisions, he dismisses conventionalism in science as monism, rather than dualism (since, if theoretical science is utterly arbitrary, it reflects only decisions and not views on facts). Popper is not very easy to follow on this, and his terminology is not too helpful. We may, perhaps, summarize his view thus. Naturalism, both in science and in society, is the monism of reducing all convention to truth or of disposing of them as falsehoods; conventionalism, both in science and in society, is seeing nothing but arbitrariness with no relation to truth or to facts or to nature. Popper recommends dualism both in science and in society. He accepts both a natural component and a conventional component as irreducible; he recommends criticism to improve the natural element; he recommends that this recommendation of criticism be incorporated in the conventional element (as a higher level convention), and he recommends that even the (higher level) conventions concerning criticism be open to criticism and improvement. But no improvement is ever the achievement of the ideal of final truth or of final ends. Convention, finally, has no anchor in the truth, even when the concept of truth is broadened to include moral truths; we cannot speak of conventions as false, merely as more or less adequate.

This, then, is as tight an amalgamation of *The Open Society* and *The Logic of Scientific Discovery* as I could achieve in my reading which I would expect Popper to endorse. Also, all this points at Popper's later work, his 'Three Views Concerning Human Knowledge', in that it considers institutions as mere instruments, but not of theories, not even of the factual elements embodied in institutions.

The question raised early in this essay is largely answered, then. The conventionalist, disregarding any factual elements in theoretical science, cannot but view theoretical science as purely an instrument. Consequently, for conventionalism the desideratum of a criterion of progress can only be satisfied by a pragmatic criterion. Both Poincaré and Duhem expressed their hope that the criterion of simplicity and of usefulness would coincide. This is either a pious hope or a theory. And so is Mach's psychological theory of simplicity as mental economy and an (implicit but quite widespread) economic theory of economy as the most useful mental commodity. Mach's psychological theory is patently false, and the economic theory could be refuted even before the day of the computer by showing that novelty, not economy, is the most highly prized mental commodity. Conventionalism, then, leads to instrumentalism, and to an obviously erroneous one at that.

Modified conventionalism, on the other hand, assuming the existence of a factual element (in science and in society), includes the assumption of an unattainable ideal of truth, which latter assumption is what has been labelled 'modified essentialism'. Modified conventionalism, then, is broader than, and includes, modified essentialism.

But, though I read all this in Popper, and though I would expect him to endorse it, I happen to know of his outright rejection of my reading; nor, on the other hand, can I overlook passages in Popper which conflict with my reading and which I find no reason to accept. Anyway, my concern is not with acceptance or rejection: I follow Faraday here, not to mention the gentle Rabbi Ḥelbo, in considering proselytizing as not the worthiest of occupations.

III. POPPER'S PROBLEMS OF DEMARCATION

The plural in the subtitle, 'Popper's problems of demarcation', is novel, and a serious contribution to the understanding of Popper's philosophy, made by his former student, W. W. Bartley, III (['*Demarcation*']). Bartley has noticed a point in Popper's philosophy indirectly already made use of in the present paper: a scientific theory, we remember, may be rendered unscientific if we decide to rescue it from refutations. As Popper notices, says Bartley, there are theories unscientific by virtue of their poor content, and there are those rescued by tempering with their content *ad hoc*, and those rescued by some inbuilt mechanism. These cases, says Bartley, though lumped together by Popper, differ greatly from each other. For my part I go further and declare (['*Flux*']) that some theories may turn out, contrary to initial expectations, to be refutable by default, some may turn out, again contrary to initial expectations, to be refutable due to unforeseen (and even extrascientific) developments. Even honest-to-goodness *bona fide* myths and superstitions, though usually as irrefutable as Popper says they should be, may turn out sometimes refutable and refuted, and sometimes refutable and rescued, and thus refute Popper's view.

I think, again with Bartley, that there is more to it than that. Let us lump all the various problems of demarcation mentioned above into one, and ignore, for example, the fact that changing (institutional) attitudes may render some or even all scientific theories unscientific. Still, the title 'the problem of demarcation', is somewhat misleading, considering that the expression is Popper's own and that he uses it from the start in two rather distinct senses, one historical and the other modern – his own sense. Usually this does not matter much, except when contrasting his view with those of his predecessors, and when attempting to decide whether Popper's criterion of demarcation legislates what one should view as scientific or whether it characterizes science in its social and historical setting. "In an as yet unpublished work" he says ([*Logic*], 55n), "I have tried to show that the problem of both the classical and the modern theory of knowledge (from Hume via Kant to Russell and Whitehead) can be traced back to the problem of demarcation, that is [*sic*], the problem of finding the criterion of the empirical character of science." There are at the very least two distinct problems here:

(a)   what makes science *empirical*?
(b)   what makes a theoretical system *scientific*?

Now, classically, the people Popper mentions, Hume, Kant, Russell, and Whitehead, answer question (b) by the answer:

(J):   scientific = certain or near-certain.

Now, (J) (for justificationism) is objectionable, as Popper explains; he replaces it by his new

(E):   scientific = empirical.

Of course (E) (for empiricism) of necessity reduces (b) to (a); but, for the tradition, (J) makes (b) and (a) possibly distinct! In other words from (b)

and (E), we can say, (a) in a sense follows; but not from (b) and (J): (J) answers (b) with no reference to (a).

Traditionally, particularly for the philosophers Popper mentions, a special status was allotted to question (b), what makes a theory scientific?, in view of the alleged obviousness of the answer to it. Indeed, this was said not to be a topic for serious consideration. (J) regularly was taken to be the answer to (b) *as a matter of course*, in spite of clear knowledge of the classical objections to (J) – which are unsurmountable. The philosophers Popper mentions have, each in his own way, tackled this situation, expressing the desire to be critically minded, while accepting (J) as a matter of course, in spite of unanswerable criticism of it. This, no doubt, is a tall order by any standard. Traditionally, the situation is even more complicated. Most philosophers of science, particularly at present, endorse two further assumptions:

(S)   a theory is certain or near-certain when and only when backed by evidence;

(C)   a theory is empirical when and only when backed by evidence.

(I ignore tautologies here, quite reluctantly, in line with the tradition on the topic.) Of course, from (S) (for sensationalism), (C) (for confirmation theory), and (J), we can deduce (E). Much ink has been spilled on these, rather small, complications, since (E), as proposed by Popper, is coupled not with (C) but with its opposite,

(R)   a theory is empirical only when it can be refuted by evidence.

In a compromising mood an embracing meaning was offered: 'empirical' should now mean, some measure of empirical decidability one way or another:

(D)   a theory is empirical if it can be backed *or* undermined by evidence.

After this change, (J) and (S), which most contemporary philosophers of science endorse, still easily entail (E). In other words, due to Popper's influence, philosophers of science who still stick to their views ((J) and (S)) now answer consistently two or three – indeed ten – problems of demarcation together:

(OMN)   scientific = (near) certain = empirical = confirmed = backed by experience.

The situation can easily be further complicated, and as it was complicated already in two ways – probability and meaning – when Popper
appeared on the scene. They have persisted, and Popper found it advisable
to attack them repeatedly.

The probability complication is the addition of one of the two following formulas, perhaps or both:

(P₁)   empirical backing = *high* probability
(P₂)   empirical backing = *increased* probability

where 'empirical backing' is backing by evidence as in (S) and where
probability is some (unspecified) additive function. (Similar formulas
can be written for undermining evidence rather than backing evidence.)
These two rules, (P₁) and (P₂), are clearly auxiliary hypotheses. Criticizing
them really amount to attacking only the periphery of the problems, and
I cannot imagine that Popper would have fallen for this digression but for
the fact that Carnap has endorsed (P₁) or (P₂).

How is one to choose between (P₁) and (P₂)? Let us accept (J) and (S),
and let us try to retain (E) as a corollary if we can. (J) tips us in favour of
(P₁), for (J) supports near-certainty which (P₁) identifies via (OMN) as
high probability. (S), however, speaking of empirical backing, would
obviously pull us the other way in favour of (P₂). The only cases where
(P₁) and (P₂) agree are those of hypotheses *a priori* improbable and *a posteriori* probable. To avoid the conflict between the choice of (P₁) and of
(P₂) one may restrict oneself to these cases (as, under Popper's pressure,
has been suggested by Bar Hillel and endorsed by Carnap). But this
restriction, other inconsistencies apart, entails Popper's demarcation:

(R)   empirical = refutable,

since, clearly, refutability is high initial improbability.

We are running in circles now, and are playing on different meanings
given to 'empirical', sliding back and forth from the classical (C) 'supportable by evidence' to Popper's (R) 'refutable by evidence' and back – via
(D) which is the inclusive disjunction of the two. To make this rather trite
exercise somewhat more sophisticated, let us add the other confusion
which was there before Popper entered the scene.

Let us consider the problem of empirical meaning and observe how
easily it effects a serious change in the problem of demarcation, and yet

does so quite incidentally. Whereas the negation of a (certainly) true statement is not true, the negation of a meaningful statement is yet meaningful. Hence, whereas for classical thinkers the negation of a scientific theory was decidedly not scientific, for the followers of Wittgenstein the negation of a scientific theory was decidedly scientific. Yet they refused to view refuted statements as scientific. (It is difficult to blame them for this, since old traditions die hard.) They therefore had to work with two concepts simultaneously, potentially-scientific and actually-scientific, to wit, verifiable and verified. This does not help: the negation of a verified existential statement is a universal statement that never was even verifiable. To allow for universal statements in science, verification was replaced by probability. Now, probability is understood here to be the degree of empirical support; yet probability is also understood here as the degree of chance or likelihood, and as such it comes as the degree of *a priori* probability, and as the degree of *a posteriori* probability. The *a posteriori* probability is in the light of some evidence, and the evidence is either potential or actual. Hence, the potential scientific character of a hypothesis may be its *a priori* probability, and it may be its *a posteriori* probability in the light of some potential evidence; so that potential and actual scientific character now have double meanings. The thesis that meaning equals probability, of course, raises again the confusion between (P_1) and (P_2), and even more strongly. It is not surprising that, whereas meaning and verifiability was discussed at length in the positivist literature, meaning and confirmation, or meaning and probability, are not: the survey which was supposed to do that, Carnap's *Testability and Meaning*, of 1936, only led to sidetracks. Carnap's own work on confirmation moved entirely away from meaning analysis, via a short and unexplained study of explication, to a probability theory of confirmation.

The confusions mentioned here, between testability and meaning, between probability and increased probability, between actual and potential increased probability, are all easy to cross-fertilize. The question, what attitude one takes towards a literature relating to one's interest when this is not up to one's standard is very tricky; concerning it there exists practically no literature. I consider Boyle's ruling, in which he proclaimed that he would ignore the publications of those chemists who would not describe their experiments in a sufficiently clear manner to render them repeatable, very wise, indeed. Even William Whewell's harsher ruling that he could not take the German philosophers seriously because they took Hegel seriously (meaning Hegel's *Naturphilosophie*, of course), I consider not unfounded. I find those philosophers who ignore Popper's criticism uninteresting; and I consider Popper's repeated attempts to appeal to their better and more critical selves equally futile. I confess even that I see little merit in my own attempts to bring Popper to engage in a public debate with his former students, myself included, and answer our criticisms of his works: this is my declared last shot.

This is not to belittle Popper's original work. It is something of a miracle – in my opinion for what it is worth – that Popper has emerged from the morass of the current philosophy of science so-called with a classical work like his *Logic of Scientific Discovery*. But I do not think he was left completely untouched by the morass. When Popper says that a refutable theory is scientific he may mean, first, that it is actually scientific: initially and in the light of any experience, corroborating or refuting; he may mean, second, that it is only potentially scientific until it be corroborated. Popper (but not his colleagues who engage in studying scientific meaning) can choose either reading; because, for Popper, as for the classical philosophers, the negation of a scientific theory is not scientific. As for myself, I had understood Popper to have meant my first reading rather than my second reading. I know now that that was mistaken: he really means to say, a theory is potentially-scientific if and only if it is refutable, and actually-scientific if and only as long as it is well-corroborated in the light of all existing evidence. In Popper's early works both my present readings are allowed; in 1960 he wrote a note at my suggestion ([*Conjectures*], 248 n), admitting that much, and further endorsing something akin to my present second reading.

Let me take up Popper's problem of demarcation following his own presentation ([*Conjectures*], Ch. 1). He had been puzzled, he tells us, by the status of certain theories which were topical at the time: Einstein's, Freud's, Marx's. Why is it, he asked, that Einstein's theory is scientific, but not Freud's? What makes the one but not the other scientific? What was fishy in Freud's claim to empirical backing?

In retrospect I find this problem somewhat embarrassing. Perhaps I should say I find that formulation embarrassing. It may mean – and I say this after having checked and rechecked the relevant texts much more closely than the normal standards require – any of the following problems.

Why should we honor theory E but not theory F? Why should we study and examine E but not F? Why should we accept the claim that E was backed by evidence but not F? Why should we accept the claim that E exhibits empirical nature but not F? Why should we allow ourselves to believe E but not F? Why must we believe E but we need not believe F? Why must we believe E and we must refrain from believing F (but either disbelieve it or suspend judgment on it)? Why should we act on the assumption that E is true but not on the assumption that F is true?

These few questions (the list is not exhausted; see above), each of which fits Popper's text well enough, may look like variants of each other – but only to the untutored eye. Let me sketch only two differences, those which happen to be substantial. First, belief and grounds for action are not the same: as Popper stresses in his discussion of instrumentalism in his 'Three Views', we knowingly *employ* false theories in technology, but we do not knowingly *believe* a false theory. The second difference is between empirical *theory* and empirical *backing*. The confusion between the demarcations of these two is the one which eluded me and troubled me for a few years. In his early autobiographical sketch ([*Conjectures*], Chapter 1) Popper discusses both the nature of empirical backing and the empirical nature of a theory. He may have suggested that an empirical theory is a refutable theory and an empirical theory may be refuted or alternatively empirically backed. He may have suggested that a theory is potentially empirical if it were refutable and actually empirical if it has been properly empirically backed. To repeat, I think Popper has acknowledged that earlier he had been unclear on the topic, and that he now considers a scientific theory proper only one which is properly empirically backed before it is empirically refuted. On top of this past confusion of mine, for a long time I found it difficult to notice, what now seems clear, that in his autobiographical sketch Popper says that Freud is in error both because Freud's theory is unempirical and because this theory cannot have been properly backed by evidence. First, says Popper, a theory is empirical only if it is refutable; second, an empirical support proper can result only from a proper test (i.e., the attempt at a refutation); third, as a corollary, only an empirical theory can be properly empirically supported; fourth, and finally, a potentially scientific theory needs some empirical support to become actually scientific.

It looks obvious to me now that this is what Popper has been saying all along. I was a devoted student of his and worked closely with him for over seven years, yet I really understood him only a few years later. I am not clear as to whether hermeneutics is ever a worthwhile activity: I liked Popper's ideas better when I understood them less. In addition to the various criticisms already alluded to, I wish to say that proper refutations and proper corroborations of both scientific and unscientific theories are quite possible. It is easy to illustrate cases of proper and improper backing and refutations of scientific and of unscientific theories, thus having (at least) 4 kinds of backing and 4 kinds of refutations.

theory test	scientific	unscientific
scientific	Michelson, Eddington	Paré, Jenner all tests in today's aerodynamics
unscientific	practically all demonstrations in classrooms and in commercials	practically all testing of oracles, diviners, etc.

The cases mentioned in this table are those of tests which ended in empirical backing and in empirical refutations. Both Paré and Jenner were scientifically testing known superstitions: Jenner found empirical backing to one, and Paré refuted many. Contemporary aerodynamics is classical and so, *qua* scientific theory, is plainly refuted; but, *qua* rational technology its forecasts are properly tested – with supervision of well known government agencies.

I do not wish to decry Popper's study of the problem of demarcation of science. I join Bartley in viewing it as an important problem which Popper's studies have rendered unimportant (see Appendix). Also, I see more value in Popper's equation of the (actually) empirical with the (potentially) refutable; though I think he disowns this very equation. What I view as the more important and most original contribution of Popper's to the philosophy of science is his solution to the problem of induction: *we gain theoretical knowledge from experience by refuting some of our theories.* (But, I equate neither refutation nor empirical character with science. See my [*Novelty*] and my [*Nature*].) Unfortunately, Popper himself, far from seeing that the problem of demarcation of science is already a conglomerate of problems, has further collapsed it with the problem of induction into one. Thus, he says, ([*Conjectures*], 52), " it took me a few years to notice that the two problems – of demarcation and of induction – were in a sense one". No doubt, in some sense any two problems are one; and so, in a sense, this statement is true. But Popper himself paraphrases this same statement (two pages later) thus: "the problem of induction is only an *instance* or facet of the problem of demarcation", which I consider patently false (see Bartley's ['*Demarcation*'], 64). I do not know how literally to take this. Elsewhere ([*Logic*], 42), he says, "the proposed criterion of demarcation also leads us to a solution of Hume's problem" – which, taken literally, may well be true. Similarly, he is cautious when he says (*ibid.*, 313), "the solution suggested here has the advantage of preparing the way also for a solution of the second and more fundamental of the two problems of the theory of knowledge (or of the theory of the empirical method)." Nonetheless, in my view, the first problem, of demarcation, is concerning *knowledge*, and the second, of induction, is concerning *method*. Popper often seems to use the terms 'methodology' and 'epistemology' almost interchangeably. This is regrettable. So long as Popper indentifies the two problems, of insuction and of demarcation, and thus, the two fields, methodology and epistemology (see Wellmer, [*Methodologie*]), the theory of learning and the theory of knowledge, he does not add to clarity. In particular, he forces us thus to identify conventionalism and instrumentalism and thus come to the very incongruity with which we started: modified instrumentalism is trite and modified conventionalism intriguing; yet they modify the same doctrine in the same way!

There are a number of questions concerning Popper's early theory of demarcation and its application to the gradualist or approximationist approach as endorsed in his later 'Three Views' and as even adumbrated at the very end of his *Logic of Scientific Discovery*. Popper assumes the following: if we take the most highly testable theory available, corroborate it, refute it; if we repeat this process of conjectures *and corroborations* and refutations, then we shall find a series of theories approximating truth. This sounds like a miracle, but is partially explained by Popper when he declares that explanatory power, degree of generality, and content, all increase with degrees of refutability. I have refuted these claims (['*Nature*']); and on this score I have suggested that harmony can be achieved better by narrowing Popper's problem of demarcation to that of empirical character, not to scientific one; and by demarcating science itself by its empirical nature as well as by its explanatory power, generality, abstractness, depth, etc.

Meanwhile I have gone further. I now find it hardly possible to characterize science even by a large set of characteristics. I can only view science as a tradition where such and such characteristics are manifest to this or that degree and even in conflict (see Appendix to Chapter 8). And, inasmuch as I endorse Popper's view of traditions as akin to myths, I cannot avoid seeing science as largely a myth. Scientism, which Popper rightly views as a myth akin to some religious myth, is rather obviously part and parcel of the Western scientific tradition, and Popper is mistaken in denying this. Consequently, it is with some surprise that I find that, though I have gone further away from Popper's teaching towards modified conventionalism as broader than his modified essentialism, my modified view is in spirit though not to the letter, more Popperian than Popper's declared view. Or should I say Bartleyan? This is a rather ironic point of departure. Naturally, it seems inevitable that at a sufficiently large distance of time wide differences of opinions vanish and historians who notice them at all wonder what the vehement disputes of the past were all about; but it is a strange sensation when it occurs to one's own colleagues and associates. But let those historians, if any, who pay any attention to details smaller than Popper's broad ideas, let them not conclude that the Popper-Bartley vanishing controversy did not feel any the less real, that by-standers did not feel caught up in its heat.

IV. THE THREE VIEWS CONCERNING HUMAN
KNOWLEDGE REVISITED

I apologize for my pedantry in advance. I shall put these matters in a table of 3×3: I shall have to present the combination of instrumentalism and conventionalism under one label – I call it the active view. Along with it I introduce two other new labels, all of which I shall presently explain. The columns present the contemplative view of science, the active view, and the new inquisitive view. The rows present the aim of science, long range or short range, dealing with the problem of its regulative principle or rationale; the method of science, dealing with the problem of induction; and the status of science, dealing with the problem of the demarcation of science. Popper's identification of the two last problems, then, is here overruled. Let me, following tradition, name the three rows 'rationality', 'methodology', and 'epistemology'.

	Contemplative	Active	Inquisitive
Rationality	True and ultimate explanation; ESSENTIALISM	Pragmatic truth or utility INSTRUMENTALISM	As in ESSENTIALISM
Methodology	Intuition of essences; methodological essentialism; divided into inductivism and apriorism; PASSIVISM	Trial and error; ACTIVISM	As in ACTIVISM
Epistemology	Essential definitions; final truth; ultimate explanations; ESSENTIALISM (Truth by nature)	Nominal definitions; divided into fictionalism and tautologism; CONVENTIONALISM (Truth by convention)	Tentative expanations, refuted and succeeded by better tentative ones; GRADUALISM: APPROXIMATIONISM (Proximity to truth) (Verisimilitude)

I hope I am allowed the indulgence of explaining my choice of terminology. My terms 'contemplative' and 'active' are self-explanatory and are due to Russell, I think, who used them throughout his career in the sense adopted here. But, actually, the terms have a long tradition, and go back to the seventeenth century. My term 'inquisitive' alludes to the original sense of 'skepticism', to Popper's *Logik der Forschung*, and to Jorge Luis Borges' *Otras Inquisiciones*. Popper prefers 'growth theory'. This is rather question-begging, since it remains to be seen whether a theory complies with the *desideratum* that it be a growth theory. For the terms 'methodology' and 'epistemology', their etymology and quite traditional meaning, I would refer the inquisitive reader to J. M. Bocheński ([*Methods*], 9–10). Of course, one can collapse epistemology to methodology by the theory that knowledge is justified by the method of its attainment. (See Popper's 'On the Sources of Knowledge and Ignorance' [*Conjectu-*

res], and my [*Historiography*].) But collapsing two fields and confusing them are different things.

(Historically it has been tacitly assumed that the problems of epistemology and methodology were intertwined in that a solution to one is almost *eo ipso* a solution to the other: if we know how we learn, then we know that we know something; and if we know how we know, then we know that we have learned something. Consequently, when studies in one field proved too rough, it looked more promising to switch to the other field. All to no avail, of course; except that the unexplained shifts were confusing and the assumptions behind them tacit. One tacit assumption was, of course, that learning means the acquisition of some knowledge, just as improving one's financial standing is the acquisition of some cash or credit. Now, no doubt, in coming closer to his target the fund raiser has raised some funds; yet the traveller comes nearer to his destination without partially being there. That learning is the acquisition of items of knowledge need not be assumed; and gradualism is the rejection of this assumption.)

My terms 'essentialism' and 'instrumentalism' come from Popper's 'Three Views', where they are used to designate views concerning the rationality or aim of science. (I prefer 'rationality' because science, as an institution, cannot, strictly speaking, have an aim but only be given an aim and thus have a rationality; see my ['*Individualism*'].)

The main disruptiveness in the table has to do with the fact that rationality may be short term or long term: the rationality of the inquisitive view is the same as that of the contemplative view only in the long run. The short-term rationality of the contemplative view is not the truth but approaching it. In his *Logic of Scientific Discovery* Popper denounces long-term ends and stresses the short-term one; in his 'Three Views' he does the opposite. Yet this is no real incongruity, it being only a shift in emphasis; such shifts are unavoidable with the change in the problem to be tackled or the situation in which it is tackled.

My term 'passivism' is new, and is not meant as an insult in any way whatever, as an examination of the passivist literature may show. Here, for example, is Bocheński's description (*op. cit.*, 19) of objectivity as reflected in Husserl's phenomenological method of intuiting essences, but which also is "an essential constituent of Western scientific method": "... the investigator... must exclude everything that comes from himself, from the subject, above all his own feelings, desires, personal attitudes, etc. What is required of him is a detached observation of the object.... The reseacher who acts in accord with this rule is a pure knowing essence, one who forgets himself completely... the rule of objectivity requires a contemplative attitude, i.e., the exclusion of utilitarian considerations..."

It is perhaps the most incredible tribute to Kant that, due to his activist apriorism, the literature has completely overlooked the fact that Descartes was an avowed and strict passivist. I owe this point to a most moving paper by Ben Scharfstein ([*Dream*]), and I can only refer the reader to that paper.

Of all the other terms, only two are new to any extent, 'gradualism' and 'approximationism'; I regret not remembering where I first met them. Perhaps they are Popper's; though I doubt it. I do not think, in particular, that the doctrines the terms designate are novel either (see my [*Novelty*]).

So much for terminology. As to the table itself, I wish to draw attention to its incompleteness, which is the same as the incompleteness of Popper's scheme in his 'Three Views'. Pointing out the incompleteness of a given schema may help orientation: it is no criticism of the schema which is not claimed to be complete. It may be useful in some cases to try and complete a schema and effect all possible permutations, etc. (see I. C. Jarvie, [*Revolution*], Ch. 3). This, however, is beyond my present purpose. To the limitations of my scheme, then.

First, Kant does not fit here at all. For, he explicitly refrained from judging the truth of his system and yet his claim for universal validity for it as higher than a claim for pragmatic truth alone. And certainly he was not a fictionalist; though an activist, he had no room for trial and error. (See [*Conjectures*], and my [*Unity*].)

Second, William Whewell does not fit here either. He fits the contemplative view of the aim of science and of its status; but for the method of science he chose the active view. This is why he is so alluring, I suppose. But the allure is uncritical to such a degree that one has to appeal to the success of Newtonianism to explain it.

Third and final exception, the majority of contemporary philosophers of science, the so-called inductive school of probability, may be viewed as an alternative modified essentialism – indeed, unlike Popper's modified essentialism it cannot consistently be incorporated within a modified conventionalism: what Popper calls "a liberal Utopia", ([*Conjectures*], 351) is what Carl L. Becker described as *The Heavenly City of the 18th Century Philosopher*, a collection of individuals governed by nothing but enlightened self-interest and free of all tradition, convention and prejudice. I think the only place where this rather charming and incredibly naive aspiration is still – inadvertently – preserved is in the literature on probability and induction. It is, indeed, this charming dream of rationality that may explain the allure of an otherwise so depressing a doctrine as one which is both utterly passivist and yet is not essentialist. (Indeed, when attacking essentialism, its adherents sometimes cross the borderline to become instrumentalists; see my ['*Flux*'].)

With these exceptions out of the way, let us glance at my table and see what it illustrates. The contemplative and the active view share the idea of scientific certitude. They differ as to the source of scientific knowledge: the contemplative view is that the source is nature, the active view is that the source is the human mind. Hence the contemplative view is the view that scientific growth is entirely a passive contemplation of nature which is the intuition of true essences (methodological essentialism). It is this passivity, shared by both the inductivist and the (pre-Kantian) apriorist, which allegedly guarantees the truth of our theories (epistemological essentialism).

Fictionalism or instrumentalism became taboo since the foundation of the Royal Society, on the authority of the fathers of modern physics and of Sir Francis Bacon. Yet, Descartes already was willing to fall back on fictionalism or instrumentalism if this proved the only way to claim scientific status (see Sabra, [*Light*], 23ff.), perhaps following even Galileo and Torricelli in this respect (see Rossi, [*Bacon*], 221). Even the arch-skeptic and popular Joseph Glanvill took the same view ([*Vanity*], 211–212). The Cartesians graciously ascribed to the Newtonian philosophy mathematical but not physical certitude (Koyré, [*Newtonian*], Ch. 3); and the Newtonians, led by Helmholtz (Duhem, [*Aim*], 99ff), in like manner viewed fields of force as a useful fiction. Poincaré (*Hypothesis*], Ch. 4) and Duhem (*op. cit.*, Pt. II, Ch. 5) developed the view of science as useful fictions or empty definitions – recommending the latter as the better alternative. Later Eddington thought fictions were inescapable but strictly private, and empty definition was objective science ([*Nature*], 319–321).

Poincaré stressed the conventional aspect of scientific theory, its being a set of empty definitions. Duhem stressed the instrumental aspect of it, the use of these definitions for practical purposes. He had an ulterior motive here, since his conservatism was not Newtonian but Aristotelian and nourished by his Roman Catholicism; he was, therefore, proud, he said, to find as an ally in adopting instrumentalism a person like Mach (333).

And so, by a switch, views of science deteriorated. First, my science was certainty-by-nature and true, whereas and your science was, as a consolation prize, certainty-by-convention and merely useful. Soon, all science was certainty-by-convention and merely useful, and certainty-by-nature became the prerogative of obscure metaphysics utterly divorced of all relation to science. (See Appendix to Chapter 8.) Men of science who view all science as instrumental and definitional find themselves with a vacuum when it comes to ideas on the nature of things.

The inquisitive approach has ancient ancestry, of course, in the Socratic philosophy. Yet it is the most recent approach to science, for which we are indebted to Popper. (See my ['Novelty'].) Its chief characteristic is the idea that a given scientific theory, say Newton's mechanics, is a description of nature, though not always the best we have, and even the best we have need not be the last word. Even those who claimed to hold the Socratic approach to science could not usually stick to that view because of a strange feature of the situation. On the one hand, as we contemplate the application of the inquisitive approach to science, we face at once the problem of demarcation of science: if scientific theory is not certain or quasi-certain, what differentiates a good theory from a bad one? It is thus not surprising that Popper has stressed the problem of demarcation and its significance throughout his philosophical career. On the other hand, the more successful he was in his application of the inquisitive approach to science, the less important this problem turned out to be; quite apart from the fact that Popper's initial solution has turned out to be either inadequate or false. In retrospect we may perhaps say, one theory is preferable to another when and only when it is a better approximation to the truth. One may object to this as a very narrow sense of preferability, as I have shown by counter-example. This may be remedied by the introduction of a theory of degrees of refutability. Popper himself has found it necessary to modify his theory of degrees of refutability: and he did so by relativizing them against some background knowledge; this cannot be viewed as more than a progress report: we do not yet have a sufficient theory of background knowledge (see my ['*Background*']). I have tried to offer elsewhere (['*Nature*']) examples illustrating cases where barely testable hypotheses are preferred to highly testable ones in some intuitive sense of 'testable' perfectly acceptable to Popper. But, I agree, the situation in its actual history or in the abstract still leaves too many questions not fully answered: I suppose we are all at sea, and may benefit from stopping to discuss old solutions to the problem of demarcation or offer new ones and ask, instead, is the problem pressing enough to make us continue investigating it?

But if we proceed, I suggest a much broader sweep and a set of problems to be arranged first. And we may study the problems within Popper's philosophy or in a comparative study. Take Popper's philosophy first.

The status of science is still not clear in Popper's theory of science. What status has a refuted theory? An untested refutable theory? A tested unrefuted theory? A rescued theory? All these are open matters, or ones viewed as open by Bartley and myself, especially in view of the fact that the major questions of traditional philosophy are answered by Popper without going into the question of status. This is the main point which my table illustrates.

We may speak of status, however, merely by contrasting Popper's view with the other two schools replacing the status of certitude with that of tentativity. This excellent move, however, is clearly not a demarcation proper: obviously, all unrefuted synthetic propositions are tentative, be they scientific or not. Dealing with the problem of demarcation proper, we may do it more historically – interpreting the history of science with the help of the gradualist view of the rationale of science. Then some degree of approximation to the truth – of verisimilitude – may perhaps be defined to demarcate science proper, and, again, we have only progress-reports here. This may perhaps be done after the incorporation of additional desiderata – which means changing our view of the rationale of science, in its actual history or in the abstract. Be our choice here what it may, I feel, we may do better to explore at first only the desiderata of degrees of explanatory power – a modification of Leibniz's view to include refutations – and see if it cannot suffice. It should follow from a good theory of verisimilitude that Newton's theory, though false, is more verisimilar than Kepler's. I suppose this will not suffice, but I cannot say. If it will not, then, obviously, our view of either rationality or methodology will have to be altered.

Popper's attitude here is still not clear to me. In his early work he repeatedly claimed that high degree of refutability is the only desideratum we need insist on. Not that he was unaware of other desiderata, like explanatory power, generality, content, simplicity, but that, presumably, he viewed them all as taken care of by refutability. He says, for example ([*Conjectures*], 135n), "The reason why I consider the argumentative and the explanatory functions as identical... are derived from a logical analysis". In other cases he claims not identity but weaker kinds of reducibility, such as monotony. For example, he distinctly presents explanatory power, content, and confirmability as different; but he derives from his definitions of these functions that they increase together with the increase of improbability. I find all this of great heuristic value, but as satisfactorily refuted by a number of instances I have provided ([*Nature*']). (I need not say the 'logical analysis' Popper speaks of is erroneous; indeed the error is not difficult to spot: Popper's analysis is of a partially ordered set, and rests on ordering as if it were not partial. A theory which entails another both explains more and is more refutable; Weyl's theory explains both Einstein's and Maxwell's, and is at least as refutable as either. The Bohr-Kramers-Slater theory was refuted without having explained a single phenomenon, much as Thales theory was. A highly *ad hoc* explanation, on the other hand, explains what it was set to explain and is quite possibly exhausted thereby, and is thus irrefutable. We may answer this last ponit by claiming, the explicanda are refutable, and so their explicans is refutable too. Taking this reply seriously we will conclude that no *ad hoc* theories are possible, only partial *ad hoc*ness exists. No doubt, then, Weyl's theory is (partially) *ad hoc* in this sense and hardly refutable, yet highly explanatory all the same; whereas Einstein's theory is less explanatory and also less *ad hoc*. Whichever way you look at it, Popper's alleged logical analysis is shoddy: even great analytic minds may falter on a rare occasion.) I consider the desiderata mentioned here quite significant, and perhaps the inquisitive view as described in my third column is already obsolete.

But there are other desiderata to be considered as well, and they may all define other kinds of rationality plus status; credibility, if one likes (I do not), usefulness, challenge, depth (this I like very much).

As to the ideas of challenge and of usefulness, they have not yet been attacked, and we do not know what incentive they constitute for what action, what is conductive to challengeability, or to use, under what conditions a challenge is taken, especially of possible usefulness of an innovation, etc. It is amazing how many obviously challenging problems are as yet unstudied. As to the idea of depth, there are hints about it in the literature, but nothing substantive or systematic. Popper claims depth to be undefinable. In a trivial sense this is true of all concepts; yet we can always try. I see no reason why we should leave this magnificent concept, and the allied one of enlightenment, unexplored. Popper's view of credibility is not always very clear. He says, ([*Logic*], 414–5), "…while it is a mistake to think that probability may be interpreted as a measure of the rationality of our beliefs, … degree of corroboration may be so interpreted… It must not be interpreted… as a degree of the rationality of our belief in the *truth of h*; indeed we know that… [the degree of corroboration of h is zero] whenever h is logically true. Rather it is a measure of the rationality of accepting, tentatively, a problematic guess…" I do not know what this may possibly mean. If when saying 'accepting' Popper means 'apply', then, as Popper himself says in his 'Three Views' ([*Conjectures*], 111–113), we do not hesitate to apply a theory which is not 'a problematic guess', but which we know to be false. If when saying 'accepting' he means 'believe', then I do not see why belief should be rational in this sense: Popper himself confesses (e.g. [*Logic*], 38) his holding metaphysical beliefs which are, of course, uncorroborable, yet held quite rationally, of course. So, what on earth does he mean by 'accept'?

Anyway, his theory of corroboration is supposed to have a wider scope than mere acceptability. I wish, however, to repeat that it is difficult to place the rôle of empirical backing or corroboration – in Popper's sense of failed refutation – here (see my ['Corroboration']). This is particularly so since a scientific theory may be corroborated scientifically – through a genuine test – or not, just as an unscientific theory may be tested scientifically. For example, a superstituon may be tested and refuted or corroborated scientifically. So can a theory not of any explanatory yet of technological import, – a theory solving a technological not a scientific problem

– be corroborated scientifically or not (we have scientific technology, pre, pseudo, etc.). For the sake of, and in the name of, social responsibility, a theory must be corroborated before it be institutionally implemented, as I have argued elsewhere (['*Confusion*', II]). Viewing scientific theory as a contemplative institution – textbook – or as technology of sorts – educational or otherwise – may resolve some of the difficulty. The difficulty, to repeat, is that the institutions fostering science are of a very different type from the institutional, or accepted, scientific 'received' opinion. One of the interesting questions which are today on the agenda of quite a few students of the topic (see my ['*Kuhn*']) is, does the changing textbook of science through the ages cover the field of the history of science? Suppose the answer is affirmative. Query: does the science textbook only endorse corroborated theories? Suppose the answer is, yes. I will then have to modify my view of the role of corroboration in Popper's philosophy of science, alter my view to one between my present one and Popper's present one, and cease to see Popper's theory as strictly a theory of conjectures and refutations. This, of course, purely on the descriptive level of discourse – studying science in its traditions. As to the prescriptive side, I may, and under the pressure of the theory of the importance of corroborations in science I would, then, call for a reform of the constitutions of science, in the hope of its moving further towards a more open and free inquiry, towards a more problem-oriented and problem-solving attitude, where solving a problem interestingly will merit highly, regardless of whether the new solution be refuted in the first test or in the second.

How shall we go about this difference in viewing science and its tradition? Do we decide to reform our view of the tradition or do we decide to reform the tradition itself? The first move in the direction of answering this question may well be the move in the direction of explaining corroborations, and in particular of their past importance in the scientific tradition. It is easy to explain their importance in technology (see my ['*Confusion*', II]). It is possible to show that corroborating evidence is at times important in science, though not as such. In reply to my claim that in science corroboration as such does not signify, Popper claims that historically corroboration meant the difference between the survival or perishing of modern science. I do not deny that. The novas of the late sixteenth century have played a similar rôle, yet we do not believe in them:

we try to explain them. Suppose we explain the rôle corroboration played in
the rise of science as the result of its competition with religion as a justification. In this case we may see in Popper the last vestige of justificationism.
Suppose, however, that without corroboration science as we know it may
perish; at least as an institution. Why? Will this lead to a new intellectual
phenomenon superior to science? Or will it be a disaster? If a measure of
greatness of a philosophy can be given, it should be its hinting beyond
itself. And I find Popper's philosophy admirable in this respect.

APPENDIX:

BARTLEY'S CRITIQUE OF POPPER

In the present chapter I rely heavily on Bartley's paper (['*Demarcation*']),
and also on his previous work ([*Retreat*]). I have not referred, however, to
Bartley's own chief departure from Popper's philosophy, his comprehensively critical rationalism which he wishes to replace Popper's critical
rationalism. On this point I am much less in agreement with Bartley than
with Popper; and I prefer an even more skeptical attitude than either.
Nevertheless, I think Bartley's effort to push a theory of rationality to its
limit is fresh and interesting and yields many new and interesting results.
One of these is Bartley's thesis that the problem or problems of demarcation are facets of the problem of rationality. In brief, he says, once we
prejudge issues and identify rationality with being scientific, the problem of
demarcating science becomes urgent; but it is preferable to pay less uncritical respect to science, i.e., to respect science to the extent that it is
rational, and to dwell on the problem of the demarcation of rationality
instead. This criticism is not very upsetting, but worthy of notice all the
same, I should have thought.

Perhaps I should refer to the critical comments on Bartley's paper which
were originally published together with it, as far as the thesis from which
I have just paraphrased it is concerned. These are by J. O. Wisdom, J.
Giedymin, A. E. Musgrave, and Sir Karl Popper himself. Wisdom agrees
with Bartley. He says, "one can draw a demarcation line very near to
where Popper drew it, not exactly in the same place, though in spirit the
same"; he also says the problem of the demarcation of science is important because it relates to the problem of rationality (66). Giedymin defends Popper. He says, it all depends on whether one accepts Popper's

statement, "nontestable systems of statements are of no interest to empirical scientists" ([*Conjectures*], 257; quoted by Bartley, ['*Demarcation*'], 54), questions Popper's concept of 'interest' and relates this to Popper's concept of 'accept', questions this concept as well, and then proceeds to analyze only the latter concept (70). Musgrave does not comment on the point at issue but registers disagreement (83, line 3 from bottom). Popper repeats his demarcation of science, and his claim that the problem of demarcation of science is very important. It is very important both historically and philosophically, he says. Bartley says, philosophically the problem of rationality is now (thanks to Popper) more important than the problem of demarcation of science. Popper is doubtful; considering the problem, which of these two problems is more important, he says he doubts that this problem "will become a very important philosophical problem; if for no other reason, because the two ... are historically extremely closely interrelated" (95). True, but historically all too many empirical scientists declared themselves not interested in unempirical statements, and by unempirical they all too often meant unverifiable. They were in error about verification as Popper claims, and they were also in error about the value of metaphysics for research, as Whewell claims. Will Popper endorse the statement Giedymin quotes from his *Conjectures*? If he will, his view is narrower than even the one he has expressed in his *Logic of Scientific Discovery* of 1935. If he will not, then, surely, he can see why Bartley is not very happy with his insistence on old problems while holding newer and more liberal views – even if he does not share Bartley's feelings. Finally, Bartley replies to Giedymin's comment: Bartley had said that due to Popper's studies the problem of demarcation of science is now less significant than Popper had initially claimed; Giedymin quotes Bartley erroneously to assert the grossly erroneous view that the problem of demarcation of science is insignificant *tout court* (104).

The rest of the fifty odd pages of comments on Bartley are devoted to other points of Bartley's important paper and to Popper's important 'Back to the Presocratics'. As Popper says in his comment on Bartley, all criticism is of some value; let me apply this maxim to the criticism (quoted in the last paragraph) of Bartley's thesis (64): "The later development of Popper's thought, and the generalization and application of his ideas outside science, have rendered his discussion of demarcation obsolete". In this thesis, the word 'obsolete' should be qualified by, "though decidedly not in retrospect and not very significantly for those who insist that 'untestable statements are of no interest to empirical scientists'". I have not checked, however, whether Bartley would agree with me on this qualification.

But I am being much too finicky. My main point is that Bartley's chief thesis, just quoted, seems to me to be of the best kind of compliment any philosopher can receive, particularly one who advocates the critical approach. Popper, however, declares Bartley's criticism of very small value and says it "came as a shock" to him (['Remarks'], 89). A few pages later (97) Popper says, "I am always grateful for any criticism". His chief thesis in his 'Remarks' (101) is a conjecture concerning the reasons for "Bartley's change of mind". A conjecture which Popper says explains "to some extent" some of Bartley's criticism. His reason for offering the conjecture, or perhaps for having the problem that it comes to solve, is this: "(By eliciting this conjecture, these criticisms may be regarded as having some intellectual value.)" I wonder, pondering this bracketted statement, not only about its own possible meaning, but also about the possible meaning of intellectual activity in general, and of intellectual value in particular.

The problem of demarcation of science, the problem of demarcation of rationality, and, for my money, the problem of demarcation of intellectual interest and of the examined life which may perhaps be worth while – this kind of activity has an extremely dangerous allure, in that the solutions it may lead to, if true, may look somewhat dictatorial. And we all think the views we hold are true or else we would not hold them. And so, I feel, first and foremost, interesting as these problems are, one prerequisite for their study, as a safety measure, may be the attention to the fact that we study them on account of our finding them interesting – that by the same token anyone may study any problem which he may find interesting, regardless of whether it fits professor X's demarcation of science or professor Y's demarcation of rationality, not to mention professor Z's demarcation of interest. I say 'safety measure' with some chagrin: When Popper and Bartley speak of importance, they have something interesting in mind yet sound a trifle banal – at least to each other. Popper's 'Remarks' center on the intellectual value of criticism. For me glamorous intellectual values have their honourable place in a broader context of spiritual values which likewise includes such unglamorous items as human understanding and concern. Perhaps these values, the intellectual and the plain, must work in an interdependent fashion – for reasons I have ventured to indicate in the body of this chapter. Let this stand as my tribute to my former teacher and as my final challenge to him. My challenge, to repeat, is that he try to consider Bartley's criticism in the manner described in his theory of the critical tradition, i.e. read it in its most forcecul reading and take it as a tribute.[4]

NOTES

[1] The motto is from p. 137 of Vaihinger's [*As If*]. (Square brackets refer to the bibliography at the end of this essay.) This essay was intended for the Schilpp Popper volume but withdrawn and replaced with a three page extract at Popper's request.

[2] Popper makes another dichotomy within naturalistic epistemology – between optimists and pessimists ([*Conjectures*], 6, 11). The optimist trend leads to liberal utopianism; the pessimist trend allows for true knowledge only by authority, this leading to authoritarian politics (for example Plato's *Laws*). It seems to me clear that the proper inquisitive approach rests on a pessimistic or distrustful view combined with a restless quest or curiosity.

I suppose that this is a highly individualist philosophy which goes back to the mediaeval skeptics, who were all highly individualistic and even loners. It is no accident that they seldom studied problems of social philosophy. See my ['Unity'], especially notes there. See also Scholem's [*Kabbalah*], especially Chapter 2, for the conservatism cum radicalism of the Kabbalist.

[3] A little contemplation might raise alarm as to the consistency of these remarks; they look alarmingly paradoxical. I need not say that it is impossible to judge the consistency of the view adumbrated here in its present state of fluidity and sketchiness.

[4] I have been disappointed on this matter already. See note 1 above.

BIBLIOGRAPHY

Agassi, Joseph, 'The Confusion between Science and Metaphysics in Standard Histories of Science' ['*Confusion*' I], in H. Geurlac [*Ithaca*]. Reprinted here.

Agassi, Joseph, 'The Confusion between Science and Technology in Standard Philosophies of Science' (['*Confusion*' II]) *Technology and Culture* 7 (1966). Reprinted here.

Agassi, Joseph, 'Changing our Background Knowledge' ['*Background*'], *Synthese* 19 (1969).

Agassi, Joseph, 'The Nature of Scientific Problems and Their Roots in Metaphysics' ['*Nature*'], in M. Bunge, [*Festschrift*]. Reprinted here.

Agassi, Joseph, 'Methodological Individualism' ['*Individualism*'], *Brit. J. Sociology* 11 (1960).

Agassi, Joseph, 'The Novelty of Popper's Philosophy of Science' ['*Novelty*'], *International Philosophical Quarterly* **8** (1968). Reprinted here.

Agassi, Joseph, 'Thomas S. Kuhn, *The Structure of Scientific Revolutions*' ['*Kuhn*'], *J. Hist. Philos.* **4** (1966).

Agassi, Joseph, 'The Role of Corroboration in Popper's Philosophy' ['*Corroboration*'], *Australasian J. Philos.* **39** (1961). Reprinted here.

Agassi, Joseph, 'Positive Evidence in Pure and Applied Science' ['*Positive*'], *Philosophy of Science* **37** (1970). Reprinted here.

Agassi, Joseph, 'Science in Flux: Footnotes to Popper' ['*Flux*'], in Cohen and Wartofsky [*Studies*, III]. Reprinted here.

Agassi, Joseph, 'Sentionalism', *Mind* (1966). Reprinted here.

Agassi, Joseph, 'Unity and Diversity in Science' ['*Unity*'], in Cohen and Wartofsky [*Studies, V*]. Reprinted here.

Agassi, Joseph, *Towards a Historiography of Science* [*Historiograhpy*], *History and Theory*, Beiheft 2, The Hague, 1963; reprint, Wesleyan University Press, 1967.

Akzin, Benjamine, Art. 'Legislation' in *Encyclopedia of the Social Sciences*, 1968 edition.

Bartley, William W., III, *The Retreat to Commitment* [*Commitment*], N.Y., 1962.

Bartley, William W., III, 'Rationality versus Theories of Rationality', ['*Rationality*'], in M. Bunge [*Festschrift*].

Bartley, William W., III, 'Theories of Demarcation Between Science and Metaphysics' ['*Demarcation*'], in Lakatos and Musgrave [*Problems*].

Becker, Carl L., *The Heavenly City of the Eighteenth Century Philosopher*, Yale University Press, 1932.

Bocheński, J. M., *Methods of Contemporary Thought* [*Methods*], Harper Torchbook, 1968.

Bradley, Francis H., *Appearance and Reality*, Oxford 1893.

Bunge, Mario (ed.), *The Critical Approach, Essays in Honor of Karl Popper* [*Festschrift*], N.Y. 1964.

Carnap, Rudolf, 'Testability and Meaning', *Philosophy of Science* **3** (1936) and **4** (1937).

Cohen, Robert S. and Wartofsky, Marx W. (eds.), *Boston Studies in the Philosophy of Science* [*Studies*], Vol. III, N.Y. 1968; Vol. IV, N.Y. 1969.

Duhem, Pierre, *The Aim and Structure of Physical Theory* [*Aim*], Atheneum, N.Y., 1962.

Eddington, Sir Arthur S., *The Nature of Physical Reality* [*Nature*], Ann Arbor paperback, 1952.

Frings, M. S. (ed.), *Heidegger and the Quest for Truth* [*Heidegger*], Chicago 1968.

Glanvill, Joseph, *The Vanity of Dogmatizing* [*Vanity*], London, 1661.

Golding, M. P. (ed.), *The Nature of Law, Readings in Legal Philosophy* [*Law*], N.Y. 1966.

Guerlac, Hency (ed.) *Ithaca* 1962 [*Ithaca*], Paris 1964.

Hart, H. L. A., *Law, Liberty, and Morality* [*Law*], N.Y. 1963. *Punishment and Responsibility, Essays in the Philosophy of Law* [*Punishment*], OUP, 1968.

Hegel, George F. W. *The Phenomenology of Mind* [*Phenomenology*]; (transl. by T. M. Knox), OUP, 1967.

Jarvie, Ian C., *The Revolution in Anthopology* [*Revolution*], London 1963.

Jarvie, Ian C. and Agassi, Joseph, 'The Problem of Magic of the Rationality of Magic' ['*Magic*'] *Brit.J. Sociology* **18** (1967).

Koyré, Alexandre, *Newtonian Studies* [*Newtonian*], Cambridge, Mass. 1965.

Kelsen, Hans, *General Theory of Law and State*, 1945. *Theorie pure du droit*, Paris 1962.

Kraft, Victor, *The Vienna Circle* [*Circle*], Transl. by Arthur Pap, N.Y. 1953.

Lakatos, Imre and Musgrave, Alan E., *Problems in the Philosophy of Science, Proceedings of the International Colloquim in the Philosophy of Science, London, 1965* [*Problems*], Amsterdam 1968.

Merleau-Ponty, Maurice, *The Primacy of Perception and Other Essays* [*Perception*], James M. Edie (ed.), Evanston 1962.

Musgrave, Alan E., See Lakatos.

Poincaré, Henri, *Science and Hypothesis* [*Hypothesis*], Dover, N.Y. 1952.

Popper, Sir Karl R., 'The Aim of Science' [*'Aim'*], *Ratio* 1 (1957).

Popper, Sir Karl R., 'Remarks on the Problems of Demarcation and of Rationality' [*'Remarks'*], in Lakatos and Musgrave, [*Problems*].

Popper, Sir Karl R., *Conjectures and Refutations* [*Conjectures*], Harper Torchbook, 1968.

Popper, Sir Karl R., *The Logic of Scientific Discovery* [*Logic*], Revised Ed., Harper Torchbook, 1969.

Popper, Sir Karl R., *The Open Society and Its Enemies* [*Open Society*], Harper Torchbook, 1963.

Popper, Sir Karl R., *The Poverty of Historicism* [*Poverty*], Harper Torchbook, 1969.

Price, Derek J. de Solla, *Science Since Babylon* [*Babylon*], Yale Paperback, 1962.

Ross, Alf, 'Directives and the "Validity" of Law', in Golding, [*Law*].

Rossi, Paulo, *Francis Bacon, From Magic to Science* [*Bacon*]. London 1968.

Sabra, A. I., *Theories of Light From Descartes to Newton*, [*Light*], London 1967.

Scharfstein, Ben-Ami, 'Descartes' Dream' [*'Dream'*] *Philosophical Forum* 1 (1969).

Sholem, Gershom, *On the Kabbalah and its Symbolism* [*Kabbalah*], (transl. by Ralph Manheim), N.Y. 1960.

Vaihinger, Hans, *The Philosophy of the As If*, [*As If*] London 1925.

Wartofsky, Marx W., See Cohen.

Wellmer, Albrecht, *Methodologie als Erkentniss Theories, Zur Wissenschaftslehre Karl R. Poppers* [*Methodologie*], Frankfurt 1967.

CHAPTER 18

UNITY AND DIVERSITY IN SCIENCE

... the value of science as metaphysic belongs ... with religion and art and love, with the pursuit of the beatific vision, with the Promethean madness that leads the greatest men to strive to become gods. Perhaps the only ultimate value of human life is to be found in this Promethean madness. But it is a value that is religious, not political, or even moral.

(Russell, *The Scientific Outlook*,
ch. iv, 'Scientific Metaphysics')

He who has science and art has religion.

(Goethe)

Abstract

The idea of the unity of science is the historically very important idea of total rationality and objectivity.

This idea is a utopian dream, and a rather dangerous one.

The Popperian view of rationality as a goal directed, i.e., as problem-solving, method of trial and error is a better view or rationality.

Solutions to problems offer the element of unification, and their criticisms offer the element of diversification.

The idea of the unity of science is a utopian dream, which is appealing because it has so many aspects – emotional, moral, social, methodological, epistemological, and ontological – especially, as we shall see, ontological. The idea has seldom been articulated, and almost never been examined – even superficially. On the face of it, at least, it may look as if the idea of the unity of science clashes with the idea of diversity of unique individual things; which latter idea has a great appeal too. Therefore, still on the face of it, there is much room for discussion. Both the idea of unity and the idea of diversity are by no means confined to science. People prefer to apply the idea of unity to science because they take science as synonymous with, or at least as a paradigm of, rationality. Moreover, behind the general idea of unity itself stands the idea of rationality. This idea of rationality, in ethics as well as in epistemology, was the idea of universalizability – of all individual phenomena in all their aspects. Furthermore, behind the idea of universalizability stands the idea of rationality again: namely, that universal laws not only cover *everything*, they can be proved by arguments that all rational *men* must accept. Still further, behind the idea of proofs acceptable to all rational men stands the idea of the rational unity of mankind.

Now this idea of utter universalizability, as it is used in the history of thought, is rather dangerous. I would contend that the dangerous element in it is precisely the theory of rationality which stands behind it: behind the idea of universalizability stands the idea that rationality is comprehensibility and comprehensibility is universalizability; hence everything is rational (i.e. universalizable). In opposition to this confident promise of complete rationality, I suggest that science is propelled not by aiming single-mindedly at universality, and not by aiming at diversity either, but by the constant conflict between these two tendencies. I further suggest that this thesis can be maintained only at the cost of replacing the old-fashioned idea of rationality as universalizability by means of proof, with the Socratic idea of rationality as a process of conflict between universality and specificity, often rooted in conflict of ideas about universality, to wit rationality as Socratic dialectic. And insomuch as science may be viewed as a paradigm of rationality – and (considering Plato's early dialogues seriously in this light) certainly not *the* paradigm – its history has to be reconstructed as that of a dialectical procedure.

Thus far the dialectic view of rationality presented here is that expressed by Popper in Chapter 24 of his *Open Society*; I shall now present an elaboration on it, which may be attributed collectively to Popper and his school.[1] In the *Open Society* Popper identifies the principle of rationality with the principle of criticism of putative views, proposals, criteria, desiderata, etc.; here I shall present rationality of thought as a special case of rational action in general. Rational action is (traditionally) viewed as conduct conducive to given ends – i.e. goal directed. In intellectual discourse, the ends are called desiderata; we begin with desiderata. Hence, rational thought is goal directed, and hence rational. (For example, a discussion within the traditional field of astronomy will at times begin with the description of phenomena to be explained, at times with difficulties stemming from empirical observations to be ironed out, at times with the aim of modifying an existing theory which is at odds with other existing theories, at times with the aim of inventing new tests of a given theory.) Whatever the situation is, we may begin with stating the agenda, the desiderata. These may be critically debated, or ideas allegedly conducive to them may be critically debated. This raises a problem: in order to criticize our standards of x, we need standards of criticism (of x); when x stands for criticism, things (seemingly, at least) get out of hand. How then, can our ideals of rational thinking – as conducive to our desiderata – be subject to critical debate? Criticism is not more fundamental than rationality, since criticism is impossible without reference to desiderata – in this case desiderata for rationality. Nor do the desiderata stand above criticism.

Let us leave this difficulty now, and go back to my historical sketch.

I. AMBIVALENCE TOWARDS UNITY:
AN IMPRESSION

Let us examine impressionistically a few intellectual clichés and their transformations – remembering that in the present case, at least, clichés are not necessarily of the vulgar. 17th- and 18th-century philosophy entertained an optimistic mood that was largely[2] shared by 19th-century philosophy, though it hardly agrees with out present apprehensive mood. We are now as a matter of course worried about tasks which may be beyond human capacity. H. G. Wells stand on the watershed, as the author of *The Shape of Things to Come*, who foresaw the possibility of a return to the Middle Ages, and as the author of *Joan and Peter* where he puts in the mouth of God the declaration, "There isn't a thing in the whole of this concern of mine, that man cannot control if only he chooses to control it. It's arranged like that. ... But ... men ... are too lazy...." It's arranged like that, he says. Today we turn to apprehensive pre-Enlightenment clichés suggestive of an inflated intellect and a deflated morality; Russell echoes Rabelais' saying "Science without conscience is damnation." Yesterday the cliché was Spinoza's: the rational man knows that it is the best policy to react to hatred with love; or Kant's: the sense of respect for the law, the desire to universalize, is what makes one moral. For Kant, even love and friendship and so on, were either derivative of rationality, or peripheral to morality since, *eo ipso*, the problem of ethics is, how should I, as a rational being, act in the present circumstances.

How far this is from the Christian disquisitions on ethics, where love was central and all else peripheral![3]

The optimistic doctrine of rationality as a doctrine of total proof – in principle of all truths – of total control – in principle of all circumstances – and of total propriety – in principle of all conduct – is not easier but harder to grasp than the old problematic idea of an omniscient, omnipotent omnibenevolent deity. Yet the idea of man-God is what I am trying now to present as historically powerful, as an idea advocated by Descartes and Spinoza, by Kant and Laplace, by Comte and Marx, as the most significant regulative idea in the post-medieval period.

Admittedly, this idea was seldom discussed fully and openly. Yet its pervasiveness is symbolized in the overpowering optimism of Condorcet and of Madame Roland even when persecuted by the French Revolution itself. The optimism Condorcet displayed was hardly a matter of temperament – it was a case of an ideology which overrode any temperament. A young man by the name of Berthollet, known merely because his father was a famous chemical philosopher, was so depressive that he was driven to suicide. He sealed himself off hermetically, retaining a candle, pen, and paper, and resolved to record his own observations of his dying self. His writing, incidentally, soon became illegible; as an act of faith it strikes us as remarkable, perhaps at the time it was less so.[4]

The optimism, then, of the idea of the unifying supremacy of reason was not in the psychological domain but in the intellectual, as was the reluctance to discuss it; indeed, both factors still persist. Even nowadays, when the idea is so much more tempered than it used to be, it is seldom clearly and lengthily discussed; it is often alluded to, but no more. We still are very ambivalent about it. In the opening paragraphs of *The Identity of Man*, Jacob Bronowski hits it exactly. The fundamental wish man has to be at home in the world, the intuitive assumption so-to-speak that man is a part of nature, he says, has a strong intuitive appeal and also clashes head-on with the equally strong intuition that man is unique, in a sense above inanimate matter and even above all other creation.

This ambivalence can be further illustrated by Freud's personal attitude. The feeling man has of being part of nature he called the oceanic feeling, and this feeling plays quite a significant role in his metapsychology. Yet in the very beginning of *Civilization and its Discontents* he expresses an ambivalence towards it. His ambivalence at least in part derives from his fear that there might be a kernel of truth in the idea that the oceanic feeling is associated with the need for religion. When he comes to technology he exults – but he first creates a barrier; he identifies oceanic feelings with narcissism and religious needs with a sense of helplessness, the need for mother's womb, for father's protection. These needs, he tells us, give rise to the religious ideas of omnipotence and omniscience. And these same ideas come impressively near to realization, he adds, in modern technology which has rendered man almost Godlike. Yet the connection between narcissism and a sense of helplessness could not escape Freud. "There may be something behind that", he admits, "but for the present it is wrapped in obscurity." "For the present" means when he discusses oceanic feelings; elsewhere he easily connects helplessness with narcissism. Freud usually allows for the fusion of psychological extremes, which may replace each other in a variety of mental mechanisms; and the same goes for the fusion of their symbols. Freud's theories of all dreams, of many neuroses, and of schizophrenia, are instances. We have the axis impotence-omnipotence, and its symbols demonology-religion; we have the axis of insecurity-security or alienation-oneness, and its variety of symbols of self-hate and of narcissism. If you allow impotence, omnipotence, demonology, and religion, to mix together, you might just as well allow impotence and insecurity to mix together, especially since, according to Freud, a sense of insecurity is what gives rise to a sense of impotence. Indeed, this mixing is the core idea of Freud's theory of (paranoid) schizophrenia (where insecurity leads to the deluded sense of omnipotence rather than of absolute security).

Freud cannot tell us why a sense of impotence, in its transformation to its opposite, may lead one person to a religious vision of omnipotence and another to a technological vision of power. Nor does his doctrine enable us to tell why a sense of insecurity and alienation, in its transformation to its opposite, to the sense of security and integration, may drive one person to a mystic dissipation of his soul, through rituals of purification, into unity with the soul of nature and another person to a technological dissipation of his body, through a control panel, with a panorama of bulldozers devouring the landscape and transforming it into a human artifact.

Once we are alerted, through Freud, to the ability of insecurity or alienation to transform itself into security and oneness, we may notice this transformation occurs fairly frequently. The connection is all too obvious even from a seemingly casual example which Wittgenstein carefully analyzes in his lectures on ethics. Consider a remark, such as, "I feel so safe, as if no harm will happen to me no matter whatever else may happen." Wittgenstein analyzes the verbal formulation of such statements, and shows that the above statement should read, "I feel so safe, as if nothing can happen which may harm me." He does not explain why he first formulates it in the objectionable manner except to say it is common. In the light of Wittgenstein's and Freud's analyses we can say, it is common because it expresses – adequately – both the oceanic feeling of oneness, and insecurity or alienation. Wittgenstein himself, however, centered on the religious aspect of the matter alone.[5]

My stress on ambivalence here is not a digression to psychology but a running explanatory theme. I wish to explain why the idea of unity which, as I claim, is so common, is so rarely expressed and never analyzed in the cold light of reason.

Let me give an example from recent philosophy before going to the history of the treatment of the topic. In the first monograph of the *Encyclopedia of Unified Science* Carnap briefly formulates the thesis of the unity of science which is substantially almost identical with Bacon's. This is a remarkable fact, particularly in view of the absence of literature on the topic – even Bacon's presentation is all too brief. Now Carnap admits that his thesis cannot be proven, but he consoles himself by saying that at least it cannot be refuted. He adds that a precondition of the idea of unity of science is the idea of the unity of the language of science and he is thus able to conclude his discussion in a few sentences. He does not even notice that his own philosophy makes it possible to have diverse languages of science developing into a unified language as they go along. All this is amazing. Even the slightest acquaintance with Carnap's philosophy, or of the philosophy of the Vienna Circle, suffices to raise an alarm: ideas which cannot be proven or refuted surely sound dangerously metaphysical and hence meaningless. The alarm may turn out to be a false alarm; my point is that Carnap does not even notice it! He really wishes to take the unity of science so much for granted as to have to say about it the barest minimum.[6]

Bacon's conduct is almost identical. For Bacon metaphysics is evil not in itself but in its methods. The proper method is inductive. From facts we conclude inductively some laws; from the laws the more general, and hence fewer, laws – the *axiomata media*; from these we conclude inductively the few laws of metaphysics; and from the few laws of metaphysics we conclude, by induction again, the single law of natural theology. This natural metaphysics and natural theology are proper, as they are based on induction and thus scientific. But they belong to the future. Traditional metaphysics is bad due to its speculative methods, due to its being based not on induction but on speculation! And yet, the very idea of the unity of science just outlined is highly speculative. And Bacon freely admits this. "And therefore the speculation was excellent in Parmenides and Plato", he says unabashed, "although but a speculation in them, that all things by scale did ascend to unity." [7]

If the idea of unity had met with a single critic, surely his task would not have been hard to perform, especially since in the Western tradition of methodology the dominant school was inductive, not deductive.[8] Yet there was hardly a critic around.

The deductive version of the idea of unity was more consistent – especially in the work of Spinoza; but it was even more optimistic in that version. To believe that already now we possess in essence all knowledge, and that knowledge is a sufficient basis for ethics, already makes us uncomfortably Godlike – in reality: in principle, Spinoza claims, he possesses the knowledge of the essence of all things. In all the literature against the Cartesian apriorist school, however, this point is never taken up: the apriorism of the school was attacked, and specific scientific views – yet what it shares with Bacon's speculations was curiously left unmentioned; on the contrary, one feels that the quarrel between inductivists and apriorists was a family quarrel just because of so much unmentioned common ground and mutual understanding.

(I should like to mention that this explains a quaint historiographic fact. However Baconian the French thinkers of the Enlightenment have declared themselves, again and again certain historians of ideas cannot avoid declaring them Cartesian. Indeed, the further away from the Age of Reason we get, the harder it becomes to distinguish between Bacon and Descartes, and the latter's private letter to Mersenne in which he declares almost full agreement with the former sounds commonplace; yet when we meet this communication immediately after the standard sophomore course in the history of philosophy from Descartes to Kant we feel numbed as though we had had intellectual shock. Incidently, Descartes does not explain his agreement with Bacon: as I have said, high pitched rationalistic optimism was seldom fully articulated.[9])

II. THE ETHICS OF SCIENCE AS A UNIFIER OF SCIENCE

Thus far I have discussed the strange fact that the unity of science was seldom fully presented and seldom criticized. I have said, it is often alluded to and then always highly favourably, and it was always conceived as a sacred point of science – but there is almost no criticism of it. Let me outline some of its appealing aspects.

I have already mentioned the enormous emotional appeal, when discussing the oceanic feeling. This, really, is only a variant of the overflowing sweetness allegedly felt by the mystic when uniting with the universe. And there is little doubt in my mind that if one traces the idea of the unity of science back to the 16th century one arrives at the mystic doctrine of the time, now called light-metaphysics, then called Pythagoreanism, and previously called Cabbalism – which is a mutant of neo-Platonism. And, if, with Bacon, one goes back to ancient Greece, one finds the same mysticism there, in the Pre-Socratics, particularly Parmenides, and even in Socrates, his skepticism notwithstanding.[10]

One need hardly say that the mystic doctrine is, as Bacon noticed, not only emotional but also, even chiefly, ontological. And rationalism, whether inductive or apriorist, always rested on the thesis of the utter comprehensibility of the universe by reason. Hence the hostility of science to religion. Hence, likewise, the call to return to religion, which stemmed from the aftermath of the crisis in physics of the turn of our century; once one claims that reason may be limited, the old conflict is gone, once and for all. This is how Maimonides put it; and this is how things still stand. Though Maimonides believed in unity and in comprehensibility, he stressed the limits of reason, as his chief quarrel with the rational unbelieving philosophers.[11]

In order to show that the utter comprehensibility of the universe drives religion out, one has to place morality within the domain of comprehensibility. This move looks incredible; at the very least it looks like the error known as naturalism, or perhaps as the naturalistic fallacy (not quite in the sense of G. E. Moore, inventor of the phrase), namely of (invalidly) deriving prescriptions from descriptions. Not necessarily so; Kant, at least, knew better, and yet he did subsume morality under rationality. There is one obvious naturalistic element in Kant, namely, the fact that we are rational (he strangely never called this highly questionable view into question); yet he did not commit a fallacy, since the link is in his claim to observe a conscience – in the form of a sense of respect for the law – which, he claims, obliges us to universalize, to take others as we take ourselves, which is the golden rule.

How is this egalitarian universalistic sentiment, one may wonder, integrated within the optimistic tradition?

This is a point much more articulated in popular philosophy and cheap science-fiction than in more serious philosophy. Its epitome may well be the mad scientist from the marvellous cheap chiller-movies of the thirties, sipping his claret, and airily, half-pensively, mumbling in perfect Oxford diction, 'Murder, you say? My dear fellow, this is a scientific experiment!' The mad scientist does not recognize morality and, in effect he claims, he has no duty to morality as long as he is fulfilling his duty to science: as long as he is a good scientist, he implies, he is good. Being a good scientist first was conditioned on being a good citizen; soon being a good scientist became a touchstone for being a good citizen (show me a good scientist and I will show you a good citizen); the last step, the perversion, is the conclusion that *eo ipso*, being a good scientist implies being a good citizen. In the beginning of this line of reasoning, morality is taken for granted, and science is declared subject to its laws; in the end of this line of reasoning, the roles have subtly been reversed (in a manner, incidentally, quite common in schizophrenia: the mad scientist is clinically nearer to the truth than the whole of Ibsen's lunatic asylum).

It ought to be stressed, if confusion is to be avoided, that the idea of the separateness of science and morality is 20th-century; traditionally we have pious science and evil science and nothing in between. Also, one must notice, the idea of evil science is as old as the idea of pious science. It is a serious mistake to think that fears of evil science were invented by Mary Shelley, or, more generally, by the 19th-century antiscientific Romantic movement. The idea is as old as the myth of the proximity of the gates of Hell to the gates of Heaven; the practical Cabbalist and the alchemist shared it, and their monster of Frankenstein was the Golem of Prague. Francis Bacon had it too. His idea that the scientist must be pure is not accidental to his philosophy, even though, as C. W. Lemmi has shown in his *Classical Deities in Bacon* of 1933, Bacon borrowed it from an alchemist by the name of Natalis Comes.

Although Bacon often spoke of the inductive method of demonstration, he never explained it; much to the frustration of his chief commentator Robert Leslie Ellis. His real view of the matter, however, had nothing to do with demonstration, but with the idea of grace as understood by the practical cabbalist and the alchemist. Speculation is putting Mother Nature in chains – we force her to conform to our blueprint. Induction is not so much collecting facts for the purpose of demonstration as the act of worshipping Mother Nature, of showing her humble obedience and respect. And, obedience is the best method of enlisting Her services; for She will reward us by revealing to us all Her secrets. You may be sure of that, since it is the prerogative of God alone to be hidden forever.[12]

Abstract speculation, says Bacon, is tyranny over Nature as well as the conquering of kingdoms of the mind, which are philosophical schools. The speculant is impatient and wishing to achieve fame he arrives at his theories by spurious methods, attaining the goods by force instead of by honest work. Most attempts at rational reconstruction of Bacon's philosophy are directed at finding some technical (inductive) rationale for his demand for hard work and attention to detail. I do agree that Bacon does provide snippets of such technical justifications of his demands to work hard and amass details without end (e.g., theories are squeezed out of facts like wine out of grapes); that he even promises much more to come. Quite apart from the fact that he rejected the most common rationale – we collect facts in order to generalize them – he is definitely baffling on this point. Justus von Liebig claims that Bacon's promises are as vain as those of his patron James I. Robert Leslie Ellis was more sympathetic and he tried hard to make sense of Bacon's diverse requirement – but only to admit defeat. Quite possibly, and in my opinion more than possibly, the large collection of facts was less the material for input into the induction-making machine which Bacon promises, and more a matter of ritual labor which, like all ordeals, cleanses its bearer and makes him worthy of success. The rationale for hard work, then, is not technical but moral. There is an element of the so-called Protestant ethic in science – nowadays still so popular under the impact of Weber and Merton. With the exception that it is not Protestant but Medieval – alchemist-cabbalist – as I have already noted.

We have three traditional characters worthy of examination, equally at home in popular literature, in traditional philosophy, and in scholarship of all sorts. We have the good scientist who is a good man, a good citizen, a pious person – a man answering all descriptions of your personal ideals. He is the hero of most so-called science-fiction literature; he is the hero of most of the trash and nonsense and willful distortions and embarrassingly obvious self-deceptions of the so-called biographies of such impossibly difficult individuals as Newton and Laplace, heavily made-up as placid, warm, and worldly-wise. Also, we have the extreme opposites of such ideals – the enemies of science, with no redeeming features – stupid and brutal all at one. The two are reminiscent of Dr. Jekyll and Mr. Hyde. (Stevenson tells us that Jekyll is the usual mixture of good and bad, who succeeded in isolating only the bad – the Hyde – but not the good. He describes Jekyll, however, as an angel, perhaps because he is a scientist.) These two are the only utterly unproblematic and straightforward stereo-types. The mad scientist is the freak, the only somewhat problematic stereotype: he is the person whose mind is bright but whose heart is dark; Dr. Faustus.

Whereas the only three traditional archetypes are Dr. Jekyll, Mr. Hyde, and Dr. Faustus, the 20th century has created a new one, best presented in the biographical movie on the life of Dr. Wernher von Braun. Whether the movie tells the truth about its hero's scientific or moral career is not of much significance here. The movie does, it is well-known, represent its hero as he himself has chosen to appear to the public: able and concerned as far as science is concerned, but rather indifferent to moral and political questions, or at least one who puts scientific considerations first and moral ones second.

Dr. Faustus is not at all Goethe's Faust. Goethe's hero is a really unified person – he loves science, youth, love, happiness, righteousness and, finally, action. Unlike Faust, Dr. Faustus comes from the same folklore to which Bacon's philosophy belongs: he is an occultist whose mind got twisted and who became evil, poor fellow. He is not a refutation of Bacon's doctrine: he is a fallen angel, a freak. He was a good scientist and hence a good citizen, but he lost his soul to the devil. Like his heir, the mad scientist, he lost his heart but not his brain; to be precise, not the storage part of his brain. It is von Braun who breaks away from the Baconian tradition. We are still ambivalent about him – emotionally, morally, philosophically. We cannot decide, on each of these three levels, whether we prefer a mad scientist to a heartless one. Emotionally, the mad scientist repels us violently, we fear empathizing too much with him; whereas the heartless scientist arouses only revulsion. We answer the heartless scientist with heartlessness, yet we can empathize with the very same hateful, sadistic mad scientist – especially when invited to empathize with his psychiatrist. Indeed, when we meet Jules Verne's mass-murderer Captain Nemo and see how unjustly and severely he had been tortured, how kind and clever he still is, we tend to forget and forgive his aberration as a bad dream. Morally, too, we vacillate; in *The Devil's Disciple*, George Bernard Shaw places opposition and hate nearer to love than cold and deep-seated indifference; Shaw speaks the same language as Buber's *I and Thou*. Buber offers the same quasi-Christian message philosophically (there is more life of dialogue in disagreement, says Buber, than in ignoring an opponent). Of course, the situation can get completely distorted and revert to the worst versions of Christianity where indifference equals hostility. This, however, is not forced on us because there is a difference between our attitude to those with whom we happen to have no ties of sympathy and our attitude to those with whom in principle we can have no ties of sympathy. It is the latter class that, morally, we do not quite know where to place. Those who have put themselves beyond morality, have made us unable to feel anything for them, and thus they make us doubt whether our morality applies to them at all. Over this ground much contemporary moral philosophy wrestles. Much of the interest in and concern about the Eichmann trial may have stemmed from this, rather than from empathy or antipathy towards him. But our ambivalence is also philosophical, and it reflects a reluctance to give up the philosophy of the unity of science.

To illustrate this, let me refer to a strange similarity between the views of the leader of the Vienna Circle, Otto Neurath, and those expressed by the great scientist and Nobel Prize Laureate Werner Heisenberg. In his contribution to the *International Encyclopedia of Unified Science* of 1936 Neurath stresses the traditional hope that unified science will bring its practitioners together from different disciplines, from different walks of life, and from different countries. There is nothing objectionable in

Neurath's view except perhaps that, especially for the mid-thirties, it is somewhat naive – and who does not regret this! The same thesis is to be found in a lecture given by Heisenberg in Göttingen University in 1946 and reprinted a few times under the impressive title 'Science as a Means of International Understanding'.[13]

The problem aired in that paper was, understandably, very much in the air where it was delivered: what should a German science-student do to remedy the moral and political situation of post-war Germany? There is perhaps a lot to be said on this question, e.g., regarding the need for expiation and the like. I shall say no more than that Heisenberg barely faces it. Yet he does give an answer which is the same perversion of Baconianism that we have already met. He argues, to use my words, that since a good scientist must be a good citizen, you should now forget as much as is reasonably possible about all your civic duties so as to concentrate maximally on your duties as science-students. Become good scientists, and *eo ipso* you will also be good citizens. So much for my words; now I shall give a critical summary of the paper in its own words, so that you may judge whether I was fair.

"It has often been said ... repeatedly been stressed, with justification" that science is international, interracial, interreligious. "At this particular time [1946] it is important that we should not make things too easy for ourselves. We must also discuss the opposite thesis, which is still fresh in our ears." Why must we?

It is no doubt the case that we need not make things too easy for ourselves, that we should listen to critics, pay serious attention to opposite opinions, be ready to learn from opponents and change our minds. But there is a limit to all this. We are not obliged to take seriously every opponent. Heisenberg goes on: "I should like [!] to discuss which of these two views", he says, namely the unity of science and the national, racial exclusive view of science (i.e. Nazism), "is correct and what are the relative merits of the arguments that can be produced in their favour." I hope the reader suspects that there is a distortion here. Fortunately, indeed, there is: Heisenberg does not quite mean what I have quoted him to say. Yet the quotation is rather thought-provoking.

"To gain clarity on this question", he says, he wants to show how science works; and he tells the story which he can tell from personal experience: how international and unified the Copenhagen school in physics was when he joined it in the twenties. He learned in Copenhagen, Heisenberg says, that race and nationality are of no use in science; he received later "further proof of the 'objectivity' of science and its independence of language, race or belief". The thesis of the unity of science, then, is proven. The argument, opened on the first page of the lecture, is over on its third page; before the opponent had the chance of a hearing, it seems.

But Heisenberg goes on, nonetheless. A digression to the story of the internationalism of science in the 17th century, including a quotation from Dilthey, and we are ready to slip into a difficulty. "It has always been considered self-evident that adherence to such an international circle would not prevent the individual scientist from devotedly serving his own people and feeling himself one of them. On the contrary", internationalism enhances one's sense of love and indebtedness to one's own country.

Is there a problem lurking here? Indeed, it soon appears. How is it, asks Heisenberg, that all these cases of international scientific collaboration "seemingly do so little in preventing animosity and war?" Note the word "seemingly". The nation of Gauss and Helmholtz and Kirchhoff and Fechner and Wundt and Koch and Planck, the nation of Kant and Goethe and Beethoven, to throw in an additional bit of international collaboration, only "seemingly" benefitted "so little" from that volume of collaboration in its attempts at "preventing animosity and war".

But I misread: the words "seemingly" and "so little" merely serve to make an understatement. Heisenberg clearly states that science is politically impotent; and for a simple reason: there are very few scientists around, and the forces of politics are much too strong, resting, as they do, on masses of people, their economic conditions, and on "a few privileged groups favoured by tradition". The Junkers and their like plus mass unemployment, in case you miss the hint. "The political influence of science has always been very small, and this is understandable enough. It does, however, place the scientist in a position which is in some ways more difficult." We remember that it is not a question of divided loyalty: the scientist loves his country as much as anyone else. The reason for the difficulty, says Heisenberg, is that politics largely depends on applied science. "Thus, the action of an individual scientist often carries far more weight than he would wish and he frequently has to decide, according to his own conscience, whether a cause is good or bad. When the difference between two nations cannot be reconciled he is therefore often faced with the painful decision" of whether to work with his colleagues or countrymen.

Not that the unity of science has been disproved, mind you: it has been proven less than two pages ago. Not that there is a conflict of loyalties either – this was established less than a page ago. Yet, on the fifth page of Heisenberg's lecture all of a sudden the poor scientist is torn by inner conflict and outer strife.

The lecture goes on. There was a profound recent change in the relations between scientists and governments. In World War I, scientists adhered to their national loyalties. Not so in World War II, when "scientists claimed the right to judge the policies of their governments independently and without ideological bias. ... Eventually scientists were sometimes even treated like prisoners in their own country and their international relations considered even immoral." In 1946 Heisenberg spoke of governments in general, as if the status of scientist in one part of the world was comparable to that of another. Historians of post-war Germany may care to take notice of the fact that in Göttingen such words were spoken in 1946.

Heisenberg's outline of the history of international collaboration is not clear to me. In the 17th century and in the twenties of the 20th, science was international. In World War I it was not; yet 'a profound change' took place in the thirties! I ask the reader to forgive me if I do not discuss this puzzlement, however, and if I now become a bit more cursory in my summary. The next paragraph speaks of the martyrs Bruno and Galileo. The next to it speaks of the bomb, chemical warfare, and the like. The paragraphs which follow speak of the moral burden this puts on the individual scientist. "Can science really contribute to" international understanding in the present circumstances? After a brief digression to the Middle Ages, we come to the history of nihilism (which "we find ... in many parts of the world today" (not only in Germany?) even in "its most unpleasant form" (they too have concentration camps?) "disguised by illusion and self-deception" (they too dream of world domination?)), and to the possibility of its reinforcement by science, if not more than that. I am again at a loss: was science ever influential? If not, then, since nihilism is influential "in many parts of the world", the accusation that science contributed to nihilism fizzles out. If yes, how can we maintain the belief in the unity of science in Germany in 1946? I am at a loss. Why does he not refute this charge?

Heisenberg's discussion is nearly over: against his critics he must reaffirm his faith. The fact is that science is the way to "the centre", which is often called "spirit", or "God", or by other names. As the highway to God, then, science, in principle, is accessible to all. And the more people learn science, the greater the numbers of people who learn science, the nearer we shall come to God, and hence to live in a happier and nicer world.

And so Werner Heisenberg begins and ends with the profession of the same traditional faith as Otto Neurath. On the way he somehow succeeds in adumbrating certain severe criticisms of the thesis of the unity of science (e.g., in an unexplained manner it has led to nihilism), and he even speaks in a vein, totally alien to the traditional spirit of the thesis of the unity of science, that above and beyond his duties to science a scientist has to consult his conscience about the ethics and politics of science. But these 'above and beyond' are transient, I understand Heisenberg to suggest; soon science shall triumph and one's duties to science and one's other duties will merge again. Hence, the advancement and spreading of science are of great importance because a scientifically orientated world is one in which science can indeed help foster international cooperation and thus minimize hostility and war. Science, you remember, is in Heisenberg's view the highway to God.[13]

Heisenberg is only an extreme example of a less extravagant but rather popular view. We are all willing to agree that the world is not black-and-white, and so our archetypes are not quite true; but they are, by and large, not too remote from attainment and so can serve as reasonable ideals. C. P. Snow, in his celebrated *The Two Cultures*, blames artists for being Luddites and commends scientists as progressivists. For the sake of balance he admits that men of science all too often have no use for Dickens, even for books sometimes; but he sees in this only an aberration. He is, by and large, committed to progress, to science, to rounded scientific education, to an art – especially literature – which is committed to both science and progress. Literature is committed to progress, science equals progress, humanism equals progress, progress is a free man's worship – these ideas are common to the tradition of the unity of science. In his dream Descartes sees a poetry book and a dictionary, and in his dream he interprets the dictionary to be science and the poetry book to praise wisdom. So felt Dryden, so felt the Moderns of the late seventeenth century England, so feels Lord Snow.

Lord Snow will admit that in fact things are not so simple; he will admit, in particular, that all too often men of science ignore the arts. Indeed, the great novelty and the importance of his work lies precisely in his recognition, which is his starting point, that scientists have no more commerce with artists than artists with scientists. Yet, finally, he is mildly critical of the scientists but he comes down hard on the artists. His bias cannot be overcome by the mere observation of symmetry.

What I have tried to illustrate with these two examples – Heisenberg and Snow – is, first, that empirical criticisms of the thesis of the unity of science can be easily dismissed, and they often are. Second, that the thesis may be employed to defend positions alien to its spirit.

The idea, then, that scientists must be pure and disinterested etc., is in many new variants still very popular; what I have tried to show previously is that this is an essential element in Bacon's philosophy, just as much as in Spinoza's or Kant's or Russell's. The comprehensibility of the world is not sufficient – there must also be the acceptance of the task or duty to comprehend it, the intellectual love of God. The fault of the Middle Ages was not that man had less capacity to comprehend, not that the world was then less comprehensible, but that man had degraded himself by intellectual laziness which made him give up his intellectual freedom of choice, just as Sartre's modern morally lazy man sells out his moral freedom of choice: in both cases the condemnation is total.

It is perhaps hard to articulate fully the moral-religious aspect of rationalism as I have presented it thus far – as the task of universalizing and comprehending universality, of moral and intellectual universalism, of the moral duty of high optimism. Let me leave all this now, however, and show how it ties in with the idea of rationality as proof.

III. PROOF AS THE UNIFIER OF SCIENCE

Bacon saw the existence of contending schools as an unmitigated evil – similar to divisions between warring empires. Had any school possessed the truth, he said, there would be neither need nor room for schools. Truth speaks for itself, truth shines, truth appears naked.

This doctrine, which Popper labels the doctrine of manifest truth or of truth as manifest, may sound as if it were a version of naive realism and thus in contradiction with the theory that Nature loves to hide. Far from it: nature does love to hide, but not forever; under some conditions, She even has to reveal Herself; but whereas when She hides She merely drops hints and clues as to Her whereabouts, when She does appear, Truth appears in her stark nakedness and dazzling beauty; and then no one need nor can doubt Her; the Naked Truth does not carry a handbag containing identification, corroboration, or letters of recommendation.

Taking what I have just said somewhat seriously may suggest that there is never any need for proof; truth is just known as truth. To this one might retort, but we must be able to distinguish the feeling of certainty from certainty proper; to use Sir John Herschel's terminology, we must differentiate psychological certitude from mathematical certitude. Indeed. But if we take seriously the idea that truth is manifest, then this aspect of truth is covered as well. If we could see the truth clearly yet doubt that the clarity is in fact and not in our minds, then there would be at least this lack of clarity in the picture. You may say, perhaps, that if we demand such a high degree of clarity from ourselves before we can claim knowledge of the truth, then we can never claim knowledge. Perhaps; it is harder however, to deny that once we do clearly see the truth, we can claim knowledge. And so, it looks as if it were analytic or near-analytic to declare that we do not need proof when we see the truth. This is what we mean when we say, it is so obvious, I do not know how to put it; it is so clear, I do not know how to explain it. To put it more congenially, knowing a truth is knowing a proof inside out – at a glance, as we say – seeing all steps at once. This idea is shared by many thinkers up to, and including, Wittgenstein – of the *Tractatus* in which truth '*shows* itself' as well as of the *Remarks on The Foundations of Mathematics*. Once you accept this idea, you see that when we speak of a truth in need of a proof, we either speak of a hypothesis which only *may* be true, of a truth revealed to *me* needing proving to *you* so that you can see it as well. So the role of proof, of proving, is inevitably that of introducing the newcomer to the truth. And the mode of introduction is, in my opinion, what differentiates – at least traditionally – science from mysticism. The mystic too – be he Oriental or European Mediaeval – introduces novices to the truth; but his introduction is only highly preliminary: most of the pilgrimage on the road to truth must be done alone, and he who has travelled it successfully is an initiate. Not so in science. This is a point endorsed even in the twentieth century by those who address themselves to the problem: it is agreed by most writers that it is the public character of science (to use Popper's expression of a common sentiment), as opposed to the esoteric character of mysticism which makes us prefer science to mysticism. And so, again, we see the universalist ethic behind science.

To approach matters differently: Russell says there must be a difference between the views of Einstein and the ravings of a madman. The difference lies precisely in the universality of the one, through rational justification, as against the caprice and idiosyncrasy of the other. What Bartley calls "the justificationist doctrine of rationality" is the view that what distinguishes the rational from the mad is the former's appeal to universally valid proof. Take as a conspicuous instance Parmenides' doctrine that the world is full, round, immobile, unalterable, etc. As a doctrine it may be declared a variant of mysticism, not much different from some Oriental or Occidental doctrines. As a doctrine it may be declared madness – as many if not all mystic doctrines were viewed by rationalists through the ages. But Parmenides presented proofs of his doctrine – indeed one may view him as the father of Western justificationism, as Popper does, and as the father of Western mathematics as Lakatos does.

The justificationist approach seems essentially different from that of truth as manifest; rationality is identified with proof rather than self-evidence. But not at all: for, self-evidence and proof are one: proof is what makes one see the objective truth, and acceptance of objective truth is rational as opposed to subjective, idiosyncratic, arbitrary. Rationality is universality, i.e. objectivity, i.e. truth; proof is the universalist method of revealing the truth. Indeed, were proof a matter for the initiate, then our whole edifice would collapse. Or, were proof of falsehood possible, then our whole edifice would collapse.

Consider the discoveries of Lakatos in this light. What has happened to Parmenides' proofs? They have been refuted, of course, and his view is not a proven theorem, of course. For, were the proof valid, we would be able to prove falsehoods. But how can one claim to have a proof without having it? How is error or deception at all possible if proof is the same as revealing a truth in all its clarity? There is no answer; such cases are freaks, they are rare, their proper place is therefore in the origins of mathematics or of a mathematical theory. But look at the history of mathematics, says Lakatos. There is almost no acceptable proof in Euclid, almost no acceptable possible proof in the calculus before the mid-19th century; there is, in brief, no proof a century old or more that an ordinary mathematician today cannot knock out.

What, then, is the role of proof? For Lakatos it is a challenge, an invitation to refute a proven theorem and a guideline for a possible refutation. This is exactly the same, I contend, as the role of every kind of proof in science, whether *a priori*, a *gedanken experiment*, or empirical support. Just as Lakatos views the world of mathematics as a system of ever increasing possible alternatives, so I view science as a system of ever increasing possible alternatives. And so, I think, every proof indicates something – and is thus quite informative – of the prover's shortcoming, of the limitations of his (intellectual) imagination.

I do not invite the reader to accept such views but to consider the difference between viewing proofs in this way from viewing proofs as establishing the truth in a manner open to all to follow and be convinced by. Indeed, once justification is rejected, the problem of objectivity arises full force all over again. Popper has stressed this point in connexion with science in his *Logic of Scientific Discovery*, and he has substituted for the universality of proof the universality of empirical criticism. In his *Open Society* he raised the problem a little more generally, but there he neither gave examples of criticisms outside the domain of empirical science, with the exception of an all-too-sketchy outline of criticism in ethics, nor did he claim that criticism pertains to all intellectual activity: he explicitly claimed, for instance, that interpretations in history, historical points of view, are immune to criticism.

The broadest presentation of the problem, how can we develop a non-justificationist philosophy of objectivity, equally applicable to science, religion, and morality, was forcefully presented by Bartley, in his *Retreat to Commitment* and elsewhere. He feels that the *tu quoque* argument of the esotericists of our century, the claim that rationalism too rests on esoteric foundations, must be answered, not to the satisfaction of the irrationalist – whose judgments may be too arbitrary anyway – but to the satisfaction of that rationalist who, though a non-justificationist, wishes to carry over from justificationist philosophy the vestiges of objectivism and universalism that can be salvaged.

For my part, although I do not agree with Bartley's solution (comprehensively critical rationalism), I find his presentation of the problem admirable. Indeed, if universalism and objectivism is to be retained, I feel, it can be retained not by a doctrine, be it Kant's or Bartley's, but by a perennial struggle towards objectivity and universality. On this point I am in full agreement with the doctrine which Gurwitsch ascribes (rightly or not, I cannot judge) to Husserl (who was, nonetheless, a justificationist). According to Gurwitsch, the most rational thing to do in the absence of a theory of rationality is to search for one (rather than sink into irrationalist despair).[14] This point is, doubtlessly, a minimum thesis that all non-justificationists would accept; it is in intention close to the attitude of the Age of Reason, but in points of technicality it is its extreme opposite. For Bacon and his followers, disagreements and splits between and within schools are anathema, the symptom of faulty justification; for Popper and his followers the opposite should obviously hold. That it does not always hold, only shows that non-justificationists can hold nothing as manifest, not even non-justificationism, not even to the arch-non-justificationist. I should therefore state tentatively, that it is always the case that we do not know that we have a satisfactory theory of rationality, so we may always try to either criticize existing doctrines, or create alternatives to them, or both. We may or may not be successful in this, but we may always try.

This point, too, is no novelty. Sextus Empiricus already notes that the consistent skeptic (etymologically, searcher) exercises a skeptical attitude even towards his own skepticism. This, he adds, leads to peace of mind. But, he adds a rider to this rider: this claim for peace of mind is not a dogma but a (psychological) hypothesis which, historically, came to his school as an after-thought.[15]

IV. MANIFEST TRUTH AS THE UNIFIER OF SCIENCE

The standard criticism of justificationism can be found very early in Sextus Empiricus' *Elements of Pyrrhonism*: all demonstration uses criteria which are either arbitrary or in need of further demonstration etc. *ad infinitum*. There are two reasons for not taking this criticism seriously; one traditional, one modern; one from science, one from commonsense. Both modern science and traditional commonsense tell us that obviously we do justify all the time, and hence that there must be a mistake in the argument to the contrary. Hence, though we should try to locate and exhibit the fault in this argument, the argument need not be taken seriously even in the meantime.

The appeal to commonsense justification is not quite up to standard; at least it is not as strong as the appeal to scientific justification. It is not up to standard for two reasons. First, one may take it as a part of commonsense that, if any truth is obvious then the fault in a criticism of it is obvious, and that, if a fault is obvious it can be detected after a relatively brief and straightforward inspection. Now, if truth is manifest, then criticisms of it, including the criticism from infinite regress, contain serious faults. And if it is obvious that there is a fault in the argument from infinite regress, then the period allotted for finding it should be brief. Yet the fact is that since the days of Hume attempts to locate the fault have failed dismally. If, however, the fault is not obvious, our whole tradition of justification may be resting on an error and hence on an arbitrary or accidental development. Hence, we must view the fault as obvious, and hence Whitehead was right in viewing our inability to locate and exhibit it – our inability to solve the problem of rational justification, which, for empiricists, is the problem of induction – as no less than 'the scandal of philosophy'.

The second reason for the weakness of the commonsense argument against Sextus' criticism of justificationism is a bit less obvious, not because of intrinsic difficulty but because it has in the past been (mis)-directed against non-justificationism. It has been claimed again and again that since in ordinary circumstances we justify our opinions and act on them, then the very fact that a non-justificationist acts proves him a justificationist. But this is only one reading of the situation, and a reading which leads one – in vain, as we shall soon see – to a search for justifications in ordinary life. An alternative reading is this. Justificationists maintain that we must justify our opinions in order to act, and hence conclude that they, as actors, do justify; but the very existence of non-justificationist actors proves that unjustified action is possible. In particular, they cannot justify their action in proclaiming themselves non-justificationists (since, say the justificationists, non-justificationism cannot be justified). Hence, the existence of non-justificationists in our midst refutes justificationism. It is only commonsense that even in cases when we all agree to justify our opinions – say in courts of law – we only agree to justify them relative to accepted fundamental propositions – say, the constitution – which we do not allow to challenge and which we shall not justify. The argument from infinite regress is commonsensically so powerful, that even if you reduce its infinity to, say, ten steps, it is still valid; in any ordinary circumstance there comes a point beyond which we bluntly refuse to justify our views and call anyone who persists in challenging us to do so a pest and a skeptic and a highfalutin philosopher. It looks as if the persistent challenger is a non-justificationist, but a moment's thought will reveal him as a justificationist.

The one extraordinary circumstance where the regress is allowed to be explored beyond what is commonly accepted as reasonable is in intellectual discourse, especially scientific debate. So let us consider, then, the argument from science against Sextus. Do we justify in modern science?

Take Descartes' justification of science, that is his theory of the natural light; as a criterion of truth it is subject to the criticism from infinite regress which is applicable to all criteria. Take the natural light to be a fact, however, i.e. that Truth imposes Herself on us, and you see that we arrive at a stalemate: the skeptic can doubt the veracity of even the most imposing ideas, whereas, Descartes will say, after having performed my duty (Protestant ethic) and having been forced by Nature to believe this or that, if even then I am not right, then God is a liar. It is not an accident that the skeptic Boyle attacked just this point of Descartes' epistemology: he said Descartes still blames God for his own errors; meaning, Descartes was in fact not forced to believe in Cartesian physics.[16]

The same holds for all the philosophers of the Age of Reason: to be able to take minimal sufficient account of Sextus they had to claim that in fact Truth has imposed Herself on us (to call this imposition revelation is in line with the religious doctrine of revelation proper: God's word imposes itself on us). Hence the enormous importance of Newton, so stressed by E. A. Burtt in his by now classical *Metaphysical Foundations of Modern Physical Science* (which is, not accidentally, 20th-century); if Truth has revealed herself we should be able to point at her, and Newton's Theory is a (the only, really) paradigm of Truth Manifest.

All this may explain a puzzling fact. There is a difficulty, in the traditional literature, of distinguishing between methodology and epistemology. Methodology and epistemology are theories concerning the road to knowledge and theories concerning existing knowledge respectively. They seem to be as easily distinguishable from each other as a road-map to Rome is distinguishable from a street-map of Rome. Of course, in each of these two cases the distinction need not be hard and fast, but they are unproblematic. Now, one would expect the distinction between epistemology and methodology to be equally unproblematic; yet this is not the case. Whereas Descartes wrote on method, not on knowledge, he counts as an epistemologist of the first order of significance, but hardly at all as a methodologist, to take but one example. Why is the distinction between methodology and epistemology so hard to make? Perhaps it is because it looks as if proof belongs to epistemology, since it supports our claim for knowledge, whereas this is not so: as long as we need proof we are groping. And so proof belongs to methodology, whereas epistemology can say nothing except that truth *shows* itself. So did Helmholtz understand matters when he observed that investigation is empirical yet knowledge is *a priori*. Helmholtz compares this to the difference between climbing up-hill while struggling through the bush and not seeing any road to the peak, and the standing on the hill-top and being able to see with ease all the shortcuts. Perhaps the best metaphor is Wittgenstein's ladder which one has no need for after having arrived at the rooftop. This is why throughout the modern age the empiricists criticized apriorists on methodological – moral, I should say – grounds, expressing contempt at their lazy armchair method. In epistemology, empiricists and apriorists do not differ.[17]

To illustrate all this consider the following passages from Victor Weisskopf's essay on Niels Bohr in the *New York Review of Books* of April 20, 1967. Perhaps casual passages are not the best illustrations of serious and important and deep-seated convictions, but since I have already invoked the name of Freud, the arch reader of deep meanings into casual passages, allow me to go on for awhile. Let me add, however, that I am quoting what is evidently the central paragraph of the essay, that the essay appears in a not so unimportant a periodical, and that Weisskopf usually does not take lightly his duty as an advocate of science. Well, then; let me quote.

During this great period of physics [the formation of the new quantum mechanics] Bohr and his small group of scientists touched the nerve of the universe. Man's eyes were opened to the inner workings of Nature which before had been obscure. Once the fundamental tenets of atomic mechanics were established, it was possible to understand and calculate almost every phenomenon in the world of atoms, including atomic radiation, the chemical bond, the structure of crystals, the metallic state. Before that time, the environment was composed of unknown forces: electric, adhesive, chemical, and elastic. All these were reduced to one – the electromagnetic force....

So much for the quotation, which I find quite fascinating. The claim that quantum theory is not a revolution but only a further insight into the electromagnetic force of a century ago is in line with the philosophy Weisskopf accepts from Rutherford and expounds early in the essay: in science there are only insights, no revolutions. The claim that the new quantum theory can explain so much and unify so much is highly questionable. No doubt it has grounding in fact and rings a familiar bell to the physicist. No doubt it also gives a misleading impression to the layman. More important, it enables Weisskopf to touch the nerve of his own faith in science: Bohr and his followers, let me repeat, touched the nerve of the universe and saw the working of nature. Once all this "was established, it was possible to understand and calculate". Not, quantum theory enabled us to understand and calculate and was thereby established, but once it was established, we could understand and calculate. Not, our calculations justify our theory, but, our justified theory enables us to calculate. Perhaps I am making too much of a piece of eulogistic writing. Let it be so; let us see what nerves do eulogies touch, what chords resonate more freely in deep sympathy. ·

To stress this point, I would refer to a less euphoric part of Weisskopf's essay. "Nuclear structure presents impressive evidence for the general validity of quantum mechanics", he says in a later part. After quantum mechanics "was established" and "Man's eyes were opened to the inner workings..." etc. one more success merely "presents impressive evidence"! Evidently this additional success resolves doubts which only it raises. Indeed, Weisskopf explains: "The weight of Bohr's personality was so great that [under his influence] for a decade the typical differences between nuclear and atomic states were at the center of interest, whereas the similarities between these states were [erroneously] pushed aside." What Weisskopf finds hard to admit is that the present state of nuclear physics is still not half as satisfactory as that of atomic physics – in spite of the latest advances which he is alluding to. Hence, clearly, "impressive evidence" is only Nature's hint that our eyes are turning towards the right "center of interest". Thus, there is a difference between turning our eyes toward the right "center of interest" (methodology) and seeing the "inner Workings of Nature" (epistemology). Though in methodology a sworn empiricist who has to turn his eyes in the right direction, in epistemology Weisskopf endorses the views of Giordano Bruno, Galileo, and Descartes that Truth reveals herself. Again, the barriers between apriorist and empiricist epistemology break down: empiricists cannot escape a touch of apriorism; only the difference in methodology remains, the empiricist methodology being inductive – hard labour and humble efforts and purity of heart – or, as Robert Merton calls it (approvingly) "Protestant".

Whether all this is true or false, it is meant to explain why the skeptical criticism from infinite regress which is so obvious and clinching has failed to dent the optimistic attitude of the classical scientists which assured them that they could find the unity of nature. For my part I prefer to combat the classical view not by the criticism from infinite regress or by any variant of it – as Popper frequently does – out by the argument from diversity which seems to me equally as appealing from any vantage viewpoint as the argument from unity. Let me show, then, that the classical view of rationality imposes on reasonable people an unreasonable degree of unanimity.

V. UNITY OF SCIENCE AS A DICTATOR OF UNANIMITY ON ALL QUESTIONS

When we distinguish the rational from the capricious, from the whimsical – even merely from the private and personal – we scarcely do more than accept the minimum program for rationality, the desideratum for a doctrine of rationality accepted by everyone. To claim that taste is entirely personal, or subjective, or arbitrary, is also to claim that taste is entirely lacking in objectivity and rationality. Hence rationality is linked with objectivity and hence with truth. From this to justificationism is but a small and reasonable step – with one most unreasonable consequence.

Justificationism, I claim, totally polarizes the subjective from the objective, the non-rational from the rational. It leaves no room for any middle ground of partial rationality of opinion, of partial reasonableness, of degrees of rationality; similarly, it leaves no middle ground for any degree of rationality of a decision, in the realm of morality and responsibility, in the realm of daring and courage and boldness, in the realm of putting political and social proposals to critical debate and attempted implementation; there is no room within justificationism for the partial justification even of a scientific project, be it a theoretical exploration or a design of an experiment or a sociological survey or anthropological fieldwork.

This contention of mine is open to a number of criticisms which I shall soon expound and deflate. First let me show how very unreasonable the polarization is, especially in that it makes rationality so uniform that no room is left for diversity. Both the Middle Ages and the 18th century, at opposite poles, suffered from this.

One of the central tenets of Mediaeval philosophy was that man is irrational; for all men sin, and the wages of sin are infinitesimal in comparison with both the punishment sin entails and the rewards of righteousness. The theory of rationality of the 18th century pointed at the very opposite. Granted that some men are beset by prejudices and are thus not rational, there is no irrationality that a person can commit once he has given up his prejudices. His ends are given, his circumstances are given, his knowledge is solid as far as it goes, and it goes no further (i.e. he is unprejudiced); how else can he act but in the light of all these? Still, the eighteenth century philosophers admitted not only the possibility of prejudice but even the fact that the multitude is prejudiced. Nowadays even this concession to irrationality has been tacitly omitted.[18]

Nowadays the leading theory of rational conduct is that of Bruno de Finetti, the theory of subjective probability and of revealed preferences. The theory is very simple. It grants, first, that there is no universal scale of assessing the probability of a given prediction in the light of given experiences; it assumes, second, that everyone adopts his own scale and everyone acts on it. If any person claims to have no judgment of the likelihood of a prediction, we may in principle show he conceals his opinion by challenging him to a wager. This will call him to act on his assessment, and since he will act rationally he will reveal it. As long as we are not concerned with the cause of the concealment, be it deception or self-deception, the theory seems to cover all possible cases – in principle, that is.

This view of de Finetti is now endorsed by Rudolf Carnap and many other leading thinkers.[19] It represents the pole that we are all rational. Contrary to it, and contrary to the Mediaeval idea, as well as the 18th-century idea, I suggest, as a matter of sheer commonsense, that we are all partially rational, that there are degrees of rationality, that there are cases of better and poorer rationale, that there are degrees of responsibility, that we are sometimes more, sometimes less reasonable, responsible, deliberate. In particular, we may be in two minds, both reasonable, or belong to controversial groups differing from each other yet respecting each other as rational, reasonable, adult, responsible. I suggest, further, that this should be treated as a hard fact (in as much as there are such things) to be contended with even if we have no theory as yet to account for it.

We have already met the opposite idea, explicitly stated by Bacon, that competing schools, doubts, controversies, are symptoms of lack of reason. Bacon also stated in the preface to his *Great Instauration*, you can't criticize a man and yet respect him. This is in character, yet fantastic. Adam Smith, in his obituary of his friend David Hume says, of his philosophy I shall say nothing since those who agree with it admire it, and those who do not, despise it. I think today most of us respect and yet dissent from Hume. Benjamin Franklin exhibits in his *Autobiography* a curious ambivalence to debate. He enjoyed debate in his youth, but took pride in not having argued with his scientific opponents who lost ground even without a battle. Much as we disagree with Adam Smith, we usually agree with Franklin. At least we have a problem here.

In a well-known paper [20] Stephen Toulmin rejects the thesis that Priestley was foolish to reject Lavoisier's doctrines and seeks to show that he was justified. The criterion of choice of a hypothesis, says Toulmin, is that of simplicity; and which of the available doctrines was simpler was an open question for quite a while.

This is a nice point. Toulmin agrees with Bacon and his followers about clear cut decisions and the inadmissibility of dissenting from them once they were computed; he only claims that the interim period of computation may be long. But Toulmin is even historically mistaken: it was Kirwan, not Priestley, who held the criterion of simplicity; and applying the criterion he soon gave up his phlogistonist views and joined Lavoisier; Priestley had different criteria, which are too complicated to present here. [21] Whatever the criteria within science, even when one accepts them within science, and even when their application is unproblematic, dissent may still be sustained, say within metaphysics.

Examples: Faraday agreed that Newtonian mechanics had been as well confirmed as possible, yet he rejected it on metaphysical grounds: he rejected action at a distance on the basis of the idea of conservation of force. Einstein agreed that quantum theory is highly satisfactory as a consistent explanation, yet his determinism made him reject it. Were these people unreasonable? By implication, Toulmin's paper comes dangerously close to saying, yes. I think it is not too much to suggest, intuitively if you like, that they were both reasonable, though most of us would side with Faraday more easily than with Einstein.

The doctrine of proof requires that all arguments be quickly terminable. The verification principle is but a variant of it. To claim that instead of verification we have only probability looks like partial rationality, but this is not so: we all have to endorse the most probable hypothesis, and to the degree that it has been rendered probable. The question is, of course: What hypothesis, among competing ones, is the most probable? To this there is no generally accepted answer, but since inductive philosophers are still struggling to find such a universal criterion, claiming to be able to find it in the relatively near future, we should not press this question now; suffice it that they claim that in science unanimity is achieved over any specific case of choice of hypothesis. Of course, the unanimity thus reported is a myth too, but let us not press this point either. Rather, we can ask: What does the endorsement of a hypothesis amount to? Endorsing the most probable hypothesis, most inductive philosophers claim, is the bare recognition of the (logical) fact that a given hypothesis is the most probable among the class of competing hypotheses in the light of experience and thus the recognition of its rational preferability.[22] Does this solve any problem? Does Einstein's recognition of the merits of quantum theory amount to endorsement of that theory? In the accepted sense it does; in fact it does not. The stress on unanimous endorsement of the most probable hypothesis is so heavy in the current literature, that there is almost no explanation of what endorsement means in terms of 'cash value'; in practical terms, that is. And the few explanations that exist have not been put to any critical examination.

But perhaps we can find two equally probable hypotheses? Will not this, at least, justify diversity of opinion?

We barely allow for diversity this way and the choice looks to us dangerously arbitrary. So we add a criterion of choice: of two equiprobable hypotheses choose the simplest. And farewell to diversity.

Ambivalence plays its tricks here; diversity looks arbitrary and hence irrational since we have contrasted caprice and idiosyncrasy with objectivity and universality and rationality. Once we have done that we cannot ameliorate matters. We have to start afresh: we must identify caprice and lack of rationality but not conclude from that that rationality is compulsive to all rational men. How can this be done is a question which I shall come to in the next paragraph. First I wish to state that – paradoxical as it sounds – we do in common experience identify caprice with lack of rationality yet refuse to view all diversity as caprice; hence we refuse to see rationality and caprice as contrasts though we identify caprice with lack of rationality.

The paradox, in other words, is a logical error; identifying the minimal points of two scales is not identifying the two scales. Even if each is composed of two points each may have only one point in common with the other. It may happen that all good money is genuine, and that all forged money is bad, yet from this it does not follow that all bad money is forgery! It *a fortiori* does not follow from all irrationality is sheer caprice that all rationality is totally binding. We may introduce leeway, and it is commonsense that we should do so.

Even on a question on which we have utterly no evidence, the classical doctrine of rationality allows no diversity! Some say, have utterly no judgement on matters concerning which you have no proof, and with no judgement engage no action. Others, notably Descartes, require conventional conduct on matters on which there is no rational judgement. Others, still, suggest consulting *a priori* probabilities. Strange as it sounds, there is disagreement on the criteria to be applied, but not on the doctrine that the criteria should eradicate all disagreement. The doctrine of induction is based on the transcendental argument that there must exist a principle of induction since scientists are in perfect agreement, and surely not on a religious authority or some such silly criterion. Yet the debate over the principle of induction indicates as wide a disagreement between inductive philosophers as can be. Why cannot they use the methods of science and determine empirically what is the principle scientists employ (as they say scientists do) and eradicate this shameful disagreement?

The intuitive clash between universality and diversity was felt in different ways by different thinkers. Let me mention one. William James felt it would be a bigotry to consider his own brand of Christianity the only rational one or to say that choice of religion is totally arbitrary. He developed pragmatism in order to permit everyone to choose a religion, that is to say, to offer more than one religion as rational choices. He ended up by a criterion which, if it were operative at all, would compel each one to choose one and only one religion – the most useful one. I suppose it is the religion of his environment, but I am not sure.[23]

The same may be said of Carnap's principle of tolerance. Carnap first claimed that metaphysical problems – and solutions to them – are meaningless; he then claimed that they may be rendered meaningful by translating them to questions concerning the language-systems we may employ. Now, it seems, the question of choice between competing metaphysical systems gets translated or transformed into questions concerning the choice of language-systems. This question Carnap has answered with a principle he called "the principle of tolerance": everyone has the complete right to choose his own language-system. Now this principle may be a principle of indifference, in which case there is little value in the tolerance employed: we do not have to display much tolerance to allow people to choose between alternatives of equal merit. If, on the contrary, one language-system is preferable to another, tolerance (which we all take for granted anyway – otherwise there would be no problem of choice) is no solution to the problem of choice: we need criteria and rules of rational discussion to facilitate rational choice – i.e. the choice of the best language-system. We are caught here in the same net. Either there is no rational choice of a framework – the framework is arbitrary whether it be a religion or a language-system (a metaphysics) – or there is a criterion of rational choice and it imposes on us a unanimity; all who share it are rational and all dissenters are irrational.

One final example, an intriguing one, and with a note of depair. R. G. Collingwood, *Speculum Mentis or the Map of Knowledge* of 1924; chapter on History, section on the World of Fact as Absolute Object. I quote:

In science the attempt is made to bridge the gulf between God and the world. Instead of God we find the concept or law, which because it is the law of the world no longer stands outside the world. Here what was transcendent has become immanent. But in this very success, failure is revealed; for it becomes plain that the immanence is a false immanence [for the following reason: in science the fiction is held that] the world is not a world of individuals but a world of particulars, and because the concept [i.e. the law] is indifferent to the various particulars in which it is embodied, their diversity remains meaningless and the world is to that extent a chaos ... the failure of religion is repeated and the quest for immanence ends in transcendence, but an abstract transcendence more intolerable than the concrete transcendence of religion. Thus [art, religion, and science] ... are attempts ... to reach ... organized individuality. ... Art comes nearest to success. ...

This conclusion of Collingwood is very impressive, but it leaves us with the problem open. And the problem, how to have unity and diversity combined, hinges on the problem of rationality. And, anyway, the problem of rationality is significant in its own right.

Anyway you look at it, there is still a significant and challenging difference between the choice of the madman and the choice of the reasonable man; only the reasonable man has a number of reasonable alternatives, some of them equally reasonable, some more than others. This should be the desired corollary to any doctrine of rationality that may claim superiority over the traditional doctrines of rationality.

VI. A THEORY OF RATIONAL DISAGREEMENT

Einstein agreed with the majority of physicists that quantum theory satisfied certain desiderata. His contention, however, was that these desiderata are not sufficient. In other words, we need not specify *a priori* the desiderata of rationality, but leave them open to rational debate. Also, given the desiderata, we may leave open to rational debate the question of which alternative satisfies them best, and if few are adequate, leave them to arbitrary choice without declaring this arbitrariness irrational.

This seems straightforward enough, and it applies to views as well as to conduct.

There are also two moves that seem widely acceptable (at least they have been proposed by Kant, and were endorsed by Neurath, Carnap, and practically all members of the Vienna Circle at one time or another), but which *within justificationism* do nothing to relieve the difficulty; within the above framework of rational debate are quite helpful. They are the idea of presenting universality as a methodological desideratum rather than as an ontological claim and the replacement of objectivity by intersubjectivity.

Let me explain.

It is agreed by practically all philosophers and historians of science that something similar to Bacon's pyramid of levels of generality takes place in science; the classical instances are Newton's unification of Kepler's and Galileo's theory, electrodynamics as a unification of theories of electricity and magneticism, and electromagnetic field theory as a unification of electrodynamics and physical optics. The question is, what is the status of this pyramid? Does the pyramid in science correspond to a factual one, to one in the real world? Or is the pyramid something to which there can be no correspondence, is it a characteristic similar to the linguistic structure of science? For justificationists this question makes little difference: they have to endorse each law of science, independently of the question on which level of the pyramid it happens to be. It is a secondary matter to them that laws are ordered in a pyramid, the primary matter being that the laws are true. This, however, raises an old difficulty: does not science rest on the *a priori* assumption that the pyramid does exist, that the next level of generality will be discovered, that laws will fit in a pyramid? In reply to this it might be claimed that whether or not the pyramid reflects the true state of affairs in the cosmos, science may try to find it, and if the pyramid is there science may even be successful. Thus, the pyramid belongs to methodology. For justificationists, who claim that science proves its doctrines, this move is of no avail. For, when we prove a higher level theory we show *a fortiori* that one more level of the pyramid is filled by true laws. If we replace proof with probability, (meaning when we say that a scientific theory is probable, that most scientific doctrines are true) then the previous sentence will have to be modified, but not in a way prejudicial to the present discussion. And so, justificationists must accept the pyramid not only *a posteriori*, but also as the *a posteriori* corollary to any *future* development of science. This means that it is a transcendental part of all science, that for science to be possible God must have created parallel pyramids, one in nature and one in the intellect or in language or in some human quality. This is not as primitive a thesis as Wittgenstein's picture theory of every statement in language, but it is identical with Bacon's claim that God has created Nature and the Intellect on a par, so as to make science possible.[24]

If, however, we claim that the lower-level theories are often refuted, or simply clash with the theories they replace even when the former are not refuted, then we can deontologize or methodologize the claim for universality in science. Moreover, in such a case, tests and crucial experiments are the particularizing effect that keeps the universalizing effect going, that keeps us dissatisfied with existing universal theories and pushes us towards further generalities.

To take an example. William Whewell claimed that Newtonian mechanics cannot be further deduced from a more general law. Why? The tendency, he said, was to explain both existing laws and the as yet unexplained facts, by newer laws which are thus more universal. But in the case of mechanics there were no such facts. Simple. But where do such facts come from? Whewell stressed that there are no uninterpreted facts, and that interpreted facts follow from the theories they verify. So there can be no theory-isolated facts to increase the generality of a theory! To my knowledge this simple criticism of Whewell has never been noticed. To forestall dismissal of Whewell as an isolated case, let me point out that on this point Whewell's view – of explaining theories and isolated facts by higher-level theories – was endorsed by Duhem and Meyerson.[25]

The reason for the general significance of the difficulty implicit in this view – namely, that isolated or uninterpreted facts are unavailable – is that more and more philosophers tend to share it. They increasingly tend to endorse the view that there are no uninterpreted or isolated observations of fact. This means, for almost all philosophers of science, that every new observation of fact confirms some known theory, which is a bit incredible; and that a new observation of fact cannot refute a theory except if we first invent a new alternative to that theory, which is even more incredible. There is little doubt that any critical examination of the current theory of interpretation, which (adumbrated by Kant, Johannes Mueller and Helmholtz) is due to Whewell and Duhem, is no satisfactory substitute for the traditionally most popular (Baconian-Lockean) doctrine of science as based on uninterpreted data. But there is an alternative theory of observation.

Both traditional theories of observation derive from their originators' views of the role of observation in theoretical science. Bacon viewed facts as the raw material which is somehow processed into theories; hence his insistence on raw data. Whewell viewed facts as verifications; hence his view of interpretations as predictions. Popper views new facts as refutations, and so his theory of interpretation allows for both predicted and counter-predicted observations, the latter constituting novelties relative to the theories which they refute.

The result of this new view of interpretation is the methodological aspect of scientific theory. A theory is designed for explanation. It is thus the refuted theory, especially one which had once served a given purpose, which cries for a replacement which may better fulfill the same purpose. And here Whewell's doctrine is rectified, as well as Duhem's – into Popper's.

Thus, I have argued, methodologizing or deontologizing universality can be effective only in a non-justificationist system. The same may be said of replacing objectivity with intersubjectivity.

Indeed, Kant replaced objectivity by intersubjectivity in order to deontologize universality, as should be obvious upon a simple reflection. But his deontology allowed no diversity, nor was it intended to. Nor did it remove the polarization of all human thought and action into the rational and the irrational. When Kant argued for the limits of reason and concluded that beyond these limits scientific intersubjectivity breaks down and can only be replaced by arbitrariness, he plainly and assuredly concluded from this that attempts to transcend the limits of reason must be irrational. This attitude of Kant's should have made him view rationality in the narrow fashion common to his time, and practice it with consistency uncommon at any time. To an extent he was, indeed, quite impatient: not in the least glad to suffer fools and quite willing to consider all opponents as fools. Yet, if there is one aspect of Kant's conduct which was grossly out of character, it was his ambivalence towards differences of opinion. I do not mean his table-talk: light-hearted controversies, even serious yet admittedly preliminary controversies, were tolerated as a matter of course. But Kant was willing to admire even opposition to the last as a symptom of a keen mind concerned with truth. It is perhaps amazing that such a prolific writer, who was so willing and able to elaborate a point, was so brief on some most crucial points of controversy, to the effect that we still know too little about his change from Leibnizianism to his own, critical, version of Newtonianism. No doubt, ambivalence often is the source of reticence – certainly not any lack of frankness in this case. It is also quite possible to see the intellectual reason for the ambivalence: Kant's critical doctrine of intersubjectivity seems to be a liberating doctrine, an idea enabling us to divorce rationality from unanimity; but it fails to do that: rather than establish possible understanding between individuals, it clinched understanding to the point that it became full consent.

The newer theory of the intersubjectivity of tests of theories likewise does not help diversify thought and depolarize rationality and irrationality. For, tests are confirmatory and, we are told, the acceptance of the confirmed theory is declared imperative. The same idea does, however, work if tests are attempted criticisms; they are then aimed at particulars and when successful they diversify – they open new areas of choice which are free for all.

In concluding this paper, which is intended more to open up issues than to close them, I wish to state three points.

First, if, as I suggest, instead of laying down desiderata we allow different contenders to state their own desiderata and debate them, then this may lead to an infinite regress. Indeed a regress there is here, just as there is the regress from one explanation to a higher level one; both these regresses may be infinite, yet they are not futile – not necessarily so; each step in the regress may offer its own interests and challenges.

Second, the desiderata may be solutions (perhaps of given characteristics) to given problems. Popper's orientation has always been towards problems, and in his post-war work he has stressed that problems are the starting-point. I hope I have introduced this idea of his in a somewhat integrated fashion. To mention one corollary: Popper speaks of the objectivity of problems and perhaps also of their relative significance; but he says nothing about the objective choice of a problem to study – sometimes he even suggests one must fall in love with a problem, which smacks (quite erroneously, of course) of subjectivism. Once we suggest that problems are chosen by more general criteria, which themselves are open to criticism, we render matters of choice of problems inter-subjective to some degree.

Finally, the matter of choice of problems seems to me to be the crux of the philosophy of the future. Problems abound; yet most of them are not interesting, or not important, at least in comparison between the expected investment in, and the expected return from, the forthcoming solution. But all this is essentially highly speculative and vague. (Mössbauer was assured by his superior that he was wasting time on a useless problem, very much as E. O. Lawrence before him, and as Planck even before.) And so this may lead to honest disagreement and even some debate. More important, the sense of significance is thus brought into research in a new way. In the classical doctrine it was brought in in the abstract, in the sense that science is important and requires dedication. Here it is introduced particularly, as a particular problem pertaining differently to any different research project. It thus leads anew to some connection between the prescriptive and the descriptive and again *via* rationality, but with a built-in diversity.

REFERENCES

[1] The development of a new theory of rationality, following the lead of Chapter 24 of Popper's *The Open Society*, seems to me to be a collective effort, but Bartley has the lion's share. The following is a partial list of the efforts in this direction.

K. R. Popper, *Logic of Scientific Discovery*, London 1959, Preface to The English Edition, 1958, and new footnotes pp. 44, 98, 206 (and compare with old note on p. 55), and new appendixes ix (introductory part), and xi. *The Open Society*, 4th ed., London and New York (Paperback), 1962, Vol. 2, Addendum. *Conjectures and Refutations*, London and New York 1963, 1965, Preface to 1st and 2nd ed., Chapters 4 (originally 1948), 5, 8, (2), 10 (v, vi). See also his 'Naturgesetze und theoretische Systeme', in in *Ratio* (Oxford) **1**, No. 1 (1957) 24–35, both reprinted in his *Objective Knowledge*, Clarendon Press, Oxford, 1972.

J. O. Wisdom, *Foundation of Inference in Natural Science*, London 1951, final chapter. 'Respect for Persons, the Pleasure Principle, and Obligation' in *Atti del XII Congress Internationale di Filosofia*, Vol. VII, G. C. Sansoni, Florence, 1958–9. 'The Refutability of "Irrefutable" Laws', *Brit. J. Phil. Sci.* **13** (1963).

J. W. N. Watkins, 'Confirmable and Influential Metaphysics', *Mind* **67** (1958), 'The Haunted Universe', *The Listener* **57** (1957), 'Epistemology and Politics', in *Proceedings of the Aristotelian Society*, London 1957.

W. W. Bartley, 'I Call Myself a Protestant', *Harper's*, May, 1959, reprinted in *Essays of Our Time* (ed. by L. Hamalian and E. L. Volpe), New York 1960. *The Retreat to Commitment*, Knopf, New York, 1962, 'How is the House of Science Built?', *Architectural Association Journal*, February 1965. 'Rationality vs. Theories of Rationality', in *The Critical Approach to Science and Philosophy. Essays in Honor of K. R. Popper* (ed. by M. Bunge), Macmillan, London, 1964. 'Theories of Demarcation and the History of the Philosophy of Science, in *Problems in the Philosophy of Science, Proceedings of the International Colloquium in the Philosophy of Science, London*, 1965 (ed. by I. Lakatos and A. Musgrave), Amsterdam 1968.

I. Lakatos, 'Proofs and Refutations (in four parts)', *Brit. J. Phil. Sci.*, 1963–4. 'Infinite Regress', *Proceedings of the Aristotelian Society* (Supplementary Volume 36, London 1962).

P. K. Feyerabend, *Knowledge Without Foundation*, Oberlin 1961. 'How to be a Good Empiricist – A Plea for Tolerance in Matters Epistemological', in *Philosophy of Science*, The Delaware Seminar (ed. by Bernard Baumrin), New York 1963. 'Problems of Empiricism', in *Beyond the Edge of Certainty* (ed. by R. G. Colodny), Prentice Hall, 1965. See also his contributions to Vols. II and III of *Boston Studies in the Philosophy of Science*.

I. C. Jarvie, *The Revolution in Anthropology*, London 1964 (see Index, art. 'Rationality'). 'The Objectivity of Criticism of the Arts', *Ratio* **9** (1967).

Jarvie and Agassi, 'The Problem of Rationality of Magic', *Brit. J. of Soc.* **18** (1967). Reprinted in R. Wilson (ed.), *Rationality*, Oxford, 1972, Harper Torchbook. Also: 'Magic and Rationality Again', *Brit. J. Soc.* **24** (1973).

M. Bunge, *The Myth of Simplicity*, Prentice Hall 1963; *Scientific Research*, 2 vols., Springer, Heidelberg-N.Y., 1967.

S. Anderson, 'Planning for Fullness', in *Planning for Diversity and Choice* (ed. by S. Anderson), MIT Press, Cambridge, Mass., 1968.

J. Agassi, 'Epistemology as an Aid to Science', *Brit. J. Phil. Sci.*, 1958. 'Science in Flux: Footnotes to Popper', *Boston Studies*, Vol. III, Reidel, Dordrecht, 1967 reprinted here. 'Rationality and the *Tu Quoque* Argument', *Inquiry* **16** (1973).

Finally, Tom Settle, Agassi, and Jarvie, 'The Grounds of Reason', *Philosophy* **46**, 1971. 'Towards a Theory of Openness to Criticism', *Phil. Soc. Sci.* **4** (1974).

[2] The pessimistic mood which crept into 19th-century philosophy is a fascinating topic which I am not qualified to discuss. Largely, I should only say in parentheses, it is the irrationalists' tribute to reason that their despair of reason led to total despair – even against their will. Walter Kaufmann is right in viewing Hegel, for example, as by and large a pessimist (*Philosophical Review* **60**, 1951) – quite contrary to Hegel's own pretended optimism and progressivism (and rationality of sorts), one should add. The Malthusian view was rational and, in at least one aspect, pessimistic; yet, the evil predicted by Malthus was a matter to overcome by a technicality; and when a technical suggestion to overcome the evil was made, it was advocated with incredible zest and great zeal. Darwinism, too, had an aspect which could be viewed pessimistically, and the advocates of German nationalism did stress it; but the rationalists saw in Darwinism the strongest support for progressivism. The first collapse of the equation of the two dichotomies – rationalist–irrationalist and optimist–pessimist – takes place with Freud's world-view, since he was both a staunch rationalist and a confirmed pessimist. This is, perhaps, one of the most traumatic aspects of Freudianism.

Norman O. Brown, in *Life against Death, the Psychoanalytical Meaning of History* (Wesleyan University, 1959, Part 5, Chapter XV, section: 'Rationality and Irrationality') says: "That the instinct of psychoanalysis – for it too has instincts which it represses – makes it want to attack the rationality of prudential calculation and quantitative science is an indubitable but not widely advertised fact. It is concealed by the use of a quite naive and traditional (therefore unpsychoanalytical) notion of the 'reality-principle' and 'reality-thinking'. Behind this naive notion of 'reality-thinking' is Freud's un-questioning (he could not question everything) attitude to science, that Comtian attitude which saw man passing through the stages of magic and religion till it finally arrives at the scientific stage, where he is at last mature – i.e., where he has abandoned the pleasure-principle, has adapted himself to reality, and has learned to direct his libido toward real objects in the outer world. Behind this scientist pose of the psychoanalyst lies the repressed problem of the psychoanalysis of psychoanalysis itself."

Here we see the ideal of rationality in the double sense of (a) objective comprehension and (b) it being used successfully in purposive action. This is optimism regarding both the descriptive and the prescriptive aspects of reason – welded so well that Brown cannot but see a conflict between Freud's rationalism and pessimism.

Kleist sounds like a precursor of Freud and Adler, not, as some presume, because of his intense obsessive personality (such abound in history), but because he despaired of reason (after reading Kant, recognizing his excellence, and not being satisfied with him), yet remained crystal clear and rational in style as well as in mode or reasoning – psychological or otherwise. Thomas Mann's claim that he was a romantic (Preface to English ed. of *The Marquise of O*-----) is based on most incredibly weak evidence (such as Kleist's making the midwife reaffirm the view that with the exception of the Holy Virgin no woman ever conceived without sexual intercourse; in the described circumstances, what else can a Christian midwife say?).

Mann's claim that Kleist is a romantic (meaning somewhat mystic, meaning ir-rationalist) is a deduction from Kleist's pessimism and Mann's equation of the two dichotomies – rationalist–irrationalist and optimist–pessimist. There are other ra-tionalistic pessimists, of course, prior to Freud, the most notable of which is a certain literary school in Russia, culminating with Chekhov (and excluding pessimistic ir-rationalists like Dostoevsky and optimists like Tolstoy). Mann's essay on Chekhov is an attempt to explain away Chekhov the rationalistic pessimist. It looks as if Mann held on to the equation of the two dichotomies in order to convert his commitment to rationalism into a much-desired but never-quite-felt optimism (see also note 15 below). See his ambivalent attitude to Freud whose views are used optimistically in *Tonio Kröger*, but not everywhere else. Indeed, here lies Mann's literary effort, in his struggle through reason towards optimism, the 20th-century fallen Faust. See Henry Hatfield, 'Religion in Thomas Mann's *Joseph and His Brothers*', *Boston University Graduate Journal* **15** (Fall 1967).

To conclude, rationalism or irrationalism can, *apriori*, be optimistic or pessimistic, depending on many factors. That irrationalism should lead to pessimism is, however, quite reasonable (in view of the high value of reason), though not very pleasant, of course. But then, pessimistic rationalism is the least pleasant choice (*Ecclesiastes*). This, however, is not to say that optimistic rationalism is easy – see the quotation from Basil Willey in the next note. The easiest, obviously, is optimistic irrationalism, also known as fools'-paradise. This exhausts the possibilities, unless we become more detailed and cautious in describing rationalism and optimism.

[3] See Spinoza, *Ethics* (trans. A. Boyle), Everyman's 1910, Pt. IV, *Prop. XLV* to *Prop. LII*. The following extracts are sufficiently indicative.

"*Prop. XLV*. Hatred can never be good.

Prop. XLVI. He who lives under the guidance of reason endeavors as much as possible to repay his fellow's hatred, rage, contempt, etc. with love and nobleness.

Prop. XLVII. The emotions of hope and fear cannot be in themselves good.

Proof. – The emotions of hope and fear are not given without pain.

Note. – To this must be added that these emotions indicate a want of knowledge and weakness of mind ...

Prop. XLVIII. The emotions of partiality and disparagement are always bad.

Prop. XLIX. Partiality easily renders the man who is overestimated proud.

Prop. LII. Self-complacency can arise from reason, and that self-complacency which arises from reason alone is the greatest.

Note. – Self-complacency is the greatest good we expect.

Prop. LXI. Desire which arises from reason can have no excess."

Spinoza is not likely to have been influenced by Bacon, yet the similarity between the above passage about the deficiency of hope and the following, from Bacon's *Meditationes Sacrae*, 'On Earthly Hope' (*Works*, ed. by J. Spedding, R. L. Ellis, and D. D. Heath, new ed., London 1870, vii, p. 248), is neither accidental nor insignificant. "Certainly ... to keep the mind tranquil and steadfast ... I hold to be the chief firma-ment of human life; but such tranquility as depends on hope I reject, as light and unsure." See Aristotle, *Metaphysics*, XII, 1972b-1073a; Moses Maimonides, *Guide for the Perplexed*, Shlomo Pines' translation, Chicago 1963, part III, Chapters 51 and 52; Gershom G. Scholem, *Major Trends in Jewish Mysticism*, paperback edition, New York 1961, pp. 131-132; cp. pp. 56-59, 345. Consult also works referred to in note 10 below. .

It is hard for me to comprehend how Kant's affinity with Spinoza has been so persistently overlooked. In his *Critique of Pure Reason* Pt. II, Ch. II: 'The Canons of Pure Reason', Section 2: 'The Ideal of the Highest Good, as a Determining Ground of the Ultimate End of Pure Reason', quotes Leibniz (A812 B840) but makes no reference to Spinoza. Yet, clearly, its sentiment (A816 B844) is Spinozist, its conclusion (A814 B842) is Spinozist, and its critique of previous attempts to arrive at the same conclusion (A815 B843) is obviously a critique of Spinoza's system. But the greatest compliment in that section, also not explicitly directed at anyone in particular (A817 B845), is perhaps the most obvious allusion to the *Ethics* of Spinoza: 'Accordingly we find, in the history of human reason, that until moral concepts were sufficiently purified and determined, and until the systematic unity of their ends [i.e. the intellectual love of God] was understood in accordance with these concepts and from necessary principles, the knowledge of nature, and even a quite considerable development of reason in many [?] other sciences, could give rise only to crude and incoherent concepts of the Deity, or ... resulted in an astonishing indifference ....'' (The observation of the indifference, I suppose, is in an allusion to Descartes.) See also Ch. III, 'The Architectonic of Pure Reason' (A840 B868): "The legislation of human reason (philosophy) has two objects, nature and freedom, and therefore contains not only the law of nature, but also the moral law, presenting them at first in two distinct systems but ultimately in one single philosophical system."

Note also that the principle of "architectonic of pure reason" (Ch. III) itself provides a whole system of metaphysics culminating with rational theology. Kant asserts that too much was claimed for metaphysics once, thus leading it to disrepute; but "we shall always return to metaphysics as to a beloved one with whom we have had a quarrel." ... "Mathematics, natural science ... have a high value [but merely] as means ... to ends that are necessary and essential to humanity. ... Metaphysics is the full and complete development of human reason ... it is an indispensable discipline ... [T] hat, as mere speculation, it serves rather to prevent error than to extend knowledge, does not detract from the value. On the contrary this gives it dignity and authority, through that censorship which secures general order and harmony, and indeed the well-being of the scientific commonwealth, preventing those who labour courageously and fruitfully on its behalf from losing sight of the supreme end, the happiness of all mankind."

Schiller criticised Kant's moral philosophy (*Über Anmuth und Würde*, 1793) as too austere – "carrying with [the sense of duty] a monastic cast of mind", as Kant puts it in his reply to it in his *Religion Within the Limits of Reason Alone* (note (in the second edition) to the first *Observation* in Book I, p. 18 of *Harper Torchbook* ed., transl. by T. M. Greene and H. H. Hudson, New York 1960). "The majesty of the moral law (as of the law on Sinai) instils awe (not dread, which repels, nor yet charm, which invites familiarity)", replies Kant. (The 'Anmuth' in Schiller's title is the charm Kant refers to.) See also F. Überweg, *A History of Philosophy* (transl. by G. S. Morris), 4th ed., London 1885, Vol. 2, p. 198.

The same view has been stated by Sir Leslie Stephen, in his *History of English Thought in the Eighteenth Century*, 3rd ed., New York 1902, 1927, Vol. I, pp. 19–20: "Newton laid down mathematical doctrines which were speedily accepted by all mathematicians. To study Newton is therefore to study the history of mathematical investigation of the time. The difference between his views and those of other inquirers is simply a difference of extent, not of substance. One thinker has more knowledge and a wider intellectual horizon; but all thinkers agree so far as their knowledge goes. If the same statement held true in philosophy, we should simply have to expound the views of Locke and Hume, and to show how those views were developed by later inquirers. The thoughts of the greatest man would include those of the less, and afford a starting point for his successors. In fact, however, we have to consider a complex process of antagonistic theorizing, where every position is in turn assumed and abandoned, instead of a simple evolution of thought. ... Men have been arguing metaphysical questions for many centuries without deciding them. Why are these studies, so apparently fruitless, so perennially fascinating? ... What is this world in which we live? What are the ultimate limits of our knowledge? How can it be increased? ... What are the rules to be deduced for the conduct of life? If we could answer [metaphysical] questions, we could satisfy the demand of the intellect for a firm basis of knowledge and a systematic coordination of all discoverable truth. But ... the true theory is reduced by blundering into every possible error ..."

See also Basil Willey, *The Eighteenth Century Background, Studies on the Idea of Nature in the Thought of the Period*, London 1940, p. 44: "Optimism at this level – the level at which Spinoza could declare that *omnis existentia est perfectio* – so far from being a facile or complacent creed, is admittedly almost impossibly hard to attain, and can never be long sustained by flesh and blood. ... With a Spinoza or a Leibniz, unquestionably, it represents the conclusion of long and arduous metaphysical reflection. But with Pangloss ... Thoreau ... or Browning ..., it generally seems to denote contentment. ... In the early and middle years of the eighteenth century the wealthy and the educated of Europe must have enjoyed almost the nearest approach to earthly felicity ever known to man. Centuries of superstition, error, and strife, lay behind. ... 'The vulgar', not yet indoctrinated with the Rights of Man, were contented with their lot. ... The universe had been explained. ..."

For further details of the 18th-century unitary views see next note.

For the difference between Christian ethics and 'scientific' ethics see Paul Hazard, *European Thought in the Eighteenth Century from Montesquieu to Lessing*, New Haven 1954, Chapters IV–VI, esp. p. 64ff.

The references to H. G. Wells and to Russell (*The Scientific Outlook*) I have borrowed from A. E. Baker, Canon of York, *Science, Christianity, and Truth*, London 1943, which I consider a charming and informative period piece, pp. 109, 104–5. See also Russell's *Religion and Science*, London 1935, p. 175: "When Canon Streeter says that 'science is not enough', he is, in one sense, uttering a truism. Science does not include art, or friendship, or various other elements in life. But of course more than that is meant." I think when Russell says this truth is a truism (i.e. widely accepted) he overlooks the fact that in the rationalist tradition of the 17th and 18th centuries (which the Canon was fighting in 1930) this was anything but a truism. Indeed, not only an apologetic Canon, but even an ally to Russell such as Schrödinger, had to repeat the truism and even at a much later day; and a truth, however obvious to Russell, which is repeated by such a polished and laconic writer as Schrödinger, is not quite a truism. See E. Schrödinger, 'On the Peculiarity of the Scientific World-View' and 'The Spirit of Science', both reprinted in *What is Life and Other Scientific Essays*, Doubleday Anchor, New York, 1956, *Mind and Matter*, Cambridge 1958, pp. 44–7: "Dear reader or, better still, dear lady reader, recall the bright, joyful eyes with which your child beams ... and then let the physicist tell you that in reality nothing emerges from these eyes. ... In reality! A strange reality! Something seems to be missing in it."

[4] For the clichés of the period see Peter Gay, *The Enlightenment: An Interpretation*, New York 1967, pp. 34–6; see also John T. Mertz, *A History of European Thought in the Nineteenth Century*, Dover, New York, 1965, I, 147.

For Condorcet's martyrdom see Henry Ellis, *The Centenary of Condorcet, an Address*, London 1894, and J. S. Schapiro, *Condorcet and the Rise of Liberalism*, New York 1934, pp. 105-9: "... a modern instance comparable to Socrates ... calmly writing on the perfectability of mankind under the shadow of the guillotine. [His] true greatness ... was ... evident ... in his attitude towards the Revolution which now threatens to destroy him. 'I have the good fortune' he declared, 'to write in a country in which neither fear, ... nor respect for national prejudice has the power to suppress or to veil any universal truth'."

For Madame Roland's martyrdom see Carl L. Becker, *The Heavenly City of the Eighteenth Century Philosophers*, New Haven 1932, Chapter 3, Section III (paperback ed., 1959, pp. 151 ff.).

For Berthollet's son's suicide see Dr. Thomas Thomson, *History of Chemistry*, London 1830, II, p. 151.

Kleist's suicide was not in spite of optimism, but due to despair – he could "neither learn nor gain anything" by staying alive, he said. Even in the deepest despair, his Enlightenment background (see note 2 above) was showing. So with Freud, who was in constant terror lest he lost his originality and thus be driven to suicide. See my 'Revolutions in Science, Occasional or Permanent?', *Organon* (Warsaw) 3 (1966).

[5] L. Wittgenstein, 'A lecture on Ethics', *Philosophical Review*, January 1965, 8-9: "I will mention another experience straight away which I also know and which others of you might be acquainted with: it is, what one might call, the experience of feeling *absolutely* safe. I mean the state of mind in which one is inclined to say 'I am safe, nothing can injure me whatever happens.' ... The first thing I have to say is, that the verbal expression which we give to these experiences is nonsense! ... We all know what it means in ordinary life to be safe. ... To be safe essentially [*sic*] means that it is physically impossible that certain things should happen to me and therefore it's nonsense to say that I am safe *whatever* happens. ... This is a misuse of the word 'safe'. ... I want to impress on you that a certain characteristic misuse of our language runs through *all* ethical and religious expression. ..."

For the connection between the senses of absolute security and of omnipotence, see, e.g., Freud, 'On Narcissism: An Introduction' (1914), *Collected Papers*, Vol. IV, London 1946, p. 98, and Thomas Freeman, John L. Cameron, and Andrew McGhie, *Chronic Schizophrenia*, Tavistock Publication, London, 1958, pp. 26-42. Cf. Norman Malcolm, *Ludwig Wittgenstein, A Memoir* (with a 'Biographical Sketch' by G. H. von Wright), O.U.P., London, 1958, pp. 70-71 for the close link Wittgenstein saw between the remarks on absolute safety and religion. See *loc. cit.*, pp. 3, 10-11, 20-21, and 32 for Wittgenstein's own character. See also L. Wittgenstein, *Remarks on the Foundations of Mathematics*, Oxford 1956 (ed. by G. H. von Wright, R. Rhees, and G. E. M. Anscombe, transl. by G. E. M. Anscombe), Pt. IV, Section 53, p. 157: "The philosopher is a man who has to cure himself of many sicknesses of the understanding before he can arrive at the notion of the sound human understanding. If in the midst of life we are in death, so in sanity we are surrounded by madness." The customary reference to Wittgenstein's use of psychotherapeutic terminology as if it were metaphorical, I suppose, is myopic or more likely mythical. See also the quotation from Wittgenstein in note 15 below.

[6] Robert McRae, *The Problem of the Unity of Sciences: Bacon to Kant*, Toronto 1961, claims (Preface) that it is not even clear what exactly is the thesis of the unity of science: "The differences in the conception of unity" he adds (viii), "are in some cases so great that one may wonder indeed whether they belong within the discussion of the same

subject." Therefore, his book is mainly an attempt "to bring them together at the onset systematically and unhistorically, to make it clear that there is in fact a common subject".

One might expect the modern followers of the new movement for the unity of science to be more explicit. The impression one gets from browsing in the literature, however, is disappointing. There is little on the topic, and that little contains complaints, such as the one launched by P. Oppenheim and H. Putnam in their 'Unity of Science as a Working Hypothesis', in *Minnesota Studies in the Philosophy of Science*, vol. II (ed. by H. Feigl, M. Scriven, and G. Maxwell), Minneapolis 1958, opening sentence: "The expression 'Unity of Science' is often encountered, but its precise content is difficult to specify in a satisfactory manner." "A concern with Unity of Science", they add, "hardly needs justification", and they justify it by the need for "counterbalancing specialization".

In their conclusion, Oppenheim and Putnam admit that the Unity of Science has thus far been adumbrated "without very deepgoing justification". "It has been our aim", in the paper here cited, they add, "first to provide precise definitions for the crucial concepts involved, and, second, reply to the frequently made accusations that belief in the attainability of unitary science is 'a mere act of faith'."

The definition of Oppenheim and Putnam, incidentally, is the theory of the reduction of all science to microphysics in a hierarchy of steps reminiscent of Comte's hierarchy of sciences. I am not clear about it all, since they do not refer to Comte and they do not explain how to reduce, say, general relativity to microphysics. Their 'justification' of the unity of science is inductive: it has been impressively successful in the past, etc.

Perhaps Herbert Feigl sketches ('Unity of Science and Unitary Science', in *Readings in the Philosophy of Science* (ed. by H. Feigl and M. Brodbeck), New York 1953) the same view – it is hard to say on account of his brevity.

The modern place for the thesis of the unity of science seems to me to be, perhaps, Wittgenstein's *Tractatus*, which, of course, nobody claims to be in full comprehension of (see note 14 below). The clearest expression of the modern view seems to me to be that of Rudolf Carnap. See his *The Logical Structure of the World and Pseudoproblems in Philosophy* (transl. by Rolf A. George), Berkeley, 1967, Introduction, p. 9: "*there is only one domain of objects and therefore only one science ...*", (p. 10) "in construction theory we sometimes speak of constructed objects, sometimes of constructed concepts, without differentiating." This is neutral monism, cf. p. 284: "177. *Construction Theory Contradicts Neither Realism, Idealism, nor Phenomenalism.*"

See also p. 290: "180. *About the Limitations of Scientific Knowledge. ... There is no question whose answer is in principle unattainable by science ...* [there exist] 'mere techni-cal obstacle[s]', not an 'obstacle insurmountable in principle'."

Nonetheless, Carnap's view is not as classical as it sounds. See pp. 292–3: "181. *Faith and Knowledge. ...* we do not here wish to make either a negative or a positive value judgment about faith and intuition (in the nonrational sense). They are areas of life just like poetry and love ... as far as their content is concerned, they are altogether different from science. Those nonrational areas ... and science ... can neither confirm nor disprove one another." See also pp. 296–7: "183. *Rationalism? ...* For us there is no '*Ignoramibus*'; nevertheless, there are perhaps unsolvable riddles of life. This is not a contradiction. *Ignoramibus* would mean: there are questions to which it is in principle impossible to find answers. *However, the 'riddles of life' are not questions but practical situations.* The 'riddle of death' ... has nothing to do with questions about death. ... These questions can be answered by biology, but these answers are of no help to a grieved person. ... Rather, the riddle consists in the task of 'getting over' this life situation. ..."

Contrast this with Spinoza's solution of the problem of death: *Ethics*, IV, *Prop. LXVII*: "A free man thinks of nothing less than of death, and his wisdom is a meditation not of death but of life.

Proof. – A free man, that is, one who lives according to the dictate of reason alone ..."

Spinoza says the extraordinary and Carnap says what is widely accepted and expressed in a multiform of manners and styles. In *The Story of Gilgamesh* Gilgamesh does not change his view of life when he loses his alter ego Enkidu; he simply loses the taste for life. In his suicide note, likewise, Stephan Zweig implies clearly that in his view life is still worth living and fighting for – if one has the strength; yet he simply lacked the strength to start all over again past middle age. All this well illustrates Carnap's view. By immense contrast, Spinoza says, science proves that life is worth living to such a high degree and in such a poignant manner, that he who sees the proof can have no emotional problem facing life, no matter under what conditions. This is the optimism which, Basil Willey observes (see note 3 above), is so hard to sustain emotionally for a thinking person with strong feelings, and yet so easy to sustain for a shallow person like Pangloss.

It is easy, then, to overcome the emotional problems of death if one does not suffer them strongly. Otherwise all philosophy is not enough of a consolation. Philosophy is not enough of a consolation, may I add, because the optimism to the extent which both Spinoza and Carnap advocate is not very convincing: perhaps the intellectual problem of death cannot exist, but the optimism with which they both dismiss it is doubtful. So Carnap's view of the emotional problem of death is true; yet it tallies less well with his view of the intellectual problem of death, than Spinoza's view of the emotional problem tallies with the same view of the intellectual problem: Spinoza's optimism is more of one cloth than Carnap's.

To conclude, it may be said that much as the thesis of the unity of science does exist and writers on it do have much in common, it is hard to make the thesis so specific as to declare that it does or does not contain reductionism of this or that sort. The unity may be of approach, of method, of language, or of subject-matter. The subject-matter may be mother-nature or atomic facts or physics. The unity may be the indifference to various existing diversities of opinions, e.g., concerning the mind-body problems. It is even hard to say whether Oppenheim and Putnam are reductionists although doubtless, reductionism of some sort is what they advocate as the thesis of the unity of science. Similarly, when Schrödinger attacks reductionism he is attacking the traditional unity thesis, just as Russell does; but unlike Russell, Schrödinger is advocating a deeper doctrine of unity.

[7] F. Bacon, *The Advancement of Learning*, Book II: Human Learning, III: Philosophy, 2: Natural, (1) science, (2) metaphysical. *Works* (ed. by J. Spedding, R. L. Ellis, and D. D. Heath), new ed., London 1870, pp. 361–2; see also *ibid.*, p. 321. Note that Bacon uses both images of the pyramid and of the ladder. See also notes 12, 17, and 24 below.

[8] I do not mean to say that the inductive method is used in science, of course, but rather that scientists often said (and believed) that they used that method regularly. And sometimes, no doubt, after sufficiently idealizing experiments, and adding to them some powerful thought-experiments, they could indeed apply the inductive method to them, and perhaps they did. Funnily enough, it is because this practice was recommended by both apriorists and inductivists that it is not easy to say where they differ.

They differ, of course, mainly in their claims for the final ground for knowledge, and this difference is somehow reflected in their different views on method, but it is not so easy to say how.

Laplace, in his *System of the World*, Book 5, reports that the victory of Newtonianism over Cartesianism signified the victory of inductivism over deductivism. For my part, I think the story is somewhat more complicated, and linked with the success of the Royal Society, the success of British politics, its link with Locke's theories of toleration and of balance of power, and with Locke's friendship with Newton – as well, of course, as the inductivism of Newton, Locke, and the Royal Society. See also P. Duhem, *The Aim and Structure of Physical Theory*, Princeton 1964, Part I, Chapter IV, Section 8: 'The Diffusion of the English Methods'. See also Paul Hazard, *European Thought, etc.* and H. B. Acton's 'The Philosophy of Language in Revolutionary France', *Proc. British Academy*, 1960, reprinted in *Studies in Philosophy* (ed. by J. N. Findlay), Oxford Paperbacks, 1966.

Nevertheless, apriorism was by no means dead, and the traditions of Euler, Kant, Oersted, and Helmholtz, kept it alive and very significant indeed yet, doubtless, it was more than somewhat disreputable to be an apriorist in the eighteenth and nineteenth centuries. And it is this last point that is pertinent to the text above, since the unity thesis obviously squares less easily with inductivism than with apriorism.

[9] Descartes, *Correspondence* (ed. by Ch. Adam and G. Milhaud), Paris 1936, Vol. I, to Mersenne, 23 December 1630 (p. 184): "About this [making useful experiment] I have nothing to say after what Verulam [Bacon] has written about it, namely, without being too curious to research into all the small particulars touching on a matter, mainly one must make general collections of all those things that are most common and which are very certain and which can be known without expenditure: such as, that all colchea are rotated in the same direction, and to know if this is the same after the equinox; that the bodies of all animals are divided into three parts, *caput, pectus*, and *ventrem*, and also other examples; since these are those which serve infallibly in the search for truth. For more detailed items, it is impossible that one does not get many superfluous ones, and even false ones, because one does not know the truth of things before one makes them [the experiments]."

Op. cit., to Mersenne, 10 May 1632 (p. 226), a follow-up on the previous letter: "You have informed me elsewhere that you know people who are pleased to work for the advancement of the sciences, even of their desire to make all arts of experiments at their own expense. If any one of this disposition would want to undertake to write a history of celestial appearances, according to the method of Verulam [Bacon], and if without putting forward neither any reason nor any hypothesis, he would describe for us exactly the heavens as it appears now, which position each fixed star is in respect to its neighbors, which difference, either of size or of color or of clarity or of more or less twinkling, etc.; *item*, if this corresponds to what the ancient astronomers have written of it and what difference he has found in it (since I do not at all doubt that the stars do not ever change their relative positions because one deems them fixed)." And he goes on about comets and their orbs, and about ecliptics and apogees of planets, "very useful to the public ... will relieve me of a lot of labour, but I do not hope that someone else will do it. ..." (My translation.)

See also Martha Ornstein, *The Role of Scientific Societies in the Seventeenth Century*, Chicago 1928, pp. 44–6, for an exposition of Descartes' Baconianism.

See also Leibniz's praise of Bacon, and its roots in the influence of Bacon's doctrine of prejudice on him, in Ellis's preface to *Thoughts on the Nature of Things* (*Works, op. cit.*, Vol. 3, pp. 70 ff.). "We do well to think highly of Verulam", i.e. Bacon, says Leibniz, "for his hard sayings have deep meaning in them." And the "hard sayings" are criticisms of the scholastics, whom Leibniz admired all his life, alone in the world of radicalists whose contempt for everything mediaeval was constantly mounting.

Spinoza, however, dismissed Bacon as confused and dogmatic, especially regarding his doctrine of prejudice (of the cause of error); see his first letter to Oldenburg, *Works*, Dover, New York, 1951, Vol. II, p. 278. Nevertheless, one should note that he dismisses both Bacon and Descartes in the same letter, and expressly at Oldenburg's invitation. His final verdict of Bacon is, that whatever valuable he has said, has since been better said by Descartes.

Concerning the dispute regarding the French philosophers and their following Bacon or Descartes, see C. C. Gillispie's and L. Pearce Williams' contributions to *Critical Problems in the History of Science* (ed. by M. Clagett), Madison 1959, and comments there; see also R. Emerson's 'Peter Gay and the Heavenly City', in *Journal Hist. Ideas* **28** (1967), and references there.

There is, of course, the idea that science is both inductive and deductive, or both compositive and resolutive, or both synthetic and analytic (these are three sets of more-or-less synonymous, though always intolerably vague, terms). This idea may be methodological, and it may be epistemological. See J. H. Randall Jr., *The School of Padua*, Padova 1961, and comments on it in J. W. N. Watkins, *Hobbes' System of Ideas*, London 1965, Section 9. See also Justus von Liebig, *Induktion und Deduktion*, 1865, and F. Engels, *Anti-Dühring*. There is a paper by G. Buchdahl on 'Descartes' Anticipation of a Logic of Scientific Discovery', in *Scientific Change* (ed. by A. C. Crombie), New York 1963, and comments on it by N. R. Hanson there. I find Buchdahl's summary of his problem in his second paragraph (p. 399) incomprehensible, as he raises the question, how shall we handle the question of scientific truth, which second question he leaves unformulated. Still, I suppose Hanson is right when he says (p. 461) Buchdahl has rendered us a service "in revealing as myth-eaten the picture of Descartes as a naive Cartesian rationalist".

To say that Descartes was not quite a Cartesian, as Hanson rightly says, is not the same as to say that Marx was not a Marxist. For, what Marx meant was that he was at liberty to change his mind, whereas what Hanson means is much more interesting. When one has a major theory which one assumes to be true, one may accept only conclusions from that major theory, or also other truths which ought to follow from, or at least harmonize with, that major theory, even though we may not know how. And, of course, two truths must harmonize; but one's major theory may turn out to be false after all. In this sense, we may say, our false major theory did not prevent us from accepting truths which contradict it: had we known these truths contradict the major theory we would simply have given up the major theory.

And so, what Hanson says to Buchdahl is that quoting instances from Descartes which are, say, inductive-deductive in character, does not prove that Cartesianism is inductive-deductive, but merely that Descartes could think out of character while believing that he was thinking in character.

Buchdahl's error is part of a thesis of A. C. Crombie's *Robert Grosseteste and the Origins of Experimental Science 1100–1700*, Oxford 1953. There Grosseteste's philosophy is linked with the view of the school of Padua, and the view of that school is declared to be the one both widely preached and widely practiced in the seventeenth century. If this were so, then the book's lengthy title would indeed read, 'Robert Grosseteste as the Origin of Experimental Science'. The thesis is expressed at the end of the Introduction (p. 13): "Grosseteste took the double, inductive-deductive procedure described by Aristotle in the *Posterior Analytics*", he says. "He found the inductive side illustrated by the writings of the medical school and the deductive side illustrated by the writings of Euclid, Ptolemy, and others. ... Grosseteste was favoured ... by unusual opportunities to make his influence felt. ... With Grosseteste, Oxford became the first centre of the methodological revolution with which modern science began. On the continent, Grosseteste's influence may be traced with certainty in several writers, and there is evidence to suggest that the methodology of the Oxford school exerted a decisive influence on European science as a whole." Paris, he admits, was also influential, and was somewhat independent of Oxford. But "Oxford took the lead. ... There is no doubt that from the time of Grosseteste the experimental science ... began to appear in centre after centre...."

In his Conclusion Crombie follows Randall in claiming that Padua was the most important 16th-century methodological centre, adding in a footnote (p. 297): "According to Randall ... the Paduan teacher Paulus Venetus was sent by his Order to Oxford in 1390 and remained there for three years ... after which he taught for two more years in Paris", and this, we are led to believe, was the world's most significant, world shaking, travel-grant. "In his various encyclopedic writings he fully though critically expounded Oxford ideas on logic and dynamics", says Crombie in a manner of proof.

Those who need criticism of such a 'proof', can consult N. Gilbert, 'Galileo and the School of Padua', *Journal of the History of Philosophy* 1 (1963) 223–31, and Benjamin Nelson's 'The Modern Revolution in Science and Philosophy', *Boston Studies*, vol. III, especially p. 10.

So much for Crombie's claim of Grosseteste's influence on Padua. As to Padua's influence on posterity, he tells us that, for instance, (p. 301), "Francis Bacon's method of discovering the form was precisely the 'double procedure' worked out during the preceding four centuries". And (p. 302), "His method was mentioned and used by more than one seventeenth century scientist, particularly in England. For example, Harvey [this is no slip of Crombie's pen but a point supported by a very scholarly reference], Hooke, and Boyle all referred to it in the midst of their investigations." Not only "Bacon's method ... was precisely the 'double procedure'", Galileo's was too, it seems (p. 303). "The originality of Galileo's method lay precisely in his effective combination of mathematics with experiment." "To connect the observation with a theory" Crombie adds (p. 305), "Galileo described precisely the double procedure of resolution and composition which his predecessors in Oxford and Padua had made familiar." Precisely Bacon, Galileo, everybody, is precisely Grosseteste.

Crombie's thesis is no less than that all important methodologists preached and all important scientists practiced the method of induction-deduction, or resolution-composition, or analysis-synthesis – including Bacon and Galileo. It is amazing to me that in the mid-20th century such a thesis should be advocated, and that the volume which contains this as part of a major thesis (the thesis being, all this goes back to Grosseteste) should be viewed as a scholarly contribution merely because it is (doubtlessly) very scholarly.

Nevertheless, incredible as the thesis that Galileo and Bacon preached the same view, there is a certain amount of truth in it. In retrospect we may differ so much from both that we may see their differences as a family quarrel, much as they would not like our viewpoint. This can be said by one who, like myself, rejects both the inductive and the deductive view (in favour of the skeptic view). But Crombie agrees with both Bacon and Galileo, which is by no means an easy feat.

In defense of Crombie one can say, it is in epistemology that Bacon and Galileo differ, not in methodology. In my view, however, the opposite is the case. The methodological differences between the inductivist and the apriorist are much easier to delineate than the epistemological ones. In method the empiricists insist that experience plays a role much more crucial than intuition, and apriorists hold the converse; in epistemology it is hard to maintain the same distinction with equal clarity and sharpness. Already Kant, in the last chapter of his *Critique of Pure Reason*, on the history of pure reason, makes the observation that though the distinction between intuitionism and sensationalism (a priorism and inductivism) is subtle, these two positions are represented by schools which go back uninterrupted to the very beginning of ancient philosophy. Unfortunately, Kant does not elaborate here. We shall return to this point in notes 17 and 24 below.

Note also in Crombie (ed.), *op. cit.*, Henry Guerlac's 'Some Historical Assumptions of the History of Science', especially Section III, on historians' failure to treat science as a unity, let alone as a part of the intellectual unity of our tradition. See, however, Koyré's comments on Guerlac there and Koyré's subtle move from unity proper to interconnectedness and its problems.

[10] See H. A. Wolfson, *Philo*, Cambridge, Mass., 1947, especially the first and final chapters, and *Crescas' Critique of Aristotle*, Cambridge, Mass., 1929, especially the first chapter. See also his 'Extradeical and Intradeical Interpretations of Platonic Ideas', in *Ideas in Cultural Perspective* (ed. by P. P. Wiener and A. Noland), New Brunswick 1962. See also J. L. Blau, *The Christian Interpretation of the Cabala in the Renaissance*, New York 1948, and 1965; Frances A. Yates, *Giordano Bruno*, Chicago 1964; P. P. Wiener, 'Problems and Methods in the History of Ideas', in Wiener and Noland, *op. cit.*, esp. p. 28.

[11] Moses Maimonides, *The Guide of the Perplexed*, Shlomo Pines' translation, University of Chicago, 1963, Pt. I, Ch. 31 on "the insufficiency of the human intellect and its having a limit at which it stops". Ch. 32: "In regard to matters that it is not in the nature of man to grasp, it is ... very harmful to occupy oneself with them." Ch. 72: "And just as in the body of man ... so are there in the world. ... Accordingly it behooves you to represent to yourself in this fashion the whole of this sphere as one living individual possessing a soul. ..." Ch. 73, the tenth premise, note to the reader: "We wish consequently to find something that would enable us to distinguish the things cognized intellectually from those imagined. For if the philosopher says, as he does: That which exists is my witness and by means of it we discern the necessary, the possible, and the impossible; the adherent of the Law says to him: The dispute between us is with regard to this point. For we claim that that which exists was made in virtue of will and was not a necessary consequence. Now if it was made in this fashion, it is admissible that it should be made in a different way, unless intellectual representation decides, as you think it decides, that something different from what exists at present is not admissible...."

The argument is repeated by Russell, see references in note 3 above.

[12] F. Bacon, *Principles and Origins According to the Fable of Cupid and Coelum etc.*, third paragraph (*Works, New Ed.*, London 1870, V, p. 465): "And certainly it is the prerogative of God alone, that when his nature is inquired by the senses, exclusion shall not end in affirmation."

See C. W. Lemmi, *Classical Deities in Bacon*, Baltimore 1933, pp. 50, 57, and 60. The note on p. 58 reads: "In other words, I think it probable that Bacon's *Thoughts on the Nature of Things* is an expansion of Comes's chapter of Cupid."

In *Novum Organum* (I, Aph. 75) Bacon speaks of systematic doubt as of "a calumny of nature herself" and (I, Aph. 129) of the arts and sciences as "that right over nature which belongs to [the human race] by divine bequest". His metaphor of speculations as chaining Nature, and his claim that we must woo Nature so that she reveals Her charms are systematic, and appear both in the Preface to his *Novum Organum* and early in that work (Book I, Aphs. 1, 3; also Aph. 102) and elsewhere.

This sentiment of Bacon's is deeply engrained in the western tradition. I shall quote only two passages illustrating it, on account of their interest.

Sir Charles Sherrington opens his classical *Man on His Nature*, Pelican 1940, thus: "As to Natural Theology and what we are to understand by it, more than one well-known statement offers us counsel. Bolingbroke, type in his way of eighteenth-century culture, wrote to Alexander Pope, the poet, 'What I understand by the first philosophy [metaphysics] is "natural theology", and I consider the constant contemplation of Nature, by which I mean the whole system of God's works as far as it lies open to us, as the common spring of all sciences, and of that', i.e. Natural Theology. There is, too, Lord Bacon's famous definition (*De Augmentis*, iii, 2), that 'spark of knowledge of God which may be had by the light of nature and the consideration of created things; and thus can be fairly held to be divine in respect of its object and natural in respect of its source of information'." (For the metaphors 'spark', 'light', etc., see Evelyn Underhill, *Mysticism*, 12th ed., London 1930, and G. G. Scholem, *op. cit.*)'

It sounds funny to quote Sherrington quoting Bolingbroke quoting Bacon, but that is the stuff traditions are made of. The following is Basil Willey's comment (*op. cit.* p. 156–7) on Holbach's comment on Clarke; do I have to say not Clarke but Clarke's comment on Bacon? It is perhaps more important to stress that according to Willey, Holbach, in a manner characteristic of the whole 18th century, "declares that all our misfortunes are due to our neglecting and departing from Nature. ... Our errors *cannot* be 'natural', are not what Nature intended." And he goes on to expose Holbach's error as a typical "eighteenth century mental habit", namely "that of honouring Nature with a reverence which, in spite of professed atheism, is in fact religious or transfused from religion". Holbach, he says (p. 161) "treats Clarke rather as Marx afterwards treated Hegel; all that Clarke says of 'God', he assures us, may truly be said, and intelligibly said of 'Matter' or 'Nature' ...".

[13] Werner Heisenberg, 'Science as a Means of International Understanding', in his *Philosophic Problems of Nuclear Science* (transl. by F. C. Hayes, London 1952), reprinted in *Great Essays by Nobel Prize Winners* (ed. by C. Hamalian and E. L. Volpe), New York 1960. See also Heisenberg's 'The Role of Modern Physics in the Present Development of Human Thinking', in his *Physics and Philosophy*, New York 1958.

Heisenberg's metaphor of the way or highway to God may be an allusion to Max Brod's *Tycho Brahe's Way to God*: of course the word 'way', 'method', 'tao', frequently had religious as well as scientific-religious connotations, and so Brod's use, as well as Heisenberg's, is quite common.

[14] L. Wittgenstein, *Tractatus Logico-Philosophicus*, London 1922, p. 67. "4.022. The proposition *shows* its sense. The proposition *shows* how things stand, if it is true. And it *says*, that they do so stand." P. 79: "4.121. ... That which expresses *itself* in language, *we* cannot express by language. The proposition *shows* the logical form of reality. They exhibit it." No less! "4.1212. What *can* be shown *cannot* be said." What this last statement *can* mean without being *obviously* false I have *no* idea. No idea at all. P. 187: "6.522. There is indeed the inexpressible. This *shows* itself; it is the mystical." I do not know how *well* it shows itself, but I *do* agree.

See also M. Black, *A Companion to Wittgenstein's Tractatus*, Ithaca 1964, pp. 190–192: "The Notion of Showing. Wittgenstein uses the verb 'to show', or its cognates, very often ... forty occurrences. ... Unfortunately, this crucial concept is most elusive."

Black says that what *shows* itself is a symbol – and that the showing is immediate – 'in a flash'. This is not very illuminating.

See also L. Wittgenstein, *Remarks on the Foundations of Mathematics, op. cit.*, Pt. V, Section 9, p. 167: "Take a theme like that of Haydn's (St. Antony Chorale), take the part of one of Brahms's variations corresponding to the first part of the theme, and set the task of constructing the second part of the variation in the style of the first part. That is a problem of the same kind as mathematical problems are. If the solution is found, say as Brahms gives it, then there is no doubt; – that is the solution." And so on. This goes for rationality, though I think it can be sharply contrasted with the following quote from Husserl which is a marvellous open defense of a new, unknown, kind of rationality!

In the concluding summary of his 'The Crisis of European Man'. Husserl characterises the 'crises' as the "seeming collapse of rationalism". It is only a crisis of a false theory of rationality, he adds, based on 'naturalism' and 'objectivism', not of "the essence of rationalism itself" (E. Husserl, *Phenomenology and the Crisis of Philosophy*, Harper Torchbook 1965, p. 191). It never occurred to Husserl to doubt essentialism and thus justificationism and certitude, any more than to Wittgenstein. And so he could not but conclude that the choice is between "the ruin of Europe ... fallen into a barbarian hatred of spirit" and "a heroism of reason" (*ibid.*, p. 192). But at least he knew and confessed inability to offer a satisfactory theory of rationality. In other words, though he stuck to his concepts of rationality and of essence, he viewed them as open-textured, and stressed this openness beautifully.

Now, obviously, at least *prima facie*, the idea of rationality had better be open-textured, but this can hardly be said of the idea of essence. See the defence of the open-textured concept of rationality in Aron Gurwitsch, 'The Last Works of Edmund Husserl' in *Philosophy and Phenomenological Research* 17, No. 3 (1957), p. 396: "*Historical forms of rationality*, however, must be distinguished from the *idea of rationalism* or rationality, a Platonic idea which is specified and approximated in those historical forms. Surmounting a certain historical form of rationalism is one thing; abandoning the very idea of rationalism is quite another. ... *Philosophy and the idea of rationalism are one and the same. ...*"

For Bartley and for Lakatos see note 1 above.

[15] Sextus Empiricus, *Outlines of Pyrrhonism* (transl. by Rev. R. G. Bury, Loeb Classical Library, London and New York 1933) I, Ch. XII, "What is the End of Skepticism? ... the Skeptic's end is quietude in respect of matters of opinion and moderate feeling in respect of things unavoidable. ... The man who determines nothing ... neither shuns nor pursues anything eagerly; and, in consequence, he is unperturbed. ...

... The Skeptics were in hopes of gaining quietude by means of a decision ... and being unable to effect this they suspended judgment; and they found that quietude, as if by chance, followed upon their suspense. ... We do not suppose, however, that the Skeptic is wholly untroubled...."

As far as inner logic is concerned, one must consider the skeptical hypothesis – non-justificationism leads to ataraxia – empirically refuted. The Greek word 'ataraxia', here translated as 'quietude' and 'unperturbedness', is somewhat hard to render, since it is a quasi-religious term overloaded with nuance and overtone, and since modern experience and modern psychology enable us to distinguish with ease a variety of states of undisturbedness, from the insensitive yet most optimistic and the sensitive yet philosophical. It is easy, for example, to contrast the Zen-Buddhist matter-of-fact calm-through-cultivated-philosophic-indifference – particularly as understood and/or misunderstood in the West (see A. Koestler, *The Lotus and the Robot*, London 1960) – with the self-assured calm-through-cultivated-philosophic-optimism of early 19th-century English writings, whether literary (Jane Austen) or scientific (Sir John Leslie, John Dalton, Dr. Thomas Thomson). Perhaps the nearest modern variant of ataraxia is to be found in the 17th-century writings – due to imitation, I think, rather than either accident or inner logic. We have in the 17th century a skeptic and justificationist, Boyle and Spinoza, exhibiting it in a marvellous fashion. In central European late 19th century we find a certain degree of irritation accompanying all intellectual discourse, manifesting, it seems, an ideological opposition to ataraxia; so I read the atmosphere emanating from works of Kirchhoff and Boltzmann, Thomas Mann, and Freud. I should even say that a certain irritation at the idea of ataraxia is present in the works of Wittgenstein and of Popper alike – again, indifferently to the justificationism of the one and the skepticism of the other. Unless Popper will repudiate my calling him a skeptic, that is to say my identifying skepticism with non-justificationism. But even the refusal to identify skepticism with non-justificationism, to wit, the preference to ally skepticism with cynicism (in the modern deprecating sense, not in the ancient austere sense) and nihilism (alluded to in Heisenberg's text cited above) – even this refusal is but a profound distaste for ataraxia; which refusal, regrettable as it is, nonetheless refutes Sextus's hypothesis that skepticism leads to ataraxia; or, if you wish to put it in a different way, a non-justificationist's refusal (such as Popper's) to identify non-justificationism with skepticism refutes Sextus's thesis that non-justificationism is identical with skepticism.

Since the word 'skeptic' means searcher, it is a bit hard to see why the desire to have ataraxia seems to exclude the desire to search for the truth. There may be a different explanation for this. Possibly this is due to Kant's discussion of the difference between Hume's censorial skepticism and his own critical skepticism, which encourages the search for arguments even beyond their legitimate limit; see *Critique of Pure Reason*, Pt. II, Ch. I, Section 2, 'Impossibility of skeptical satisfaction of the Pure Reason that is in conflict with itself' (A760 B788). Another possible line is expressed by Ludwig Edelstein (Crombie, *op. cit.*, p. 34). "A Platonic metaphor expresses the same thought in a different way. He who does not learn to work like a slave for the possession of the truth will never reach it" (*Republic* IV, 494D). This sentiment is expressed by Schopenhauer in his attack on Hegel, approvingly quoted by Popper (*Open Society*, II, Chapter 12), "Who can really believe that truth also will come to light, just as a by-product?"

The answer to this rhetoric question is, all those people who are able to maintain any measure of a critical attitude towards Protestant ethics. And, let me repeat, Protestant ethics so-called, is Cabbalistic-alchemist ritualistic apologetics. As a matter of fact, unpleasant to the labour pietists, truth ever so often does come to light as a by-product. Whatever is the proper method of science, be it the one described by Popper or not, clearly it is not the one consciously adopted by the fathers of the scientific revolution: if they have adopted it, they did so unintentionally – as the unintended consequence of their conscious and different intentions – and so the proper method, and the truths it had revealed, came quite as by-products.

It is amazing how similar in Baconian – 'Protestant', if you will – sentiment, are Popper and Wittgenstein. See Ludwig Wittgenstein, *Philosophische Bemerkungen* (ed. by R. Rhees), Oxford 1964, Foreword: "... I would like to say 'this book has been written for the greater glory of God', but nowadays this would be contemptible, i.e., it would not be understood aright. Namely, it has been written in good will and as far as it has not been written in good will but out of vanity etc., so far the author would like to have it condemmed. He cannot purge it further of these ingredients than he is himself clean of them." (My translation.) Clearly, according to this passage, vanity is a sinful intent that cannot possibly lead to the discovery of the truth. And, according to this passage, any motive other than "the greater glory of God" is vain or sinful: Of course, this does nbt mean that Wittgenstein – and his like – would oppose ataraxia, but ataraxia, so it seems, must only come at the very end of a very hard day's work: this is the ambivalent irritability about ataraxia here-and-now which I have alluded to above. I join Feyerabend in saying, a good intellectual debate over a glass of beer makes the beer taste better – even though beer is not exactly my cup of tea.

Thus, the question, does ataraxia help learning must be answered with, sometimes. It can only be reopened after developing a better theory of both peace of mind and work. There are varieties of peace of mind; and varieties of hard work; and of the virtues and defects of either. One might then work out a view of the impact of philosophy on states of mind – somewhat similarly to, but more critically than, Erik Erikson's study of William James. As he shows, James was depressed by determinism, and was finally able to overcome his depression by refuting determinism to his own satisfaction. (See Erikson's introduction, to *Emotional Problems of the Student* (ed. by G. B. Blaine Jr. and C. C. McArthur), Appleton-Century-Croft, New York, 1961). Or, in reverse, a state of mind may appear to influence philosophy. See John Stuart Mill's story in his *Autobiography*, where he blames his early utilitarianism for his depression and where he narrates how his emergence from the depression led him to a new version of his utilitarianism. (See, however, Ruth Borchard's *Mill, The Man*, London 1957.)

Finally, Bartley has drawn my attention to the possibility that not all ancient skeptics shared Sextus's view of ataraxia as a by-product, that for some of them possibly the end of all discourse should be ataraxia – thus making them more akin to Zen Buddhists then to Socrates. See John Owen, *Evenings with the Skeptics, or Free Discussion on Free Thinkers*, London 1881, "Arkesilaos might easily have taken his own test of, and ideas concerning, truth,. as possessing not only subjective but an objective validity" (Vol. I, p. 307), and remarks on "the pursuit of ataraxia" (309), "the probability of Karneades is ... a compromise between dogmatists and absolute Skeptics" (318, 319), "the perpetual appeal Ataraxia [had for Sextus] in its accurate definition ... [is] opposed to a [truly skeptic] philosophy which makes non-definition the chief principle in its method" (338-9).

[16] R. Boyle, *Works* (ed. by Th. Birch), 1st ed., 1744, Vol. III, p. 432a ff.: "... the study of physick has one prerogative, (above divinity). ... I mean the certainty and clearness, and the resulting satisfactoriness of our knowledge of physical, in comparison with any we can have of theological, matters, whose being dark and uncertain, the nature of the things themselves, and the numerous controversies of differing sects about them, sufficiently manifest.

... Cartesius was [so] sensible of a dependence of physical demonstration upon metaphysical truths, that he would not allow any certainty not only to them, but even to geometrical demonstrations, until he had evinced that there is a God, and that he cannot deceive men, that make use of their faculties aright.

[But] when ... Descartes ... demonstrate[s] ..., the presumed physico-mathematical demonstration can produce in a wary mind but a moral certainty, and not the greatest ... that is possible to be attained...."

In the above paragraph the demonstration referred to pertains to comets, and comets were at the time the biggest trouble for Cartesians since the 1661 comet moved in the wrong direction along its orbit thus destroying the vortices. (See Laplace, *System of the World*, Book V.)

See also *op. cit.*, Vol. IV, p. 346a: "... And for the rule ... that there is always the same quantity of motion ... the proof he [Descartes] offers, being drawn from the immutability of God, seems very metaphysical, and not very cogent to me, who fear, that the properties and extent of the divine immutability are not so well known to us mortals, as to allow Cartesius to make it, in our case, an argument *apriori*."

B. Scharfstein stresses in his 'Descartes' Dream' (*Philosophical Forum*, Vol. I, 1968–69), the fact that Descartes thought active curiosity a sin and tried very hard to be a passive student; that, indeed, Descartes expressed this sentiment strongly on his death-bed. This is not merely a psychological factor, certainly not in the 17th century when receptivity was contrasted with (sinful) willfulness by Bacon (doctrine of idola, end of *Sylva*), Locke (*Conduct of the Understanding*), Descartes (*Discourse*) and Spinoza (*The Improvement of the Understanding*, and letter to Oldenburg cited above). [17] The idea that development can be not from the little satisfactory to the much satisfactory (Bacon) but from the more unsatisfactory through the less unsatisfactory to the perfect, can be found in Plato's *Symposium* (211c), "starting from individual beauties, the quest for the universal beauty must find him ever mounting the heavenly ladder, stepping from rung to rung ... until at last he comes to know what beauty is ... and once you have seen it, you will never be seduced again" by (lesser) beauties of particular things, but stick to the highest "vision" of "the heavenly beauty" itself, i.e. the ideal of beauty, thus achieving knowledge, virtue, and perhaps even immortality.

The very same rejection of the ladder can be found in Aristotle, more didactically stated and with its edges trimmed: "the premisses of demonstrative knowledge must be true, primary, immediate, more knowable than, and prior to, the conclusion" where the premiss is a theory and the conclusion an observed fact (*Post. Anal.*, i, 2, 71b). It is quite clear that to Aristotle empirical experience is the ladder as far as demonstrable knowledge is concerned but the real basis as far as probable knowledge is concerned.

Whereas the empirical background to demonstrable knowledge is, according to Aristotle, removable – it is not a justification – as a part of the hierarchy of causal explanation it is no more removable than a corollary of a mathematical theorem is removable from mathematics: the first principles are the very causes of all facts; once we know them we have a causal hierarchy of causes and effects, each step being the effect of the higher step and the cause of the lower step, as well as being the conclusion from the higher step and the premiss to the lower step – except for the first principles which are causes or premisses alone and the (general) facts which are the effects or conclusion alone. Here is a ladder, or a pyramid, which was transmitted in the Renaissance by Sir Francis Bacon to the modern world, and reinforced by thinkers like August Comte. (See note 6 above.)

It is important, if confusion is to be avoided, to note that there are two ladders here. One is for justifying our theories and hence, as I am arguing, methodological; and this ladder can be thrown away once the goal – knowledge – has been achieved. The other ladder is imbedded in knowledge itself, and is, thus, epistemological; it cannot be removed. (See also the ladder in Boethius, *Consolatione*, I, 1 and in Frances A. Yates, *The Art of Memory*, Chicago, 1966, Index, Art. Ladder.)

To return to Bacon, he uses the metaphor of removing the ladder, but only when he clearly deals with methodology, not with epistemology. When he insists that ancient knowledge was attained by induction, he worries about the absence of historical evidence for this. The evidence, he claims (*Novum Organum*, I, Aph. 125), has been removed – just in the manner in which builders remove scaffoldings and ladders out of sight.

Goethe reverted the metaphor: the scaffolding cannot usually be evidence as evidence is legitimately a part of science, and so he viewed hypotheses as scaffoldings. See his *Maximen und Reflexionen* (Goethe, *Gedenkausgabe der Werke, Briefe, und Gespräche*, Zürich 1949, Vol. 9, p. 653, para. 1222): "Hypotheses are scaffoldings which one puts up before building and which one tears down once the building is complete. They are indispensable for the worker: only one should not take the scaffolding for the building." (My translation.) The same idea is put more tersely in R. T. H. Laennec's *Traité de l'ausculation médiate et des maladies de poumons et du cœur*, Paris 1826, I, 280 (see E. H. Ackerknecht, *Medicine at the Paris Hospital, 1794–1848*, Baltimore 1967, p. 9) and also in Herschel's *Preliminary Discourse on the Study of Natural Philosophy*, London and Philadelphia 1830–31, Pt. II, Ch. 7, para. 216, p. 153: ..." to lay any great stress on hypotheses ... except inasmuch as they serve as a scaffold for the erection of general laws is to quite mistake the scaffold for the pile."

Let us not inquire as to the loss of the empirical data by the ancients, and overlook the questionability of Bacon's simile of movable ladders and scaffoldings here. Can an empiricist treat hypotheses as scaffoldings? (Not that Goethe was an empiricist; the above quoted empiricist maxim is followed by a few anti-empiricist ones. But Herschel was as much of an empiricist as any thinker, and his metaphor stuck.) Admittedly, it seems quite innocuous for an empiricist to say that hypotheses are mere scaffoldings, meaning, finally the basis for science is empirical fact not conjectures; but if epistemologically we remove empirical evidence too – as both Plato and Aristotle maintain – then where is the difference between apriorism and inductivism? We see here, again, how right Kant was to claim that the difference is subtle (see note 9 above).

One may defend Sir John Herschel by claiming that he was no Platonist-Aristotelian; he removes hypotheses as scaffoldings but never the empirical basis. But this is overshooting the target. Herschel insists on the certitude of Newtonian mechanics, but not on its empirical foundation: he quite permits *a priori* proof of it as much as *a posteriori* proofs. He does not really care, and so we cannot care more than he and insist on declaring him apriorist or inductivist.

Those who did insist on inductivism while rejecting apriorism, for example Ampère, had to be more subtle: for them both hypothesis and experience are ladders to be removed, but hypothesis must be removed first. This is not merely an order of chronology but also an order of priority, even of emotional priority. As G. E. M. de Ste. Croix suggests (Crombie, ed., *op. cit.*, p. 84), Plato's idea of removing the ladder presents not only apriorism, but also a characteristic aprioristic contempt towards experience, which Plato sometimes exhibited (in the *Republic* and later works). It is very interesting that Bacon quotes (*Advancement*, II: '*History of Nature Wrought Mechanical*') Plato's other discussion of the ladder, also leading to the first principles of beauty, not the *Symposium* (where neither contempt nor praise for facts is registered) but *Greater Hippias*, where Socrates pokes ironic fun at those who are impatiently contemptuous of lower forms of beauty. Bacon reads this to say, you can remove the ladder, but only after having used it. (Attitudes to empirical knowledge in Plato's works may be used as means of attacking the Socratic problem.)

There is also the question of the vulnerability of the ladder here. The empiricist as well as the apriorist view hypotheses as highly vulnerable and so, of course, something that must be sooner or later removed "out of sight" (Bacon). But whereas apriorists view evidence of the senses as vulnerable temporary means (Socrates of Plato's early dialogues, Bruno, Galileo, Descartes), empiricists view evidence of the senses as final even if removable (Aristotle, Bacon, Herschel, Whewell).

Thus, it is possible to view empiricism or inductivism as a double-secure system, where our hierarchy of final knowledge is based both on immediate intuitions and on empirical facts. Perhaps this represents Herschel's view better.

In my view, not only Plato and his followers, but, clearly Socrates too, felt that in principle apriorism should do.

This may explain why it was so hard to penetrate the classical arguments for scientific certitude and criticize them. For a discussion of the 20th-century background to E. A. Burtt's *Metaphysical Foundations of Modern Physical Science*, London 1925, see my 'Science in Flux', *Boston Studies*, III, p. 304, this volume, pp. 20, 39.

It would be unfair to Burtt, however, to suggest that in our day and age his idea (there is no finality in science) is common property. Some writers fail to adhere to it even while advocating it. I shall give one example here: Heisenberg's 'Recent Changes in the Foundation of Exact Science' (*op. cit.*). Heisenberg attempts to criticize the view that there is finality in science. "Even Kant's philosophy, intended as a critique of premature dogmatizing in scientific concepts, could not prevent the torpescence of the scientific concept of the universe – it may even be said that it encouraged it" (p. 22). And Heisenberg blames this on Kant's apriorism. But Heisenberg's explanation may be false. How well Heisenberg himself – not an apriorist – has succeeded in preventing "premature dogmatizing and torpescence" may be surmised from the following sentence, two paragraphs later: "... modern physics has shown that the structure of classical physics – as that of modern physics – is complete in itself." What this exactly means I do not know, as I do not know how a theory can declare itself complete or incomplete, in itself or besides itself. But it sounds to me to be not exactly devoid of the attitude condemned in Kant. It sounds to me that Heisenberg's 'complete' means for modern physics what Kant's "premature dogmatizing" means for classical natural philosophy (see my 'Is Physics Complete?', *Synthese*, 1958). Indeed, one may ask, does Heisenberg disagree with Kant about the finality of classical physics? The answer is, no: "Columbus's discoveries were immaterial to the geography of the Mediterranean countries, and it would be quite wrong to claim that ... [he] had made obsolete the positive geographical knowledge of the day. It is equally wrong to speak today of a revolution in physics. Modern physics has changed nothing in the great classical disciplines ... only the conception of hitherto unexplored regions, formed prematurely ... has undergone a decisive transformation" (p. 18).

Now, the idea that Columbus's discoveries were immaterial to Mediterranean geography is such a folly, that Heisenberg himself has to modify it, replacing the word 'geography' with 'positive geography'. All that remains now to do is define 'positive'. 'Positive geography' evidently has nothing to do with views of the Mediterranean basin as the centre of the earth. Indeed the concept of positive geography, that is to say, of cartography of a refined form acceptable to Heisenberg's taste, did not exist in that day; nor were there at the time maps accurate enough to be viewed as positive; nor is positive cartography of the Mediterranean basin indifferent to the question of the curvature of the earth. At most Heisenberg could say that as a result of the Columbian revolution, the scientific revolution, etc., we can now say in retrospect that an idealized and improved version of pre-Columbian Mediterranean cartography is a very good approximation to post-Columbian cartography.

One may say that the idealization and correction of past theories have to be accepted anyway; and that then the difference between their being first approximation to, and being parts and parcels of, present day theories, is so negligible it may be ignored. First, this sounds dogmatic to me; second, when one insists on there being no correction at all, no change, and even in violent language ("it is equally wrong"), than it should be stressed that corrections were made, however small; third, the whole picture of science is rendered hyper-positivistic in order to render this idea plausible – which is a high cost for a very small return.

I cannot escape the impression that even Heisenberg feels uncomfortable about his own position and its all too seemingly dogmatic character. For, I think he proceeds from the above quoted passage to an attack on Kant's dogmatism as an expression of some unease. As if to draw attention to Kant's alleged even worse dogmatism. But Kant lived before Einstein, and so his error is easier to sympathise with than Heisenberg's.

[18] Robespierre's famous speech, his *Report on the Relations Between Religious and Moral Ideas and Republican Principles* of May 7, 1794, expresses all this very well. The art of government, he says, "has hitherto been the art of cheating and corrupting men, but ... ought to be that of enlightening and improving them". He proves the existence of God and the immortality of the soul thus. Man needs enlightenment and improvement. In order to achieve these man needs "more respect for himself and his fellowmen". In order to achieve these he needs faith in God and in the immortality of the soul. Finally, "I cannot see how Nature can have suggested to man fictions that were more useful than reality." This is fascinating, and a shrewd combination of a Cartesian and a Kantian mode of arguing. A Cartesian argument is from God's veracity and its form is, if an inquiry went properly yet the outcome was not the truth, then God would be a liar; which is absurd. A Kantian argument is transcendental and its form is, if such and such were not the case then knowledge would be impossible; but knowledge exists. Robespierre argues from the existence of knowledge and morality to the existence of religion and from the existence of religion in a proper manner to its truth – a combined Kantian-cum-Cartesian argument. This is not a small achievement for an allegedly verbose second-hand-ideas-spouting mere politician.

Since the above argument is rational, Robespierre continues, it is binding yet non-partisan: opponents to it are merely irrational. "You fanatics have nothing to hope from us. To recall men to the worship of the Supreme Being is to deal fanaticism a mortal blow. All follies fall to the ground before Reason; all fictions fade away in the light of truth. *Without compulsion, and without persecution, all sects are to be merged* in the universal religion of virtue." (Italics mine.) (J. M. Thompson, *Leaders of the French Revolution*, London 1929; New York 1967, pp. 236–7.)

In his second speech of July 26, 1794, he is even sharper. "I know but two parties, that of good citizens and that of the bad. ... There does exist a generous ambition ... an egoism of enlightened men. ... What then are we to do? ... to establish a single control ... and thus to crush all factions under the weight of national authority, and to build on their ruins the power of justice and freedom" (*ibid.*, pp. 139–141).

See also J. L. Talmon, *The Origins of Totalitarian Democracy*, London 1952, particularly the quote from Lemercier (p. 36) on the beneficial 'despotism of evidence', and on the "'natural and irresistible force of evidence' which rules out any arbitrary action on the part of the administration", and the quote from Dubois (p. 167): "Either there exists no demonstrable ethics at all, or there should exist only one – just as there exists only one geometry." (Translations mine.)

See also H. B. Acton, 'Prejudice', *Revue internationale de philosophie* 21 (1952);

J. W. N. Watkins, 'Milton's Vision of a Reformed England', *The Listener*, January 22, 1959; and F. A. Hayek, *The Constitution of Liberty*, London 1960, p. 527, n. 15 on Jefferson's opposition to academic freedom.

As to the sentiment prevalent today, Popper claims that the slogan 'Vox populi – vox dei', i.e. the idea that public opinion has already achieved a state of near perfection, is very widespread in western democracies, he even says it is the official ideology of western liberalism, and he wishes to combat it. See Popper's 'Public Opinion and Liberal Principles' in his *Conjectures and Refutations*. The same view, or a similar one, is criticized by Michael Oakeshott in his 'Rationalism in Politics', though from a very conservative viewpoint; M. Oakeshott, *Rationalism in Politics*, London 1962.

[19] De Finetti's view was foreshadowed by Wollaston, as reported by Faraday in 'Observations on Mental Education', alternatively, 'Observations on the Education of the Judgment', *Lectures on Education Delivered at the Royal Institution of Great Britain*, London 1854, reprint, 1855; also in *Modern Culture, etc.* (ed. by E. L. Youman), London 1867; also in *Science and Education: Lectures Delivered in the Royal Institution* (ed. by Sir E. Ray Lankester), London (1917); also in M. Faraday, *Experimental Researches in Chemistry and Physics*, London 1859.

For De Finetti's views, see *Studies in Subjective Probability* (ed. by H. E. Kyburg Jr. and H. E. Smokler), New York 1964, and bibliography there. See also H. E. Kyburg, 'Recent Work in Inductive Logic', *American Philosophical Quarterly* 1, No. 4 (1964), and bibliography there.

The sentiment is well expressed in J. W. N. Watkins, 'Decision and Uncertainty' *Brit. J. Phil. Sci.* 6 (1955), which is a review-article of Shackle's *Expectation in Economics*. See his summary (p. 78): "Although there are ineradicable elements of uncertainty in human life, and although it is the most significant decisions whose outcomes tend to be the most uncertain, the classical theory of decision-taking, common to both the philosophical and the economic utilitarians of the 19th century, presupposed foreknowledge of the decision's outcome. In the 20th century foreknowledge was reduced to knowledge of the probabilities of the possible outcomes of a decision, in line with the tendency to substitute probability for certainty."

[20] S. E. Toulmin, 'Crucial Experiments: Priestley and Lavoisier', *J. Hist. Ideas* 18 (1957), reprinted in *Roots of Scientific Thought: A Cultural Perspective* (ed. by P. P. Wiener and A. Noland), New York 1957.

[21] Priestley's case is discussed in my *Towards An Historiography of Science*, Mouton, The Hague, 1963, '12. Priestley's Dissent', and in my 'Revolutions in Science, Occasional or Permanent?', *Organon* 3 (1966).

[22] The best description I know of all this is in F. L. Will, 'The Preferability of Probable Beliefs', *Journal of Philosophy* 62 (1965). I do not pretend to understand Will's discussion – nor do I consider it interesting – but I do think his conclusion (pp. 66–7) is striking. There is, he says, "no logical room for such questions" as why prefer probable beliefs. To say probably *A* often means, to say *A* is preferable; or else it "serves ... to express and appraise the grounds of possible assertions. ... And in some uses the specification of the strength of the grounds for a proposition may not, for somewhat special reasons, close the question of the credibility of that proposition. ... In some circumstances the appraisal ... may be made in such a way that the acceptability or credibility of the proposition ... [also depends] upon whether the person ... is, in fixing [!] his beliefs, acting within what may be referred to collectively as the institution of human knowledge." Whether Will has in mind *only* people who are contemptuous of science – who are at liberty to reject credible views and "what might be referred to

collectively as the institution of human knowledge" – I do not know. Interestingly, his description fits Einstein as well: "I live in that solitude which is painful in youth, but delicious in the years of maturity" ('Self Portrait', *Out of My Late Years*, London 1950, p. 5).

Does Will recommend alliance with "the institutions of human knowledge?" He does not say. It is not a matter upon which "the institution of human knowledge has a confirmed belief"; in the west that institution tolerates dissent and rebellion and the rejection of the most confirmed belief. See my 'The Confusion Between Physics and Metaphysics in the Standard Histories of Science', in *Proceedings of the Tenth International Congress for the History of Science, Ithaca, 1962*, Paris, 1964, reprinted here.

²³ The case of William James is rather complicated. On the one hand James was prone to assertions which are quite exasperating in their obvious unacceptability. On the other hand, James's desiderata were interesting though highly problematic. This may explain how come Russell wrote, in reply to the criticism that he had "caricatured pragmatism by saying that, according to it, truth is what pays"; merely "but this is a verbal quotation from William James" (Russell, 'Reply to Critics', in *The Philosophy of Bertrand Russell* (ed. by P. A. Schilpp), Evanston 1944, p. 731). James's desiderata were the defense of pluralism as well as of extreme empiricism, while being a reformist. See R. B. Perry, *The Thought and Character of William James*, Harper Torchbook 1964, chapters 24, 27, 30, 32, and especially the rather moving Conclusion.

²⁴ See Ellis' comment on *Novum Organum*, II, Aph. 36 (Experimentum Crucis), *Works*, new ed., p. 297, note 2: "Nothing shows better than an instance of this kind, the impossibility of reducing philosophical reasoning to a uniform method of exclusion. Bacon seems to recognize as the only true form of induction .. that ... which proceeds by exclusion. ... The argument depends on a wholly non-logical element, the conviction of the unity and harmony of nature."

For Bacon's pyramid of nature and science, see notes 7 and 17 above. The most explicit expressions of these sentiments concerning parallelism between mind and nature are to be found in Bacon's frankly mythological writings, such as his *Wisdom of The Ancients* (myths of Pan, Echo, and Proserpine), *Thoughts on the Nature of Things, De Augmentis Scientiarum*, Book II, Ch. XIII, as well as his preface to his *Great Instauration*, Preface and Conclusion of *Novum Organum, Parasceve*, etc. Those who wish to take Bacon's mythology lightly should consult Spedding's introductions to both *The Wisdom of the Ancients* and *The New Atlantis*, as well as C. W. Lemmi's *Classical Deities in Bacon*, Baltimore 1931. See also *De Augmentis* Book 7, Ch. III, where the parallel is more boldly stated as the topics are "the culture of the mind" and "moral knowledge" (*Works, op. cit.*, Vol. IV, p. 28). Moral knowledge, however, is declared in the opening of *Valerius Terminus, The Advancement*, and *De Augmentist*, to be the highest end of all learning. These are the most cabbalistic passages of Bacon (esp. *Valerius*, Ch. I), which adumbrate the intellectual love of God most clearly.

There is no doubt that one way or another all inductivist philosophers felt that God must have put some constraint on Nature so as to render Her comprehensible by inductive means (see quote from Bacon in note 12). It was J. M. Keynes who, in his *Treatise on Probability*, Cambridge 1921, Chapter 23, claimed that both Bacon and Mill were rather vague about it, and he, wishing to be clear and explicit, postulated his famous principle of limited variety. The principle is clearly enough stated by Ellis, and quoted in the beginning of this note.

The most subtle attitude towards this principle and similar ones is Kant's, need one say? On one hand, he says, any such principle is concerning reason and nature and so could only be judged from above both nature and reason – namely not by us. He even goes further and rejects violently any such principle even as the merest hypothesis. Transcendental hypotheses, as he calls these, "such as the appeal to a divine Author" (A773 B801), are not explanatory, and lead the mind to unhealthy self-satisfaction (A772 B800). "Order and purposiveness in nature must themselves be explained from natural grounds and according to natural laws; and the wildest hypotheses, if only they are physical, are here more tolerable than a hyperphysical [transcendental] hypothesis, such as an appeal to a divine Author, assumed simply in order that we may have an explanation."

On the other hand, Kant does not like Hume's 'censorship', and claims that transcendental hypotheses may be stated; that he always likes to read new books defending them. He adds that transcendental hypotheses are useful to combat contrary transcendental hypotheses – the result of the combat should be a draw. That even after the draw there is use – one may take such a hypothesis as an ideal, or as a regulative idea or principle, which should guide one's research (rather than elicit false satisfaction). What is unclear to me is whether the ideal is inter-subjective or private. It can be nothing else; if it were inter-subjective it would be apriori demonstrable, and if private they do not belong to the metaphysics of morals, namely, they are not rational guiding principles (of research). In a sense Kant says they are private (A782 B810), in a sense he stresses the universality of the ideals or regulative ideas or principles of pure reason. I cannot follow his subtlety there. I can only say, surely he wants research, the intellectual love of God, to be a supreme universal principle, but he may perhaps claim that research is not necessarily bound to ideals. This is not satisfactory, but I leave it at that.

Much has been said in comment on Carl Becker's *Heavenly City of the Eighteenth Century Philosophers* and its claim that there is so much in common between the medieval and the 18th-century philosophers. Yet one might sum up the intellectually relevant similarities. First, the main difference between the Medieval thinkers and their heirs lay in the formers' sense of utter impotence. The medieval philosophers shared with the Renaissance philosophers the dream of the recapture of antiquity, but before Brunelleschi's success in constructing a dome in an ancient manner, followed by his disciples' success in sculpting, painting, and building, like the ancients, the general sense was that of impotence. Similarly, the idea of the hierarchy of knowledge and the knowability of the world, potent in the post-Baconian era, existed before. Indeed, the main message of Bacon, as he himself stresses, is a message of hope and of cajoling people to do research.

The second important difference is that the medieval writers are much more confused than the modern writers. This point is hard to divine from secondary sources, such as Crombie's in particular, because authors usually quote only what they think is clear or what they wish to clarify, or some such. Confusion in the Middle Ages much depends on slavery to diverse ancient authorities; but, as Galileo showed, the elimination of error is an arduous task, and of confusion even more so.

So much for the differences; as to the similarities, they lay in the confusions between Plato and Aristotle, between induction and deduction, between learning (methodology) and knowledge (epistemology), all as means for supporting the optimistic thesis of the knowability of the universe, to be found both in the Middle Ages and later. Yet, on the whole, in the 18th century this led to an ode to nature and to man's mastery of nature, whereas earlier it often sounded like a wail over a paradise lost.

(On the unity of science in the Middle Ages see M. DeWulf, *Philosophy and Civilization in the Middle Ages*, Princeton 1922, end of Ch. 4, vi and 5, i, ii; DeWulf is a Catholic apologist worse than Crombie, but he is still rather informative.)

And so, again, Crombie may score a point in viewing Grosseteste as a predecessor after a fashion to Bacon and to Galileo. But in a way I do not think he would like. In his second chapter he admits (p. 32 note) that the 12th-century philosophers received these ideas from Aristotle via Boethius; yet he calls them Platonists and complains of their mistrust of the senses – forgetting Kepler's and Galileo's mistrust of the senses; forgetting that Grosseteste himself does too (p. 73)!

And even Francis Bacon who is among the Moderns almost the only Quixotic believer in the senses, even he constantly and systematically refuses to side with Plato or Aristotle about the hierarchy of knowledge, speaking of 'axioms and definitions' in one breath, more systematically then Grosseteste and Roger Bacon, and even claiming explicitly that Plato is the only one who tried the inductive method, but only after corrupting science by mixing it with theology (*Novum Organum*, I, Aph. 105). (The widespread Ellis translation says, "Plato, who does indeed employ ... induction ... for the purpose of discussing definitions and ideas." The expression "of discussing" is a rendering of "executiendas"; in Khintchin's translation, Oxford 1855, it is rendered "of formation"; why not "of executing"? Ellis also translates "dialectics" as "logic". See my dissertation, University of London, 1956, unpublished.)

All this, of course, raises again the question of the difference between apriorism and inductivism, both in the Middle Ages and later. Now in the Middle Ages the hierarchy of causal theories was identified with Jacob's ladder, namely with divine illumination. During the late Renaissance, as the outcome of some measure of growth of both rationalism and optimism of sorts, two attitudes developed towards divine illumination. One was the neo-Platonist or light-mystic, i.e. Cabbalistic attitude run so optimistic as to expect illumination more or less here and now. The other, more academic, attitudes excluded divine illumination altogether and replaced it with the authority of the senses. And so, whereas in Grosseteste knowledge by divine illumination has no recourse to empirical basis but defective knowledge is empirical, in the late Renaissance academies both kinds of knowledge tend to merge and raise the serious problems *which are still with us*. And to think that it all starts with Aristotle's two systems – of certain and of probable knowledge!

[25] For the connection between Whewell and Duhem, see my 'Duhem Versus Galileo', *Brit. J. Phil. Sci.* 8 (1957), as well as my *Towards An Historiography of Science, op. cit.*, Section 10.

For Duhem's influence on Meyerson see Meyerson's preface to his *Identity and Reality* and references there.

See also my 'Sensationalism', *Mind*, 1966, reprinted here, concerning the difficulty involved in the view that there are uninterpreted data.

APPENDIX ON KANT

One of the facts I have attempted to explain in the notes to this chapter is the attitude which the rationalists of the classical school, of the Age of Reason proper, showed towards their critics – an ambivalence: an attitude of respect, of respectful disagreement and even of appreciation, mingled with an attitude of contempt. I have singled out Immanuel Kant, the cleverest philosopher and the peak of that age, as well as Baruch Spinoza, the profoundest and one of the pioneers of that age. I wish to enlarge a bit on Kant, using Arnulf Zweig's edition, in his own translation, of Kant's philosophical correspondence (University of Chicago Press, 1967), from which the reader may get a more comprehensive picture. Here I only wish to present a few snippets.

Kant expressed his attitude of respect and appreciation towards criticism in various letters. "I am not offended by your criticism" he says in one letter (p. 234). In another, more complicated one, he says (p. 158), "I have always thought it my duty to show respect for men of talent, science, and justice, no metter how far our opinion may differ. You will, I hope", he wrote privately to a friend whom he was attacking in print, "appraise my essay... from this perspective." That is to say, seeing himself as a man of talent, science and justice, Kant hoped that the target of his criticism will respect him. He goes on thus: "I was requested by various people to cleanse myself of the suspicion of Spinozism, and therefore, contrary to my inclination, I wrote this essay. I hope you will find in it no trace of deviation from the principle I have just affirmed."

Of course, Kant was very much of a Spinozist, as I have explained elsewhere in this chapter. And, yet, as he says, he could cleanse himself of the suspicion of Spinozism nonetheless (though, personally, the word 'cleanse' makes me uneasy), because, as he explains in a letter of a few weeks earlier (p. 152), he was less of a Spinozist than Solomon Maimon. At the distance of two centuries the distinctions Kant draws between his views and those of his associates may look totally insignificant, but this is neither here nor there. It is also amusing that Kant finds controversy "contrary to his inclination" even when he debates with "no trace of deviation from the principle" of respectful disagreement; and he even hides behind friends. It may sound strange to the modern ear, but in fact it was the style of the period; already Robert Boyle did it; they somehow mentioned friends to make the act more social and less personal: even when one goes by the book one may secretly enjoy the stab at an opponent, which is not very nice.

That Kant was rather nice is amply evident. In the letter about Maimon addressed to a mutual friend, which I have already mentioned, he says (p. 155) very openly that "Maimon's book contains... so many acute observations, that he could have it published at any time with no small advantage to his reputation, and without offending me thereby, though he takes a very different path than I do." He also wrote to Maimon himself (Maimon, *Autobiography*, London, 1888, p. 282), assuring him of his entertaining "no feeling of disparagement" toward Maimon's "earnest efforts in rational inquiries", which "betray no common talent for the profounder sciences." All this is so very nice that I wish I could leave things at that. But in his letter to the mutual friend Kant goes on to make two qualifications. First, he stresses that Maimon's disagreements with him are secondary: the letter continues thus: "Still, he agrees with me that a reform must be undertaken, if the principles of metaphysics are to be made firm, and few men are willing to be convinced that this is necessary." The word 'still' indicates, I suppose, Kant's satisfaction with the fact that Maimon agrees with him on a fundamental and highly controversial point. "But, dearest friend," he continues with a second qualification, "your request for a recommendation from me, to accompany the publication of this work, would not be feasible, since it is after all largely directed *against me*." And Kant goes on, neither commenting on Maimon's disagreements with him nor leaving the point alone: He notices the incompleteness of Maimon's work and makes some suggestions.

Again, I feel, the logic of the situation is so very different in our own age that the reader is apt to misconstrue Kant's reasoning. Today we take it as a matter of course that Einstein could publish a recommendation of a work directed against him. That is because Einstein was quite skeptical about the truth of his own theories; he was in search of the truth, but he doubted that his search was ever successful. Solomon Maimon expressed the same view in a letter to Kant about Kant's views (p. 175). Kant did not respond. I think it is clear that Kant could hardly respond without losing his temper. At least when he did respond – to Fichte, not to Maimon – he felt justified in losing his temper. He said against Fichte (p. 254)

"May God protect us from our friends, and we shall watch out from our enemies ourselves;" and he said against Fichte (p. 254), "... I took the completeness of pure philosophy within the *Critique of Pure Reason* to be the best indication of the truth of my work ... I declare again that the *Critique* is to be understood by considering exactly what it says and that it requires only the common standpoint that any cultivated mind will bring to such abstract investigation ... the system of the *Critique* rests on a fully secured foundation, established forever; it will be indispensable too for the noblest ends of mankind in all future ages." When one sincerely views a book in such a light – complete, fairly easy, and established forever – then one cannot suffer dissent gladly. True, Kant had other reasons to be angry with Fichte, and so he spoke freely in a public declaration. But he had no reason to be angry with Maimon, except that Maimon was not in full agreement. And so, though he said Maimon's disagreement was not offensive, we remember, nevertheless, in a very revealing passage in another private letter, not even concerning Maimon or disagreement, he failed to control his temper when he remembered Maimon's disagreement.

The revealing passage intrigues me on a few counts (particularly since it concerns senescence). Here it is (pp. 211–212): "... age has affected my thinking ... I feel an inexplicable difficulty when I try to project myself into other people's ideas, so that I seem unable really to grasp anyone else's system and to form a mature judgement on it. (Merely general praise or blame does no one any good.) This is the reason why I can turn out essays of my own, but, for example, as regards the 'improvement' of the critical philosophy [of Kant] by Maimon (Jews always like to do that sort of thing, to gain an air of importance for themselves at someone else's expense), I have never really understood what he is after and must leave the reproof to others.... Otherwise I am quite healthy, for a man of 70."

Antisemitism apart – no one in his senses would take Kant for an antisemite – what strikes one about all this is the odd mixture of astute self-knowledge, with coarse dogmatism, as well as of resignation with self-irritation. Without being able, by his own admission, to penetrate Maimon's ideas, Kant was able to remain convinced of their worthlessness. This is out of character: he generally took it for granted that there is no point in condemning an opponent before being able to empathize with him to quite a high degree. Nevertheless, since he could not empathise with his critics well enough, he decided that he was more efficient going on developing his own ideas further than criticizing Maimon. He was resigned to do so, yet it irked him very much. He must have noticed his own rigidity, and he could not derive sufficient comfort from the completeness and finality of the system which Maimon was so conceited as to try to improve. Perhaps he was not sure of the finality and completeness after all; who knows? He must also have noticed that going on developing the same ideas was following the road of diminishing returns, getting oneself into a rut. But this is an inevitable result of finality plus completeness. Laplace, in the end of his *System of the World*, having declared Newton the greatest and luckiest, asks, what is there for us, his lesser successor, to do by the way of improving the state of the science? And, he answers, merely to develop the application of Newton's ideas a step of two further on.

No doubt, completeness and finality lead to diminished returns; but we do not need these for diminished returns – we may get them anyway. A man can get there when he is too old to break from his own system – I have explained elsewhere that this is what makes one old, the inability to follow the struggle for improvement. But perhaps Maimon's improvements were really too small. Let us glance, then, at Kant's change that Maimon was a parasite.

When is an author's work to be considered an attempted improvement on another author's? The obvious answer is, an improvement which ceases to be marginal should be considered on its own. This inhibits one from declaring one's work more than a mere improvement; and so a measure of humility is all too possible a cover-up for something rather parasitic. On can speak of intellectual indebtedness, perhaps, instead of parasitism. No matter: however greatly one author may or may not deviate from his predecessor, he may wish to stress his indebtedness – as I do towards Popper. Or one can take the other's works as a point of departure and care little as to how far one has departed. Why, then, does it matter whether one is attempting a small improvement or a revolution, whether one is an intellectual parasite or a megalomaniac?

The only possible answer that may signify relates to an author's choice of his possible audiences. And for this what matters is his point of departure. For, what characterizes one's audience is primarily, their interest and level of comprehension or background knowledge. There are secondary qualities of audiences, such as boldness and good taste, but we need not discuss these. If there was an audience concerned only with the broad outlines of Kant's philosophy, then Maimon's offering to them but a slight variant of the same may be of a questionable value. If, however, there were readers concerned with more rigorous details, matters could stand differently. Moreover, the skeptical component which Maimon tried to introduce to Kantianism was by no means a small matter. Indeed, it was so significant a change that it got lost: Maimon had nobody to carry on his skeptical ideas and he created no skeptical tradition within rationalism. Rationalism remained rigid and this, no doubt, facilitated the rise of romanticism.

Perhaps it is too much trouble to go into details of philosophical schools. But major philosophical problems – particularly of rationality – do deserve the growth of philosophic traditions. Maimon could give rise to a tradition of critical rationalism; he did not. It would be sad if such a venture should fail again.

CHAPTER 19

CAN RELIGION GO BEYOND REASON?

> Real mystics don't hide mysteries, they reveal them. They
> set a thing up in broad daylight, and when you've seen it
> it's still a mystery. But the mystagogues hide a thing in
> darkness and secrecy, and when you find it, it's a platitude.
>
> G. K. CHESTERTON, 'The Arrow of Heaven'

Let me first state my views on salient points, so as to declare my hand.
Explanations will come later.

Religion in its traditional forms is a thing of the past – largely due
to the development of science and to the discrediting by the sciences of
religion's archaic views of the world and of man. There are constant
attempts to retain and revitalize parts or aspects of traditional religion
in the new conditions. These are transformed ritual, transformed faith,
and meaning, where meaning is meant to retain aspects of salvation.
It turns out that these three aspects rather hang together, and that faith
still seems to clash with science. It is my observation that these days
see the growth of a new silent avant-garde of able and civic-minded
religious scientists. They belong to various denominations and hold a
new version of religious philosophy which follows Duhem, Buber, and
Polanyi. It is compatible with science and revives ritual and faith in a
desperate effort to find meaning. I oppose this avant-garde philosophy
as one which makes its holders more living-dead than is bearable, as
one which empties both science and religion of their significance. Following Arthur Edward Waite, I find quest to be more significant in
religion than faith, or ritual, or salvation. Like Russell in his less belli-
cose and more pensive moods, I find quest to be the heart of research,
and I find it full of religious overtones. The true religion, the quest,
seems to be in science now as in Spinoza's days.

I. RELIGION AND REASON

We do not know what constitutes religion and we do not know what constitutes reason. Since reason regularly allies itself with science, we may just as well confess right away our ignorance of what constitutes science. Had we been in possession of theories of religion, of reason, and of science, we would then try to use these theories to answer the question in our title: Can religion go beyond reason? But we are not in possession of such theories; one who wished to answer this question, nonetheless, may first offer such theories. However, it is bad business to start with a full-fledged theory rather than with a problem and a problem situation.[1] For the sake of the problem at hand, I shall begin with traditional religions, traditional theories of rationality, and the corpus of scientific knowledge; I shall suggest that the traditional views on these matters are defective; I shall then discuss some modern modifications of these – always with an eye on our question, of course.

Traditionally – that is to say, in the Western tradition – faith and reason meet and immediately clash when Jew (faith) and Greek (reason) meet in the Hellenistic world. Traditionally, for neither pre-Hellenistic Greek nor pre-Hellenistic Hebrew does the problem arise, since the problem rests on the specific conflicts between the two traditions so symbolized and which neither knows before they meet – not even Job, not even Ecclesiastes. Julius Guttmann says in his classic *Philosophies of Judaism* that there is no (rational or critical) philosophy in pre-Hellenistic Judaism. As many writers have suggested, pre-Hellenistic Greek philosophy is singularly free of religious problems proper – its theology and ethics being only very loosely linked with any specific religions. In Hellenism this changed drastically: the Philonic tradition tried to harmonize faith – a specific faith, that is – and reason. The Talmudic tradition, just as much as the Tertullianic and Augustinian, claimed that reason is, and should be, limited. One may not ask the unanswerable question! The differences between traditions were differences of commitment as to which was the true faith. The agreement was regarding the claim that beyond the limit of reason stood one specific true faith. So were matters understood by Talmudists and Cabbalists, Scholastics and alchemists. To quote Alfred Weber's *History of Philosophy* (also quoted in *St. Anselm's Basic Writings*), "The Second Augustine, as Saint Anselmus had been called, starts out from the same principle as the first; he holds that faith precedes all reflection and all discussion concerning religious things...." Maimonides indicates in his

Guide for the Perplexed that there existed in the Middle Ages a school of unbelievers. They tried to show, he more or less reports, that there is no room for faith, and their proof was based on the claim that there is no limit on reason. That is, as a matter of principle; nobody ever as yet denied that reason is, in fact, limited. He agrees that this is the best way for the unbeliever to destroy faith: The only way, were it possible, to destroy faith, is to deny that reason is limited. Here, Maimonides says, the unbelievers and I understand each other very well.

Change came in the Renaissance, and with it the problem, the conflict, as we know it today. Admittedly, in Medieval philosophy one can find passages suggestive – but not much more – of modern controversies; yet, the modern concern with the peace between faith and reason belongs to the Renaissance of science. The problem – "does reason conflict with faith?" – became central. Not much room was left even then to the question, assuming that there is no possible conflict between reason and faith, "can they cooperate or not?" This question is characteristically of our age. The concern of the Renaissance remained, like that of Maimonides, to answer those who wished to reject faith in the name of reason. And, let us be clear: Maimonides' line was – still is – very strong: The way to dispose of faith is to make unlimited claims for reason. In the Renaissance people wished to make unlimited claims for reason, but without thereby wishing to dispose of faith. "Can this be done?" was really their question. The question was put in cold storage by the Royal Society at the end of the Renaissance.

When the Royal Society of London was founded in the mid-seventeenth century, its chief concern was to separate faith from reason in order to prevent any possible conflict. This is all well and good. The question is, however, can we separate the two? One can always say faith and reason are inherently separate, since they share no problem. This, however, may mean that there is no limit to reason, that reason handles all questions, and faith handless all quests (but not the questions). And, we remember, Maimonides viewed this as the only foundation for the antireligious philosophy which a believer should seriously criticize.

The tradition based on this claim is very important. Therefore, many authors, from Maimonides to Marx to Marcel, agree: Either reason is unlimited, or is it not. If it is unlimited – meaning if on principle reason can solve all problems – then there is no room for religion in any significant sense; it is then relegated to poetry and emotion and such. If reason is limited, and we may assume it to be limited in any way, then certain problems are beyond reason. Perhaps even reason depends on other agents for success; perhaps there is no reason and no sicence without intuition, instinct, belief, or whatever you call it. In such cases room is made for religion. In such cases, perhaps, there may be room for rational theology, perhaps for the application of reason to the study of the intuitive faculty.

How much of all this was acceptable is hard to say. What seems to have been widely accepted is that either reason is unlimited and excludes religion, or it is limited and calls for religion. Soon one side of the dichotomy got the upper hand. Ever since the foundation of the Royal Society, the claim that beyond reason stands faith has been questioned. This naturally led to the conclusion that reason is not limited. The success of Newtonian science led spokesmen of science increasingly to the bold expression of the view that reason is unlimited, and when they became bolder they openly concluded from this that reason must be hostile toward faith. To be more precise or to put it in modern parlance, the religion of science became increasingly hostile to all established religions, Christianity and Judaism in particular. Individual religious scientists found their positions increasingly uncomfortable. As Michael Polanyi puts it in his *Personal Knowledge*, John Locke, as the spokesman for the new scientific community, kicked religion upstairs, made it like the lords and kings of England – venerable but powerless. Authority went to the House of Commons.

Since the crisis in physics at the turn of the present century, a new breed of religious scientists has developed. Members of the breed tend to endorse an instrumentalist philosophy of science, one similar to and very often influenced by, that of Pierre Duhem. Such religious philosophies of science strip science of its claim to know about the nature of things. The new religious scientists also endorse Buber's quasi-existentialist philosophy of religion which considers religion a private matter between a man and his god. And they endorse Buber's and Polanyi's traditionalist philosophy, according to which there is no rationality without prior commitment; and commitment, though not entirely arbitrary, is arbitrary within the limit of coice between existing traditions and the philosophies they endorse. In a distinct sense this group of scientists plays the role of the religious avant-garde of our days; the taste makers and molders of educated opinions and attitudes in the religious sphere; a quiet avant-garde of professional scientists of a traditionalist inclination who prefer to operate within their religious, social, and political institutional frameworks, rather than use open public platforms for open debates.

To conclude the present introduction, let me present the broad outline of the situation as I see it. The conflict between faith and reason has occupied much of the literature on the relation between the two, including the theological writings of Kepler and Galileo, whose chief concern was to prevent any such clash. Accepting their view, most philosophers now agree that either reason is limited, thus making room for faith, or not, thus rendering faith a matter of mere psychological or poetic interest. Very few writers, notably Kant, thought differently: Reason is limited yet the claims of faith have to be carefully checked. Somewhat in line with this philosophy, I suggest we depart from traditional polarizations, and even from traditional equations.

> Tradition has (falsely) equated:
> religion = faith;
> as contrasted with:
> reason = science.

I find both equations unacceptable and the polarization between reason and religion even more objectionable. Still following Kant, and more so in accord with more recent views, particularly of Popper and of Bartley, I suggest that reason is limited, and that hence there is room for faith within reason: Such a faith conflicts with the modern versions of traditional religion as advocated by the avant-garde religious scientists as described here.

In the present chapter I shall try to present the background of this avant-garde movement. Here is an outline of my presentation (the numbers indicate the subsequent sections).

II. There is a dissatisfaction with both science and religion.

III. Once science was a handmaiden of established religion. When science freed itself of the authority of established religion, claims for science were made which later proved to be exaggerated. Both reason and faith seem now to be courting one another.

IV. But one must examine carefully the question: In what sense is it possible, and in what sense desirable, that science and religion supplement or complement each other?

V. What each expects from the other is that it complement the other's intellectual weakness. Otherwise the intellectual dissatisfaction with both will not be removed.

VI. The idea of cooperation, then, is that of intellectual supplementation between science and religion, which idea emerges from the intellectual disappointments in both.

VII. Hence, these disappointments should be the first indications of ways leading to remedy.

VIII. To evaluate these we need standards of rational thought and of rational action more general than hitherto available.

IX. In view of the failure of credulity and naïve hopes, new rational standards must, first and foremost, be those of utter self-reliance – perhaps merely out of despair – much as expressed by Jorge Luis Borges in almost every essay of his.

X. But the inevitable dose of despair need not be as large as that contained in the pragmatism of the new avant-garde.

XI. The religious aspect of science offers a better remedy of existing defects – of both religion and science – than the uneasy merger of old-fashioned philosophies of science and of religion. The honest religion of science, the true agnostic religion, does easily what other blends cannot possibly achieve.

II. DISSATISFACTION WITH SCIENCE AND RELIGION

A. One may approach the situation from the scientific or rationalistic tradition. We have been looking for something – knowledge, power, happiness; success for the human race. We had expected to attain it with no outside help. It is this attitude which we call variously reason, science, humanistic agnosticism, mature self-reliance, rational responsibility. This attitude embodies a certain contempt toward those who rely on people whom they cannot or would not question (priests or party leaders) or on ideas they cannot or would not present and examine critically (the catechism or party line).

Is this self-reliance rationality? Or is it empirical science? It is hard to tell. As long as one is pleased with this attitude, with any attitude for that matter, one need not bother to clarify it and to nail down fine distinctions concerning it and related attitudes. But something may go wrong. Some of our expectations may meet with deep disappointments. What should be radically modified? It may be reason, science, or self-reliance.

Alternatively, we may try to keep our old attitudes substantially intact, and modify them only to the extent necessitated by the addition of a new ally – whose task should be to undo the disappointment. This attempt is plausible and shows great respect for the old attitudes, even though they proved to be less potent than previously hoped. As it turns out, however, the intruder, like a cuckoo, soon outgrows and expels the older inhabitants.

B. Let us now approach matters religiously. There is an imperfection in man which science promises to remove but fails. And the question is, can religion succeed here? The perfection sought is what religious thinkers call grace. This is the meaning of the word grace (at least in this context; but, I suggest, even more generally). Grace, as we are told, is never a right; although we are not entitled to it, we may be granted it, especially if we fulfill certain conditions. Those who believe in grace – especially those who believe that they have attained grace – are different from those passionately engaged in the search for perfection. The searchers are troubled; the blessed are not. The searchers may not quite know what they are looking for: they may merely feel the need for some support, for some meaning in life, for some improvement. At first, it is true, they had expected it from religion, and then from science, and now they are bewildered and may even look again toward religion. It does not matter so much in the first place what the source and history of the dissatisfaction is – rather, what matters in the very first place is that there is, indeed, dissatisfaction. Once religion has given us the support we crave, then the primary dissatisfaction is removed and the situation is thus radically altered.

Not only one who has attained grace, but even one who listens to him in the hope of emulating him – regardless of how and why – has nothing to do with our discussion, even if the latter never will attain grace. The situation is similar to that of those theories of reason which have never fulfilled their promises of certitude or near-certitude in science; so long as one accepts the promise no problem arises. To be drawn to our present discussion, the religious person must be dissatisfied, disappointed, frustrated. He may, then, look to reason for consolation. And, taking a dose of reason to support his religion, he may, indeed, all too easily destroy his religion. But this alone will not do. He has to be doubly frustrated: Reason destroys his religion and fails to replace it.

C. The problem, then, is whether religion and science are complementary. Assuming that neither religion nor science alone is a sufficient means of attaining perfection in man, perhaps a combination of them would be. And even if a combination of the two would not perfect man, perhaps it would bring him nearer to perfection more rapidly than either component alone. This idea – of reconciling science with religion, the view that science and religion are complementary – must nowadays be quite popular, since it is peddled in all sorts of literature, from philosophy, history of religion, and science to sheer science fiction. Also popular, of course, is the traditional idea of separating science from religion, a result of the view that mixing science with religion destroys both. One might reconcile these two ideas of complementation and of separation in the following way: Science and religion may help each other perfect man – but only if use is made of each in its separate place. This view, that science and religion should be separate but complementary, is the one now coming into vogue within the scientific community. It is the chief aim of the present chapter to argue that this idea destroys the vitality of both science and religion, and is thus doubly objectionable.

III. REASON AND FAITH

The dual dissatisfaction with science and religion is rooted in Western history: Once science was the handmaiden of established religion; now reason and faith both seem to be courting each other.

A. In the second half of the thirteenth century, the Church openly attempted to suppress rationality or self-reliance; and one assertion which illustrates the mood of the avant-garde of the age was condemned, namely, that there is no need to accept as a matter of faith a thesis which can be accepted as a matter of reason.

In general, self-reliance was presumably viewed as such an obstacle to the endorsement of faith that even its specific employment in support of faith was feared. Efforts to destroy self-reliance abounded; the cleverest and most appealing to the self-reliant is the effort to do so philosophically. The Christian philosopher could destroy the self-reliant's self-reliance by proving to him the truth of Christian faith. But this is not all. Not only in Thomas Aquinas's *Summa Contra Gentiles,* but also in his *Summa Theologica,* the idea that science is subordinate to religion reigns supreme: Aquinas proves that without sacred doctrine there is no science. It is incredible: The man whose direct intellectual ancestry was Jewish and Muslim (Maimonides and Averroës), and whose intellectual heritage was secular and pagan (Plato and Aristotle), the same man said, in effect, if you are not a Christian you cannot be a scientist!

The view of science as handmaiden of religion has been retained after a fashion even in modern times. Thus, Descartes could still pretend that his philosophy, though it started in skepticism, ultimately reinforced religion. A few decades earlier, Kepler, Galileo, and Bacon tried to present science and religion as noncompetitors – on condition, of course, that under the pressure of reason the claims made by established religion will sometimes have to be modified. With the official institution of the scientific revolution, with the rise of the Royal Society and its scientific code, things changed even more radically.

B. After the foundation of the Royal Society in 1661, and prior to the Einsteinian revolution of 1905, the relations between science and religion seem to have been progressively those of polite hostility. Frequently the leaders of science were irreligious, or else they tried to conceal the fact that they were religious. There are exceptions, to be sure; but even since the formation of the Royal Society, the exceptions have been rare. Boyle and Newton were leading scientists and they were both religious – each in his own very peculiar and highly unorthodox way. And yet the forceful leader of science at the time was Edmund Halley (of the Halley comet), and he was an agnostic, though a less aggressive one than some or his successors. As is well known, Bishop Berkeley developed his philosophy in reaction to Halley's agnosticism. He also attacked Newtonian self-reliance as the cause of Halley's agnosticism. His views adumbrate the much more modern views, to be discussed below. Joining the scientific or rationalistic movement at that period often was tantamount to leaving behind established traditional religion. It was customary at that period to conceal the fact that one came to rationalism and science after disappointment with traditional religion: it doesn't matter what one thought previously, we all look alike before science. And so, after the foundation of the Royal Society, and prior to the Einsteinian revolution, reason, namely science, surreptitiously won over religion, while officially it was not hostile to established religion. Reason was not supposed, however, to prevent the study, critical or otherwise, of religion as an intellectual and social phenomenon; and the study was quite critical in part – and that part had quite a devastating effect. The official policy was expressed openly only in the latter part of the period, particularly within the Marxist movement: Let established religion live in peace; help rational education (both scientific and political) to develop; and subsequently religion will quietly fade away.

For three centuries, the seventeenth through the nineteenth, religion has been on the retreat; scientists often had no part in established religion, or belonged to a church but felt awkward about it. Even the religious scientists, including the pious Robert Boyle himself, openly preached against any religious idea which clashed with their own reason. Loyalty to science came first, and so in every conflict between science and religion, science invariably won. Bible criticism, archaeology, geology, Darwinian biology, social anthropology, every field which developed scientifically, led to new retreats for religion. A well-known instance of this is Albert Schweitzer's work early in this century – his *The Quest of the Historical Jesus*, and his doctoral dissertation on the psychiatry of Jesus; they are frankly apologetic, but only to the extent permitted by reason or science and scholarship.

C. Meanwhile, the dissatisfaction with late nineteenth-century science bred a new attitude toward religion. The dissatisfaction became almost universal among literary intellectuals, and affected many scientists. The trend had finally reversed, and scientists started courting religion. Russell's *Religion and Science* of the 1930s records symptoms of that transition, and in it he expresses his surprise at the phenomenon. It has occurred time and again, of course, that a philosopher who had attacked religion when young endorsed it when older (Heine, *Religion and Philosophy in Germany*), that a rationalist, with an anticlerical career almost completed, called the priest to his deathbed. But, as a public phenomenon what Russell narrates was obviously a novelty: A movement of religious scientists is a twentieth-century product.

IV. THE QUESTION OF COMPLEMENTARY RELATIONSHIP

Let us, then, try to examine the question: In what sense is it possible, and in what sense it is desirable, that science and religion supplement or complement each other?

A. The antireligious thinkers, such as Russell, readily acknowledge that, for fulfillment of life, science is not enough. We need friendship, and we need arts; we need all sorts of things apart from sicence. Religion was claimed to be a substitute for sex, for instance. Science was never seriously claimed to be such a substitute. At most, the claim of science was that of a means by which to achieve enlightenment – plus the subsidiary claim that religion cannot bring enlightenment. For instance, this is how Kant puts it in his *Religion within the Limits of Reason Alone:* It is only when religion claims to fulfill the function that science or reason also claims to fulfill, that the clash arises. And in each such clash, says Kant, science must win and religion lose.

Kant, in his desire to write a very liberal and tolerant book, was even willing to concede that one can be reasonable while believing in virgin birth, which he personally considered idiotic. Yet at the end of the book he says that whenever religion claims to do what reason claims to do, it is phoney. Against his intentions, against his temperament, when it came to the relation of religion and science, Kant was an enemy of established religion. Religion can only win in cases in which science cannot even start to compete! This was the situation until very recently.

For example, Michael Faraday, whose life was dedicated to science, was also a profoundly religious man, who (outside his church) rarely alluded to his religion; the rare exceptions were cases of enormous pressure. And yet he found no difficulty in alluding to all other nonreligious complements to science. In a moving passage in a letter to a friend (Schönbein), he says, "After all, though your science is much to me, we are not friends for science sake only, but for something better in man, something more important in his nature, affection, kindness, good feelings, moral worth."

B. Religious assertions were often dismissed by scientific leaders, if not as superstition, then at least as highly problematic and in need of much interpretation. This did not exclude even Robert Boyle, who was the most religious leader of the scientific community, of the commonwealth of learning, to use his phrase. He was, for instance, the man who instituted the rule that in scientific circles people should not argue about religion. He was deeply religious – he gave most of his money for religious purposes of missionary works (especially spreading the Bible) and of charity – and he was said always to have paused for about one minute after he used the Sacred Name in speech, with the result that he tried to avoid using it because it became a burden on his audience. Also, one must add, he was an unusually honest, frank, and sincere man. His philosophic doctrine was that there are two faculties of mind – reason and emotion; that justice belongs to reason and mercy to emotion: and that God gave us the ability to comprehend him by reason alone, as justice requires, but that out of pity for those of us who are a bit dumb or stubborn, he created miracles, which have a merely emotional appeal. It might seem strange that a man so deeply religious should have thought so, but so he did, and he was heard by many. He has also written that theological questions are beyond reason, and, therefore, that we should leave them to religion; but he never meant this to express the idea that religion is a complement to science. This would have been quite impossible anyhow, sine in his view religion is not another form of understanding. Perhaps he meant to suggest that since religion is ot a form of understanding, let religion try to handle the incomprehensible.

Until the twentieth century, the rule was that whenever science and religion disagreed, science won and religion proved to be wrong. Russell still held this view in 1935. The Bible says that the hare chews the cud and the biologists say it doesn't chew the cud; and, of course, it doesn't. I don't know why it is so important to Russell to insist that the Bible says that the hare chews the cud, that the biologists disagree, and that the biologists are right. I suppose it is a remnant of the reaction to medieval science. It is not uncommon to hear even nowadays such claims; we are still told repeatedly that the religious leadership was against inoculation which scientists recommended. In truth, many scientists were against many inoculations – sometimes correctly, sometimes not – and many religious people, as missionaries who went to the bush and administered inoculation, were for most advanced medicine, and even contributed to medical science in their small ways. It is, in my opinion, often difficult to know when and how and why established religion clashed with science, and precisely on what issues. And science is not always right in such clashes. It is even often unclear what is meant by the claim that science is always right in such clashes. Even the greatest clash – between the Church of Rome and Galileo – has turned out to be not half as obvious a case as most writers a century and two ago would have us believe.

C. Somehow, the question of how, exactly, do science and religion compete, belongs to history. By now established religion in the West has entirely capitulated on this issue. By now no religious leader in the West, not even the fundamentalists whose parents forbade the teaching of Darwinism in their public schools, not even the pope, would dare clash openly with science on any issue. What they offer is, they claim, what science cannot offer.

V. TOWARD INTELLECTUAL COMPLEMENTATION

What each side – the established religions and the rationalistic scientific movement – expects from the other, is that each complement the other in the intellectual arena and in the area of each other's weakness. Otherwise the intellectual dissatisfaction in either case will not be removed.

A. The cycle, then, is complete. Science once dared not contradict religion, and posed as an ancillary to religion. Now the opposite is the case. When Kepler, Galileo, and Bacon said that the book of nature cannot contradict the Bible, they meant to mollify opposition. When Pope Pius XII made the same idea into the guideline for his policy, it was an admission of defeat. You may reinterpret the biblical story of the creation, said the pope, if you believe Einstein's story of creation. You may believe Darwin's theory that man descended from apes only if you believe this occurred just once in history.

B. The development of a sophisticated view of religion as ancillary to science was brought about by developing the antireligous scientific view to its extreme. Extremes touch, we are told. When one takes any viewpoint to its extreme, said Samuel Butler, one sees its absurdity. The viewpoint in question, then, may be either discarded or complemented. There are those who push a viewpoint to its extreme in order to force its replacement (the scientific avant-garde), and others who do so in order to force its complementation (the rear guard of science, but also the avant-garde of religion).

Both extreme mechanism ("man is a machine") and extreme positivism ("only science makes any sense") are the paradigms here. They are so very narrow that they make it almost undeniable that there is more to life than science. When extreme mechanism presents the world as utterly dehumanized and aimless, it may suggest[2] to us that there is depth and meaning to the world, but outside science: the body belongs to science, and the soul to religion. Alternatively, a sensitive religious soul entering science may be drawn to mechanism in order to arrive at such a conclusion. He would say, "science does not capture meaning, but I do experience meaning"; hence some experience is extrascientific – let us call it religious.

The extreme positivist sees religion as a refuge for ignorance and a bastion for superstition. Apart from this, he may see nothing in religion; he may even refuse to comprehend the meaning of a proper name like "God" and declare it meaningless, wanting not only in denotation or designation or reference, but even in connotation or sense; it may declare theology proper as less than false, as sheer meaningless gibberish. The philosophers G. E. M. Anscombe and Frederick Copleston, S.J., have endorsed extreme positivism in order to advocate a move which is extremely easy to implement, which is nothing but the tacking of a small rider onto extreme positivism, and which is becoming increasingly popular in certain circles – as the magic solution to hosts of troublesome problems. They favor some version of extreme positivism just because it evidently requires complementation; any proliferation of the meaning of the word "meaning" or "sense" will permit this. "God" does not make *scientific* or cognitive sense, but can it make *artistic* sense, or *religious* sense, or perhaps *social* or *political* sense, etc., etc.? Proliferation and compartmentalization "sense" put an end to strife. When the same string of words appears in both a scientific and a religious context, then they do not necessarily possess the same sense; and hence, obviously, scientific discourse need never conflict with religious discourse; everybody is happy now – separate, but equal.

C. Yet, what we have achieved is complementation like that which love and friendship as well as arts and ceremonies offer, not intellectual complementation. The language of music is not the language of science, we all agree; and even the literal meaning of a ceremonial declaration is not of much import, at least according to the sophisticated modern bride who seemingly promises in church to love and honor (or even to obey), not only in the foreseeable future, but "until death do us part" – yet without meaning to disclaim her legal rights to equality and to divorce. All this was and is accepted – perhaps regrettably – without any contestng or debating. In such cases, no doubt, the ceremonial promise differs from a verbal contract and therefore the word "promise" may signify ceremonially something utterly divorced from what it signifies in business. We do not need extreme positivism to arrive at such conclusions, and such conclusions do not offer new complementations. Anscombe and Copleston offer us stale cakes instead of fresh bread.

VI. POSSIBILITIES OF COOPERATION

It is from dissatisfaction with both religion and rationalism that the new idea of cooperation emerges.

A. A lot of old intellectual rubbish is still extent; and sometimes the crudest arguments impress the sophisticated. Some religious writers try to show that some wise old sages may sound Freudian; that the taboos of Leviticus are hygienic, etc. The critics of this kind of intellectual rubbish fall into a trap when they expose the old sages as unFreudian fools, and when they explode all alleged connection between Leviticus and modern hygiene. For all this belongs to the quarrel between science and religion of the middle of the nineteenth century. Inasmuch as religious doctrines were concerned with matters of fact, they were often significant precursors to present-day doctrines, but one must take it for granted that they are now superseded. Religion now claims only historical respectability for some of its old doctrines, but no more; it does not claim that any of them is true. Contrary to the eighteenth-century mood, present-day scientists and, to a greater degree, present-day historians of science, often show sympathy toward such claims for historical respectability. In the wake of religious cultural historians such as Arthur Edward Waite, of medievalists like Huizinga, and, above all, of the illustrious historian of medieval science, Pierre Duhem, practically all historians of science today agree that, superstitious as astrology and alchemy surely are, in the Middle Ages they were part and parcel of learning, and as such belong to the history of science proper. Erroneous as biblical medicine may be, it was no worse (perhaps better) than all its competitiors at the time it was recorded; hence, it deserves our appreciation. Here we see a tremendous shift in the rationalist attitude toward old religious doctrines: as a result of a historical perspective, the attitude has become less contemptuous and more tolerant and even respectful toward old errors. Does this improve the relation between the religious and rationalist today?

B. The shocking fact for the rationalist is not that some religious doctrines are now respected as once reasonable but now superseded; but rather that some scientific doctrines are now in exactly the same category of respected as once reasonable, but now superseded. When scientific theories can be superseded, it is no longer feasible for the rationalistic advocates of science to hold religious theories in contempt for the same reason. The gospels are not gospel-true, but neither are the books of Newton. Any scientist who denies the last sentence should be told to read Newton in order to shake his dogmatism. No doubt, there is a matter of degree here – explicable by the fact that the Bible is older than Newton's *Opticks*. From a historical perspective there may be little difference in validity between the doctrines of science and of religion of one given period; in some cases the two doctrines are identical (Aristotle). What the one hostile to science forgets is that theories can be superseded only when they go with claims for ultimate truth;

he often stresses that in some sense scientific theories have not been superseded, as when they go with claims for useful application and when they are somehow absorbed in newer theories. In this sense theories cannot be superseded, but in this sense theories are not as rational as they were once claimed to be. The aspects of science that cannot be superseded, namely, usefulness and incorporation into later views, have never caused any hostility or clash with religion; the clash concerns aspects or interpretations of science – its claim for final theoretical knowledge – in which science definitely can be superseded. The classical claim was that scientific theories are absolutely true, that Newton's theory is the last word in mechanics, that Newton has achieved what the gospels failed to achieve.

Some religious thinkers stress that Newtonian mechanics or Daltonian chemistry have been superseded, though they were once claimed to be the undeniable literal truth, the demonstrated last word. These religious thinkers stress that such claims are no longer made, and their memory deliberately obliterated. It seems, then, that these religious thinkers are debunking science. It looks as if, in revenge for the scientist's debunking of religion, now the religious are debunking science. But this is a very gross error. We are not speaking here of irrationalists rejoicing in the inability of the rationalists to keep believing in what only yesterday they were claiming to be the dictates of reason. We are speaking here of scientists who yesterday were themselves such rationalists; who were disappointed in their rationalistic view of science; who subsequently ceased viewing scientific theories as the dictates of reason and who started to see in them more technology than enlightenment; who are returning to their church or synagogue.

And so, it is not that the religious are now debunking science; it is not tit for tat. The religious scientists, who seemingly debunk science by reminding us of old defunct promises, are sophisticated leaders of sophisticated communities; they debunk the old rationalist view of science, not science itself; they are moved by a sense of disappointment, not of hostility. Hence, they come to religion, at least in part, from science and old-fashioned rationalism.

C. We can now see what kind of enlightenment religion is offering; how that enlightenment can be claimed to be complementary to, but not competing with, science. What religion offers is intellectual commitment; faith in certain doctrines which are not amenable to scientific treatment, and which can be adhered to safely. Empirical facts and metaphysical doctrines are permanent. Scientific doctrines as doctrines proper, and religious doctrines which can clash with science, are both highly transitory and should be totally dispensed with. This, I contend, is the view now endorsed by the religious scientists who are the religious avant-garde; and though unacceptable, it is a very serious view and merits close examination.

This view is, in a definite sense, quite existentialist; but it is already fully articulated in the works of Pierre Duhem, the philosopher and historian of science of the period of the crisis in physics, the link between the Newtonian and the post-Newtonian era. He was a philosopher's philosopher and a historian's historian. He did not gain much recognition in his day. Just now he is becoming popular enough to be the leader of the twentieth-century intellectual avant-garde. Very soon he will be superseded, and then his doctrines may become accepted by the vulgar.

VII. DEFECTS OF BOTH RATIONALISM AND RELIGION

The disappointments leading to the new situations, both in old-fashioned rationalism and in old-fashioned religion, should be the first indication of ways leading to a remedy.

A. Ignorance, even in the best scientists, is not something new, nor was it left unexploited by the enemies of reason. But it did not impress scientists until recently. Indeed, we may represent the traditional viewpoint by a sharp quotation from Kant's *Religion within the Limits of Reason Alone:* "It is the commonest subterfuge of those who deceive the gullible to appeal to the scientists' confession of their ignorance."

The reason that scientists could so easily confess ignorance and yet be unmoved by proposals from the religious to seek enlightenment elsewhere is fairly clear: Scientists were arguing from a position of strength. It is not how much they knew, but their ability to know, the very idea of self-reliance through knowledge, that offered them more hope than all religion could.

This very idea was contested by Pierre Duhem. Science must be devoid of all pretense to theoretical knowledge, he said, because science can never prove its theories empirically. Unproven theories are more likely to be erroneous than true, and hence it is better to view science, not as a system of theories, but as a system of mathematical definitions used to correlate empirical data. Thus, if we think Newtonian mechanics is an empirical theory about the behavior of planets and stars, then we may be disappointed by the subsequent need to revise our theories. However, if we view Newtonian mechanics as a system of second-order total differential equations to correlate observations, then these are immutable. True, the domain of application of Newtonian mechanics, the range of correlated facts, is changeable: We constantly try to apply the equations to new situations or with increasing precision, until we are stopped by experience from doing so indefinitely. When our attempt to extend the range of applicability of our equations is thus frustrated, we may look for new equations.

In this manner Duhem succeeded in rescuing science from the state of permanent revolution to which it might have been thrown when it turned out that empirical proof of scientific theories is impossible. But there was a price to pay: The informative content of scientific theory was gone. Theoretical science had to be viewed as a mere mathematical system to pigeonhole and correlate empirical data, and its aim had to be viewed as mere convenience and usefulness. This view of the status of theoretical science (conventionalism) as mathematical, and of the aim of science (instrumentalism) as technological, sharply contrasts with the classical view of science as enlightenment – as chiefly the knowledge of the true laws of nature, as the true explanation of the empirical phenomena, with technology as a mere by-product of true knowledge.

In advocating this change Duhem had the support of antireligious philosophers like Poincaré and Mach, and he was very proud to count these as allies. But he alone went further, and argued that the new philosophy of science permits, perhaps even requires, an adjustment of our philosophies of religion and of enlightenment.

Duhem was an orthodox Roman Catholic; his philosophy helped him harmonize his religious and his scientific commitments. He readily admitted all this, but he stressed that he advocated his view of science, his conventionalism (theoretical science belongs to mathematics), and his instrumentalism (theoretical science belongs to applied mathematics), not only for the sake of religion, but chiefly for the sake of science itself.

B. The conclusion that religion has won, that it has not capitulated to rationalism in any way, is thus being pressed. Indeed, my own teacher, Sir Karl Popper, in his classic "Three Views concerning Human Knowledge" (*Conjectures and Refutations*, 1964), accepts it. I wish to explain my dissent from it.

Popper's argument is this. In the late Middle Ages, instrumentalism was the current philosophy of science, as Duhem has observed. The argument between the Church and the Copernican heretics was not scientific but philosophical. The same Jesuits who attacked Galileo's philosophy used Copernicus in their astronomic calculations. Saint Robert, Cardinal Bellarmine, S.J., accepted the Copernican hypothesis *mathematically*, and insisted that Galileo was transgressing his rights as a Catholic when accepting the Copernican hypothesis *philosophically:* As long as the Copernican hypothesis was unproven it was not the duty or even the right of Galileo, as a scientist or as a Catholic, to assert that Copernicanism was really or philosophically true. Strangely, not only Duhem, the Catholic, but even Poincaré, the free-thinker, endorsed Bellarmine's position; with Niels Bohr, says Popper, instrumentalism became the accepted fashion; and so science capitulated and the Church won.

In the tradition of science, it was taken for granted, as Giorgio de Santillana illustrates in a detailed study, that Bellarmine stood for obscurantism and Galileo for enlightenment. In the late nineteenth century, as Popper shows, Mach and Poincaré tacitly, and Duhem openly, endorsed Bellarmine's view of Copernicanism and rejected Galileo's. Copernicanism, they said, was not true information about the universe, about the center of the universe; rather, they said, it was a system of applied mathematics. Copernicanism, they said, does not tell us what and why, but how. Some unsophisticated historians of science still repeat the nineteenth-century story, according to which the debate between Galileo and his opponents is now dead, since Copernicanism has won; the observations of stellar parallaxes (shifts of the scenery caused by the motion of the observer), which Galileo could not observe (for want of a strong telescope), are by now established as the facts that prove Copernicanism to everybody's satisfaction. So claim most historians of science. But this claim is contestable; Poincaré and Duhem did contest it. Copernicanism is the statement that the sun is the center of the universe. (Even Newton read Copernicus so; he thought that the center of the solar system is the center of the universe, and was troubled by the discrepancy rooted in the fact that the center of the solar system is not identical with the center of the sun.) This assertion has been superseded. Taken literally it must be proclaimed false; of course, as a very powerful mathematical tool it is still useful within its limitation, and a much better tool it is than any of its predecessors. Thus, Bellarmine won.

This is misleading when taken as the overall picture. Bellarmine has won, but on a technicality and concerning a minor point. When Bellarmine argued that Galileo had no right to proclaim Copernicanism philosophically true, he was not debating with Galileo. Rather, he was threatening Galileo; more precisely, he was defending the Church's authority over the scientists. His argument concerning Copernicanism was a rider to explain his threat – a rider concerning just a point at issue, not the main issue itself. And Galileo agreed with Bellarmine on what the issue was. In his *Letter to the Grand Duchess*, he defended nothing less than the scientists' freedom from the authority of the Church, suggesting that the Church has no business telling scientist (as scientists) anything at all. It was the self-reliance of reason, of the individual's ability to read the book of nature without the aid of authority, tradition, and priests, that Galileo was defending. Bellarmine darkly and menacingly had hinted that Galileo was siding with the Protestants, and Galileo darkly and vehemently repudiated the charges. John Watkins, Paul Feyerabend, and other followers of Popper, have recently sided with Bellarmine on this: Protestants were self-reliant when reading the Book of God, and scientists when reading the book of nature. The Spinozist formula, *Deus sive Natura*, God equals Nature, the Copernican claim that the two books cannot contradict each other as they are both true, or any other kind of correlation, will make Bellarmine's hint very plausible. And though his hint may be merely plausible, already his fear that science, just like Protestantism, weakens the Church's authority, is amply justified. Indeed, science did undermine religious authority. Yet the Catholic Church has finally allowed men of science to be almost as self-reliant as their irreligious colleagues. And self-reliance spread both with the spread of scientific education and with the spread of the scope of science. Subsequently, some Catholic leaders declare openly that many moral problems have now become matters for individuals to decide in accord with their own consciences.

Bellarmine showed great insight: when self-reliance is allowed to any extent, there may be no stopping it. This insight is, indeed, Platonic – as explained in Popper's *The Open Society and Its Enemies*. Popper himself endorses it. The choice, he says, is between self-reliance and the return to the apes (return to complete dependence). Bellarmine lost to an extent that would have alarmed him – that indeed alarms many a Catholic leader today, leaders such as Cardinal Ottaviani, who, in protest, resigned his position as head of the Vatican Congregation in charge of faith and morals in 1967.

Protestants almost unanimously capitualed to science much earlier; the Protestant equivalents of Ottaviani are a handful of fundamentalists. Even branches of the most orthodox sections of Orthodox Jewry have accepted science. It is not merely that religion yields to science what is due to science. With few exceptions the religious these days allow the rationalists to spread the gospel of self-reliance even in the midst of religion. Bellarmine has lost as few valiant fighters ever have.

C. Symmetry between the defects of rationalism and of established religion is hard to advocate. What the religious are losing to science and to the scientific tradition is viewed as progress by most people. Whether the same can be said of the scientific tradition or the rationalist tradition is highly debatable, for the tradition of science lost aspirations for theoretical knowledge when it accepted Bellarmine's instrumentalism. True, Bellarmine's instrumentalism, his view of science as applied mathematics, makes room for the freedom to accept any metaphysical commitment. But this is hardly a gain; it is the loss of the hope, of the ideal, to develop a scientific metaphysics or a scientific world view; it is thus a catastrophic loss of self-reliance, or at least loss of the hope of self-reliance, or at the very least loss of the precious illusion of self-reliance. When religion loses, self-reliance gains; when rationalism loses, self-reliance loses too. Hence there is hardly any place for any symmetry. Things look bleak.

VIII. STANDARDS OF RATIONAL THOUGHT AND ACTION

We need standards of rational thought and of rational action more general than either of the older standards. This claim is contrary to the tradition one, according to which science is autonomous and hence cannot abide by external standards. Traditionally the standards of science were equated with the standards of rationality; the new religious avant-garde merely adds that there are different standards of rationality in the different fields of human thought and action – science being a prominent one, but not the only one.

Would we be better off giving up the illusion of self-reliance and settling down with a commitment? It is here that criteria diverge. Common sense is usually – but certainly not always and not on principle – against both illusion and commitment, against self-deception and dogmatism. Yet, it seems that we must give in to one or the other. Say which, and you have decided whether to enter the rationalistic tradition of unadulterated science or to switch to the new tradition of reconciliation with religion.

Let us first see clearly why the commitment and the illusion are so very inimical to each other.

A. Let us approach things first from the rationalistic point of view. To be precise, the issues discussed here are not scientific. Should we approach them scientifically, and, if so, how? The viewpoint traditionally endorsed by scientists, the scientific attitude, so called, is the readiness to apply the method of science to all intellectual activity, to all intellectual problems, to the attempted solutions to them, and to the examination and the application of the better of these solutions. Such a viewpoint, of course, prejudges the issue of a possible complement to science in the sense discussed here – the intellectual or enlightening sense. Let us examine this viewpoint for awhile, even though it is prejudiced.

Can we apply the scientific attitude outside the usual domain of science? To answer this we ask, "What is the application of the scientific attitude?" The traditional answer is that the scientific attitude is the application of scientific method. Now we must ask, "What is scientific method?" The traditional answer is vague, except on one point: Whatever scientific method is, it is an empirical method – and in that it involves the quest for empirical evidence. Now we can reformulate the question, "Can

we apply the scientific attitude outside the usual domain of science?"
into the question, "Is scientific method essentially empirical, namely, is
there any intellectual activity to which we can apply scientific method of
inquiry without thereby rendering the inquiry empirical?" The traditional
answer is, "No: enlightenment = rationality = science = empiricalness."

This is classical rationalism and classical positivism, of course.
The most obvious criticism of it is that logic and mathematics are
rational yet unempirical. As late as 1922 this criticism worried Ludwig
Wittgenstein; his *Tractatus Logico Philosophicus* contains a supreme
effort to do away with logic (including mathematics) by declaring it a
peripheral, unintended by-product, and as such, rather senseless. A sentence can be meaningless like the sentences of theology, metaphysics,
and ethics – which, strictly speaking, are no sentences at all (just as
poetry contains no elements of arithmetic, even when appearances give
contrary impressions) – and a sentence can also be properly framed, but
simply say nothing, just as zero is a number, and as a map can be mapped
onto itself, and as a mill can grind water. For Wittgenstein, logic and
mathematics were rather senseless freaks because he wished to accept the
equation at the end of the preceding paragraph in all its narrowness.
They really did not exist for him any more than ordinarily zero counts as a
number. The totality of true propositions, he said, is the total natural
science!

When we break away from such a narrow rationalistic attitude,
namely from strict classical positivism, we feel the great need to distinguish between the scientific and the rational, and we feel the strong
urge to define the rational as broader than the scientific.

More than that – much more. We may wish to explain the desirability
of science, of the application of scientific method, in terms of rationality. And this amounts to the wish to have the criteria of rationality
put limits to the applicability of scientific method. But this means
that the field of rationality should be wider than the field of science – in
the sense of the existence of instances of discourse which are rational
and yet nonscientific (and, to keep matters neat, nonmathematical,
nonlogical as well). For, if we want to decide rationally where to apply
scientific method, we also want to decide where to refrain from such application and so allow, a priori, such possibilities. The criterion of rationality would, consequently, be deeper than the criterion of scientific charac-

ter, at least as an instrument for making decisions when to apply scientific method and when not. All this is impossible within the old positivist framework where the identity of rationality and empiricalness is determined *a priori*.

B. But why should such a fruitful and useful activity as science be in need of justification in terms outside its own? Pierre Duhem stressed that science ought to be autonomous, that is, not judged by any external criteria. For, he said, scientists need not share any external criteria, yet they support the unanimity which all scientific activities enjoy. If science be judged by external standards, divergence concerning these will destroy unanimity in science.

The autonomy of science, as advocated by Duhem, is a dual autonomy; first, concerning the rationality of its method, and second, concerning the lack of metaphysical commitment of its content. The threat to either kind of autonomy is a threat to the unanimity observed to rule science. The autonomy of method is the undesirability of judging scientific standards by external standards of rationality. The autonomy based on freedom from any metaphysical commitment prevents metaphysical disagreements from leaking into science. Thus, we cannot judge the acceptability of the continuum theory of elasticity and of the atomic theory of thermodynamics, by either a metaphysical commitment to an Aristotelean antiatomistic process metaphysics or to a Democritean atomistic-mechanistic metaphysics. Both Aristotelean and Democritean agree about both elasticity and thermodynamics; hence physics and metaphysics cannot clash. Metaphysics alone pertains to reality; physics (i.e., empirical science in general) handles only phenomena and prediction, only economy and usefulness, perhaps aesthetic value to boot, but not truth and not finality. Thus spake Duhem.

Duhem's concept of the autonomy of science is very close to the classical rationalistic conception as advocated by the Royal Society, except that the old rationalists forbade commitment to any metaphysics prior to its having gained scientific status, and Duhem took it for granted that everyone has a metaphysics and not all scientists are in agreement about metaphysics. This seems a very plausible and congenial modification of the view of the autonomy of science. Yet its main thrust concerns not the autonomy of science but the autonomy of metaphysics – and in the sense that metaphysics need not be troubled or constrained by science. That is, the autonomy of metaphysics is secured by depriving theoretical science of its informativeness.

From the autonomy of metaphysics to the autonomy of religion there is but one step. It is no accident that the new positivists prefer to debate at length the autonomy of science. They so act on the presumption that once the autonomy of science is decided, the rest follows easily: first the autonomy of metaphysics, and then of religion. If science is judged by universal standards of rationality, these have to apply elsewhere, thus leading to the (Russellian) ideal of applying the scientific attitude wherever possible. This, they claim, makes no sense. Where one applies the scientific attitude one applies the empirical method and achieves science, so one cannot apply the scientific attitude elsewhere; elsewhere meaning metaphysics, which either does not exist (old rationalism) or has its own standards (new positivism). The claim for the autonomy of science thus leads the new positivists to claim autonomy for other fields as well. When the practitioner of metaphysics fears science, what he really fears is that *he* transgresses and trespasses; and, indeed, science tolerates no trespass. But then, neither does metaphysics. Bellarmine was right in denying Galileo the right of trespass into metaphysics and theology, but wrong in that he permitted himself to trespass into the domain of science. So both Bellarmine and Galileo were wrong in the clash (the one was naïve and the other was overconfident, they add, but let things rest in generalities); the correct attitude, then, should be that of autonomy, of no possible clash. This is the new positivism of the religious avant-garde (which I wish to combat).

Duhem and Wittgenstein (when young) were both narrow positivists, recognizing no standard and beyond those of science. But, whereas Wittgenstein endorsed science and only science, Duhem endorsed science when dealing with matters scientific, and religion when dealing with matters religious. Doubtless, Duhem's position is superior to Wittgenstein's. Between the two, the new is more tolerant of metaphysical commitment and in this it is preferable to the old.

C. The decisive argument of the previous paragraphs is defective on two accounts. The first is this. Unanimity in science was classically endorsed as a corollary of the idea of certitude. Now that certitude has been given up as a bad job anyhow, we need not endorse unanimity. Certainty was indeed overthrown because unanimity was exploded. Twentieth-century physicists disagree with practically all nineteenth-century physicists about atoms, about action at a distance, and about geometry. Duhem speaks of a new brand of unanimity – the agreement of living scientists amongst themselves, to say the most. Even this agreement is highly questionable. Faraday, for example, attacked the accepted views on atomicity and of action at a distance. In his day he could be dismissed as a small minority; yet today the majority sides with him against the majority of his day; so today it is harder to dismiss him. Thus, we cannot declare even unanimity among nineteenth-centure physicists!

Anyway, what is the value of unanimity? The criteria of science must be subject to external criteria, and extreme traditional positivism must be avoided in order to avoid dogmatism. To say that the criteria of science are valuable for the sake of unanimity, or fruitfulness, or what have you, is to value what have you, and thus to judge the criteria of science by some external criteria. Duhem knew all this; he did not commend unanimity but observed its existence. But, of course, his observation is false: He himself as a physicist followed Ampère and Weber and attacked Faraday and Maxwell.

The same holds for religion. To say that the family that prays together stays together is to value religion as an instrument; it is a positivistic attitude toward religion; religious positivism so called. It is likewise based on a false observation. Religious positivism is empirically refuted yet it is consistent with religion proper: One may hold that religion is both right and useful. However, most religious positivists do not care whether religion is right or wrong, and even are prepared to concede that it is partly wrong, partly meaningless mumbo-jumbo. Alternatively, if one holds that religion is both right and useful, one has to say which of these two characteristics one values more; that is, one has to say this, for the sake of critical debate. For, if we try to make a person alter his views, we wish to know which criterion he employs in his act of changing his views.

Let us elaborate on this point for a moment. Of course, if one's defense of religion because of its usefulness collapses, one may indeed attempt to defend it on theological and metaphysical grounds. This, however, may be a reflection on one's intellectual makeup. Perhaps one is using delaying tactics, perhaps one is attached to a position and will not relinquish it until all defenses of it fail, perhaps one is not attached but is loyal to one's position, and perhaps one simply thinks one ought not to alter frivolously one's view under insufficient pressure. One's opponent may feel frustrated, but then one's opponent was in error when pursuing a line of attack in depth before attempting even a superficial survey of the opponent's defenses. Will the opponent capitulate more easily when this defense of his collapses, or that, or both? and so on. Taking things intellectually rather than personally, the same question reads: Which criterion is more significant? Any attempt to sidestep this question may amount to the claim that religion is both truthful and useful because of some necessary link between truth and utility. This claim, then, should be explored first as the deepest. Even further, will the deepest claim about such a link be true, useful, or what? Will it, Heaven forbid, be empirical and thus lead the believer to mix his theology and his science?

On this line of attack one can expose the new religious avant-garde as religious positivists who base their philosophy on empirically unexamined but easily examinable claims about the roles of religion.

IX. ENLIGHTENMENT AND SELF-RELIANCE

We now come back to our chief question, "Can religion complement science as enlightenment?" with the attempt to demarcate enlightenment in an improved fashion. What, then, is enlightenment? What prescription do we make when we advocate enlightenment? Enlightenment is, first and foremost, being self-reliant as opposed to being guided, blindly obedient, and servile. Is commitment independence of mind, or is it servitude to a principle? Is the committed person more or less self-reliant? Can we answer this question without being empirical, or, if we must be empirical, without mixing science with metaphysics?

A. Traditionally, enlightenment was viewed as independence and as freedom, as opposed to voluntary servitude, as the refusal to be led by the nose – essentially, as self-reliance. The self-reliant person, the enlightened person, consults his own judgment and taste, or, if need be, opens his own judgment or taste to criticism; he will not simply take it from the Vatican or Moscow. He may take it, but not *simply* take it; that is, if and when he does take it, he will have a reason, *he* will find it acceptable.

Old-fashioned rationalists are ready with an arsenal full of the weapons of psychoanalysis and of socioanalysis and of political analysis. They will sneer ta the Catholic (Communist) who accepts the judgment of the Vatican (Kremlin) only after consideration. They will call such considerations rationalizations, ratiocinations, self-deceptions, pseudo-rationality, pseudo-intellectualism, fake self-reliance, mere gloss and varnish of rationality on old and defunct irrationality. It is not unusual to find such people compared to a dog on a leash who divines the direction his master is heading for and keeps ahead of the master in that direction. These people are, says the old-fashioned rationalist, the most useful arm of the Vatican (Kremlin) for the purpose of deceiving the half-sophisticated and the half-dissatisfied.

There are such people. But here we are talking of the true avant-garde, of a small portion of the self-declared avant-garde, which is for the Vatican (Kremlin) much headache and many a sleepless night; and which is for the rationalists a living refutation of the equation of the committed-obedient with the slavish-guided. We should ignore the slavish-guided and their ratiocinations and discuss the committed who claim to be self-reliant in the sense that they claim responsibility for their own commitment.

The new positivists are thus a serious faction. Though the Establishment may wish to dismiss it as too small and too intellectual, time and again this turns out to be not so easy. Such an inner strenght rightly becomes influential and does deserve close study.

B. The religious person who claims that in his very commitment he is self-reliant has, above all, to explain his allegiance to one religious establishment among many competing ones. Affiliation entails some sort of a package deal, of course, which includes some less desirable aspects of any given set; and so the religious who claim self-reliance have to explain not only their allegiance, but also their affiliation to a given sect, their readiness to take the rough with the smooth.

No; they will declare affiliation to be a practical matter, subject to empirical study. If they can do more to reform the sect by a struggle from without, they say, they would leave their sect even if this would take them to Hell. I do not know how seriously this terrible boast is made, but I have heard it made frequently. On second thought, perhaps the boast is not so serious after all: Where is the Hell referred to by the party boasting readiness to land there for the sake of the common welfare?

The answer depends not on affiliation – which in this stage of the discussion is subject to some superior pragmatic criteria – but on allegiance: how much of the official doctrine of the sect does the self-reliant religious person really endorse? We do not know. He may be committed to one doctrine really and to a somewhat different doctrine demonstratively – as loyalty prevents broadcasting one's criticism of the doctrine of one's sect. So it is hard to rely on empirical evidence and so the debate must close here, or we may venture an a priori reconstruction instead.

First, consider the believer who accepts the metaphysical doctrine of his sect about the nature of God and his world, about nature and morality; and then the positivist who has no faith, no metaphysics, only a moral sense and a sense of ritual (aesthetic and/or social).

Commitment, say those who claim to be self-reliant believers, is the necessary conclusion to all discourse on self-reliance and on rationality. Justifying one's own view by one's own arguments makes one hop from one defense to the defense of that defense, from one criterion to the criterion which leads to its choice, and so on ad infinitium. This is the well-known critique of classical rationalism from infinite regress. One takes one's fate in one's own hands, says the modern postrationalist, and makes a decision which prevents this: One commits oneself to a standard and acts in the light of the standard one has chosen oneself. This is the only possible road to true self-reliance.

C. Here we come back to the old theme of disappointment: The religious avant-garde come to religion from the classical view of science as rational; in the sense of justified; in the sense of its resting firmly on final evidence by a final criterion. Sir Karl Popper has treated the irrationality of this "despair of reason" in chapter 24 of his well-known work, *The Open Society and Its Enemies*. Popper suggests that the minimal standard of rationality should be openness to criticism, and that the commitment, necessitated by the breakdown of the classical overoptimistic view, should be minimized into the minimal faith in reason, to wit, the faith in the fruitfulness of criticism. Popper calls the classical view *uncritical* rationalism, and the one he proposes in its stead, *critical* rationalism.

A similar, though much more detailed, and epistemologically more sophisticated, view was developed by Bartley in his *Retreat to Commitment* (1962) which, though a few years old, is not yet the bestseller it deserves to be. Bartley centers on the *tu quoque* argument. The irrationalist says to the rationalist – you too are committed to some dogma. And Bartley answers, no; whatever view of mine you criticize effectively I shall reject – even my view of enlightenment as the fruit of critical debate; if you, too, are willing to accept criticism of your commitment, it may thereby cease to be an irrational commitment, it may cease to be a commitment at all, and merely become a tentative opinion. This is how I read Bartley.

It is important both for Popper and for Bartley to distinguish between the standards of science and the standards of rationality, and to argue that the latter support the former. Popper, Bartley, and others, have tried to develop this point; I shall only briefly state it here in two parts.[3]

The first part is that criticism may be of diverse kinds; in empirical science criticism is ideally a new experiment which criticizes a good theory. The second part tells us what is a good theory. It tells us, first, that rational action is directed toward diverse ends, even that enlightenment can be constituted of diverse aims; and that, second, the end of science is comprehension of the world, so that a good scientific theory – which is what we wish to characterize – is that which explains much. And so, empirical science concerns a very special kind of enlightenment, namely, comprehension of the empirical world. The aim of science is the search for new theories which explain empirical facts and for newer empirical facts to criticize these theories. Enlightenment, however, may be of diverse kinds, aimed at diverse ends, criticized in diverse manners, with the proper correlations between the intended aims and the construed criticisms.

So much for the Popperian or neo-Popperian view. For my part, I prefer a new variant of the classical idea of self-reliance. Classically,

a self-reliant person accepts an idea on its own merits as he under-
stands it with his own mind. Classically, and subsequently to the previous claim, a self-reliant person accepts a view after considering its
proof satisfactory. This, we saw, leads either to infinite regress or to
commitment! Alternatively, a self-reliant person has forever to cope
not so much with proof and acceptance as with quest, trials, criticism,
rejection, modification, new quest, and so on, as long as life permits.
Subsequently, justification cannot rest on proof or knowledge, and likewise – when these fail – justification cannot rest on commitment; suffice
it if justification be given by showing that certain criteria lead to the
conclusion that such and such a theory is the best available. Thus,
whereas the classical justification is that Newton's mechanics was demonstrably true, the new justification of the fact that Newton's mechanics was endorsed at the time is that at the time it was the best available
explanation of the then-known phenomena, and one which remained,
then, impervious to criticism. This, of course, may lead to doubting
the criteria; but we may claim that thus far the criteria, too, are the
best available; that when they have serious competitors we may well
reject them too, as, indeed, I think we should.

   Self-reliance is the reliance on one's own judgment, on one's own
criteria, etc. Now, in judging quantum theory, Einstein and Heisenberg had the same criteria and reached the same conclusions. Yet Einstein,
but not Heisenberg, rejected quantum theory out of a metaphysical
commitment. Commitment enters science very forcefully: He who
is committed to causal metaphysics conducts one research program,
he who is committed to chance metaphysics conducts quite a different
research program.[4] Hence, if we wish to avoid a retreat to commitment, we had better attempt various commitments. Admittedly, life
is short, and too many possible commitments, then, have to be ignored as not very promising – of course, with no proper justification!
We must go back, I fear, to the philosophers of commitment and consult
them on such matters.

X. THE SOPHISTICATED RELIGIONISTS: BUBER AND POLANYI

The new religious positivist, the self-reliant believer, has certain dogmas about the value of self-reliance which should be taken as tentative

opinions to be criticized. Self-reliance must be taken as a fundamental point of departure, as a primary principle; however, not the optimistic (and thus questionable) principle of the old rationalist, but rather the desperate one (rooted in admitted failure, not in questionable success), akin to that of the new religious positivist.

A. Again, we see how clouded simple issues may become. What is self-reliance, commitment, metaphysics? In science? In religion? Historically, much of the study of philosophy has centered on the truth and falsity of certain philosophical propositions, which, indeed, enter reliance and commitment in a large way; but, consequently, the studies of attitudes, programs, and ways of life, have regrettably suffered too much neglect. Thus speaks the modern enlightened theologian.

There is here a quaint mixture of something apologetic, almost dishonorable, and something noble and admirable. The apologetic aspect is to minimize the various doctrinal differences between various sects and to see these as mere reflections of a variety of ways of life. This is religious positivism pure and simple, and no enlightened theologian is free of it; not even writers who, like Martin Buber, frankly reject parts of their traditions, doctrinal or ritualistic, which they view as superstitious and magical.

I wish to quote here one of the more popular introductions to the philosophy of religion, Frederick Ferré's *Basic Modern Philosophy of Religion* (1967). The thesis of the volume is summed up on page 371, at the introduction to the discussion of "the cognitive possibilities of theistic language," which simply adumbrates the author's worry that unless his theology is empirical it may be arbitrary and thus irrational. Ferré's thesis: "The two primary functions of theism's logically primary images are (1) expressing and influencing basic life styles, and (2) reflecting and shaping ultimate 'ways of seeing.' It is hard to come to grips with either of these functions."

The obvious criticism of Ferré's thesis is that what is "hard to come to grips with" is not enlightenment to be led by, but ignorance to be tackled. It is here, however, that most rationalistically or scientifically oriented critics are slightly in error, attacking the religious not head-on, but off-tangent – a tangent which leads straight to old-fashioned or classical or positivistic or uncritical rationalism. The two "primary images"

of Ferré are not to be rejected, since they are quite correct; also they
are indeed "hard to come to grips with"; but they are not "theistic" in
the least, especially not when "theistic" is a euphenism for Ferré's
own denomination. Rather, "the two primary images" are religious in
the skeptical sense in which Einstein and even Russell must be regarded
as religious. The ideas of "life style" and of "ways of seeing" are admittedly religious in some traditional sense, but they are not identical
with traditional religion, much less with that of a given traditional sect.

B. The best defenders of religion as the new enlightenment, as a "style of
life," are Martin Buber and Michael Polanyi. Both are Jews by descent.
The one advocates a refined version of Hassidism (which is the way of life
of a neo-Cabbalistic Jewish sect) but is opposed to all mandatory traditional Jewish ritual; the other advocates a refined version of Catholicism.
The one draws his analogies from the social sciences and the humanities
and the finest of the fine arts; the other from the natural sciences and their
philosophy and history. Both advocate the new ideas of intensified or
heightened mode of cultured and civilized "way of life" – gracious living
if you wish, but not of isolated individuals as much as of prefectionists in
interpersonal relations, I-Thou relations – and both advocate the connoisseurship of the style of life; both see here a commitment and a refinement
of education, both see here a new lease of life for the best in traditional
religion on condition that the worst in it be frankly jettisoned. In this they
are, of course, a part of a larger movement, Bultmannism or the demythol-
ogization of religion. What is important in their works, however, is more
the positive aspect of their religiosity, their readiness to justify their
unwillingness to demythologize religion so far as to let it vanish completely
(the death of God).
     The works of Martin Buber do not yield to summary, even a brief
and superficial one, within the limits available here. Indeed, much of
his contribution is part of a process of reviving a lost past and a lost
education which is both religious and meaningful for modern Jews.
I shall not discuss all this. Two points of his work should suffice. First,
his *Two Types of Faith*, which is a brilliant piece of linguistic analysis
in the wake of Georg Simmel but supporting an existentialist philosophy,
and a strange piece of pseudohistory supporting a quaint pseudo-
Judaism. We can believe, or have faith, or the like, says Buber, in two

different senses, one indicated in the preposition "*in x*," the other "*that y*" or "*concerning y*", etc., where *x* is a *person*, and *y* is a *proposition*. The Jew, the Psalmist, has faith in the Lord – not in his existence; rather, the Jew trusts that the Lord will not let him down, not the proposition that he exists. Saint Paul required that the Christian have faith in the Greek sense, in the sense of accepting the truth fo a proposition, so as to be saved – thus mixing Greek and Jewish elements of his religion.

Of course, considering the first two of the Ten Commandments in the light of Buber's analysis will lead to blasphemy, since these no longer declare, respectively, the existence of the one deity and the non-existence of any other deities, but rather the trustworthiness of him and the untrustworthiness of the others. Interestingly, historians of biblical theology may very well be so blasphemous and read the Ten Commandments this way: originally, some say, Israel's deity was "jealous" and this led him to become overdeity, and only still later did he become the one and only.

Be this so; nevertheless, traditional Judaism obviously and most emphatically opposes this reading. Hence, traditional Judaism has strong fundamental articles of faith in the very sense which Buber declares very Greek and very un-Jewish. This Buber's piece of arbitrary apologetics has to be jettisoned. What remains is the advocacy of a way of life of trusting, of faith, and of hope.

The arbitrary apologetic streak in Buber's philosophy need not concern us overmuch. Buber himself was willing to see in traditional Judaism much that he rejected as distasteful. In his "Reply to Critics," in Schilpp's volume *The Philosophy of Martin Buber* (Library of Living Philosophers), (1963), 1967, in a section published in *Commentary* and elsewhere in 1964, Buber ends on this point: that the Hassidim proposed to intensify religiosity, to intensify life, but by a magical formula; that we should try to avoid the formula, yet to go on pursuing the same end, said Buber, was his chief message. However, Buber never really liked to admit that he had a message. Again and again he said he was pointing the way and no more (something, one might add, an art critic has to do, or a good educator), he only wished to help people find their own way, not prescribe (art critics cannot prescribe taste or proclamations of taste).

What Buber's philosophy amounts to in terms of religion in the strict and traditional sense is not clear to experts, let alone laymen. There are two or three biographies of him and innumerable studies, all inconclusive. It seems clear, however, that Buber shifts emphasis from doctrine to prayer. Prayer enriches life, especially when done in a way somewhat more sophisticated than going to the synagogue to recite some dead words just to fulfill a duty. If one wishes to pray, says Buber, one need not raise problems and enter discussions about the existence of the deity; when one really has to pray, one just prays.

There are two obvious criticisms of Buber, the one hostile and shallow, the other sympathetic and serious. The hostile critic will see in God a father substitute and in prayer a regression to a child's desire for protection and comfort. All this is true but irrelevant: an orphan in need of a father substitute may reasonably adopt a human stepfather, and unreasonably adopt a totem pole. The psychological need is the same in both cases of the advisable and of the inadvisable conduct. The question we should ask is, "Is there a deity which listens to my prayers?" and not, "What is my need for prayer?" Assuming that God does not exist, is prayer advisable? The sympathetic critic will draw attention to the fact that many who do wish to pray cannot honestly do so since they cannot come to the conclusion that some personal deity listens. The sympathetic critic will thus reject Buber's proposal to ignore faith of the Greek type. Buber, on his part, cannot and would not meet this criticism by an attempt to prove that there is a deity listening to our prayers. Rather, he would point out that those who refuse to assume too much about God do assume too much about their own selves – in particular about their own abilities, intellectual and moral. Excessive self-reliance is to Buber not so much the sin of hubris as the error of hubris – the overestimate of one's own abilities and resourcefulness.

Buber has a baffling point here. Once we equated reasonableness with proof and became antireligious. Now we gave up this idea of reasonableness, and Buber at least challenges us to reconsider all our ideas: Perhaps we should invent a new criterion of reasonableness and apply it to religion; perhaps we should act intuitively and simply pray, both when we feel like praying and when we feel that it is reasonable to pray; and perhaps we should simply become irrational. These are the alternatives and this is Buber's challenge. Buber is not alone in challenging us, but *his* challenge, particularly, appeals to the religious scientific avant-garde, the subject of the present study. And the particularly appeals, I think, because of his attitude to the above alternatives. Contrary to certain allegations, Buber flatly rejects the irrationalist option, and even severely criticizes Heidegger for accepting it. He also leaves the choice between intuitive reasonableness and possible new criteria for reasonableness; he merely stresses that there is no basis for the claim that prayer is unreasonable except in a defunct philosophy which suffered from excessive self-reliance.

To consolidate all this and to be somewhat more convincing, Buber has undertaken to perform two tasks, both of which he has executed. The first is to argue that the limit of self-reliance is the reliance of the individual on his specific background – social, cultural, religious, and even scientific. The second is to sift the reasonable from the unreasonable in religion.

I shall not discuss his execution of these two tasks, especially since I am studying here the phenomenon from the viewpoint of one more interested in science than in any specific religious commitment. But I have to state one outcome of these labors. The resulting philosophy will not be irrationalist, though decidedly not rationalist in the sense of classical uncritical rationalism. It will be a philosophy of commitment – of commitment as a precondition of rationality, not as the outcome of a rational decision. Yet the commitment is not arbitrary either – at least not as arbitrary as the average existentialist would have it. For the choice is not between a set of possible commitments but between a set of existing and living commitments of communities which traditionally practice them.

Buber's refusal to fit any categorization is systematic and stubborn. I have opened this chapter by dividing men not into scientists and priests but into knowers and seekers, and I view the religion of the knower as essentially different from that of the seeker. Buber fits neither category. His is not a knower, and preaches no doctrine; but he seeks not a doctrine, not the truth, but the community of those who seek communion in God. It is very hard to say, for those who demand that quest be a component of religion, whether Buber qualifies this way or not. He suggests getting rid of magic, but no new rationality; he suggests intensifying life and the quest for God, but says nothing on the quest for enlightenment. He is against excessive self-reliance but does not say to what degree one should allow, indeed strive for, integration in one's community, without thereby losing one's independence and self-reliance.

Here we see how important it is to view religion, as well as science, as a living tradition. And this is why here Polanyi comes to complement Buber: Prayer or any other religious conduct is part and parcel of the religious way of life, which is utterly parallel to the scientific way of life. Banish one, and you may just as easily banish the other; it all is a matter of initial and frankly arbitrary choice of an existing tradition practicing a given way of life. One chooses science, one religion, and Polanyi chooses (*The Logic of Liberty*) both: They both enrich his life.

Polanyi attacks the traditional rationalist philosophy of science as enlightenment. There is no rationality in the old sense; there is no proof in science; complete objectivity is impossible. Yet there is some objectivity within, and – as a matter of brute fact, if you will – to, a given scientific society (*Science, Faith, and Society*).

Comprehension, says Polanyi in the preface to his *Personal Knowledge*, "is neither an arbitrary act not a passive experience, but a responsible act claiming universal validity." This sounds surprisingly Popperian, until one remembers that Polanyi, in the same volume, flatly rejects the correspondence or absolutist theory of truth. "*X* is true," he says, really means, "I believe *X* to be true." And to see what belief means is to see what it entails in actual life, in praxis. Thus, Polanyi advocates a version of Duhem's instrumentalist philosophy of science; but he shows that Duhem's own statement of his view is too old-fashionedly rationalistic. It is less than Duhem had claimed it to be: within the Duhemian rules of science some alternatives are excluded, but the rules do not narrow the alternatives in each case to precisely one. Hence, moves in the history of science often had to be made by individuals prominent within the scientific tradition – moves not fully characterizable according to any articulated set of known rules (*The Tacit Dimension*). Here an element of authority enters the philosophy of science.

This point has been repeated by Thomas S. Kuhn, who is thus far utterly neutral in the debate concerning religion, in his *The Structure of Scientific Revolutions* and elsewhere. And he acknowledges his debt to Polanyi. The exact timing of a scientific revolution, says Kuhn, is not determined by any rules; the rules only prescribe a vague feeling of the approaching revolution. The exact timing is declared by the acknowledged leadership of the scientific community. So much for Kuhn's position.

Again we come to an important element in all this, Polanyi's doctrine of connoisseurship (*Personal Knowledge*). If you want to be a scientist – or an artist, or a theologian – you start neither a priori (by thinking) nor a posteriori (by observing); you start by going to the best available master and becoming his apprentice. The method, the style of life, is tacit and inarticulate; you learn it by apprenticeship. You cannot criticize religion or science from the outside, not do you become an insider by merely endorsing a doctrine; commitment is an existential affair; one learns the meaning of a commitment by practicing it.

C. Both Bubber and Polanyi become slightly authoritarian in places – out of the inner logic of their situation. As the limit of self-reliance is social, so transcending self-reliance lands one on the reliance on the authority of one's leader. The task of sifting the reasonable from the unreasonable in any given religion is left to the leader to perform; and the outcome of that performance has to be accepted. A crucial instance is Buber's *Moses*, which includes a critical introduction and an uncritical text. In his introduction, Buber rejects both the literal acceptance of the Bible as a historical document and its total overcritical rejection. The Bible, he says, should be treated as a distorted racial memory. If one can ascribe this idea to any single author, one may well accept R. G. Collingwood's (*The Idea of History*) attribution of it to F. H. Bradley. This idea is admirable, and should be appreciated even though it has been practiced before (say, by Schweitzer), and even though its explicit formulation is by now commonplace. But how do we rectify such memory? Buber does not say; he illustrates. He simply retells Moses's story as he sees it. Willy-nilly he thus plays the role of a leader.

The admitted ineffability of the essential and vital and valuable elements of tradition is an enormous source of strength here. One cannot specify the tradition as well as one can convey the feel for it. On almost any significant question Buber or Polanyi offers an elusive, refined answer. Sociology should follow neither individualism nor traditionalism, but a sort of middle course. Social and political innovations must take place, but fall on fertile grounds. Religion is not merely a private faith and not merely a social and cultural way of life, but a sort of blend of the two. Sicence is neither inductive nor deductive, but a blend.

This, then, is the Buber-Polanyi intriguing doctrine of self-reliance

which stands behind their religious doctrines. A self-reliant person
develops his sensibilities to the utmost, even to a point beyond his
ability to articulate them. He may learn to be critical on the way, but
he must evolve to the postcritical level of intuitive expert judgments.
These are not final: Sooner or later criticism may shake them; but they
are above and beyond criticism when they rule the day. The classical
rationalistic idea was that one must prove, and of course, thereby
articulate, one's views. Proof is too much to expect. We have, then,
daring based on intuitive feelings, on expert touch. This is not only un-
provable but even not articulated.

   Now, what is not articulated can hardly be criticized, it seems. And
so the new philosophy seems to land in blatant irrationalism. Yet this is
an error. Strange as it may sound, the differences between the views of
Buber and Polanyi and those of Popper and Bartley are secondary.

   There are two important ingredients to judge, intuition and criticism.
Now, readily or reluctantly, all major thinkers today agree about intuition, its value, the difficulty of articulating it, etc. Similarly, those
who do not appreciate criticism are dismissed as irrationalists.

   There are two secondary ingredients to judge: connoisseurship and
the choice of a style of life. Now, the Popperian will say, it is preferable
to attempt to articulate one's intuitions in order to open them to critical examination and evaluation. Even the connoisseurs and artists may
benefit from attempts to articulate, though they cannot ever be entirely
successful. The Buberite, however, will defend the working ineffable
residue which may be destroyed by overarticulation. This, however, can be
studied empirically – like the many tenets of positivist commitment.
Buber and Polanyi are, when all is said and done, religious positivists:
Whatever body of religious doctrine is articulated, they will relegate to a
secondary position; it is the living working practice which they stress.
Martin Buber was immensely consistent here and, where he could more
easily point the way to the perplexed, regularly and repeatedly debunked
his written words as poor substitutes for teaching, for live conversation.
And, true enough, there is some distance between this and the old religious
positivism, which in the works of some disciples of Malinowski identifies
religion – including the affirmation of the faith – with mere ritual.
Buberite positivism evades the intellectual question: How much of the
doctrine of any religion is acceptable? This evasion, however, is not

enough of a novelty to make a difference, even when Buber adds myriads of sophisticated and exquisite cultural and historical studies to his philosophical works.

The difference between the two avant-garde groups, the critical rationalists and the sophisticated religionists, becomes even smaller; yet the religionists are not rationalists; ever so often they skate dangerously close to authoritarianism and hence to irrationalism, because they advocate the connoisseur's intuition as best and hence binding – even though it keeps changing (improving?).

XI. SCIENCE AND UNIVERSALISTIC RELIGION

What is most obviously missing from the new refined religion of the religious scientific avant-garde whom I have ventured to describe, and who follow, I think, Duhem, Buber, and Polanyi, is very odd indeed: What is missing from their religion is the religious aspect of science, the scientific quest. Their religion incorporates beautifully some religious aspects of the arts, of Fra Angelico, of or Bach. They admit the aesthetic value of science – indeed they stress it. But the religious value of science, or the intellectual love of God, short circuits their philosophy. The separation of religion from science is essential to them. It is essential to them in order to prevent science from ousting religion, for (like Pascal) they have found science without religion intolerable.

What I recommend here, then, is exactly such a short circuit, an intensification of all that is good in this avant-garde philosophy. The result will be "A Free Man's Worship," much as Russell envisaged it half a century ago.

A. The religious scientific avant-garde I have discussed here are not professional philosophers or theologians. Also, they have high standards of professional ethics. Hence, they can seldom express their views on religion, whether in writing of in public speaking. Even in private, when conversation turns really intellectual, they prefer to talk on subjects they know: mathematics, pure and applied; science, natural and social, pure and applied – real hardware by the most severe positivist standards. They are also inhibited from discussing religion due to loyalty – both to their religious denominations and to science. They are hardly heard. But they have clear and strong opinions, and these often show in action, usually in committee or in private consultation.

The picture of the behavior of the religious scientific avant-garde then – quite unintentionally – is that of a conspiracy of silence about religious doubts; of a grimness on the part of desperate intellectuals who try sheer tenacity as their last effort at adjustment – a seeming conspiracy very much akin to the traditional seeming conspiracy of silence on the same topic though from the rationalistic side of the barricade. The disappointment in science was too painful; the new positivist view is precarious and really depends on keeping life as gracious as possible and on not tearing one's hair and shouting at each other.

The cool, suave attitude of the religious scientific avant-garde to science is the not really cool and rather stiff expression of past disappointment, of the once burnt twice shy. Also, their mood fits Duhem's despair of ever attaining informative theoretical science very well. Dedicated they are; enthusiastic they fear to be. Buber has done a lot to bring religion to life for them (for they often lack religious upbringing), at least to the Protestants and Jews among them, but in a significantly mutilated form – at least one can show this with respect to Judaism. For Buber, Judaism is a faith, not a scholarly way of life; the love of learning, the respect for scholarship, and all that, so characteristically Jweish, do not appear in Buber's picture of Judaism. The most important Jewish ritual – study – is totally bypassed by Buber.

This is no oversight: Pragmatism is a shift away from intellectualism. Duhem has debunked the intellectual values which rationalism preaches, and rendered science a part of technology; likewise Buber has ignored the intellectual values which religion preaches, and rendered religion a part of a way of life. What members of the new religious scientific avant-garde desire most is intellectual, and what they want reinforced is the moral and religious value of intellectual activity. Exactly this they lost to commitment. In a grim determination to hold to the commitment, they have lost sight of their own quest.

Much as I do not like this grim determination, I do not think I can show that it is redundant. For all I know, the grim determination may work; perhaps because it is a mode of convincing the divine authority of our goodwill and sincerity (the new Kierkegaardians!) so as to induce the authority to spare us. The grim determination may not suit my character, or yours, or yet the suave mode of gracious living as cultivated by the new religious scientific avant-garde. Psychology may be against it. Yet, philosophically, grim determination has the upper hand. Religion, in the late Middle Ages, preached pessimism, even in the high Renaissance (R. F. Jones, *Ancients and Moderns*). Science competed hard, and offered certainty and optimism. Now that science has disappointed us, it should no longer offer mock certainty and shallow technological optimism. The members of the religious scientific avant-garde refuse to let go of the last shred of the promise of science, and in desperation they try then to complement it with religious promises. The desperate mood of such a move is not canceled by the outcome. Moreover, contrary to all their intentions, their philosophy remains irrationalist both in that it disallows scientific examination of their pragmatist tenets and in their loss of their own main objective: the intellectual love of God.

B. My own view of the matter is this. Science is better off not competing with religion concerning promises, but competing with old sectarian religions frankly as a new universalistic religion. The pretense that science was no competitor to established religion has led science to engage in fierce hostilities of a positivist and radical nature. As Comte has already noticed, positivism *is* a religion; yet he was not boldly positivist enough, to reject this religion. Competing as a religion, science may, however, appear as what the avant-garde religious thinkers do wish religion itself to appear: Not anti-religious, but rather religion properly modernized, a way of life with as little dogma and taboo as possible, and with as much possible enrichment by varied traditions as desired.

The idea sketched here is not different, I hope, from what Einstein, in his "Religion and Science," referred to as: "the cosmic religious feelings" that he felt. For more details on the religious aspect of this philosophy, I would recommend consulting R. Robinson's woefully neglected *An Atheist's Value* of 1964 (though I definitely intend not to advocate atheism). But in order to integrate this into a more coherent philosophy, one must devise a philosophy of science and of rationality to compete with Duhem's philosophy of science. Here, I think, a modification of the views of Popper and of his former students – especially Bartley – seems to come in rather handy, though as yet their views are far from having the finish a serviceable doctrine may need.

C. Let me conclude with two quotations from Bertrand Russell's *Autobiography*, volume 1. Anticlerical as he always was, in his deeper moments he expressed his quest in thoroughly religious terms. The first and brief quotation is from a letter written on April 22, 1906, exhibiting Russell's views of the religious aspect of intellectual activity. The other, from a letter written on July 16, 1908, exhibits his view of religion which, though half a century old, seems to me to remain the real avant-garde attitude in these matters.

And another thing I greatly value is the kind of communion with past and future discoverers. I often have imaginary conversations with Leibniz, in which I tell him how fruitful his ideas have proved, and how much more beautiful the result is than he could have foreseen; and in moments of self-confidence, I imagine students hereafter having similar thoughts about me. There is a "communion of philosophers" as well as a "communion of saints," and it is largely that that keeps me from feeling lonely [p. 280].

I am glad you are writing on Religion. It is quite time to have things said that all of us know, but that are not generally known. It seems to me that our attitude on religious subjects is one which we ought as far as possible to preach, and which is not the same as that of any of the Voltaire tradition, which makes fun of the whole thing from a common-sense, semi-historical, semi-literary point of view; this, of course, is hopelessly inadequate, because it only gets hold of the accidents and excrescences of historical systems. Then there is the scientific, Darwin-Huxley attitude, which seems to me perfectly true, and quite fatal, if rightly carried out, to all the usual arguments for religion. But it is too external, too coldly critical, too remote from the emotions; moreover, it cannot get to the root of the matter without the help of philosophy. Then there are the philosophers, like Bradley, who keep a shadow of religion, too little for comfort, but quite enough to ruin their systems intellectually. But what we have to do, and what privately we do do, is to treat the religious instinct with profound respect, but to insist that there is no shred or particle of truth in any of the metaphysics it has suggested; to palliate this by trying to bring out the beauty of the world and of life, so far as it exists, and above all to insist upon preserving the seriousness of the religious attitude and its habit of asking ultimate questions. And if good lives are the best thing we know, the loss of religion gives new scope for courage and fortitude, and so may make good lives better than any that there was room for while religion afforded a drug in misfortune.

And often I feel that religion, like the sun, has extinguished the stars of less brilliancy but not less beauty, which shine upon us out of the darkness of a godless universe. The splendour of human life, I feel sure, is greater to those who are not dazzled by the divine radiance; and human comradeship seems to grow more intimate and more tender from the sense that we are all exiles on an inhospitable shore [pp. 285–86].

NOTES

[1] For the importance of problems, see M. Bunge, *Scientific Research* Springer-Verlag, (Berlin, 1967), chap. 4. For the importance of problem situations, see W. W. Bartley III, 'Approaches to Science and Skepticism', *Philosophical Forum* **1** (1969), 318–31.

[2] The word 'suggest' is too mild here: the depressive qualities of these doctrines bring to mind the brainwash techniques discussed in William Sargant's *Battle for the Mind*, Harper and Row, N.Y., 1957, 1959.

[3] For more detail, see my 'Science in Flux,' and 'Unity and Diversity in Science,' both reprinted here.

[4] See Einstein's 'Reply to Criticism' in P. A. Schilpp (ed.), *Albert Einstein: Philosopher-Scientist*, 1949, 1959, p. 675, "... there is, strictly speaking, today no such thing as a classical field-theory.... Nevertheless, field-theory does exist as a program...". For more details, see my 'The Nature of Scientific Problems and Their Roots in Metaphysics,' reprinted here.

[5] Charles Scribner's Sons, New York, 1967.

[6] Little, Brown & Co., Boston, 1967.

APPENDIX ON BUBER

Professor Nahum N. Glatzer has drawn my attention to a number of passages in works of Martin Buber concerning the importance of studies and urged that I was unfair when I said Buber had completely bypassed the Jewish commandment to invest in studies as much time as possible. I stand corrected. Though I mentioned that Buber did much to revive Judaism and its meaning for modern Jews with little or no religious background, I did not say, what I think I sould have said, that Buber's contribution in the direction of adult education was a remarkable one. As far as German Jewry is concerned, the only effective response to Nazism was the Jewish adult education movement in which Buber was one of the leading figures; it was a movement which started in earnest only when the Nazis were already in power; for many German Jews who survived the holocaust, their physical survival was the result of this spiritual revival. But this is merely a dramatic and impressive chain of events; it is the cold philosophy behind it which concerns us here. It is a quaint mixture of a new secular Judaism, with accent on heritage and history and on moral values, with a tinge of Jewish religion. Come to think of it, Buber's religion in this respect is extremely similar to that of Robert Boyle, which I have already discussed. Except that Boyle, though he believed that miracles had merely symbolic value, also believed they had ocurred. Unlike Buber, Boyle was no Bultmannian. And, of course, he had no need to add the idea that his own Protestant religion is an ingredient in the proper blend of cultural refinement. (He granted this as far as Catholicism was concerned and he preached the preservation of Catholic statues, etc.)

To return to Glatzer's criticism, one could try to answer it thus:

Buber's public activity aside, in his philosophy, such as expressed in his *I and Thou*, there is no room for scholarship. This reply is erroneous, however, since Buber wants his reader to be a gentleman and a scholar. One could reply to this, however, that it is not important to be a gentleman, only a scholar. In another chapter in this volume, in the notes on 'Unity and Diversity in Science' to be precise, I have described the atmosphere of philosophy and of science in central Europe around Buber's time as advocating just such a view (be a scholar and not much of a gentleman). This makes me reconsider my assessment of Buber: his recommentation to his reader to be a gentleman and a scholar is more original and bold than I have acknowledged. Nevertheless, and very reluctantly, I do have to conclude without full agreement with Glatzer: for Buber and Polanyi alike, the gentleman is a scholar only on condition that his critical faculties are not too sharp; whereas according to most others in Buber's environment (Mach, Boltzmann, Planck, Freud) you should be as critical as you can though this is most likely to reduce you standing as a gentleman as it could lead to your being rather venomous towards some of your colleagues. Contrary to both, I say, there is no conflict between the sharpest criticism and the highest cultivation of the art of living – I mean, there need be no conflict, and thus no place for venom; indeed the gentleman and the scholar can reinforce each other. It is a historical question whether this is the idea expressed in the Talmudic tradition; I think it is very much so. One can reject the casuistic aspect of Talmudism yet endorse much of its other aspects, particularly that of friendly controversies for the greater glory. There is a complete parallel here with Buber's suggestion that we reject the magical aspect of Hassidism while endorsing its advocacy that we intensify and enhance the meaning of life. Though the Kabbalistic aspect of Hassidism was significant, and was studied by Buber, its Talmudic aspect was no less significant, and Buber never did it justice. He accepted a part of the Hassidic Kabbalism, and rejected another part. Inasmuch as Hassidism rebelled against excessive Talmudism, Buber fully accepted the rebellion. Of the rest of the Hassidic Talmudism, at best Buber preferred to ignore it.

CHAPTER 20

ASSURANCE AND AGNOSTICISM

The scientific ethos of the modern Western world contains two conflicting moods, of utter lack of assurance and of utter assurance. The question posed here is how do the two go together? The answer I offer is, all assurance is contingent on the supposition that our system as a whole still survives, but there is no assurance for the survival of the system as a whole.

I. THE COMPLEAT AGNOSTIC

The mood of agnosticism, expressed in the music of Schoenberg and Webern (even though the latter, at least, was profoundly religious), in the art of Jackson Pollock and Franz Kline, in Buñuel's 'Los Olvidados', and in ever so many other manifestations of the modern world, is philosophically best expressed in Bertrand Russell's 'Free Man's Worship'.

This work, Russell's 'Free Man's Worship' is mainly a manifesto. (The general idea in it he inherited from T. H. Huxley and H. G. Wells.) Russell himself said he did not like, later in life, its pseudo-Miltonian prose; but he never withdrew its content, and his biographer Alan Wood testified that he felt strongly, later in life, that the view and mood expressed there is correct.

It is a true sleeper: there are few references to it in the philosophic literature, none of any import. And perhaps there is, indeed, no need to refer to it. Yet, any survey, even a superficial one, will show that it is immensely popular – students of philosophy read it and fall under its spell, people unfamiliar with it will assent to anyone who quotes to them crucial passages from it. It truly expresses both the view and the mood of most Western intellectuals who profess agnosticism and even of some of those who profess some religious denomination or other.

Free Man's Worship – and Free Man is both a species and an individual, so that the reader gets a whiff of a sense of eternity from the mere style of the work – is the erect pose with which he walks over a narrow bridge, not knowing where that bridge ends, if anywhere, not having any assurance that it leads anywhere: the bridge is only seen for a brief distance and then disappears in the fog, or, still better, in the void. The meaninglessness and cruelty we see around is inescapable, yet Free Man does not succumb. He goes on regardless, hardly even hoping for the best, but acting as if he does.

II. THE IMAGE OF INDUCTIVE SCIENCE

The traditional religious agnosticism is usually coupled with a traditional scientific assurance. Again, I am making a broad and sweeping empirical observation. Let me, then, describe scientific assurance. I wish to state at once, however, that whereas the image of religious agnosticism described above is congenial to me, the image of scientific assurance I wish to draw now is personally most disagressable to me. But bitter experiences led me to observe repeatedly that I was, on this matter, the odd man out; bitter, since often my being intellectually and temperamentally out on this issue was driven home to me by those who felt they could not easily associate with such a rare bird. What helped me overcome the bitterness was my otherwise successful social life, my professional success regardless of its enormous obstacles, and my effort to understand, rather than complain about, the mood I wish now to describe.

The mood, the inductive mood for short, is that of a self-sufficient intelligent agent. Self-sufficiency is a terrific feel, one which is described in many stories and ballades and plays and movies, from Robinson Crusoe to Citizen Kane. It has diverse components, it has diversity; but usually it contains the idea of self-assurance, of self-reliance, of self-confidence and, above all, of confidence and optimism. When C. P. Snow – now Lord Snow – described the scientific temperament, meaning the temperament given to the inductive mood, he described it as optimistic, progressive, future-oriented. For Snow this will do; for us here it will not.

The specific thing to the inductive mood is the rational grounding of the self-reliant, optimistic mood. Anyone can be self-reliant and optimistic, say, by living in a fool's paradise, by refusing to look far enough into the future to be able to forecast disasters and willing to try to take proper measure against them. Not so the optimist in his inductive mood: he is rational and his optimism is grounded in his rationality.

It does not matter much who should serve as an example, but I wish to take, for the sake of convenience, as my example, none other than

Bertrand Russell, the author of "Free Man's Worship". Russell's grim vision of Man was no mere abstraction. In his late *Has Man A Future?* Russell seriously considers the possibility that we shall destroy the earth and all its inhabitants. He does not express any optimism or hope – he pleads with us not to destroy. Even in his most inductivist and celebrated *Human Knowledge, Its Scope and Limits* he start by the image science has of man; trapped in a very narrow strip of barely livable and barely stable corner of the universe, most of which is utterly uninhabitable for Man; in a hostile universe we have to make do as best we can. Yet this making do, which is largely science and scientific technology, is based on induction, in the idea that science validates its theories and assures us – not absolutely but quite very probably – that our expectations based on our scientific knowledge will indeed come true.

The idea that science has no validity at all, that it offers no assurance whatsoever, is dismissed by Russell off-hand as too defeatist. In his preface to the new edition of Nicod's classic book on induction Russell has one line on Popper's view of science: it is defeatist and our duty is to try and do better than he.

Are Russell's pessimism and optimism reconcilable? I think not. Before we examine that we can ask, do we really need assurance? The answer is, alas, yes.

III. EMPIRICAL FACTS ABOUT ASSURANCE

It is, I fear, an empirical fact that we need assurance constantly. If any one idea of the whole of the Freudian corpus has become commonplace in the modern world, it is that a child needs its mother's assurance, that cuddling and cooing to one's baby, that physical proximity to it, are of extreme importance to the baby, who would soon panic without it. Whether we call this infant-sexuality or the need for security matters here not at all: no one denies that the mother offers her child assurance at every step and that this is essential for the child's well-being.

Experience shows that industrial workers are in need for assurance too, that this need can be easily undermined by minor but noticeable modifications of their surroundings which lead to repeated disappointments of their expectations.

It is an empirical fact that by regularly disappointing an individual's expectations we can render him a nervous wreck: people who blow hot and cold, who seemingly erratically offer good or bad will to their associates and dependants are known as holy terrors. Inquisitors and Brainwashers of all sorts tighten their screws on their victims by making their violent shifts more systematic and damaging. These cruel instances show how much stability is important for normal human beings if they are to lead normal smooth lives.

Stability is not here any inflexibility. The disappointed expectations that fall into pattern and elicit well prepared responses are in no way unnerving since a higher-level pattern of satisfaction and disappointment can help. And patterns can also be disappointed and altered, but not so rapidly as to shatter the process of readjustment. That is to say, a most important aspect of stability is not only that of satisfied expectation, but also of rational adjustment to unsatisfied ones. Hence both stability and rationality are important for human well-being, as far as experience can tell us. All this, to repeat, is widely accepted as empirical facts.

Hence, the inductive mood seems to be of the essence. And it seems to be there, to this degree or that, in stable modern enclaves. Hence, the question, can we reconcile this stability with the utter lack of expectation on the large scale, the utter agnosticism with which we have started?

IV. THE NON-JUSTIFICATIONIST MOOD

There are quite a few components in our complex picture, yet one is still missing. We have the agnostic mood which is defiant of its own desperation; the inductive mood which is the optimistic, progressive and assured; and the need for assurance which is a matter of empirical fact, psychological, sociological, and common sense. The missing component is that of moral and psychological independence or autonomy or self-sufficiency. I have already mentioned it. I also noted that its importance lies in the fact that one can have stability and flexibility, a readiness to cope with counterexpectation: this very readiness is autonomy. The inability to cope pushes one towards heteronomy, namely towards emotional and moral and intellectual dependence.

Again the drastic instance, however rare or common it may be, is easy to grasp and it helps make the point sharply. The candidate for conversion, as William Sargant tells us in his *Battle for the Mind*, is ruthlessly broken down by his guide: the guide shatters the candidate's expectations; watches new ones arise and shatters these; helps the candidate build still newer ones and then shatters them as well. The candidate is bewildered, lost, helpless, destitute. In other words, his sense of dependence is heightened to a peak. Then the guide offers love and solace and a new set of expectations to go with all of these.

The crux of the present study may indeed be the contrast or the mood of the brainwashed with the mood of an independent person – politician, philosopher, scientist – who makes a great and decisive switch. For Sargant all switch is brain-wash. This is (unintended) relativism at its rather ludicrous end. For me the opposite of brainwash is the simply deliberate and autonomous decision of the well-composed flexible individual who faces certain alternative routes from the junction of his own understanding of his previous mistake, refuted expectation, or misguided aspirations. There is no hysteria or disturbed spirit here, simply the opening up of new possibilities hitherto unnoticed.

But the two extremes, the brainwashed and the serene deliberator, are not such that middle positions are mixtures of both. We have, admittedly, measures of emotional dependence and intellectual dependence; but we also have degrees of responsibility, and we can contrast responsibility with the mood of justificationism, to use Bartley's apt term.

When in civil society I stand before a judge properly accused, I am obliged to justify myself. Otherwise I am obliged not to. Justifying myself to me peers I in fact appoint them as my judges; and so long as I do so without their expressed consent I have imposed on them; I have made myself dependent on them.

The fact that people are all too often guilty of such immorality is the central theme of all of Kafka's works. Kafka also showed that this guilty act is felt by the guilty party as nothing short of a supreme effort of expiation of guilt and/or of proof of innocence or uprightness or guiltlessness – at least by comparison. But, of course, Kafka adds, the act is self-defeating.

Most common people in common situations are, again, not so extreme. They explain themselves, they show that their actions, their positions, their views, are eminently reasonable; that indeed their peers before whom they expound their own reasonableness cannot but endorse their views, justify their actions, share their positions. They are not behaving quite like the accused at the dock as Kafka's hero does; but they mention witnesses, they describe significant facts and experiences. They appeal to the supreme authority of religion, of social norms, or of science.

Here the inductive mood and brainwash come extremely close. Many thinkers, I suppose, have felt uneasy about it and tried to throw a wedge between the two: brainwashed people rely on other people's say so, scientists rely on their own experiences. But this means that I, as a scientist, rely on the experiences of other scientists – science becomes my social norm. If and when the scientific world happens to have a dogma, then I am an easy victim of it, with no redemption!

The only way out is to realize that my peers are not my judges: my actions, my expectations, my views, are not such that I need justify them except under very specific and legally well-specified circumstances. And then I have the right for legal advice.

V. CONVERSION TO AUTONOMISM

There is a final and cruel twist to all this. The justificationist mood, to repeat, is extremely common, and it fosters one's dependence under the guise of one's acceptance of the authority of science; the justificationist mood is disguised as the inductive mood. This permits the freakish cruel – if unintended – twist I now wish to illustrate by way of a real empirical example.

In recent years the philosophy of Sir Karl Popper and his followers has gained increasing popularity. It is no secret that I am a product of that school, disciple or apostate, I cannot say. I have observed a process akin to brainwash there. When a philosopher, expecially a novice, comes to meet with members of the Popper camp, usually from the inductivist camp, he tries the Popperian idiom for size and checks his expressions with the adept – as is quite common. He is used to the inductive terminology which refers to validation of scientific theories, positive evidence, supporting evidence or argument, confirmation, etc. etc. Here he finds a blank wall. In desperation he tries to translate some of what he thinks are truisms of the philosophy of science to the Popperian jargon; he learns not to use the word 'confirmation' but uses the word 'corroboration' to the same end (in line with one of the Master's most regrettable moves. I say 'regrettable' because it is not words but their imports that matter, as Popper has repeatedly declared, yet he has opened the door to a change that is merely terminological: I wish to report that in a draft of one of my papers I used the word 'verfication' with the Popperian import of positive evidence, yet my Popperian colleagues were shocked by my use of a word so branded as belonging to the justificationist camp – as if words can be monopolized. Yet words are monopolized though they cannot be, and 'corroboration' is today a Popperian monopoly. This is regrettable.)

If our novice's interlocutors in the Popperian camp are any good they are not fooled by his switch from the word 'confirmation' ot its cognate 'corroboration'. Indeed they are expecting him to do just this and are waiting around the next corner to trip him. At this point, unwittingly, they act like our brainwasher from a previous section. No matter how our novice tries to express a justificationist view, his guide sees that it is justificationist and so not autonomous and so otiose. Our novice does not thereby learn to become autonomous; he attempts, no doubt, to express quasi-autonomy; and his expectation of sympathy and understanding and at least partial approval is cruelly shattered. Hence his measure of dependence is heightened, and his conversion may amount to nothing short of brainwash, mild or severe.

Of course, once he enters the circle of the elect – and Popperianism is still not yet so vulgar as to be devoid of this booster – the moment he is boosted, our novice is told that of course there is positive evidence, support or corroboration, that his need to have some measure of stability is recognized. That the Master's social philosophy is, indeed and alas!, quite conservative!

Does this help our not very autonomous novice become critical and independent or does he become a fanatic upholder of the dogma of the new critical philosophy? Truth to tell I have observed it happen both ways. Now, apart from the educational aspects of our novice's autonomy, he has a genuine theoretical question, that his guide is not quite frank about: how do we combine agnosticism and assurance?

VI. THE ASSURED AGNOSTIC

I do not know the true answer. I have devised an answer that satisfies me for the time being, and I shall repeat it here in the briefest outline. But I wish to impress upon my reader my impression that this is a problem that has troubled many in the past and does trouble many at present. Many thinkers have felt that a total global agnosticism, a total skepticism on the large scale, precludes all assurance on the small scale. That, conversely, the existence of any assurance on the small scale is evidence enough (perhaps a transcendental proof, even) that somewhere, somehow, assurance exists also on the large scale. And I wish to deny just this.

I think the starting point is Kant's synthetic a priori, which is now better known by such words as the conceptual framework, an overview, a general view, even a welt-anschauung or world view. We have to perform two operations on the conceptual framework to de-Kantianize it. It has to be socialized: it has to be identified with what the sociologist says a member of a given society has internalized in order to share his peers' outlook. And it has to be completely deprived of all privileged status of validity, knowledge, or whatever else. The second step has been taken already by Solomon Maimon; the first is new in the philosophy of science though it has Marxian origins. (Marx himself blew it: he was a relativist.)

Once we deprive our conceptual framework of its finality, we can add to it incentives for change in the form of canons for internal criticism. Among these we should count not only those listed by diverse authors from Plato to Popper but also the canons of commensurability. Let me explain.

Some frameworks are, or at least are meant to be, incommensurable: they include the rule: prefer me over all others. Traditionally, I suppose, Judaism is the paradigm here. Other canons differ. For example, both Newtonians and Einsteinians declare Einsteinism better: the Newtonians had to convert to Einsteinism while following their own canons. Oh, yes, some philosophers have declared that no two conceptual frameworks are commensurable. They illustrate their view with historical facts that conflicts with their views. But at least their conceptual frameworks contain no canons of criticism, so that they can stick to their view: the doctrine of incommensurability (of Duhem, Evans-Pritchard, Polanyi, etc. etc.) is, indeed, incommensurable, very much like Judaism (though the leading thinkers just mentioned were Catholics of diverse sorts).

So much for the alterability of the framework. But, as it is an institution, it cannot alter so fast as to deprive citizens of their need for assurance. And within any conceptual framework there are all sorts of standards of assurance. There are commerical assurances, including banks and banking insurance, and insurance companies and federal reserves; there are social assurances, there are government agencies to test patent medicines and all sorts of inventions, including, in particular, safety gadgets; travel safety, expecially by air; etc. etc. Hence, Popper's theory of corroboration is patently false: it does not operate within any specific conceptual or legal framework but in a social and scientific vacuum.

Finally there is the great old wisdom: diversify; do not put all your eggs in one basket. It means that if my system collapses at a point and, say, my electric system goes bust, I can depend on yours; but we may go down together electricity-wise and have a total blackout. In such a case we need a substitute which should be available right how. We thus construct our system so that each detail is connected to other detail and with the framework too. This does insure that the detail is kept afloat, but at the cost of making the whole framework heavy and capable of sinking all in all.

This was realized recently with the series of ecological crises which have led to mass hysteria and to attempts to go back to Mother Nature.

But, as I have explained, this runs counter to Free Man's Worship, which is still the modern enlightened man's manifesto. We can try to criticize and improve the conceptual framework; we can only seek assurance within it while realizing that there is nothing by which to justify our enterprise as a whole.

INDEX OF WORKS CITED

See also notes and references at ends of Chapters 6, 8, 12, and 17

INDEX OF NAMES

INDEX OF SUBJECTS

SYNTHESE LIBRARY

Monographs on Epistemology, Logic, Methodology,
Philosophy of Science, Sociology of Science and of Knowledge, and on the
Mathematical Methods of Social and Behavioral Sciences

Managing Editor:

JAAKKO HINTIKKA (Academy of Finland and Stanford University)

Editors:

ROBERT S. COHEN (Boston University)
DONALD DAVIDSON (The Rockefeller University and Princeton University)
GABRIËL NUCHELMANS (University of Leyden)
WESLEY C. SALMON (University of Arizona)

14. ROBERT S. COHEN and MARX W. WARTOFSKY (eds.), *Proceedings of the Boston Colloquium for the Philosophy of Science 1964–1966, in Memory of Norwood Russell Hanson*, Boston Studies in the Philosophy of Science (ed. by Robert S. Cohen and Marx W. Wartofsky), Volume III. 1967, XLIX + 489 pp.
15. C. D. BROAD, *Induction, Probability, and Causation. Selected Papers*. 1968, XI + 296 pp.
16. GÜNTHER PATZIG, *Aristotle's Theory of the Syllogism. A Logical-Philosophical Study of Book A of the Prior Analytics*. 1968, XVII + 215 pp.
17. NICHOLAS RESCHER, *Topics in Philosophical Logic*. 1968, XIV + 347 pp.
18. ROBERT S. COHEN and MARX W. WARTOFSKY (eds.), *Proceedings of the Boston Colloquium for the Philosophy of Science 1966–1968*, Boston Studies in the Philosophy of Science (ed. by Robert S. Cohen and Marx W. Wartofsky), Volume IV. 1969, VIII + 537 pp.
19. ROBERT S. COHEN and MARX W. WARTOFSKY (eds.), *Proceedings of the Boston Colloquium for the Philosophy of Science 1966–1968*, Boston Studies in the Philosophy of Science (ed. by Robert S. Cohen and Marx W. Wartofsky), Volume V. 1969, VIII + 482 pp.
20. J. W. DAVIS, D. J. HOCKNEY, and W. K. WILSON (eds.), *Philosophical Logic*. 1969, VIII + 277 pp.
21. D. DAVIDSON and J. HINTIKKA (eds.), *Words and Objections: Essays on the Work of W. V. Quine*. 1969, VIII + 366 pp.
22. PATRICK SUPPES, *Studies in the Methodology and Foundations of Science. Selected. Papers from 1911 to 1969*, XII + 473 pp.
23. JAAKKO HINTIKKA, *Models for Modalities. Selected Essays*. 1969, IX + 220 pp.
24. NICHOLAS RESCHER *et al.* (eds.), *Essays in Honor of Carl G. Hempel. A Tribute on the Occasion of his Sixty-Fifth Birthday*. 1969, VII + 272 pp.
25. P. V. TAVANEC (ed.), *Problems of the Logic of Scientific Knowledge*. 1969, XII + 429 pp.
26. MARSHALL SWAIN (ed.), *Induction, Acceptance, and Rational Belief*. 1970, VII + 232 pp.
27. ROBERT S. COHEN and RAYMOND J. SEEGER (eds.), *Ernst Mach; Physicist and Philosopher*, Boston Studies in the Philosophy of Science (ed. by Robert S. Cohen and Marx W. Wartofsky), Volume VI. 1970, VIII + 295 pp.
28. JAAKKO HINTIKKA and PATRICK SUPPES, *Information and Inference*. 1970, X + 336 pp.
29. KAREL LAMBERT, *Philosophical Problems in Logic. Some Recent Developments*. 1970, VII + 176 pp.
30. ROLF A. EBERLE, *Nominalistic Systems*. 1970, IX + 217 pp.
31. PAUL WEINGARTNER and GERHARD ZECHA (eds.), *Induction, Physics, and Ethics, Proceedings and Discussions of the 1968 Salzburg Colloquium in the Philosophy of Science*. 1970, X + 382 pp.
32. EVERT W. BETH, *Aspects of Modern Logic*. 1970, XI + 176 pp.
33. RISTO HILPINEN (ed.), *Deontic Logic: Introductory and Systematic Readings*. 1971, VII + 182 pp.
34. JEAN-LOUIS KRIVINE, *Introduction to Axiomatic Set Theory*. 1971, VII + 98 pp.
35. JOSEPH D. SNEED, *The Logical Structure of Mathematical Physics*. 1971, XV + 311 pp.
36. CARL R. KORDIG, *The Justification of Scientific Change*. 1971, XIV + 119 pp.

37. MILIČ ČAPEK, *Bergson and Modern Physics*, Boston Studies in the Philosophy of Science (ed. by Robert S. Cohen and Marx W. Wartofsky), Volume VII, 1971, XV + 414 pp.

38. NORWOOD RUSSELL HANSON, *What I do not Believe, and other Essays* (ed. by Stephen Toulmin and Harry Woolf), 1971, XII + 390 pp.

39. ROGER C. BUCK and ROBERT S. COHEN (eds.), *PSA 1970. In Memory of Rudolf Carnap*, Boston Studies in the Philosophy of Science (ed. by Robert S. Cohen and Marx W. Wartofsky, Volume VIII. 1971, LXVI + 615 pp. Also available as a paperback.

40. DONALD DAVIDSON and GILBERT HARMAN (eds.), *Semantics of Natural Language*. 1972, X + 769 pp. Also available as a paperback.

41. YEHOSUA BAR-HILLEL (ed.)., *Pragmatics of Natural Languages*. 1971, VII + 231 pp.

42. SÖREN STENLUND, *Combinators, λ-Terms and Proof Theory*. 1972, 184 pp.

43. MARTIN STRAUSS, *Modern Physics and Its Philosophy. Selected Papers in the Logic, History, and Philosophy of Science*. 1972, X + 297 pp.

44. MARIO BUNGE, *Method, Model and Matter*. 1973, VII + 196 pp.

45. MARIO BUNGE, *Philosophy of Physics*. 1973, IX + 248 pp.

46. A. A. ZINOV'EV, *Foundations of the Logical Theory of Scientific Knowledge (Complex Logic)*, Boston Studies in the Philosophy of Science (ed. by Robert S. Cohen and Marx W. Wartofsky), Volume IX. Revised and enlarged English edition with an appendix, by G. A. Smirnov, E. A. Sidorenka, A. M. Fedina, and L. A. Bobrova 1973, XXII + 301 pp. Also available as a paperback.

47. LADISLAV TONDL, *Scientific Procedures*, Boston Studies in the Philosophy of Science (ed. by Robert S. Cohen and Marx W. Wartofsky), Volume X. 1973, XII + 268 pp. Also available as a paperback.

48. NORWOOD RUSSELL HANSON, *Constellations and Conjectures* (ed. by Willard C. Humphreys, Jr.), 1973, X + 282 pp.

49. K. J. J. HINTIKKA, J. M. E. MORAVCSIK, and P. SUPPES (eds.), *Approaches to Natural Language. Proceedings of the 1970 Stanford Workshop on Grammar and Semantics*. 1973, VIII + 526 pp. Also available as a paperback.

50. MARIO BUNGE (ed.), *Exact Philosophy – Problems, Tools, and Goals*. 1973, X + 214 pp.

51. RADU J. BOGDAN and ILKKA NIINILUOTO (eds.), *Logic, Language, and Probability*. A selection of papers contributed to Sections IV, VI, and XI of the Fourth International Congress for Logic, Methodology, and Philosophy of Science, Bucharest, September 1971. 1973, X + 323 pp.

52. GLENN PEARCE and PATRICK MAYNARD (eds.), *Conceptual Chance*. 1973, XII + 282 pp.

53. ILKKA NIINILUOTO and RAIMO TUOMELA, *Theoretical Concepts and Hypothetico-Inductive Inference*. 1973, VII + 264 pp.

54. ROLAND FRAÏSSÉ, *Course of Mathematical Logic – Volume I: Relation and Logical Formula*. 1973, XVI + 186 pp. Also available as a paperback.

55. ADOLF GRÜNBAUM, *Philosophical Problems of Space and Time*. Second, enlarged edition, Boston Studies in the Philosophy of Science (ed. by Robert S. Cohen and Marx W. Wartofsky), Volume XII. 1973, XXIII + 884 pp. Also available as a paperback.

56. PATRICK SUPPES (ed.), *Space, Time, and Geometry*. 1973, XI + 424 pp.

57. HANS KELSEN, *Essays in Legal and Moral Philosophy*, selected and introduced by Ota Weinberger. 1973, XXVIII + 300 pp.

58. R. J. SEEGER and ROBERT S. COHEN (eds.), *Philosophical Foundations of Science. Proceedings of an AAAS Program, 1969*. Boston Studies in the Philosophy of Science (ed. by Robert S. Cohen and Marx W. Wartofsky), Volume XI. 1974, X + 545 pp. Also available as paperback.

59. ROBERT S. COHEN and MARX W. WARTOFSKY (eds.), *Logical and Epistemological Studies in Contemporary Physics*, Boston Studies in the Philosophy of Science (ed. by Robert S. Cohen and Marx W. Wartofsky), Volume XIII. 1973, VIII + 462 pp. Also available as paperback.

60. ROBERT S. COHEN and MARX W. WARTOFSKY (eds.), *Methodological and Historical Essays in the Natural and Social Sciences. Proceedings of the Boston Colloquium for the Philosophy of Science, 1969–1972*, Boston Studies in the Philosophy of Science (ed. by Robert S. Cohen and Marx W. Wartofsky), Volume XIV. 1974, VIII + 405 pp. Also available as paperback.

61. ROBERT S. COHEN, J. J. STACHEL, and MARX W. WARTOFSKY (eds.), *For Dirk Struik. Scientific, Historical and Political Essays in Honor of Dirk J. Struik*, Boston Studies in the Philosophy of Science (ed. by Robert S. Cohen and Marx W. Wartofsky), Volume XV. 1974, XXVII + 652 pp. Also available as paperback.

62. KAZIMIERZ AJDUKIEWICZ, *Pragmatic Logic*, transl. from the Polish by Olgierd Wojtasiewicz. 1974, XV + 460 pp.

63. SÖREN STENLUND (ed.), *Logical Theory and Semantic Analysis. Essays Dedicated to Stig Kanger on His Fiftieth Birthday*. 1974, V + 217 pp.

64. KENNETH F. SCHAFFNER and ROBERT S. COHEN (eds.), *Proceedings of the 1972 Biennial Meeting, Philosophy of Science Association*, Boston Studies in the Philosophy of Science (ed. by Robert S. Cohen and Marx W. Wartofsky), Volume XX. 1974, IX + 444 pp. Also available as paperback.

65. HENRY E. KYBURG, JR., *The Logical Foundations of Statistical Inference*. 1974, IX + 421 pp.

66. MARJORIE GRENE, *The Understanding of Nature: Essays in the Philosophy of Biology*, Boston Studies in the Philosophy of Science (ed. by Robert S. Cohen and Marx W. Wartofsky), Volume XXIII. 1974, XII + 360 pp. Also available as paperback.

67. JAN M. BROEKMAN, *Structuralism: Moscow, Prague, Paris*. 1974, IX + 117 pp.

68. NORMAN GESCHWIND, *Selected Papers on Language and the Brain*, Boston Studies in the Philosophy of Science (ed. by Robert S. Cohen and Marx W. Wartofsky), Volume XVI. 1974, XII + 549 pp. Also available as paperback.

69. ROLAND FRAÏSSÉ. *Course of Mathematical Logic* – Volume II: *Model Theory*. 1974, XIX + 192 pp.

70. ANDRZEJ GRZEGORCZYK, *An Outline of Mathematical Logic. Fundamental Results and Notions Explained with all Details*. 1974, X + 596 pp.

SYNTHESE HISTORICAL LIBRARY

Texts and Studies
in the History of Logic and Philosophy

Editors:

N. Kretzmann (Cornell University)
G. Nuchelmans (University of Leyden)
L. M. de Rijk (University of Leyden)

1. M. T. Beonio-Brocchieri Fumagalli, *The Logic of Abelard*. Translated from the Italian. 1969, IX + 101 pp.

2. Gottfried Wilhelm Leibnitz, *Philosophical Papers and Letters*. A selection translated and edited, with an introduction, by Leroy E. Loemker. 1969, XII + 736 pp.

3. Ernst Mally, *Logische Schriften*, ed. by Karl Wolf and Paul Weingartner. 1971, X + 340 pp.

4. Lewis White Beck (ed.), *Proceedings of the Third International Kant Congress*. 1972, XI + 718 pp.

5. Bernard Bolzano, *Theory of Science*, ed. by Jan Berg. 1973, XV + 398 pp.

6. J. M. E. Moravcsik (ed.), *Patterns in Plato's Thought. Papers arising out of the 1971 West Coast Greek Philosophy Conference*. 1973, VIII + 212 pp.

7. Nabil Shehaby, *The Propositional Logic of Avicenna: A Translation from al-Shifā':al-Qiyās*, with Introduction, Commentary and Glossary. 1973, XIII + 296 pp.

8. Desmond Paul Henry, *Commentary on De Grammatico: The Historical-Logical Dimensions of a Dialogue of St. Anselm's*. 1974, IX + 345 pp.

9. John Corcoran, *Ancient Logic and Its Modern Interpretations*. 1974. X + 208 pp.

10. E. M. Barth, *The Logic of the Articles in Traditional Philosophy*. 1974, XXVII + 533 pp.

11. Jaakko Hintikka, *Knowledge and the Known. Historical Perspectives in Epistemology*. 1974, XII + 243 pp.

12. E. J. Ashworth, *Language and Logic in the Post-Medieval Period*. 1974, XIII + 304 pp.